Engineering in Textile Coloration

Edited by C Duckworth

Worshipful Company of Dyers

Society of Dyers and Colourists

1983
Dyers Company Publications Trust

Published by The Dyers Company Publications Trust, and distributed by The Society of Dyers and Colourists, Perkin House, 82 Grattan Road, Bradford, West Yorkshire BD1 2JB, England.

Printed by Joseph Ward & Co. (Printers) Ltd, Wellington Road, Dewsbury, West Yorkshire, WF13 1HR, England.

ISBN 0 901956 31 7

Contents

CHAPTER 2

Fluid Flow, Heat Transfer and Mass Transfer, *by R Murphy*

CHAPTER 3

Corrosion, *by P Elliott*

CHAPTER 7

Plant Engineering Services, *by W Ronald*

CHAPTER 8

Economics – Investment in Plant and Machinery, *by H Catling*

Dyers Company Publications Trust

THE DYERS COMPANY PUBLICATIONS TRUST was instituted by The Worshipful Company of Dyers of the City of London in 1971 to encourage the publication of textbooks and other aids to learning in the science and technology of colour and coloration and related fields. The Society of Dyers and Colourists acts as trustee for the fund.

Contributors

H Catling BSc(Eng) PhD FIMechE
Consultant. Formerly Head of Technical Economy, Shirley Institute, Didsbury, Manchester

P Elliott MSc PhD FICorrST
Lecturer in Corrosion Science, Corrosion & Protection Centre, University of Manchester Institute of Science and Technology, Manchester

J Heaton BSc AMCT AMBIM
Formerly Technical Executive, Carrington Viyella Ltd

W R Murphy BSc CEng FIChemE CChem FRSC
Senior Lecturer in Chemical Engineering, Queen's University, Belfast

M G Mylroi MA PhD FIMC
Professor of Control Engineering, University of Bradford

W Ronald
Formerly Engineering Director, The Standish Company, Worthington, Wigan

S S Smith CCol ASDC
Managing Director, Roaches (Engineering) Ltd, Upperhulme, Leek

D H Wyles BSc ARCS CText ATI CCol FSDC
Consultant. Formerly Head of Central Dyeing Department, Courtaulds plc, Spondon, Derby

Preface

In the preface to the first two books of this series an account was given of the background to the production of a comprehensive series of textbooks relating to textile coloration, and the encouragement and financial support given by the Worshipful Company of Dyers. The Society accepted a recommendation that a separate textbook be published dealing with engineering in textile coloration, partly to include in the one volume many of the specialized topics embraced by the term 'engineering', but in particular to provide a definitive textbook for candidates studying for the Society's Associateship examinations and who have chosen to sit for the 'engineering' paper as an option under Branch 1 (Textiles).

As a general policy it was intended that the separate volumes dealing with specific groups of fibres should not include detailed descriptions of machinery. One exception is to be seen in the recent publication of the textbook on textile printing, in which specialized printing machinery is described along with the process technology.

Of the importance of engineering in textile coloration there can be no doubt. In the categories: plant, services and buildings, it accounts for almost all the capital expenditure involved. In a dyeworks trading account, the sum of all engineering costs, including depreciation, is certainly the largest grouping. Insofar as the process technology of dyes and fibres cannot be commercially implemented without plant and machinery, it follows that financially and technically successful operation is dependent on the correct choice of plant and machinery, and the correct functioning, reliable operation and good maintenance of that equipment.

Certainly, this must be the province of the specialist, the chief engineer of the company, but higher management is almost invariably appointed from the ranks of non-engineers, be they colourists, chemists, accountants or salesmen. It is increasingly important that management, at all levels, is more familiar with basic engineering principles, is aware of their complexities and is sympathetic to the problems that are often encountered.

It is hoped that the present textbook will assist in providing this background, not only to budding colourists, but to those already engaged in industry, research and education.

A glance at the contents pages will indicate the spread of subjects, although there are notable omissions. There is no chapter on electrical engineering. This is excused on the grounds that primers on electrical technology are readily available for those who find this subject of particular interest, and that its practice is less intimately bound to that close relationship between machine and process. It was decided, therefore, that appropriate discussion of electrical technology would be included in the chapter on plant engineering services. The first five chapters of the book deal mainly with matters relating to actual design, manufacture and use of machinery, whereas the rest may be considered as supporting knowledge once machines and plant are in operation.

Once the scope of the book had thus been determined, it was inevitable that some chapters would mainly discuss the fundamental teachings involved, whereas others would rely upon practical exposition.

S I units have been used in Chapter 2, dealing with fundamental aspects of engineering, whereas in the more practical chapters imperial and metric units have been retained. This is justified on the grounds that much present works practice is based on the older units and will continue so for some years. It follows that all works staff should be familiar with the necessary conversions.

In a book of this nature some overlap of subject between authors is inevitable and in some cases has been encouraged, so that readers may benefit from different authors' experience, and also to provide a different interpretation of works practice.

I take this opportunity to express grateful thanks to all those who have collaborated in the production of the book – to authors, referees, members of the Dyers Company Publications Trust Technical Subcommittee and Society staff, especially Paul Dinsdale, the Society's Editor. Thanks are also due to those bodies who were generous in their permission to use copyright material.

C DUCKWORTH

Chapter 1

Functional Design of Coloration Machines

D H WYLES

INTRODUCTION

The purpose of this chapter is to describe the principles of operation of the major types of machinery used by the dyer, including those machines used for essential operations before and after dyeing. It is not intended to be a catalogue of dyeing machines nor an attempt to include the latest models and accessories offered by the machinery maker – reviews of machinery developments are more topical sources of such information.

The choice of machinery to be included has been difficult, owing to the ingenuity and inventiveness of the textile machinery engineers who, in the last twenty years, have brought about dramatic improvements in scope and performance of practically all types of dyeing machine. A compromise has had to be made between including more types of machines and increasing the chapter disproportionately with the rest of the book.

The aim is to stress the fundamentals of well-proven machinery and to indicate interesting variants produced for particular purposes. Literature references have in general been restricted to those sources readily accessible to students.

The help of colleagues in checking the script, and the kind permission of machinery manufacturers to use their diagrams and illustrations are gratefully acknowledged.

1 MACHINES FOR DYEING TOW, STAPLE, TOPS AND YARN

1.1 PACKAGE-DYEING MACHINES FOR YARN

Most yarn dyeing is carried out after the yarn has been wound into suitable packages, e.g. cheeses, cones, or cakes, which are then mounted onto perforated spindles; the yarn is dyed by forcing liquor through the package. At one time these machines were commonly called ‘pressure dyeing’ machines, but this nomenclature is not recommended as it can lead to confusion with ‘pressurized’ machines for high-temperature dyeing, i.e. dyeing at temperatures above 100°C.

The first package-dyeing machine was patented by Otto Obermaier in 1882. The earliest machines had the dyebath open to the atmosphere and permitted the dye-liquor to flow only from inside to outside of the package. Liquor was withdrawn from the main dyebath by the pump and recirculated. Later the dyebath was totally enclosed and, by reversing the pump, it became possible to force liquor from the outside of the package inwards. By alternating the direction of flow of liquor (two-way flow), better levelness of dyeing was obtained. The pressurized system improved the performance of the pump and enabled higher temperatures to be used.

Virtually all modern package-dyeing machines are pressurized and capable of operating at temperatures up to 135°C. Recent developments in package dyeing and the relevant machinery have been excellently reviewed (*1–4*).

The Dye Vessel

The pressure vessel is cylindrical with domed ends, the upper being the lid or cover, which is securely locked by a series of bolts (on older machines) or by a sliding ring or similar quick-locking device. Machine diameters go up to just over 2 m and the vessel height is usually similar, although a few exceptionally tall machines have been built. The overall height of the machine is increased by the pump, fitted immediately below the main vessel, which is supported by three or four stanchions welded to the sides.

The height of the machine and the need to load it by an overhead travelling crane necessitates either plenty of headroom or setting the machine in a pit below the level of the dyehouse floor. Where several machines are serviced by an overhead hoist on a monorail, the centres of the machines must be carefully aligned below the monorail. The capacity of the hoist must not be under-estimated – as well as the weight of the yarn and the carrier there will be a similar amount of water; a 5 tonne hoist is usually sufficient for a machine of 2 m diameter. Where the machines are mounted at ground level, a raised platform with surrounding safety rail is required for the operative; by UK law the difference in height between the platform and the top of the open vessel must be at least 3 ft (92 cm).

The lid or cover of a large-diameter machine is very heavy and needs to be counterbalanced. It is raised by a pneumatically or hydraulically operated cylinder at the back of the machine. Each pressure vessel must be fitted with a plate giving details of manufacture, name, identification number, date of production and test, the safe working temperature and pressure, etc. Pressure vessels must also have a safety valve and gauges clearly indicating the pressure and temperature inside.

Safety interlocks are fitted to prevent the vessel being opened whilst there is internal residual pressure or when the temperature exceeds 85°C. An air release/liquor bleed pipe is fitted in the lid to enable air to be expelled from the vessel during filling and to permit a small quantity of dye-liquor to pass through a water-cooled condenser before discharge into an expansion tank mounted to one side. At the start of each dyeing a visual check should be made to ensure that liquor is flowing through the overflow into the expansion tank.

Filling with water and draining are carried out via a junction in the main liquor circulation system at the bottom of the machine just outside the pressure vessel. In a few dyehouses, particularly those handling polyester yarns, exhausted dye-liquor may be discharged at temperatures around 120°C to specially pressurized drains in order to eliminate oligomers from the system.

Frame and Spindles

The packages of yarn are mounted on vertical hollow spindles, perforated to permit the flow of dye-liquor. The spindles are up to about 125 cm long and usually circular in cross-section (7 cm diameter), although some have a 'Y' cross-section to permit easier longitudinal flow of dye-liquor. The spindles are screwed firmly into holes set in a frame with a circular hollow base and a centre pillar with an eye at the top for lifting. At the centre bottom of the frame is an inlet for dye-liquor and this inlet fits onto the pump discharge when the frame is lowered to the bottom of the dye vessel. The layout is shown in Figure 1.1.

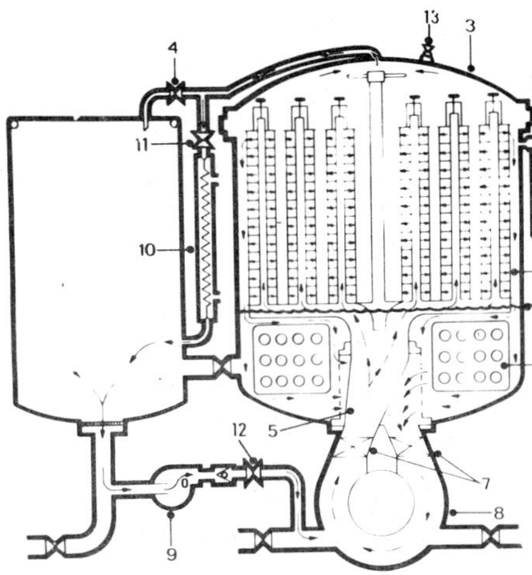

1. Directional flow shown in to out through packages for low level
2. Pressure lock sample device
3. Quick acting pressure cover
4. Pressure release
5. Flow reversal unit (for fully flooded operation only)
6. Closed coil heat exchanger
7. Flow control valves
8. Directly coupled pump unit
9. Secondary pump (addition pump)
10. Bleed cooler (isolatable)
11. Bleed cooler isolating valve
12. Addition tank isolate valve
13. Compressed air pressurisation point

Figure 1.1 – Platt Longclose package-dyeing machine

Figure 1.2 – Platt Longclose package carrier loaded with cheeses and spacers; insert shows cheeses (without spacers), dye springs and perforated spindle

The maximum frame diameter is just over 2 m and the spacing of the spindle holes depends on the maximum diameter of the packages to be dyed – the packages on adjacent spindles should not touch, but obviously the more spindles that can be accommodated, the greater the maximum load.

At the top of each spindle a screw thread is provided to enable an end-plate and end-cap to be fitted; these prevent dye-liquor from streaming out and keep the packages pressed closely together, thus maintaining a seal along the entire length of the spindle.

The number of dye packages per spindle is usually around 8–10 and, when compressible (e.g. when centre-less or on dye-springs), the column is pressed down firmly (usually mechanically) and the end-cap is screwed on tightly to complete the seal (Figure 1.2). Some textured-yarn packages tend to shrink when wetted for the first time and it may be necessary to wet out the dye-load for 5 min and then re-open the vessel and add an additional package to each column to maintain uniform permeability. Compressing the packages lengthways not only provides a good seal between adjacent packages and increases the dye load, but also tends to compact some areas of low density and high permeability, thus providing more uniform liquor flow along the column.

Non-textured continuous-filament yarns are sometimes wound or placed on rigid formers which either interlock or are fitted with spacers to provide a seal against excessive liquor flow.

Where high rates of liquor flow are required, long spindles present problems of pressure drop along the column. This drop can be reduced by increasing the spindle diameter, or by modifying the spindle cross-section; some manufacturers provide a split-level frame, although this arrangement presents complications if mechanical compression is desired.

Pump

The main pump is fitted to the bottom of the vessel, usually as close to the centre as possible in order to minimize frictional losses in flow. The pumps on early dyeing machines lost efficiency at temperatures nearing the boil due to 'cavitation', i.e. formation of steam bubbles on the reverse side of the blade where the pressure is low. First efforts to cure this fault consisted in locating the pump at the base of the machine so as to have the maximum head of liquor above it. Secondly, the adoption of completely pressurized systems prevented the formation of bubbles with the resulting loss of efficiency.

Much argument has taken place over the merits and shortcomings of centrifugal and axial pumps and this subject has been excellently summarized by Horn(5). Briefly, the pump selected must produce sufficient pressure to overcome the resistance to liquor flow and must have adequate capacity to maintain the volume of flow. The differential pressure across the package, i.e. the difference in pressure inside the spindle and in the main dyebath, may range from 25 to 85 kPa. The pump will have to produce a slightly higher pressure in order to overcome the frictional forces of the pipework.

Whichever type of pump is used, the pressure generated should be fairly constant for a moderate change in volume of flow, so that any leaks in the columns of packages cause little difference in overall pressure.

Axial pumps can easily be reversed by simply reversing the motor, but this stop–start system is not efficient if frequent reversals are required. As axial pumps operate in both directions they have two sets of flow characteristics, and thrust on each set of bearings in turn can result in excessive wear. To operate efficiently, the propeller of an axial pump must be only slightly smaller than the diameter of the surround and this narrow gap is susceptible to fouling with bits of loose yarn, etc.

Centrifugal pumps, in which the liquor is forced outwards from the centre into pairs of ducts or ports, revolve in one direction. To reverse the direction of flow, a valve connecting the pairs of ports is moved through an angle, the extremes reversing the direction of flow and intermediate positions controlling the pressure and volume of flow. This four-way valve arrangement simplifies the operation of the motor and pump, which is run at constant speed, but demands a reliable sealing mechanism for distributing dye-liquor at temperatures up to 130°C.

Most manufacturers now use pumps designed to combine the merits of both centrifugal and axial pumps.

The pump is driven via a V-belt arrangement by an electric motor; the motor size ranges from 5 to 15 kW per 100 kg of yarn, depending on maximum flow rate required.

Secondary Pump

Nearly all high-temperature package-dyeing machines are pressurized by means of a small pump operating on the inlet side of the main pump, thus pressurizing the whole system. The secondary pump takes liquor from the bottom of the expansion tank and injects it to feed the main pump. This system is also used to add dyes or auxiliaries to the dyebath.

The secondary pump is usually of the centrifugal type and operates continuously; the design of pump should be such that the maximum pressure generated cannot exceed the maximum safe working pressure of the vessel. Most machines normally operate at static pressures of around 275–310 kPa (40–55 lbf/in^2) and it is customary to pressurize even when dyeing at 100°C in order to avoid cavitation at the main pump.

Heating

The conventional method of heating is to feed high-pressure steam (550–700 kPa, 80–100 lbf/in^2) into horizontal coils located in the lower part of the dye vessel just below the frame carrying the yarn packages. This arrangement provides a large surface area for heat exchange (so permitting rapid heating) and offers negligible resistance to the flow of liquor. Being within the dye vessel itself, the system is compact and minimizes heat losses. The main disadvantage of internal heating coils is the lack of accessibility should the development of slight cracks, build-up of dye or oligomer contamination demand attention.

Normally the steam coils drain to a steam trap leading to a condensate line feeding back to the boilerhouse. In most cases coils primarily intended for heating can also be used for cooling; when the steam inlet valve and condensate valve are shut off, cold water can be passed through the coils to cool the dye-liquor circulating around the coils.

Heating/cooling coils are subjected to severe thermal and mechanical shocks when steam is first applied to heat a cold dyebath, or when cold water is fed in to cool a bath at 130°C. The coils should be well-braced to minimize movement likely to lead to cracks that are difficult to get at and expensive to rectify.

A few machines have been built with separate heating and cooling coils; this arrangement eliminates the problems of switching from heating to cooling, but the additional coil increases the cost.

Liquor Flow

The main pump forces liquor into the base of the frame, up the spindles, through the yarn packages into the main dyebath, and down through the heating coils back to the pump. This direction of flow is 'in-to-out' through the packages and when reversed is 'out-to-in'.

Care must be taken not to distort the packages during dyeing and sudden pressure changes should be avoided. Packages of hydrophilic fibres should be allowed time to wet out and swell before pressure is applied. Any region of gross distortion, e.g. a 'blown' package, leads to a flood of dye-liquor through the gap with consequent liquor starvation of other packages on the spindle.

Out-to-in liquor flow always tends to compress the package so reducing the size of pores in the structure and reducing the liquor flow. In-to-out flow usually causes slight enlargement of the packages (especially if of textured continuous-filament synthetic yarn), reducing the package density and facilitating liquor flow. However, excessive in-to-out pressure may flatten non-stretch yarns, so increasing the resistance to flow.

Most package-dyeing machines offer two-way liquor flow controlled by a timer; 3 min in each direction is a commonly used cycle, although for rapid dyeing more frequent reversal is recommended(*3,4*).

The piping conveying the liquor should be as short in length and as large in diameter as possible in order to reduce frictional losses. A reduction in the bore of the piping by a half increases the liquor velocity by the power of two and the frictional losses by the fifth power(*5*).

The pattern and extent of flow of dye-liquor through yarn packages has been a subject of much speculation and investigation(*6–14*).

In the ideal system it can be shown that

$$V = \frac{K P n \pi d^4}{i w}$$

where V = rate of flow of liquor per unit area
K = constant
P = pressure difference between inside and outside of package
n = number of channels per unit cross-section
d = diameter of channels per unit cross-section
i = viscosity of the liquor
w = wall thickness of the package.

The above finding assumes that the package is stable; if the pumping pressure is raised to a level at which yarns are flattened so reducing the diameter of the channels, then the rate of flow will be markedly reduced. A decrease in liquor viscosity (on heating) increases the rate of flow; increasing the wall thickness decreases the flow.

Package permeability varies tremendously from continuous filament (c.f.) viscose yarns (low) through staple yarn packages to textured c.f. synthetics (high). In the case of non-textured c.f. yarns the following factors have an important bearing on package permeability:

(a) Yarn denier; the permeability increases with the denier
(b) Filament denier; fine filament yarns tend to flatten and give low permeability
(c) Yarn twist; filament yarns with low twist (less than 100 turns/m), adopt a flat rather than rounded shape and give packages of low permeability
(d) Yarn traverse of the package; a low rate of traverse gives a more dense package with low permeability
(e) Yarn swelling; cellulosic yarns tend to swell in alkaline solutions, e.g. when applying vat dyes, and this reduces the permeability (some carriers have a similar effect on polyester yarn)
(f) Yarn tension; an increase in tension during winding gives a denser, harder package with lower permeability.

Non-textured yarns must have dimensional stability during dyeing. Nylon non-textured filament yarn must be wound onto collapsible bobbins, steamed (to induce relaxation) and re-wound onto the formers prior to dyeing, otherwise the strains developed in the yarn as the dyebath temperature is increased will generate very large contracting forces sufficient to damage both yarn and former.

With textured yarns package permeability is of a very much higher order and usually insensitive to factors such as yarn denier, twist and the presence of carrier. The comparative softness of the yarn package enables any shrinkage of yarn during dyeing to be accommodated without damage.

With packages of staple fibre yarns the use of fine (as opposed to coarse) denier fibres, fine counts, low rates of traverse and high yarn tension all contribute to increasing the density and reducing the permeability of the package. Additional twist usually helps to provide a rounded compact structure of the yarn and increases the package permeability.

Package densities are usually in the range 0.25–0.5 kg/l for staple yarns, the lower end of the range generally being easier for adequate liquor flow and level dyeing, but the higher end of the range offering benefits in terms of machine loading, productivity, and economies resulting from using a lower liquor ratio. With textured synthetic filament yarns, packages of around 400 g are wound and compressed by between 20 and 60% when mounted on the dyeing spindle.

Up to the mid-1960s most package-dyeing machines operated with liquor flow rates in the range 2–10 l/kg/min. The introduction of bulk quantities of textured synthetic yarns in dyeing packages of high porosity enabled comparatively high flow rates to be achieved and this permitted faster heating and cooling cycles to be used and processing times to be reduced substantially (rapid dyeing).

In conventional package-dyeing machines, flow rates are around 30 l/kg/min, giving complete dyebath circulation in about 30 s. In rapid-dyeing machines, flow rates are of the order 50–150 l/kg/min, the latter providing circulation of the complete dyebath every 6 s. Rates of up to 200 l/kg/min have been claimed(*13*) together with 20 dyebath cycles per min (i.e. 3 s per circulation), but these conditions are exceptional(*15*).

The effect of different rates of liquor flow on level dyeing has been analysed by Hoffman and Mueller(*16*), who included the effect of dye diffusion into the fibre. Care must be taken to avoid the formation of foam in the system, as this markedly restricts the flow of liquor through packages(*17*).

Liquor Ratio

In conventional package-dyeing machines, the volume of the dyebath is fairly constant and the liquor ratio depends on the weight of yarn which can crammed in. Most machines operate at around 10:1 when fully loaded.

Recently, demands for energy conservation and faster dyeing times have led to techniques in which varying amounts of dye-liquor are used:

(a) The fully flooded state, as described above, giving liquor ratios of about 10:1

(b) The partially flooded state, in which liquor covers all the packages but does not fill the dome of the machine; this enables liquor ratios of 8:1 to be used

(c) The low-liquor-level state in which only the base of the carrier is covered with dye-liquor. Obviously, only in-to-out flow through the packages can be used and the packages must be sufficiently rigid not to distort under these conditions (high-density packages of sewing thread can be dyed by this method). Liquor ratios can be reduced down to 4:1 or even 3:1 by this method.

Reducing the liquor ratio provides economies in water, heat, chemicals and dyes and, by reducing the heat capacity of the system, enables rapid heating (8–10°C per min) and cooling to be achieved without making exceptional demands on steam and cooling water services. However, the economies derived from reducing the liquor ratio are subject to the law of diminishing returns, and at very low liquor ratios there can be difficulties with dye solubility or dispersion stability, particularly where deep dyeings are concerned.

Sampling

Many package-dyeing machines are fitted with a small external pot designed to take one dye package. A bleed is taken from one side of the main pump to the sample pot to force dye-liquor through the package and then from the sample pot to the main dyebath.

For sampling purposes, the flow of liquor to and from the sample pot is shut off, the pot cooled (by passing cold water) to a safe temperature, depressurized and opened.

Unfortunately, despite liquor-flow regulators, it is extremely difficult to reproduce bulk dyeing conditions closely in a sample pot containing a single package, and the best a dyer can hope for is that the sample will be consistently slightly lighter or darker than the bulk.

An alternative system is to have a metal probe wound with samples of yarn extending into the dyebath. The probe can be withdrawn into a pressure lock which can be isolated and depressurized to allow a sample to be removed for examination. Here again it is extremely difficult to reproduce accurately the substrate/liquor flow interchange.

A third method is to insert small swatches of yarn ('flags') between two dye packages usually at the top of a column. The flag is attached to a small winding reel located in a pressure lock and can be withdrawn from the package, wound into the lock which is then isolated, cooled, depressurized and opened. Care must be taken to locate the flags so as to cause least disturbance to normal liquor flow and to avoid any tangling.

When colour matching is critical, the dyer may insist in taking a package from the bulk, but this is a very time- and energy-consuming process.

Expansion Tank

The expansion tank fitted alongside the pressure vessel accommodates the increased volume of the dye-liquor as the dyebath is heated. It also serves for the addition of dyes and chemicals to the system. With each addition of dye the total volume of the system increases and this has to be accommodated within the expansion tank.

In some cases a large expansion tank (reserve tank) is fitted in which virtually all the dye-liquor can be contained. This can be useful for retaining all the dye-liquor when sampling from the bulk goods or when keeping a standing bath. More elaborate machines provide heating coils in the reserve tank or even a pressurized reserve tank.

Coupled Machines

Despite taking all reasonable precautions, the package-to-package colour variation is such that most commission dyers of packaged yarn will not accept liability for this when the yarn is used in block colours. Likewise, batch-to-batch colour matching may show variation when yarns from different dye batches are woven or knitted immediately adjacent to each other. It is therefore desirable, from the viewpoint of colour uniformity, to dye as large a quantity of yarn as practicable in one operation and two or more machines have been linked or coupled together for this purpose. The dye vessels are doubly linked by pressure pipes to enable dye-liquor to circulate around both vessels. However, the pipework, complete with isolation valves (to enable both vessels to be used independently), is fairly complex and expensive and it has proved very difficult to obtain identical dyeings in each vessel (probably due to differences in the rate of liquor flow).

The general increase in size and load of package-dyeing machines has lessened the need for coupling systems, which are rarely used these days.

Horizontal-spindle Machines

A few horizontal beam dyeing machines, primarily intended for dyeing fabric, have been fitted with modified internal carriages containing perforated spindles mounted horizontally. The spindles are supported by the base plate, through which dye-liquor is forced, and a locking plate at the end nearer the door. Two-way flow may be used. The packages must resist any tendency to sag during dyeing otherwise top/bottom colour differences will develop.

The carriages of machines of this type are cumbersome and awkward to load, even with beams 2 m long, and their development must be regarded as primarily a means of increasing the versatility of what is basically a fabric-dyeing machine.

In the 1940s and 1950s much effort was expended in developing a horizontal-arm machine for dyeing viscose cakes. Several hundred machines of Courtaulds design were built and used in several countries to dye thousands of tonnes of viscose cakes for furnishings, corsetry, and the narrow-fabric trade. The machines were not pressurized, and operated on inside-to-outside flow only, so were cheap and simple to construct and operate.

Each arm accommodated about eight cakes with from two to 36 sample arms per machine (288 cakes, 185 kg); a set of four linked or coupled machines enabled nearly 750 kg of cakes to be dyed in a common dyebath so minimizing lot-to-lot differences in large orders for one colour.

A considerable technology was built up on methods of applying firstly direct dyes and subsequently vat dyes to viscose rayon cakes(*17–20*). However, the introduction of spun-dyed viscose filament yarn with excellent fastness in a wide range of colours at a price with which the dyer could not compete, started an irresistible decline (and the eventual extinction) of the rayon cake dyeing business, which accelerated as the synthetic fibres steadily replaced filament viscose yarn in one end-use after another.

Types of Dyeing Package

Several different types of yarn package can be produced for dyeing on spindles, the precise type of package depending on the properties of the yarn and its subsequent use.

All yarn packages for dyeing must provide:

(a) Adequate stability to withstand handling both on and off the dyeing spindle and to withstand the flow of hot dye-liquor, especially the pressure changes occurring when the liquor flow is reversed. The package must be able to accommodate both a small amount of potential shrinkage in some yarns and the effect of fibre swelling when wetted out.

(b) Adequate presentation to allow the yarn to be back-wound off the dye package for subsequent weaving or knitting. This implies that the yarn must maintain its position and not become trapped.

(c) Fairly uniform and sufficient resistance to liquor flow at different temperatures and pressures. This applies both within packages and between packages in the same dye-lot. Any areas of high porosity (low density, soft spots) will permit dye-liquor to stream through excessively, causing relative starvation of its neighbours on the same spindle. Packages of high resistance to flow will inevitably suffer a shortage of dye-liquor and be unevenly dyed.

Packages for dyeing on spindles include cheeses, cones, cakes, muffs and rockets.

Non-textured nylon or polyester filament yarns must normally be relaxed before winding onto cheese, otherwise shrinkage under dyeing conditions will cause damage to the yarn and crush the central former. Winding onto radially compressible sleeves or formers has met only moderate success.

Cheeses

Cheeses for dyeing can be sub-divided into those wound onto rigid formers and those wound onto longitudinally compressible formers.

Rigid formers (BS2547/7: 1979, ISO 3914) for in-house winding, dyeing and back-winding are usually made from stainless steel which, although expensive, has indefinite life. For external work, particularly for export sales, the tube is made from impregnated paper or plastic and generally intended only for one dyeing operation.

The most commonly used size is 55 mm in internal diameter, 170 mm long, with holes commencing 16 mm from each end.

Compressible formers have similar length and internal diameter, but are made either of stainless steel wire interlacing a helix of coarser wire (springs), or of plastic extruded in a criss-cross pattern. In both cases the objective is to permit lengthways compression of about 30% without noticeably increasing the diameter; indeed, the openness of the former permits a small degree of radial compression. Stainless steel springs have a long life and may be used for both textured and non-textured yarns and for staple. Plastic formers are usually cheaper but susceptible to dye-staining depending upon the class of dye, the temperature and time of application and the presence of carrier.

During winding the packages may be surface- or centre-driven, the linear speed being kept as constant as possible in order to maintain uniform tension in the yarn. Surface-driven systems are widely used for staple yarns, the rate of traverse (from top to bottom and return) being kept constant. Therefore, as the diameter of the package increases, the number of 'winds' (wraps round the package during one traverse) is reduced, but the wind angle (the angle formed by the yarn traversing the surface of the package and the radius) remains constant. As the package diameter increases there will be periods when the wind will nearly coincide with the traverse repeat and under these conditions the successive layers of yarn will be laid down almost in parallel thereby packing closely, to give a dense or hard place with poor permeability. To avoid wrap coincidence, an overriding mechanism (a wind gainer) has to be incorporated in the winding control.

Although the rate of traverse is nominally constant over most of the range, it drops rapidly to zero at the end of the limit. At this point the yarn briefly runs parallel to the edge of the package before starting the return traverse. This leads to an excess of yarn at the edges giving rise to 'hard edges' or, even worse, 'bulging edges'. This trouble is particularly pronounced with packages of small internal diameter; with increasing diameter (and circumference) there is more space to accommodate excess yarn.

With centre-spindle drive mechanisms the rate of traverse is related to the angular velocity of the spindle and so decreases as the package diameter increases.

Well-formed cheeses usually need sealing plates only at the top and bottom of the spindle. The adjacent cheese ends should, when pressed firmly together, form an adequate seal. After winding, the corners of firmly-wound cheeses should be 'bumped' to round them off, otherwise the corners may suffer from low liquor flow and be under-dyed.

Where practicable, a cotton stockinette or paper filter is provided between the former and the yarn. The ends of the stockinette are folded over the yarn package after winding to protect it during handling. The inside layer filters out dirt and undissolved/poorly dispersed dye before it gets onto the yarn.

Recent developments have included the use of bi-conical (BI-KO) polypropylene formers, reinforced with glass-fibre, which interlock to allow press-packing(*21*).

Cones

Cones facilitate the back-winding of yarn after dyeing and usually have a half-angle of 4° 20′ (BS 2547/5: 1978). A 17 cm long cone tapers from 81 down to 55 mm in diameter.

As with cheese formers, cones are made from stainless steel, or from plastic or impregnated paper, the latter intended only for 'one-trip' use.

Cones are generally used for staple fibre yarns, which tend to cling to each other at moderate winding tensions more so than non-textured filament yarns. Spun rayon yarns, which swell considerably when wetted, are often difficult to backwind after dyeing as cheeses and are better dyed as cones.

When mounted onto spindles, adjacent cones are usually separated by a circular plate designed to restrict liquor loss through the base of each cone.

Although cone packages are wound to ensure that the external shape is parallel to the cone former (to provide uniform wall thickness), the flow of dye-liquor is not as uniform as through parallel-sided cheeses. Cones also suffer the disadvantage that they do not accept compression lengthways when mounted on spindles.

Cakes

Cakes are parallel-sided packages without centres (most viscose filament yarn was formerly spun into cakes in a rotating cylinder – the Topham spinning box). Many thousands of tonnes of viscose cakes, covered by a cotton stockinette as protection, were dyed in the 1940s and 1950s, but this is now a forgotten art.

Cake dyeing has recently been revived for textured synthetic filament yarns. The yarn is wound onto formers about 7 cm in diameter and then the former is removed, allowing the inner yarns to collapse. The yarn can hence be dyed in a more relaxed form to give greater stretch properties.

Of more interest is the dyeing of high-bulk acrylic yarns in cake form as an alternative to hank dyeing. The yarn is first allowed to contract by passing it continuously over a steam bath, e.g. on a Superba machine, and then, after cooling, to be wound at low tension into a parallel-sided package. The spindle is then removed and replaced with a support of non-woven polypropylene. The cakes can then be mounted on dyeing spindles and compressed lengthways to give densities of 350–400 g/l at which they are sufficiently stable to be dyed using two-way flow(*15*). Care must be taken during winding to try to preserve an elliptical cross-section rather than a flat ribbon.

The advantages over conventional hank dyeing are the achievement of about double the weight in machine loading (so halving the liquor ratio), more rapid dyeing due to increased liquor flow, more rapid back-winding as the off-winding package is stationary instead of rotating, with fewer stoppages and less wastage.

Some woollen carpet yarn is now being dyed as large cakes rather than as large bulky hanks.

Muffs

Muffs are very soft, open-wound packages prepared by winding stretch yarns onto a frame of about 15 cm diameter and then, after removal, allowing them to contract. The packages are usually wound with a traverse to give a honeycomb structure and lease-strings are inserted to preserve the structure and facilitate back-winding, which can be slow and difficult as the fully crimped yarns are prone to snag with those adjacent.

Muffs may be mounted on spindles and compressed 50% or more in length or may be dyed in a pack-dyeing machine.

Rockets

The Delerue rocket winder was developed to provide fairly large packages of staple yarns which could be used directly on circular-knitting machines. The wind commenced on a conical base mounted on a spindle and traversed rapidly from periphery to centre and back again building the package vertically to a height of 25 cm or more. The package was parallel-sided with a conical top enabling packages to fit neatly on top of each other and providing excellent presentation during unwinding.

Rockets suffered from being rather fragile and were not amenable to centrifuging to remove excess water after dyeing. Commission dyers were reluctant to despatch yarn which they had not examined and lubricated during backwinding, and knitters were even more reluctant to risk producing faulty fabric through using unchecked yarns.

Bobbins
Small quantities of rigid (i.e. non-stretch) narrow fabrics (webbing) are wound onto perforated cylinders to form packages about 15–20 cm in diameter and 3–5 cm in height. Stainless steel circular plates separate the packages when on the spindles.

1.2 PACK-DYEING MACHINES FOR STAPLE AND TOW

'Pack-dyeing' refers to the dyeing of loose fibres (usually called 'loose stock' or 'raw stock'), slubbing and continuous tow, when formed into suitably shaped packs for dyeing.

The machines used for pack-dyeing are virtually identical to those used for dyeing packages, i.e. a cylindrical vessel capable of being pressurized and with means of circulating hot dye-liquor through a central point in the base of the machine, through the substrate, and back to the pump. However, the pack-dyeing machine is usually fitted with an annular basket or cage with perforated sides; the loose stock, tow, etc. is packed firmly into this cage and dye-liquor forced through horizontally.

A typical cage is about 150 cm in overall diameter with a race (annulus) of 40 cm and about 80 cm deep. The bottom of the cage is held firmly to the bottom of the vessel and a top plate fitted to prevent dye-liquor flowing over the top of the substrate. The cage is lifted in and out of the vessel by overhead crane.

The cage may be packed by hand or, in some cases, mechanically. To obtain a uniform and stable pack it is often desirable to wet the fibres with water and then compress them; pack densities are normally around 300 g/l, but can go up to 480 g/l in the case of acrylic continuous tow, which tends to deform and consolidate during dyeing(*22,23*).

When loading staple a bottom plate with lifting chains is sometimes put in the cage so that, after dyeing, the dyed fibres can be lifted out and conveyed to the hydro-extractor or dryer (Figure 1.3).

When continuous tow is loaded into cages(*23*) care must be taken to avoid inserting twist as this can cause problems in subsequent processing operations such as stretch-breaking. The cage is mounted on a turntable and rotated very slowly whilst the tow is run through a light nip, sprayed with hot water and mechanically 'stamped' to consolidate the pack. After dyeing the tow is carefully withdrawn whilst the cage is rotated in the opposite direction to avoid inserting twist. Despite these precautions, the condition of the dyed tow is markedly inferior to the original and, faced with competi-

Figure 1.3 – Platt Longclose pack of staple fibres being unloaded from its cage

tion from producer-coloured acrylic and polyester tows and a few continuous tow-dyeing machines, batchwise tow-dyeing has virtually disappeared in the UK.

Cages can also be used for dyeing hanks of yarn or slubbing, bump tops, cakes or muffs of yarn packed brick-fashion, or for dyeing terry-type towelling plaited in rope form.

A 165 cm diameter machine, which has a liquor capacity of 3 200 litres, has a nominal loading of about 365 kg of loose stock or 550 kg of continuous tow packed into cages. Two-way liquor flow is normally employed during dyeing, and dirt and undissolved dye, etc. must be rigorously excluded as the pack serves as a very fine filter.

A few loose-stock-dyeing machines are in use in which the flow of dye liquor is vertical. The loose stock rests on a perforated base and dye-liquor is forced upwards through the fibres and out via a perforated top plate to be returned to the pump. The machine is relatively cheap and adequate for many purposes.

Slubbing Dyeing
Most slubbing is dyed in containers about 60 cm in diameter and 1 m tall fitted with a central, 7 cm diameter perforated spindle for lifting and for circulating dye-liquor. The sides of the container are perforated and the tops can be either packed into the container or coiled in, the load being wetted and finally compressed in height to give a more uniform and denser pack (about 300 g/l). A large machine will accommodate four canisters, each mounted on a liquor delivery point fitted to the base of the machine.

A 1.6 m diameter machine with a liquor content of 3 200 litres will take about 320 kg of coiled tops, giving a liquor ratio of 10:1.

Warp-beam Dyeing
Warps of yarn for weaving are dyed by winding them onto a perforated beam, usually about 15 cm in diameter. Steel flanges are fitted at the precise width before winding and the beam barrel wrapped with non-absorbent, non-woven fabric to form a good 'bed' for the warp yarn, to permit diffusion from the perforations in the beam barrel and to serve as a filter. The warp must be provided with leases and a slight traverse motion is desirable in order to provide a uniform build-up with adequate porosity. The winding tension must be carefully controlled, as a marked tendency for the yarn to shrink lengthways will produce very high pressures on its inner layers and on the beam. Likewise, the effect of fibre swelling on wetting must be anticipated and the flanges must be sufficiently robust to withstand sideways pressure.

Warp beams are normally dyed in the vertical position, up to four beams being mounted onto a carrier and secured at the top. The carrier, complete with beams, is loaded into the dye vessel and dye-liquor fed to the centre of each beam; in-to-out flow through the beam is employed for most of the dyeing cycle.

This method of dyeing is used to produce dyed cotton warps for weaving with undyed weft into denim fabric.

1.3 HANK-DYEING MACHINES FOR YARN

Hank dyeing (also called skein dyeing) is probably the earliest method of yarn coloration when a number of hanks of yarn were suspended in a dyebath on two horizontal poles which were lifted manually. Later mechanically operated systems were employed, such as the use of an elliptical cross-section arm that rotated slowly immersing different parts of the hank in the dyebath. Fifty or more arms were loaded with hanks and then gently lowered into the dyebath beneath. Such machines occupied a large floor area and were prone to mechanical breakdown.

Nowadays only bulky yarns are dyed in hank form, these include woollen and high-bulk acrylic yarns for machine knitting of outerwear, hand-knitting yarns and carpet yarns (especially wool).

Recently these hank-dyeing operations have been invaded by the package dyer using very carefully prepared packages that give a product, which, although not as bulky as the hank-dyed article, is considered to be commercially acceptable especially because of the savings in water, energy, chemicals and time and, above all, the increased productivity(*15*).

Circular Machines

Circular pack- or package-dyeing machines can be used to dye hanks of yarn or narrow fabric. The hank carrier consists of perforated plates at both top and bottom with a series of quadrantal rods of different lengths to support the weight of the hanks at the top and to restrain them from movement at the bottom. The flow of dye-liquor is vertical, i.e. along the length of the hanks which must be carefully arranged to fill the vessel uniformly. Although the entire vessel is sealed and the internal pressure is 135 kPa, the differential pressure across the hanks is probably only 5-10 kPa, and the pump pressure is far less than that required in package dyeing. The liquor passes up the centre spindle and overflows onto the upper perforated plate.

With downwards liquor flow the underside of the hanks in contact with the upper quadrants may well be dyed paler than the bulk. The flow should be reversed regularly to permit upwards flow, which will tend to lift the top of the hank away from the upper quadrant.

The distance between top and bottom quadrants should be such that about a 3 cm gap is left between the bottom quadrant and the bottom of the hank. With hanks of good dimensional stability the correct spacing is easy to establish, but with high-bulk acrylic hanks, intended to shrink 20% in length when relaxed in hot water, preliminary trials may be required.

Although used more widely in Europe, circular hank-dyeing machines have not gained general acceptance in the UK, mainly because they are cumbersome in loading and unloading and tend to have excessively vigorous liquor flow for delicate, high-bulk acrylic yarns.

Nevertheless, a hank carrier offers increased versatility for the commission dyer of yarn packages.

Overflow-type Machines

The traditional type of hank-dyeing machine (often called a 'Hussong') consisted of a rectangular dyebath with a weir at one end. The yarn in hanks was suspended from an assembly of parallel rods or sticks held in a detachable frame; an identical series of rods was fitted just inside the bottom of the hanks to provide control of the load.

By means of a propeller, dye-liquor was circulated over a weir and through perforated plates fitted horizontally above and below the load to diffuse the liquor flow. The direction of flow was reversed frequently in order to avoid under-dyed places where the yarn was in direct contact with the rods. The lower rods were located about 2–3 cm above the bottom of the hanks, and, when high-bulk acrylic yarns were introduced, an allowance

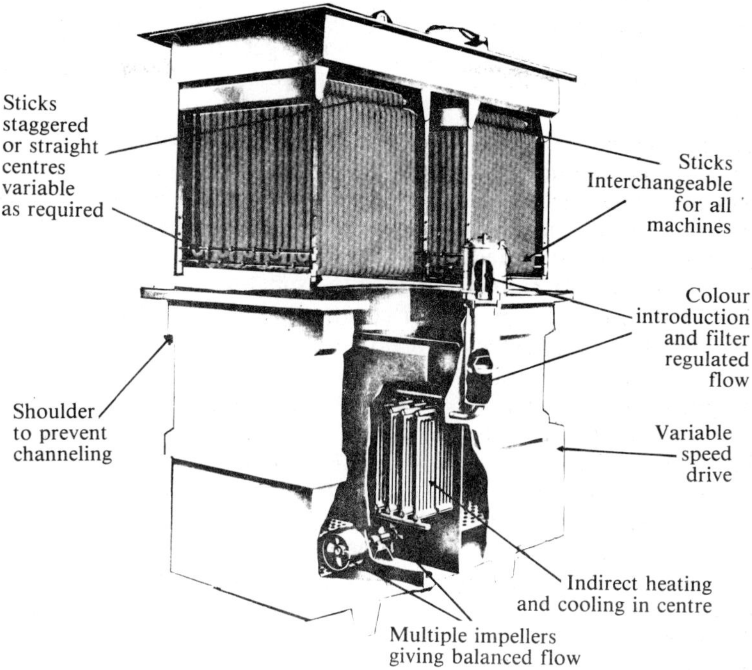

Figure 1.4 – Leemetals hank-dyeing machine

had to be made for yarn shrinkage whilst in the dyebath. A more refined machine operating on the same principles is shown in Figure 1.4.

A big advantage of this type of machine was that a set of spare sticks was cheap and, whilst one lot was being processed, a second could be set up on sticks ready to be loaded into the carrier to reduce down-time between dye lots.

This type of machine is comparatively cheap to buy and, in principle, very simple to operate. It fulfils the basis of hank dyeing, which is essentially to enable a large volume of dye-liquor to pass gently through a substrate of low resistance at low pressure. This system demands very uniform packing of the entire load with no open areas through which dye-liquor will channel.

A perpetual problem when dyeing in this type of machine is that the hanks at the outside of the carrier frame are often in direct contact with the sides of the dyebath (sometimes a perforated polypropylene fabric is used as a side cover to minimize direct contact with the metal sides). At the very low liquor pressures employed flow at the sides of the vessel can be reduced to an inadequate level. The slow flow of dye-liquor through the hanks does not usually permit a rapid rate of temperature rise to be employed. Probably due to the side-wall effect it is very difficult to get uniform dyeings when using small machines. The pattern of flow of dye-liquor has been discussed in several papers(*24–26*).

The machine is filled with water through an inlet beneath the false bottom; the drain is similarly located, as filling and emptying should cause minimum disturbance to the arrangement of the hanks. Heating is by a heat exchanger either mounted in the centre section (into which dyes are added), or (in large machines) beneath the false bottom.

Because of the voluminous nature of the substrate, liquor ratios are high, generally approaching 30:1. Power consumption is low, the electric motors driving the propellers usually being 1–2 kW. The dye vessel must be kept clear of any rough places or projections which may catch the yarn; the temperature probe is usually located below the false bottom of the machine.

The rectangular shape of the machine does not facilitate pressurization for high-temperature dyeing, but a few machines have been adapted to operate normally at 105°C (maximum 108°C) to increase the rate of levelling of basic dyes on acrylic fibres.

The machines are loaded and unloaded by an overhead crane and, as unloading is a messy business, the machine surrounds should be well-drained.

Pegg GSH Machine

This machine, introduced commercially in the 1960s, uses a yarn carrier in which the sticks are fixed on each side of the main longitudinal member; dye-liquor is circulated through the sticks and discharged upwards through perforations. The carrier fits onto a liquor entry point on the false bottom at the centre of the machine. The liquor circulates in one direction only: up through the carrier, through the sticks, downwards through the hanks, through the false bottom and back to the propeller (Figure 1.5).

Heating and cooling pipes, usually in a serpentine arrangement, are fitted underneath the false bottom, where there is ample room, and so provide negligible resistance to flow. The machine is filled so that the sticks are immersed to a depth of 5–8 cm. A row of bottom sticks is provided to prevent excessive movement of the hanks, but the use of one-way flow (top-to-bottom) and practical experience in loading made these bottom sticks unnecessary and many dyers operate in single-stick fashion.

The use of perforated sticks to discharge dye-liquor provides a more uniform distribution than the traditional overflow method and, by discharging upwards from the sticks, the hanks are lifted slightly so eliminating flattening/glazing or even undyed patches where the hank rests on the stick. The distribution of liquor flow in the carrier is controlled by internal valves on the top of the frame. A machine loaded with 200 kg of wool, and having a liquor capacity of 4 800 litres, was shown to have a cir-

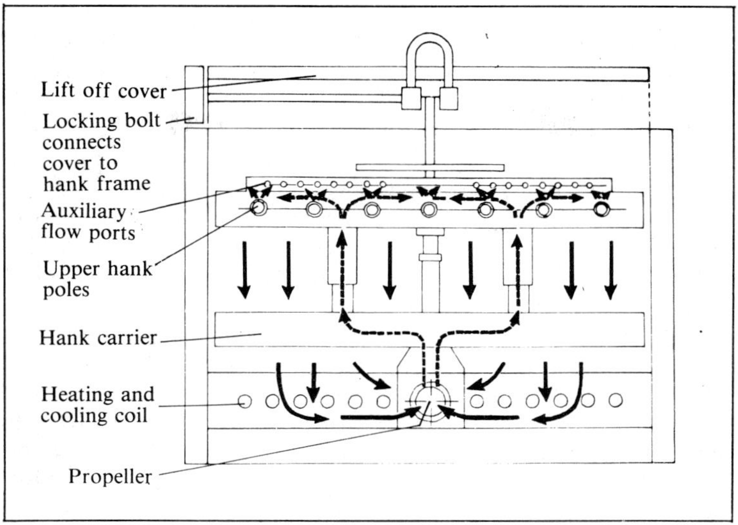

Figure 1.5 – Side view of a Pegg GSH hank-dyeing machine showing liquor circulation

culation of 3 200 l per minute, i.e. 0·73 bath vol./min or 16 l/kg/min(*27*).

Additions of dye are made underneath the false bottom so that they pass through the propellers before being distributed through the carrier. The propellers are driven through a V-belt arrangement by a 5 kW motor fitted at the side. Large carriers holding 200 kg of yarn are supplied by two propellers at the base, each driven by a separate motor.

The water inlet is fitted beneath the false bottom to avoid disturbing the arrangement of hanks. In the days when water and energy were cheap, dyers of hanks preferred to flood-cool after dyeing by running in cold water and allowing the hot dyebath to overflow to waste. This combined cool/rinse operation was justified in terms of productivity and economics.

The need to dye larger weights of yarn led to the development of a machine of 400 kg capacity in which two carriers were mounted side-by-side and a 'jumbo' model in which three carriers, each with 200 kg of yarn lay together in the dyebath. These machines performed very well, demonstrating that large machines of this type give better liquor flow than small machines, presumably because the side-wall effect is minimized.

The size of hank-dyeing machines presented some problems, particularly in regard to covers. Originally detachable metal covers were supplied which were intended to be installed using the overhead crane after the carrier with its load had been safely located in the vessel. Dyers complained that this operation took excessive time and trouble, that the lids, when not actually on the machine, occupied a lot of valuable room in the dyehouse, and that, when installed, any untoward movement of hanks during dyeing could not be seen. To meet these objections, some later machines were fitted with counterbalanced lids which were pivoted to give minimum obstruction during loading and unloading(*28*).

Despite great improvements in quality and productivity in dyeing hanks of yarn (including the stepwise increase in hank size from 150 to 450 g), hanks suffered from being extremely bulky and requiring to be rotated on a frame during backwinding which was a labour-intensive operation requiring much floor space (one establishment employed five times as many winders as dyehouse operatives!). The development of winding bulked yarns into dyeable packages which could subsequently be backwound from a stationary package at twice the speed of hanks signalled a big decline in hank dyeing, particularly when it was demonstrated that machine loadings could be doubled with consequent economies in water, steam and chemicals. Some GSH machines were converted with carriers designed to take the new mock-cakes on horizontal spindles and more powerful motors were fitted to provide the additional pressure required.

The economic benefits of this technique(*15*) are such that it appears that the importance of hank dyeing will continue to diminish except in those few areas where the product of package dyeing is not commercially acceptable due to lack of bulk.

1.4 CONTINUOUS DYEING OF FIBRES, YARNS, STAPLE AND TOW

Comparatively few systems have been developed for dyeing fibres and yarns continuously. As well as the mechanical difficulties involved, few classes of dye are amenable to being applied and fixed rapidly and many of these suffer limitations in terms of depth and subsequent fastness to washing. Furthermore, continuous dyeing ranges demand large weights per colour to achieve non-stop running, a requirement commercial conditions rarely provide.

In the heyday of cheap tufted carpeting, enormous quantities of 15 denier, matt-crimped viscose staple were required in a variety of colours for spinning into yarn, and three Fleissner continuous dyeing and drying ranges were in operation(*29*). This machine consisted of:

(a) A hopper feeding an opener to give a uniform lap of fibres.
(b) A dyebath of fairly small capacity in which the lap was taken round the underside of a wire mesh cylinder almost completely immersed in the dyebath containing direct dye and salt at 95°C. The drum was fitted with an internal propeller removing dye-liquor from inside the drum and feeding it back to the main dyebath. The liquor flow into the drum held the lap of fibres to the surface.
(c) A squeeze mangle onto which the lap was fed by a conveyor belt on leaving the drum.

(d) Two suction drums through which a steam–air mixture was circulated to provide a 2 min fixation period.
(e) A third and final drum for drying and cooling.

Despite early mechanical difficulties, mainly due to the highly corrosive condition of the dyebath, the machines worked well and many hundreds of tonnes of fibre were dyed satisfactorily and economically. However, the use of direct dyes imposed restrictions on the subsequent wet fastness and limited production to pale colours for carpeting.

The continuous dyeing of loose wool using a fibre-bowl scouring machine has been described(*30*), the first two bowls being used for scouring, the next two for rinsing, and the final bowl for applying milling acid dyes at pH 2.5–4 at 96–99°C for 60–90 s. Weights of at least 2 000 kg per colour (equivalent to 4 h running) were recommended and the short build-up time restricted the depth of dyeing attainable.

A recent development has been the incorporation of a radio-frequency unit to facilitate dye fixation. The so-called Smith machine consists of:
(a) An automatic hopper feeder fitted with spiked brattices to provide a uniform web of fibres.
(b) A feeder conveyor for delivering the web to the top of a metal slide leading to the dye padder.
(c) A dye padder consisting of two rollers mounted one above the other, the pneumatic pressure being applied to the top roller (maximum nip load 500 kg). A horizontal spray-bar is fitted to the front of the mangle and sprays the web before it enters the mangle, a padding conveyor belt assists the web into the mangle
(d) A ram section to which the web passes on a conveyor belt on emerging from the mangle.
(e) A radio-frequency (27 MHz) unit, with two 70 kW generators. The fibres are forced periodically through an inclined barrel 7–8 m long and 30 cm in diameter which passes through the heating unit.
(f) A final spray washing and squeeze nip section to remove unfixed and surface dye.

In a machine of this type the fibre remains in the barrel for about 30 min being heated for the first two-thirds of its passage. The maximum internal temperature is only slightly above 100°C and overheating will result in excessive steam loss through the exit end and drying of the substrate with consequent poor fixation. The production rate is around 500 kg/h, rather higher for acrylic fibres than for wool owing to higher packing densities and lower pick-up of water.

The Continuous Dyeing of Synthetic-fibre Tows

The increasing use of polyester and polyacrylic fibres in worsted spinning led to increasing demand for coloured tows. Spin dyeing (mass coloration) presents many problems, particularly in melt-spinning polyester, and batchwise dyeing in cages adversely affects the condition of the tow leading to

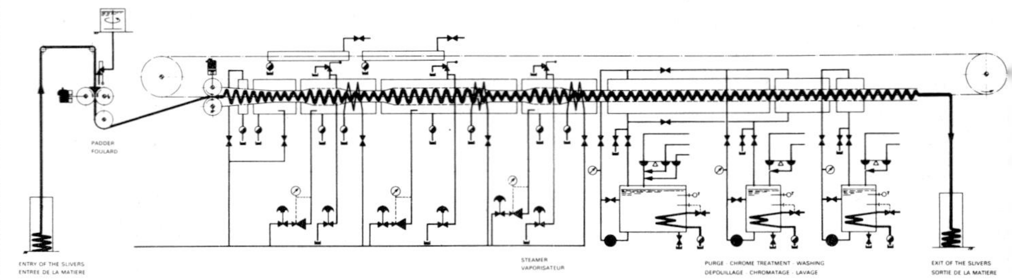

Valves shown control the flow of air, water, steam and chemicals.

Figure 1.6 – Serracant Tunel tow-dyeing machine

poor performances in subsequent processing.

The Spanish Serracant 'Tunel' machine for continuously dyeing polyester and acrylic tows is an excellent example of the machinery manufacturer's response to the dyer's needs. Introduced in the late 1960s, the machine takes two tows and, after applying dye by padding, feeds them through a tunnel with an internal diameter of about 20 cm and 15 m long. The tows are carried through the tunnel by a heavy chain with horizontal spikes on each side which support the tows. At the entrance to the tunnel two pairs of rollers are mounted on either side of the chain (Figure 1.6). The tows pass between the rollers and then converge and lie on the spikes throughout their passage through the tunnel. This arrangement serves as an effective seal enabling saturated-steam temperatures of up to 160°C to be attained in the tunnel. Generally the first half of the tunnel is used for dye fixation, the second half for washing in up to three separate zones. The packing of the tow in the tube helps to maintain the pressure in the high-temperature zone. The chain is driven by a d.c. motor and returns through a non-pressurized tunnel, steam jacketed to maintain heat in the chain.

Production capacities are rated at up to 250 kg/h for polyester or 300 kg/h for acrylic tow, and steam consumption is quoted at 1 kg steam/1 kg tow. The Serracant machine is used in many countries.

1.5 DRYING

The traditional method of drying packages of yarn after dyeing was to remove most of the extraneous liquor by centrifuging in a basket-type centrifuge hydro-extractor followed by final drying in a warm air stove, an operation often taking 24 h or more and occupying considerable space.

With increasing size of loads which can be dyed in a package-dyeing machine, dryers were developed for handling the complete load whilst still on the carrier frame.

The vessel (Figure 1.7) is of similar diameter to the largest dyeing machine and the carrier frame is lowered to fit onto a central boss. The kier is closed and securely locked and compressed air introduced to give a pressure of 480 kPa (70 lbf/in^2). The blower motor (80–100 kW) is started and air forced through the packages from outside to inside squeezing out water which is removed via a separator to drain. Out-to-in flow is applied for about 10 min during which time heat is applied via a steam coil in the heater battery.

The flow of hot air is then switched to in-to-out which tends to expand the packages and permit a greater flow of air for drying. Automatic flow reversal can be used if required. Water vapour is condensed in the condenser unit which provides a source of hot water.

The ducting is of large diameter in order to minimize frictional losses and the blower fan should be well-muffled in order to restrict the level of noise. The temperature of the air entering the vessel can be controlled as appropriate to the fibre being dried, cold air is passed to cool the yarn after drying.

The process, in terms of direction of flow, time and temperature, can be programmed so that, once set and started, no further attention is required before unloading. As a refinement, a relative humidity sensor can be incorporated to indicate when the yarn has reached adequate dryness.

The load and time per cycle required depends on the yarn, wind and size and number of packages, but a 160 cm diameter machine loaded with about

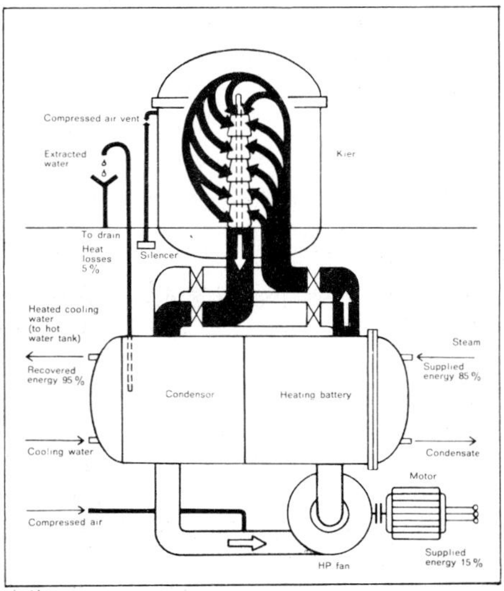

Figure 1.7 – Avesta yarn dryer

200 kg of cotton needs about 50 min per cycle; wool packages take around 30 min and textured synthetics appreciably less.

Earlier package-drying machines operated at around 200 kPa (30 lbf/in^2), but many suffered from poor reliability due to cracks and leaks in the heating/condensing units. Care should be taken to ensure that both materials and construction are of adequate standard to withstand commercial operating conditions.

Dyed staple is usually dried by centrifuging followed by drying on a slow-moving, perforated conveyor whilst passing very slowly through a stove in which hot air is gently forced downwards through the bed of fibre (usually called a brattice dryer). A spiked feeder from a hopper is important to open the mass of wet fibres and to present a uniform bed on the conveyor. The hot air is re-circulated from the bottom to the top of the machine(*31*).

Although the brattice has low power requirements, it is a large machine and presents problems of contamination when dye-lots of different colours have to be dried in succession. A more compact and effective dryer is the Fleissner suction drum in which hot air is drawn from outside through the bed of fibres held to the perforated surface of the drums. The rate of drying is very much greater than on a brattice and the time of drying reduced to about 1 min. The drums are usually made in units of two, four or six and fitted with internal baffles to promote the desired air flow.

Although centrifuging is a very efficient method of removing extraneous water from textile fibres, it is a batchwise process often requiring wet and messy handling during loading and unloading, although loading in net bags using a hoist is now used.

The centrifugal force acting on a body in a spinning hydro-extractor is proportional to the weight of the body, the diameter of the hydro-extractor basket, and the square of the number of revolutions per minute.

$$\text{The 'g' value} = 5.6 \times N^2 \times D \times 10^{-6}$$

where N = number of revolutions per min
D = diameter of basket in cm.

A 120 cm diameter basket rotating at 1 440 rev./min will produce a 'g' value of about 1 392, which will reduce the moisture contents of acrylic yarns and tops down to 6–8% and of wool down to about 27%.

One centrifuge especially suitable for yarn in hanks and staple has a basket diameter of just over a metre and a depth of 735 mm (volume = 51 dm^3). The basket is made of galvanized steel and is fitted with a removable top designed to be lifted with the load by a hoist. The top of the machine has a hinged and balanced safety cover with a sight glass. The cover is operated pneumatically and interlocked so that it cannot be raised while the basket is moving.

The basket is mounted on a central spindle carried by ball and roller bear-

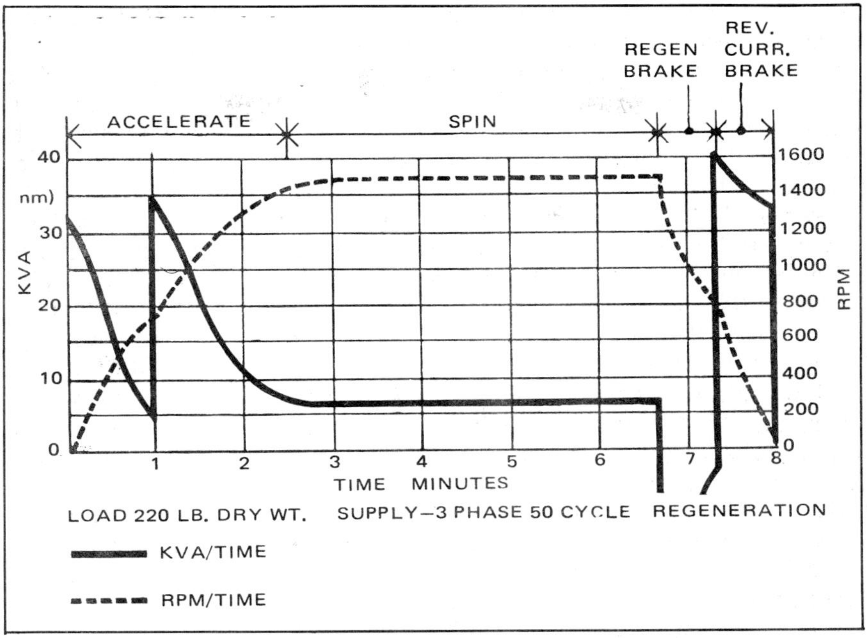

Figure 1.8 – Power and rev./min pattern for a Broadbent centrifuge

ings. The electric motor is built round the spindle eliminating the need for belt drives or coupling. Rated at 415V, 50 Hz, 3 phase, it takes 50 A at initial start-up and 56 A whilst braking. The speed attained and the current demanded during the cycle are shown in Figure 1.8. Due to the high current requirements for starting and stopping, it is recommended that the number of cycles per h should not exceed eight.

The machinery is heavy – about 2.5 tonnes – and a strong concrete floor, 150 mm thick, is required; if this is not available, a 300 mm thick foundation block is recommended. A flexible spring-suspension system is incorporated to absorb vibration, and an out-of-balance limit switch to knock off the motor if the load is badly balanced.

Progress in designing openers to provide reasonably consistent fibre beds for hot-air drying led to the introduction of squeeze mangles in front of hot-air dryers. The efficiency of squeeze mangles is dependent on the uniformity of the web of wet fibres, and moisture retention is higher than that following centrifuging. Nevertheless, as an operation coupled with continuous drying, mangling has obvious advantages over centrifuging. Care must be taken to try to avoid clumps of wet fibre passing through the nip as hard lumps are liable to damage the rubber rollers.

Hanks of yarn are usually centrifuged and then suspended from poles mounted on side-rails or from hooks on an overhead track whilst passing through a hot-air stove. In the case of synthetic yarns, e.g. high-bulk acrylic yarns, high-speed hydro-extraction reduces the retained water content down to about 7%, so making hot-air drying unnecessary.

In recent years the use of radio frequency or dielectric energy for drying textiles has been developed to the stage of industrial use(*32*). When placed in a rapidly alternating electromagnetic field, dipoles – such as water molecules – oscillate and generate heat. A major attraction of the principle is that the amount of heat generated is related to the amount of water present, so that the wettest portions heat up at the same rate as those with low moisture content. Further, as water is lost during drying, the amount of heat generated decreases, permitting a small amount of water to remain at the end of the operation. Finally, whereas conventional drying depends on hot-air drying the surface and on capillary movement of liquor from inside to the surface, dielectric heating generates heat wherever the water is, whether on the surface of the fibre or imbibed.

In the UK only a limited number of frequencies are permitted for industrial use, and most dielectric heaters for textile drying use 27.12 MHz. Dielectric drying machines have a conveyor belt moving slowly through a rectangular tunnel in which an electromagnetic field is applied. Most units are of 50 kW output and will evaporate about 70 kg of water per h giving an output ranging from 150 kg cotton (moisture content reduced from 48 to 8%) to 400 kg acrylic yarn (18 to 3%).

Total energy input is 80 kVA, and the conversion of electrical to r.f. energy produces about 30 kW of waste heat, some of which may be used to heat the air required to help remove the steam liberated in the tunnel. Water cooling may be required for the r.f. generator. Care must be taken in the design of the unit to ensure that steam does not condense on the cold surrounds and then drip back onto the goods.

For package drying it is desirable to remove the majority of water mechanically before applying dielectric heating; one machine operates on a carousel principle in which at the loading station a number of packages are placed vertically on plastic bases. As the carousel revolves, suction is applied to the base of the packages and dielectric heating applied overall. The time between loading and unloading is 5–8 minutes for polyester and acrylic yarns to 10–18 minutes for cotton and viscose rayon. Throughput ranges from 130 kg/h for rayon up to 400 kg/h for synthetic fibres. A fairly recent development is the drying of yarn packages on dye-springs, provided the latter lie parallel to the electrical field.

Considerable care must be taken concerning the practical operation of a machine supplied almost continuously by one person. Not only should the overall rate of working be carefully planned (both for the novice and for the experienced operative), but the frequency of attention is important. Operating cycles which permit a short period between successive cycles offer much greater freedom for the operative to perform ancillary tasks than in the case of cycles repeating in seconds.

Although the frequency of 27.12 MHz is selected to avoid interference with broadcasting equipment, care should be taken to minimize the leakage of stray radiation.

The development of radio-frequency dryers for textiles has advanced slowly during the last twenty years. Fundamentally, the technique uses a comparatively expensive source of energy (electricity) and only about half is converted into useful radio-frequency energy. Advantages are the elimination of large quantities of hot air to transfer heat to the substrate (along with fans, heat exchangers and ducting) together with excellent flexibility in terms of short start-up and run-down times. Future trends point towards systems in which mechanical methods are used to remove water down to a low level followed by radio-frequency drying to the desired moisture content.

2 MACHINES FOR DYEING FABRICS IN BATCHES

2.1 JIG DYEING MACHINES

The jig or jigger dyeing machine is one of the oldest types of machine for dyeing woven fabrics which must not be creased during dyeing, e.g. most taffetas, satins, poplins, ducks, suitings. Jigs have been tried for dyeing certain types of knitted fabrics, e.g. nylon shirtings, but with limited success; however, knitted power net fabrics with nylon warp and elastomeric weft are dyed commercially on both jig and pressure beam dyeing machines. The principle of the machine is very simple, the fabric being wound into a roll and then run through the dyebath at the bottom of the machine and onto a second roll and then, when the second roll is full, reversed (Figure 2.1).

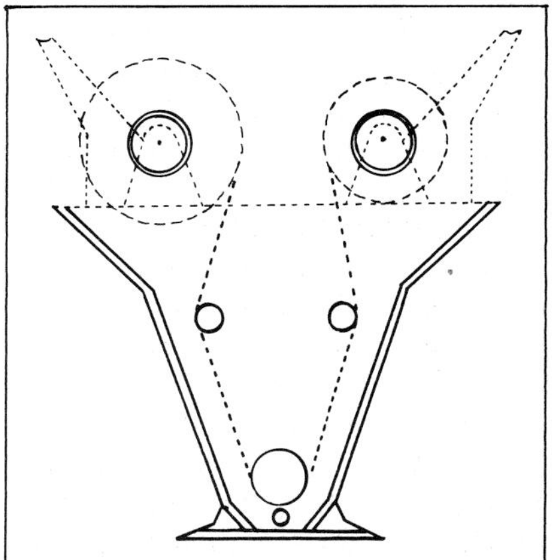

Figure 2.1 – Cross-section of a simple dye jig, showing the method of threading

Although wider jigs have been produced, most modern jigs have rollers 1.8 or 2.0 m long.

For dyeing polyester woven fabrics, a handful of experimental pressure jigs have been built to dye at temperatures up to 130°C, but with very limited success.

The fabric roll is usually formed or 'run-up' off the jig and several metres of cheap end-cloth are sewn to each end of the roll. The ends of each length of fabric are trimmed straight across the width and then butt-end sewn to avoid any overlap. The ends of the fabric should fit together as closely as possible in order to minimize the amount of dye-liquor held in the join which will mark-off onto adjacent layers of fabric in the roll.

On old jigs, roll lengths of 500 or 1 000 m were customary, but recent developments have enabled rolls of up to 1 m diameter to be accommodated, representing around 5 000 m of lightweight taffeta cloth.

Old jigs with small fabric rolls used dyebaths of around 200 l with a relatively small surface area, which was advantageous in minimizing oxidation when applying reduced vat dyes to cotton. The demand for jigs with larger load capacities, necessitating bigger diameter rolls, has led to the development of machines which have a greater dyebath capacity.

It is essential for all the fabric in the roll to have the same width and for the fabric to be run-up very straight so that the roll has sides perpendicular to its axis. On most jigs one of the rollers can be traversed to a limited extent to assist in straightening a roll which has run out of true. Some jigs have the facility to traverse over a small range automatically to minimize selvedge build-up on certain cloths.

For loading, the leading end of fabric on the roll is threaded under two free-running rollers located at the bottom of the machine and then led upwards and around the draw roller. When loading is complete, the trailing end is then secured around the second draw roller.

The Drive

On primitive jigs, the draw rollers were driven alternately, a weighted strap being used to apply tension to the unwinding roll. Consequently tension in the warp direction of the fabric was considerable, leading to lengthways extension (an additional 5% in the length of finished cloth was always very gratifying to the merchant!), but usually meant increased lengthways shrinkage of the fabric in washing, lower crimp in the warp yarns and a reduction in fabric width. Furthermore, as the angular velocity of the take-up roller was constant, the fabric started at a low linear speed, which increased as the diameter and circumference of the roll built up. Assuming that the roll diameter at the start was 20 cm and built up to 60 cm, fabric speed would increase from about 30 m/min to 90 m/min, giving in some cases, an unacceptable variation in dyeing conditions. Further, at speeds in excess of around 110 m/min, on small diameter rolls there is a tendency for dye-liquor to be thrown out of the rotating roll by centrifugal force.

Another limitation of 'non-automatic' jigs was that the dyer had to be on hand to stop and reverse the motion at the end of each passage.

In modern jigs, the length of run is set after the first passage through the machine and thereafter the direction of travel is reversed automatically. Both take-up and unwinding rollers are driven (so reducing lengthways tension on the fabric) and jig manufacturers have devoted much time and effort to devising means of driving the rollers so as to achieve fairly constant fabric speed at a moderate tension.

A standard mechanism for driving the draw rollers of a Vald Henriksen jig, widely used in the UK, is based on a friction-gear system. The motor (usually about 1.5–2 kW), through a variable V-belt pulley, drives a hardened steel roller which runs between two friction discs whose axes are slightly off-set. Each disc has a chain drive to one of the draw rollers. By means of a spring (which can be set at predetermined tension) and roller bearings the roller is normally held in a balanced position providing equal drive to each friction disc. A change in tension displaces the roller so altering the gearing between the two discs and varying the drive on the rollers.

The pressure of the two discs is automatically adjusted so that the roller does not slip when starting to drive a large roll of fabric and the V-belt pulley has an overload coupling to protect the motor and drive against overloading.

Other jig manufacturers have developed different methods of driving the jig rollers. Benninger use a d.c. motor controlled by a semi-conductor thyratron for varying the armature voltage. The motor drives the rollers through a bevel-gear stage, spherical rolling elements and overload clutches. The cloth tension is regulated by a pneumatically loaded oscillating roller. The differential drive system is complex and comparatively expensive.

Another drive mechanism, employed on Farmer Norton jigs, involves a separate d.c. motor for each draw roller. The motors have 'shunt' characteristics with the armatures connected in series and the fields in parallel. The cloth speed is varied by altering the armature voltage and the tension is regulated by varying resistance in parallel with the armature of the motor driving the let-off roller, which acts as a brake.

Another method for transmitting drive to the rollers involves the use of hydraulics and, recently, this has been coupled with electronic controls. In the Asisi 'Noujigg' machine, control of fabric movement is achieved by having a sensing roller constantly in contact with the fabric. As the fabric is being run into the jig, the length is metered by the contact roller which will thereafter control movement in each direction to that length. The control system will stop the jig at a set point in the run to permit a sample to be taken.

The tension of the fabric is measured by the contact roller which then regulates the hydraulic pressure to each roller drive, giving constant tension and fabric speed. Hydraulic pressure is used to raise and lower the covers and to drive an A-frame take-off.

Although many fabrics are adversely affected by substantial warp-ways tension under hot, wet conditions, some fabrics, particularly heavy cottons, can withstand considerable tension and most jigs can operate over a tension range up to 100 kg.

The drive control system, with its start and stop buttons, usually has facilities for setting the following variables:

Length of the first run
Cloth speed (usually between 40–100 m/min)
Cloth tension
Number of ends to be run.

At the end of the cycle the jig will stop and liquor will immediately start settling to the lower parts of the fabric roll, so creating marked imbalance in the roll, leading to gross tension variations when starting to unroll. In order to obviate this problem many jigs are fitted with an oscillating device usually introduced automatically when the programmed number of ends is completed. A few metres of end-cloth are drawn slowly through the liquor and wound onto the roll before reversing. This not only keeps the liquor uniformly distributed within the fabric roll, but, before dyeing, also keeps the 'empty' draw roller at bath temperature, so reducing any tendency to 'tailing' in which the end of the roll is dyed to a lighter shade due to a drop in temperature as heat is lost to the draw roller.

The Dyebath

The liquor capacity of a typical production jig capable of dyeing fabric up to 1.8 m wide is about 750 l for which one drain valve is sufficient. An overflow and a level indicator tube are fitted along with two or three horizontal spray pipes running the full width of the dyebath in order to expedite fabric rinsing.

Usually the predissolved and strained concentrated dye-liquor is added directly to the dyebath and thoroughly distributed throughout the bath. Some dyers prefer to split the concentrated dye-liquor into two equal portions and add half before the first dyeing run and the remainder before starting the second run; the benefit over the all-in technique is marginal. The liquor is agitated by the movement of the fabric through the dyebath and mechanical liquor-circulation systems are generally unnecessary unless dye-addition tanks are fitted (good distribution is essential) and in the case of very wide machines or where temperature uniformity is critical.

Heating

Heating is usually by low-pressure steam (300 kPa) through a perforated pipe running the width of the machine at the bottom of the dyebath. The perforations must be on the lower side of the pipe and must not impinge on the fabric.

Closed-coil heating (using higher steam pressures) may be used, but is generally slower than direct steam injection. It does, however, offer better control at dyeing temperatures, and some jigs are supplied with a dual heating system – direct steam injection for rapid heating up to near the dyeing temperature and then closed-coil heating for maintaining that temperature.

It is now standard practice to supply jigs with top covers which should be hinged to facilitate accessibility and fitted with rubber-framed safety-glass windows. The advantages of using top covers on jigs have been well-demonstrated and can be summarized as:

(a) Maintaining high temperature on the fabric roll; with some man-made fibres, the drop in temperature on leaving a boiling dyebath and being wound onto the roll in an open jig reduces dye uptake and penetration considerably.
(b) Minimizing the temperature differences on the roll; in an open jig the edges of the roll cool rapidly leading to a reduction in dye uptake (listing).
(c) Minimizing exposure to air; this is of particular importance when applying vat dyes in the reduced state.
(d) Reducing the amount of heat consumed in maintaining the dyebath and goods at the specified temperature.
(e) Reducing the loss of steam to atmosphere where some may well condense on the internal roof structure and subsequently drip, to the detriment of anyone or anything directly beneath.

Top covers should be fitted with insulated handles and be designed internally so that any condensed liquor runs to the side of the machine and back to the dyebath. The cover should also be provided with a steam exhaust valve, and many jigs are fitted with direct steam injection nozzles at the side of the machine to maintain a high internal temperature.

Temperature Controls

A thermometer is fitted to indicate the temperature of the dyebath and a thermostat to control the bath temperature via pneumatic valves on the steam line. A separate thermostat controls the injection of steam under the hood.

Expanders

When running wet fabric from roll to roll there is always a tendency to form lengthways creases which must be eliminated before winding onto the roll. Expander bars are fitted so that their convex surface is in direct contact with the fabric shortly before winding onto the roll. In order to allow for the increase in roll diameter, expander bars are usually mounted in pairs on a frame which pivots about a position which is dependent upon the tension in the fabric and the diameter of the roll.

Log Rolls

During the dyeing of lightweight filament fabrics, the surface of the take-up roll frequently develops lengthways ripples which may lead to lengthways streaks of light/dark colour. The Althouse division of Crompton and Knowles Corporation developed the log roll technique for overcoming this problem. This consists of mounting a free-running roller resting on the top of each roll of fabric. Each log roll has a deep spiral thread cut into the surface and this is claimed to flatten out the corrugations whilst permitting an even flow of liquor between the roller and the fabric. It is probable that the roller promotes lateral movement of the dye-liquor within the roll. A smooth-faced roller is ineffective as it produces a nip action, forcing liquor back along the fabric and causing blotchiness.

Sampling

For colour matching it is standard practice to stop the machine and take small samples adjacent to a fabric join near the middle of the total length.

Unloading

When jig processing is completed, the fabric is run onto a debatching roller or, in the case of large loads, onto A-frames. A few jigs are fitted with a nip or suction device in order to remove extraneous water during unloading. Wet rolls are very heavy and mechanical lifting gear is necessary for large ones.

Some jigs have been modified to handle velvets and other woven-pile fabrics. These fabrics are always run face down so that the expander rolls act only on the back of the fabric, not on the face. On conventional machines the rollers always rotate in the same direction and, when the fabric is threaded up normally i.e. the take-up is near the centre of the machine, the pile side of the fabric is always on the inside of the roll. Pile fabrics take up a large amount of dye-liquor which, once in the roll, is forced to the base of the pile by centrifugal force, leading to marked differences in shade between root and tip of the pile fibres. To equalize the centrifugal effect, jigs have been modified for velvets so that the fabric is taken-up pile inwards on one roll and pile outwards on the other. This is achieved by having contra-rotating rollers, the fabric being wound on the inner side of one roller, the opposite side of the other (Figure 2.2). As the fabric passes through the dyebath the pile is slightly laid back in the direction of travel and runs smoothly into the fabric nip formed on the roll. A modified, shortened expander bar is fitted underneath the roller with the modified drive.

Principles of Jig Dyeing

A detailed study of the fundamental principles involved in jigger processing was carried out at Courtaulds Research Dyehouse at Bocking in the period 1955–64. Much of this work has been published in the Journal of the Society of Dyers and Colourists(*33–35*) but regrettably a further paper by France and Morton describing much of the practical work carried out has not, so

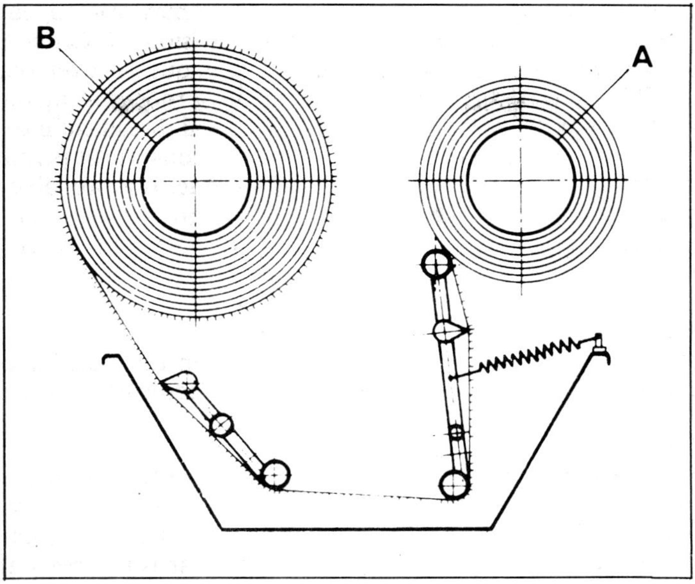

Figure 2.2 – Threading arrangement for velvets

far, been published. A later paper by Janas(*36*) described the dyeing of nylon fabric on jigs and contains useful information on the reasons for such faults as tailing, ending and listing.

Dyeing on the jigger may be regarded as a series of padding operations followed by dwell periods on the roll during which dye absorption and diffusion take place.

The factors controlling the rate of dyeing are:

(a) The amount of liquor held in the interstices of the fabric (fabric interstitial liquor)
(b) The exhaustion of the liquor in the time between successive immersions
(c) The degree of interchange of exhausted dye-liquor with fresh dye-liquor during one immersion (interchange factor).

The fabric interstitial liquor may be quoted in volume per linear metre of fabric or related to the total weight of fabric (interstitial ratio). Experimental findings for typical filament fabrics are shown in Table 2.1. These figures show that the effective liquor ratio in the fabric on the roll is about 1:1 and, under these conditions and at normal dyeing temperatures, exhaustion takes place very rapidly – usually most of the dye is absorbed in 1–5 s after the fabric has left the dyebath (diffusion into the fibre will take much longer). If slow-dyeing or temperature-sensitive disperse dyes are used, the fabric near the ends of the roll (which receives two immersions in quick succession) will be dyed lighter than the middle of the roll due to incomplete exhaustion before the second immersion; this is known as 'ending'.

The interchange factor measures the proportion of liquor exchanged during one immersion in which the fabric strikes, then passes through and out of the dyebath. Values for filament fabrics are listed in the table and, under normal conditions of dyeing, appear to be remarkably independent of fabric speed, tension, time in dyebath, temperature and whether passing round one or two immersed rollers. With heavier and thicker fabrics the interchange factor decreases.

TABLE 2.1

Fabric type	Fabric weight (g/m^2)	Weight of interstitial volume (g/m^2)	Interstitial ratio (%)	Interchange factor
Acetate poult	110	102	93	0.63
Triacetate taffeta	126	185	147	0.59
Nylon taffeta	62	59	96	0.77

Within the rotating roll of fabric the combination of gravity and centrifugal force generates a vertical pulsating motion of the interstitial liquor, the magnitude of which is highly dependent on the porosity of the fabric. It is this liquor movement which results in the 'marking off' of several repeats of imperfect joins in the fabric, which take up excessive amounts of dye-liquor.

In a classic paper(*33*), describing an investigation into the formation of watering (moiré) at the end of the rolls of acetate poult, it was shown that the fault is due to the localized flattening of the rib structure due to pressure build-up on the inside of the roll during the first run through liquor in the jig. The fault can be minimized by using large draw rollers, e.g. 30 cm in diameter, running the first passage (or 'end') under low tension and using a substantial cotton end-cloth to provide a soft bed for the roll on the stainless steel roller. If a cotton end-cloth is used, about three yards of low grade acetate taffeta should be stitched between the cotton and the fabric to be dyed, otherwise the dye-liquor picked up by the cotton (for which the dye has no affinity) will dye the inner layers of fabric to a darker shade than the bulk.

Due to lack of mechanical action, the jig provides only limited scouring action. However it enables a range of preparation and dyeing operations to be carried out on a single machine that is comparatively economical in terms of capital cost, water, energy, labour and space.

2.2 WINCHES

The winch dyeing machine (also known as the beck, vessel, or wince) is one of the oldest dyeing machines in practical use. It has the merit of being comparatively simple, and therefore economical to purchase and to operate, and, at the same time, is versatile in its range. Desizing, scouring, bleaching, dyeing, washing-off and softening can be carried out in the winch on practically all types of fabric which will normally withstand creasing when in rope form, e.g. woollen fabrics, loosely woven cotton and man-made fibre fabrics, nearly all circular knitted cloths and some types of warp knits. Compared with the jig machine the winch imposes far less lengthways tension on the fabric and the constant reformation of lengthways folds provides much greater mechanical action, so improving scouring efficiency. The mechanical action combined with low tension encourages the development of crimp in the yarns of woven fabrics and in the loop length of knitted cloths, so providing thicker fabrics with a fuller handle, more cover and better crease-recovery. Likewise, when compared with fabrics dyed on the pressure beam, although winch-dyed cloths lack the very smooth and sheer appearance of the former, they benefit from the mechanical action and increased opportunity for relaxation and have a markedly fuller and resilient handle.

Principle

The basic principle of all winch machines is to have a number (from one to forty) of loops or ropes of fabric of equal length mainly immersed in the dyebath, but with the upper part of each rope running over two reels mounted above the dyebath. At the front of the machine, about 35 cm above the top of the dye-liquor, is a 15 cm diameter reel called the jockey or fly roller which is free-wheeling; at the back of the winch is a much larger driven reel (winch reel) which pulls the ropes of fabric round.

The Dyebath

In old winches wood was used for the lower half of the machine, but nowadays stainless steel is used with welded joints, the corners being radiused and the internal welds being ground and polished to eliminate any roughness which may cause snagging of delicate filament fabrics. Good quality stainless steel should be used, AISI type 316 or equivalent, and, if sodium chlorite bleaching is to be carried out, the vessel should be given a preparatory passivating treatment with nitric acid.

The shape and dimensions of the vessel and reel vary considerably depending on the type of fabric to be dyed. Most fabric winches are about 2.5 m from the back to front and range in width from 50 cm (for a single rope) up to about 4.5 m for about 40 ropes of lightweight fabric. By law in the UK, the top of the vessel must be at least 36 in (92 cm) above the floor or working platform.

For woollen and heavy cotton fabrics a deep-draught vessel with a sloping back is preferred (Figure 2.3). The depth enables commercial lengths of bulky fabric to be accommodated and the sloping back enables piled-up fabric to be pushed easily towards the front of the machine. The dye-liquor

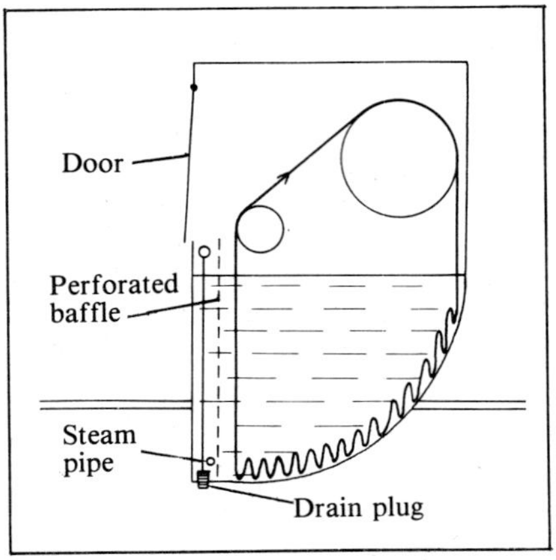

Figure 2.3 – Deep-draught winch

is usually about 1 m deep and the comparatively small surface area is an advantage when applying vat dyes to cotton.

The introduction of filament viscose and acetate fabrics necessitated the development of winches designed to minimize creasing of these types of fabric and shallow-draught winches evolved (Figure 2.4). To reduce the

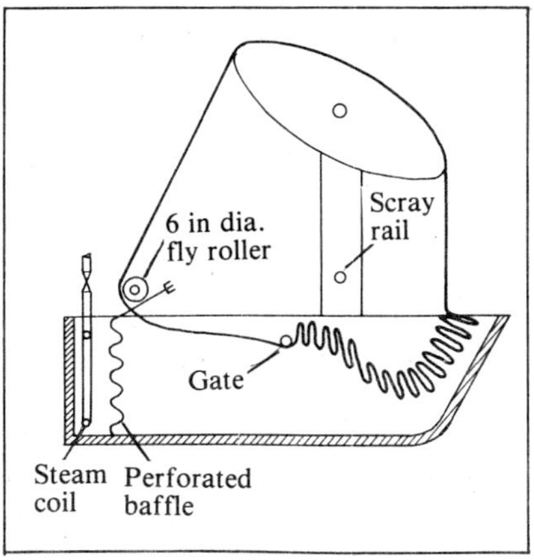

Figure 2.4 – Shallow-draught winch

pressure resulting from piling the fabric at the back of the winch, the depth of liquor was substantially reduced (to about 75 cm) and the bottom of the machine flattened.

Most winches are fitted with overflow ducts at the back so that rinsing can be carried out in liquor flow from front to back.

The Winch Reel

Much thought has gone into finding the optimum material for winch reels together with their shape and size, as not only does the reel control the rate of movement of the fabric rope but also its configuration whilst in the dyebath. The reel does not positively grip the fabric, but relies upon the weight of the wet cloth and the friction between the reel and the fabric to prevent slippage which can cause chafing(*37*). Originally, wooden-slatted reels were used, but they had limited lives and presented problems in cleaning and so were replaced with stainless steel. However, this reduced the frictional forces (particularly if soap were used in the dyebath) and, although most stainless steel reels are made in corrugated and broken surface designs to increase friction, it is customary to cover the reel with polyester or polypropylene tape wound at a narrow angle across the reel. Where tape is used, dye contamination from lot to lot must be avoided by cleaning or by careful production planning.

It was found by practical experience that not only did the presence of soap reduce the reel/fabric friction, but that this effect was increased as the liquor temperature was raised. Tape-covered reels retain a much greater degree of traction even at high temperatures.

Deep-draught winches usually have circular or slightly elliptical reels with diameters ranging from 20 to 50 cm and provide a steady pull on the fabric lifting it out of the dyebath and over the jockey roller. The fabric, on leaving the reel, falls straight into the dyebath with very little plaiting action. For vat dyeing the reel is set just above the dyebath to minimize oxidation.

Shallow-draught winches handling lighter fabrics have elliptical winch reels usually in the proportion 2:1 and about 2 m in circumference, although for acetate warp-knits reels of major axis 150 cm, minor axis 60 cm have been recommended. Elliptical reels give a fluctuating tension and speed in the fabric and provide more mechanical action than circular reels. Elliptical reels provide a plaiting action as the fabric falls into the dyebath, the size of the plait increasing with reel size. The larger the reel, the less mechanical action on the fabric – an advantage on delicate fabrics, a disadvantage on some staple cloths where improved cover and bulk is required.

Reels with adjustable cross-section have been produced and a few reels have enlarged ends designed to prevent the outside ropes from slipping off the reel; with proper control this risk is very small.

The winch reel is driven from one side of the machine by a 1–3 kW motor, depending on the size. The drive is usually through a V-belt system permitting a variation in reel speed, usually between 40 and 80 m/min, the former for delicate fabrics, the latter to increase the rate of dyeing.

The Jockey or Fly Roller

The roller at the front of the machine is mounted in free-running bearings and is pulled round by the fabric. Usually about 15 cm in diameter, most are circular in cross-section, but some are polygonal and give more mechanical action on the fabric.

As well as lifting the fabric from the dyebath the jockey roller expresses some of the surplus spent dye-liquor from the fabric. A few winches used for applying vat dyes have a second jockey roll mounted behind and above the first in order to exert a mild squeezing action on the fabric and increasing the rate of interchange of dye-liquor.

The Peg Rail

As the fabric ropes are lifted from the dyebath they pass through a peg rail which acts as a comb. The rail runs the full width of the machine and is located about 20 cm above the top of the liquor and below the jockey roll. The spaces between the pegs are usually about 12 cm in width and ensure good separation of the ropes before advancing to the jockey roll. The pegs in the rail may be fixed, but usually rotate with the movement of the fabric so minimizing tension and chafing.

The pegging of the rail controls the fabric distribution widthways in the dyebath and the ropes should be uniformly spread across the width of the vessel. The extreme end sections of the rail are never used.

A useful development is the hinging of one side of the rail so that, in the event of a tangle of ropes below the rail, one edge of it is lifted and operates a knock-off motion on the winch reel so stopping fabric movement. Otherwise the winch reel will continue to rotate and abrade the stationary fabric.

The Salting Box

On most conventional machines a perforated sheet of stainless steel is fitted vertically across the full width of the machine and about 30 cm in from the front of the dyebath. This partition forms the salting box and is used for adding concentrated dye solution and auxiliaries, e.g. salt in the case of direct dyes. The additions must be made across the full width and the contents of the salting box stirred to achieve uniformity.

The salting box also serves as a useful location for a water inlet and drain as liquor can easily diffuse into or out of the dyebath with minimal disturbance to the fabric ropes at the bottom of the vessel. Machines up to 3 m^3 capacity usually have a single drain valve operated by a lift rod; larger machines have two 12 cm diameter outlets, one at each side of the salting box.

For filling, one or two 5–7 cm water pipes feed directly into the salting box.

A refinement in some winches is a dye-addition tank fitted at one side feeding into a pipe with graduated frequency of perforations to give uniform distribution across the salting box.

Direct steam injection into a dyebath is usually carried out by means of a perforated pipe (often called a ‘sparge’ or ‘sparge pipe’) mounted horizontally at the front and near the bottom of the salting box. The equipment is simple and requires little maintenance; heating up is rapid, and forcing steam bubbles into the liquor at pressures of around 250 kPa provides vigorous agitation in the salting box. A condensate return system is not required but some dilution of the dye-liquor must be accepted.

For all the energy in the steam to be used, the steam bubbles must condense before they can reach the surface of the dye-liquor and escape. The factors controlling condensation of the bubbles are:

(a) The head of the liquor above the sparge pipe; the greater the height of the liquor, the more time there is available for condensation.

(b) The temperature of the dye-liquor; obviously the hotter the liquor, the slower the rate of heat transfer from the bubble.

(c) The size of the bubbles; the rate of heat transfer is related to the surface area of the bubble and therefore to the square of its radius. However, the volume of the bubble is dependent on the cube of its radius. Therefore, for a given volume of steam, reducing the bubble diameter by half will increase the number of bubbles by eight, and will double the total surface area. Although small bubbles are desirable, a balance has to be struck between the fineness of the hole and the number of holes required. Hole sizes are usually in the range 2 mm (for a small vessel) up to 5 mm, with 3 mm as an average.

On emerging from the sparge pipe, the size of the bubbles will immediately increase depending on the steam pressure applied and the head of liquor. Operating at around 250 kPa steam pressure and a depth of 90 cm, the volume of steam will increase threefold when transferred from inside the steam pipe to the bath. Therefore, the steam pressure applied should be kept as low as practicable, consistent with giving adequate flow.

(d) The degree of superheat in the steam; this should be kept to a minimum, as it retards condensation of steam in the bubble, especially in a hot dye-liquor.

(e) The path of a steam bubble in the liquor; this should be as long as possible. In steam pipes velocities generally range up to 24 m/s in straight pipes and 15 m/s in pipes with bends and valves, but when bubbles leave a sparge pipe they travel at very high speeds (about 300 m/s) due to the difference in pressure inside and outside the pipe, a distance of only about 3 mm. At these very high speeds it is important to locate the holes so as to give the longest path and maximum time before the bubbles reach the surface The holes are drilled in the lower half of the pipe to give the bubbles an initial direction 45° downwards towards the middle of the bath. It is inadvisable to drill holes at the bottom of the pipe or near the side of the salting box as this may lead to bubbles striking the bottom or sides of the winch at high speed, creating both noise and erosion.

Excessive noise may also result if the bubbles condense very quickly and collapse with a bang (this usually occurs when starting up a cold bath). If serious vibration results, it may be necessary to alter the location of the pipe, the size of the holes or the steam pressure.

In very wide winches two steam pipes may be employed, one emitting steam across the first half of the width, the second having no perforation for the first half, but supplying the further half of the machine.

Some shallow-draught machines have a raised front shelf and no salting box (Figure 2.5). Heating is by heat exchanger located underneath the perforated shelf along with water inlet and outlet. Dyes and chemicals are added by means of a perforated pipe running across the machine.

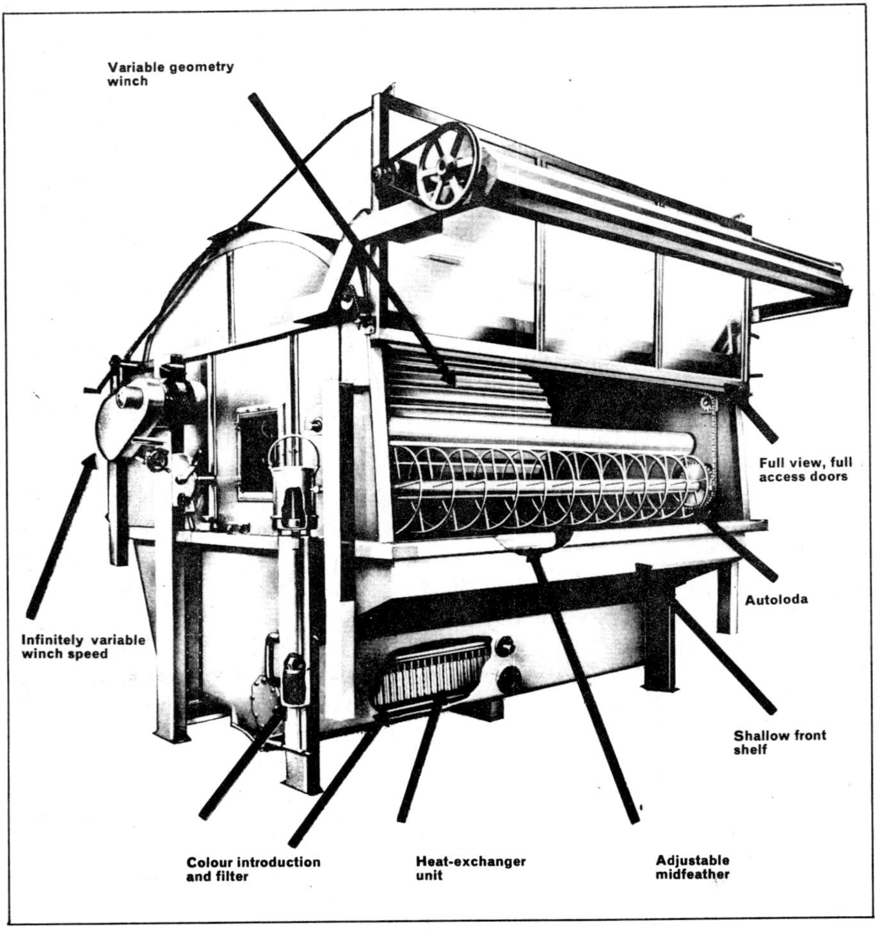

Figure 2.5 – Leemetals winch with Autoloda

Ballooning

Fabrics which have a tendency to curl at the edges, e.g. warp-knits, are chain-stitched to join the opposite edges along the length so that they are dyed in tubular form. Wherever possible circular-knitted fabrics should be dyed this way and slit lengthways after dyeing.

When loading the winch care should be taken not to twist the fabric; the join should be completely round the fabric.

When winch dyeing tubular fabrics it is found that an air bubble collects inside the tube from immediately below the jockey roll, through the peg rail and down towards the dye-liquor. This bubble serves the very useful purpose of inflating the fabric tube and dispersing lengthways creasing. Although creases are again formed when the fabric passes over the jockey roll, they are usually in a different position and change with each circuit of the rope (usually less than 1 min). The bubble is continually pushed back along the rope by the jockey roll and the pegs tend to constrict the fabric and prevent air from escaping. Fairly tight warp-knit structures ballon easily, open structures do not as they lose air readily. Practical observation indicates that ballooning is at its maximum in the 70–80°C range and diminishes appreciably above 90°C (probably due to viscosity and surface tension changes). The presence of some foam in the dyebath helps and dyers have been known to bubble air into the dyebath to encourage the formation of a balloon.

The origin of the air inside the tube is not clear and may be due partly to air originally entrapped in interstices of the fabric or air induced into the fabric as it passes over the reels or as it is plunged back into the dyebath.

Fabric Ropes

The length of fabric in each rope is largely dependent on the length as received from the weaver or knitter. Heavy woven fabrics are usually supplied in lengths of 50 m, but lighter weights around 100 m. Circular-knitted cloths range from 50 to 100 m in length and warp-knits are usually 100 m long.

Where the winch reel speeds are acceptable, winch dyers will use ropes of about 100 m, but for delicate fabrics necessitating slow speeds 50 m long ropes are used, otherwise unlevelness may result particularly if poor-levelling dyes are used.

The Mid-feather or Gate

Many shallow-draught winches are fitted with an adjustable horizontal bar half-way back in the winch and usually located about 15 cm below the top of the liquor. In such winches the bulk of the plaited fabric is kept submerged at the back of the dyebath and slowly rises to touch the gate which restrains the bulk from moving forward. The fabric is then drawn forward in a nearly horizontal sweep through the dyebath and then up to the jockey roll. This sweep rearranges the creases in the fabric so avoiding permanent lengthways creasing.

The ideal condition in winch dyeing is when the plaits at the back are moving rhythmically in the dyebath with a few of the leading plaits riding free and rising up to the back of the gate. To achieve slight positive buoyancy of the fabric, soap may be added to the dyebath.

Cover

All modern winches are covered in order to maintain good uniformity of temperature, to minimize heat loss and to prevent external condensation(*38*). To provide good accessibility to front and back, toughened glass doors are fitted to lift vertically, assisted by counterbalances at the sides. Care must be taken when opening a near-boiling winch to ensure that there is no outrush of steam. Improved designs of doors ensure that, on opening, the top of the door opens first so that any steam escapes upwards from a safe height and not towards the operative.

At the top of the winch a vent is fitted to extract steam and fumes (e.g. from sodium chlorite in bleaching, carrier in polyester dyeing). The vent is usually coupled to an extraction duct supplying a slight suction. The vent is normally at one end of the machine to ensure that any condensation running back does not fall onto the fabric.

Loading and Unloading

For loading, the leading end of one or more ropes is taken from a barrow and thrown over the winch reel, drawn forward and tied to the peg rail. The winch reel is started and draws the fabric in. Care must be taken to avoid twisting the fabric during loading.

When unloading, the joins in all the ropes are broken, one end resting over the peg rail. The ends of one or two ropes are then thrown over the unloading reel mounted high above the front of the winch and this discharges the fabric into a barrow with a perforated bottom to allow drainage.

The Principles of Winch Dyeing

In winch dyeing the liquor-to-goods ratio is usually around 25:1, although 15:1 has been achieved; small winches provide longer liquor ratios, i.e. 40:1. When the fabric is lifted from the dyebath, liquor immediately starts to drain back and this removal of spent liquor is increased by the mild squeezing action applied by first the jockey roller and then the winch reel.

Experimental work has shown that, under normal dyeing conditions, the concentration of dye in the liquor leaving the fabric as it passes over the jockey roll is appreciably less than that in the dyebath as a whole. As the fabric is plunged back into the dyebath it immediately acquires additional dye-liquor which stays closely associated with that part of the fabric as it travels slowly towards the front of the winch to complete the cycle.

The rate of dyeing is therefore controlled by the number of fabric cycles in a given time, i.e. increasing the winch reel speed will increase the rate of dyeing. The liquor interchange factor, i.e. the proportion of dye-liquor in the fabric replaced in one passage over jockey and winch reels, appears to be independent of fabric speed and the time to complete one cycle (about 1 min).

There is very little interchange of liquor after the fabric has been in the dyebath a few seconds. The fabric ropes lie side by side in the dyebath with no appreciable lateral movement of the ropes otherwise tangling may result.

There is considerable movement of liquor in the dyebath, although it is not immediately apparent. However, when dye is added to the salting box it can be seen that diffusion throughout the vessel takes place very quickly.

Levelling of dye usually takes place along the length of the fabric opposite to the direction of travel; obviously the downwards flow of the liquor as the fabric is lifted from the bath plays a major role in migration. The proportion of longitudinal to lateral migration increases as the speed of the rope increases(*39*).

Spiral Dyeing

One of the drawbacks of winch dyeing is that fabrics are processed in individual lengths, necessitating much trimming and sewing during loading, and join locating and breaking-out prior to unloading.

Many attempts have been made over the years to dye continuous lengths of rope, the fabric very slowly spiralling across the width of the winch. The major problems arose in arranging a method of feeding back across the winch, a sloping transporter trough or tube filled with dye-liquor often being employed. Such devices were used with varying success, the major difficulty being lack of flexibility with different fabric types.

A major step forward came in 1964 when British Silk Dyers at Balloch developed a system later known as the 'Spiraloda' or 'Autoloda'(*40*). In this system the leading end of the rope is attached to a loading arm which is advanced slowly by a chain drive at one side of the winch (see Figure 2.6). The time taken for the loading arm/leading end to complete one circuit is about 1 min so that if the winch reel speed is 70 m/min the loop length will be 70 m. The second modification is the replacement of the peg rail by a continuous helix roll which is rotated away from the leading end at one revolution for each revolution of the loading arm, in order to maintain equilibrium in the dyebath. The pitch of the helix determines the spacing of the ropes in the winch and interchangeable spiral rollers of different pitch are available to enable fabrics of different bulk to be handled effectively. In many cases it is preferable to use every other space in the spiral roller and accordingly the rate of revolution of this roller is doubled.

Apart from the obvious advantages such as a considerable saving in time when loading and unloading, less sewing and less fabric wastage, the Autoloda technique enables the dyer to select the most convenient loop length for the machine rather than having to accept that provided by suppliers. More uniform spacing provides higher machine loadings with lower liquor ratios, and this type of device will be used more widely in the future.

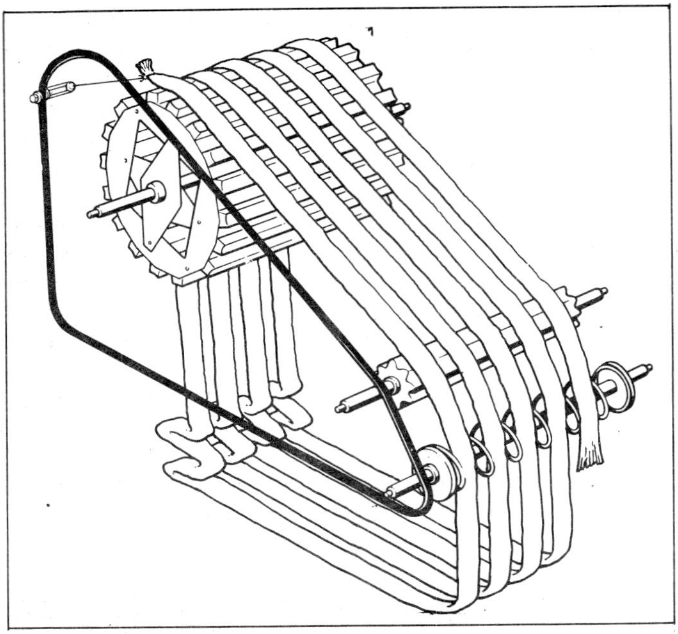

Figure 2.6 – The Autoloda; the chain drive is on the left

High-temperature Winches

The demand for high-temperature rope dyeing machines to handle textured polyester double jersey fabrics in the mid-1960s led to the development of pressure winches capable of operating at around 130°C. The machines were very heavy, comparatively expensive and their small doors made accessibility difficult. The introduction of the first jet dyeing machine to the UK in 1967 suddenly made pressure winches obsolete and few remained in use into the 1970s.

2.3 PRESSURE BEAMS

Pressure beams for dyeing fabric came into general use in the UK in the latter half of the 1950s in response to the demand for a machine to dye substantial weights of lightweight, delicate, warp-knitted nylon and acetate fabrics. Many of these fabrics when winch-dyed suffered creasing and occasionally mechanical damage, whereas when dyed on the beam the resulting fabric was completely smooth, albeit somewhat thinner and possibly slightly glazed. Lightweight woven fabrics containing filament yarns and some polyester/cellulosic cloths are also dyed on the beam, as are a very limited range of weft-knits which are slit lengthways and opened sufficiently to remove the resulting centre crease before dyeing.

The first beam dyeing machines consisted of a roll of fabric wound round a perforated beam into which dye-liquor was forced, the whole being immersed horizontally in an open dye tank. This arrangement permitted a maximum dye-liquor temperature of around 90°C and only one-way flow, i.e. inside-to-out through the roll of fabric. These basically simple machines were soon followed by high-temperature models in which the open dye tank was replaced by a circular cross-section pressure vessel and a more complex pumping circuit to permit two-way flow. The high-temperature machine, although much more expensive and complex to operate and maintain, gave greatly improved results and versatility and has largely replaced the atmospheric pressure version. The earliest machines were top-loaded using an overhead crane, but these have mostly been superseded by front-opening models (Figures 2.7 and 2.8).

Pressure vessels vary from 30 cm (for samples) up to 200 cm in internal diameter for large batches of lightweight warp-knitted fabrics. The length varies with the type of fabric to be accommodated, from just under 2 m for wovens up to around 4.5 m for wide warp-knits. At the centre of the back of the machine is the dye-liquor inlet feeding directly into the barrel of the

Figure 2.7 – Pegg beam dyeing machine for knitted fabric, showing the beam on its cradle being loaded into the machine; the sample probe and control console are on the right

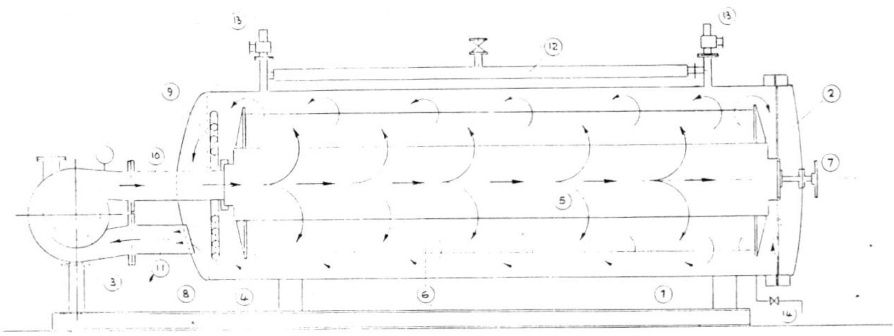

Key:
1 Main cylindrical autoclave
2 Hinged door
3 Main circulating pump
4 Machine support frame
5 Stainless steel perforated beam shell
6 Fabric wound on beam shell
7 Dye-beam locking plate and hand wheel in door
8 Stainless steel and blanking plates
9 Stainless steel closed-coil heat exchanger
10 Pump flow discharge into beam shell
11 Pump suction return from autoclave
12 Connecting pipe with vent valve on return to expansion tank
13 Stainless steel safety valves
14 Test drain valve

Note: a stainless steel water-cooled condenser is normally located adjacent to pipe (12) allowing dye-liquor to flow back to the expansion tank when machine is under pressure

Figure 2.8 – Liquor flow in a high-temperature beam dyeing machine at Spray & Burgass Ltd; this machine is fitted with circular heating coils (9) and flanges at the end of the beam

perforated beam, and, at the back of the shell, just below the centre, is the dye-liquor outlet to the pump. Two rails, running the full length of the vessel, are mounted at the sides about a quarter of the way above the bottom; these rails support the weight of the roll and accompanying cradle during the dyeing operation.

The door surround at the front of the machine is fitted with a rubber seal and the door itself is securely locked by either a series of bolts (on older machines) or a ring-locking device to speed locking and unlocking.

Other fittings to the pressure shell include:

(a) Safety valve to blow off should the pressure exceed the safety limit (usually 379 kPa, 55 lbf/in^2)
(b) Markings to show the maker and number of the vessel, date of manufacture and test, the safe working pressure and temperature
(c) Pressure gauge, clearly visible from the front of the machine
(d) Thermometer, likewise clearly visible
(e) Air/water outlet valve at the top of the machine to permit air to be ex-

pelled during filling with liquid; when the vessel is full the outlet feeds liquor via a condenser into an open expansion dye addition tank

(f) Sampling device to enable samples to be removed while the main vessel is still under pressure

(g) Test cock just below the door to ensure that the vessel is drained before opening.

A 2 m long, 1 m diameter machine has a nominal liquor capacity of about 3 000 l and effective load of about 220 kg of knitted fabric. A jumbo machine 4 m long by 160 cm diameter will take up to 8 000 m of 20 denier warp-knitted fabric weighing 700 kg; with a denser fabric a weight of 1 500 kg can be accommodated, although the fabric length is shorter. With a liquor capacity of 10 000 l the minimum liquor ratio is 7.5:1, but usually the ratio is 12:1 or 15:1.

Heating and Cooling

Heating is usually effected by high-pressure steam in pipes, although some dyehouses use superheated water under pressure or hot oil. In order to obtain a large surface area for heating and to provide minimum resistance to liquor flow, the pipes are arranged in parallel, longitudinally along the bottom of the machine. They are joined at both ends and arranged in an arc so that condensate drains into the lowest pipe and out of the vessel at the back to the condensate return pipe.

Recently a few machines, especially jumbo machines, have been fitted with external heat exchangers.

Cooling is usually carried out through the same system. After the steam and condensate valves are closed, cold water is passed through the pipes in the reverse direction, preferably to a hot-water storage tank for subsequent use.

Liquor Circulation

Adequate circulation of liquor is essential for successful dyeing on the beam and systems are designed to minimize restrictions to the rapid flow of liquor. The pump is usually fitted as close as possible to the entry point at the back of the pressure vessel to provide maximum delivery to the centre of the beam and to collect the outflow.

Although pressure systems permit two-way flow, in-to-out flow through the roll of fabric is greatly preferred. Out-to-in flow suffers the disadvantage of tending to compress the fabric, so flattening and glazing the cloth (particularly on the inside layers) hence restricting flow. Even a modest pressure on the outside of a larger diameter roll of fabric can result in comparatively high pressures on the inner layers of fabric compressed against the rigid beam barrel.

Nearly all machines are fitted with automatic flow reversal controls, the period of flow in each direction being set independently, but, as stated previously, in-to-out flow through the roll predominates.

Although the contents of the vessel are pressurized by a secondary pump (even when dyeing at temperatures below100°C) to around 310–345 kPa (45–50 lbf/in^2) the pressure drop across a roll of knitted fabric is usually only 15–20 kPa (2–3 lbf/in^2); with woven fabrics of much lower porosity and higher pump pressures, e.g. 100 kPa (15 lbf/in^2), smaller diameters of beam roll are employed. Separate gauges indicate the pressure both inside and outside the roll.

During operation, a small quantity of liquor leaves the vessel and passes through a cold-water condenser into an expansion tank alongside the main vessel. This open tank is also used for adding dissolved dye and chemicals. From the bottom of this tank, liquor is fed by a small-capacity high-pressure pump back into the main system. As the temperature of the liquor in the system is raised so the level of the liquor in the expansion tank rises.

Specific information regarding the rate of liquor flow through the beam is almost impossible to acquire as it is impracticable to fit a flow meter due to the lack of space, and the effect an orifice would have on the flow rate. Information from machinery manufacturers quoting pressure/flow characteristics of pumps indicates complete dyebath volume circulation times in the range of 35–45 s, corresponding to around 20 l/kg goods/min.

Practical observation on a commercial machine shows that the time taken for dye to pass from the addition tank through the secondary pump, main pump, roll of fabric and out through the overflow is about 5 min at 40°C. This is probably not very different from the time taken to circulate all the liquor once round the system. At 100°C the viscosity of water is about half that at 40°C and liquor circulation is doubled (assuming that the pump efficiency is constant). At 120°C (and provided the fabric resistance is constant) the liquor flow is two and a half times that at 40°C, i.e. around 2 min for a complete cycle, corresponding to about 5 l/kg/min.

Most fabric beam dyeing is carried out using liquor cycle times in the range 45 s to 3 min, but obviously there is a wide variation in rates of flow.

The Main Pump

The pump used for circulating liquor is required to deliver a large volume rather than to produce a high pressure (the main pressure being produced by a small-capacity pump). Two main types of pump may be used, axial or centrifugal and these have been fully described in the section on package dyeing.

The demand for larger machines and increased flow has led to larger capacity pumps and motors. A 2 m long, 1 m diameter machine usually has a 19 kW motor, while a 4 m wide jumbo machine has a motor rated up to 56 kW.

Beam Making

The making of a good beam of fabric is an essential prerequisite for uniform dyeing and it is regrettable that very little information has been published on this important craft.

Beam barrels are made in various diameters to fit the particular size of the machine, for instance, 48 cm diameter to fit a 128 cm diameter vessel and 61 cm diameter to fit a vessel 163 cm in diameter. The beam barrel must be sufficiently robust to withstand the weight of a full roll of wet fabric both during dyeing and in subsequent dewatering and drying operations. Apart from 20 cm at each end which are smooth, the beam barrel has numerous holes drilled to permit the liquor to flow. On a long beam there are several thousand holes, usually about 5 mm in diameter. One end of the beam is closed, the other has an aperture to fit closely on the liquor delivery pipe in the centre at the back of the pressure vessel.

In order to accommodate fabrics of less than the maximum width, excess holes towards the ends of the beam have to be blanked off and this is done by using flexible stainless steel sheets held in position by a narrower band and securely bolted. The positioning of these collars is fairly critical, as the inner edges should be inside the edges of the roll of fabric. If inadequate overlap is allowed excess dye-liquor will be forced out at the sides of the fabric near the beam barrel; if there is excess overlap the edges of the roll near the barrel will be starved of dye-liquor and therefore under-dyed. The amount of overlap is dependent on the type of fabric and the size of the roll, but between 3–5 cm is usual for most warp-knits except for very large rolls. For woven filament fabrics a similar amount of overlap is allowed, but the overlap is increased for woven staple fabrics with lower resistance to lateral liquor flow.

Before winding on the fabric to be dyed, the beam barrel is wrapped with a few layers of cotton fabric or non-woven fabric containing cellulosic fibres (the bonding agent must be capable of resisting the maximum dyeing temperature). This inner layer provides a good bed for the roll of fabric and a good medium for the dye-liquor to diffuse through (otherwise the perforations in the beam would show as overdyed marks on the inner layers of fabric); it also serves as a fine filter to remove any solid matter (including undissolved or poorly dispersed dye particles) from the dye-liquor.

Fabrics to be beam-dyed must have reasonably good dimensional stability throughout the dyeing operation. If they should extend lengthways under the influence of hot dye-liquor being forced through layers of fabric the liquor may find a channel out at one side with resulting gross unlevelness. If the fabric should tend to contract (e.g. unset, unrelaxed nylon) high pressures can build up on the inner layers and on the beam barrel. Under such conditions thermoplastic fibres are flattened, giving rise to ‘watering’ due to different light reflection; this fault is particularly pronounced on woven nylon fabrics made from bright yarn and dyed to dark colours. Excessive shrinkage widthways can cause gross flow variation near the beam barrel, and widthways extension will cause ballooning and reduced flow.

It is customary, therefore, particularly with nylon fabrics, either to give a dry heat-setting treatment under tension or a hot scour to relax some of the inherent shrinkage within the fabric prior to winding onto the beam.

Circular-knitted fabrics are slit lengthways (preferably along one crease) and opened out to full width prior to setting/relaxing and beam making.

Successive lengths of fabric are stitched together before beam making, the stitching and overlap being kept to a minimum in order to reduce marking-off onto adjacent layers of fabric on the roll. Each join means fabric wastage and some knitters produce long (300 m) continuous lengths of fabric for subsequent beam dyeing.

Fabrics for subsequent raising, or with a knitted loop, should always be wound onto the beam with the pile side out so that the structure is maintained by the in-to-out flow during dyeing. Balanced fabrics should be wound such that the edge curl tends inwards on the roll.

The basic equipment for beam winding consists of a mechanism to rotate the beam and a good fabric delivery system – obviously the fabric when wound must be free from creases. The beam is usually surface-driven and mounted on a pair of parallel rubber-covered rollers. These are driven at constant speed and control the rotation of the beam and take-up. Alternatively, a tension-controlled centre-wind take-up drive may be used. The feeding mechanism may consist of scroll rollers and uncurlers (as on a stenter) or may include a short pin-stenter frame with the beam mounted above the delivery end. The latter type of beam winder is more expensive than the former, but provides much better control of fabric width and tension.

The tension during winding must be kept to the minimum consistent with producing a stable beam.

Beam-winding mechanisms are often fitted at the delivery end of a heat-setting stenter or even a continuous wet-relaxing range, so telescoping two operations into one.

When all the fabric has been wound onto the beam, a short length of calico or polypropylene net, slightly wider than the fabric, is stitched on and wrapped round the roll and re-sewn. This provides additional stability (restraining the centre portion of the roll from ballooning excessively) and gives protection against soiling both before and during dyeing.

In the early days of beam dyeing a perforated steel corset was sometimes fixed to the outside of the roll to give stability, but this was found to be both unwieldly and unnecessary. Likewise, some dyers fitted solid flanges at each side of the roll to promote flow through the fabric rather than at the sides. These, too, were of doubtful benefit and are rarely used.

Where watering of the fabric on the inside of the roll is a serious hazard, some dyers interleave the fabric to be dyed with polyester net to absorb the build-up of pressure on the inner layers. As this practice reduces the useful load on the beam and complicates winding and unwinding, it is used only as a last resort.

Beam Loading

The full beam is loaded onto a shallow cradle provided with a shallow V at the top, the ends of the beam barrel resting on the centre of the V. At the bottom of the cradle are small (10 cm) diameter stainless steel wheels which fit into narrow rails on a trolley used for manoeuvring the beam in the dyehouse. For loading, the trolley, cradle and beam of fabric are pushed up to the opened door of the vessel so that the rails of the trolley are aligned with the rails inside the vessel. The trolley is locked in position and the cradle and fabric are pushed fully into the vessel so that the inside end of the barrel engages with the tapered end of the liquor inlet pipe. The trolley is then removed and the door closed. A hand-wheel on the door is then operated to move a spindle and internal flange against the closed end of the beam, so that good contact is made between the opposite open end and the dye-liquor inlet pipe.

Sequence of Operations

The door must be securely locked and safety devices checked, the drain valve closed and the outlet valve at the top opened before the water inlet valve is opened. As soon as the water starts flowing through the outlet valve leading into the expansion tank the water supply is throttled back and the pump started on in-to-out flow, reversing the flow every few minutes. A proprietary de-aerating agent is added to the expansion tank and the secondary pump started to pressurize the system. It is crucial at this stage to remove all air from the fabric as bubbles of air trapped in the roll result in undyed patches. Outside-to-in flow at the start also helps to compact the layers of fabric giving improved stability in subsequent operations.

Normal scouring and dyeing processes then follow, care being taken always to ensure that complete depressurization (stopping secondary pump, ensuring temperature below 85°C, and opening air inlet valve) has taken place before discharging to drain.

The pressure beam is a poor scouring machine, as there is negligible mechanical action and only soluble contamination is removed – solid matter remains trapped in the fabric. Soluble finishes may be applied to the fabric, but uniform exhaustion takes a long time and most dyers prefer to apply a finish by padding immediately prior to drying.

Sampling for Colour

Sampling for colour is a problem still requiring an entirely satisfactory solution on pressure beam machines. On most machines provision is made to insert up to three undyed samples of fabric on a steel probe immersed in the dye-liquor. The probe can be partially withdrawn into a pressure lock, which, when isolated from the pressure vessel, can be opened to remove the sample for examination. The basic problem is that the liquor flow around the sample is not the same as through the roll of fabric and therefore the colour of the sample serves only as a guide.

On some machines, small pieces of fabric attached to strings can be inserted into the edge of the roll and can be withdrawn through a pressure lock built into the door.

Where precise matching is required, some dyers have had to resort to taking the sample from the outside of the fabric roll. This entails cooling to a safe temperature (below 85°C), depressurizing and emptying the vessel, opening and withdrawing the roll from the vessel. Rather than rejecting the dye-liquor, provision is made for temporary storage of the dye-liquor in an overhead tank so that any addition of dye can be made to the exhausted dyebath rather than make up a fresh dyebath. This is an excellent example of the additional production cost incurred when very accurate colour matching is demanded.

Instrumentation and Controls

The range of instruments normally fitted to the control panel of a high-temperature pressure machine include:

(a) Gauges measuring the pressure inside the beam barrel and the pressure outside the fabric roll
(b) A temperature gauge
(c) Rate of temperature rise, temperature holding and rate of cooling controls
(d) Direction of liquor flow control, with both manual and automatic reversal for set periods.

Control of Uniformity of Dyeing

Non-uniform dyeing faults may generally be classified as:

(a) Gross unlevelness near an edge, resulting from a 'blown' beam having been wound too slackly
(b) Undyed or light patches due to air bubbles
(c) Variation near the edge on inside layers due to misjudged overlap of the blanking collar
(d) Variation in depth of dyeing through the roll – the inside layers invariably being deeper than the outside.

The last phenomenon is a perpetual hazard when using in-to-out flow in beam dyeing.

Factors Controlling Dyeing

The mechanisms controlling the exhaustion of dye from dye-liquor circulating through a package have been discussed in many papers, but almost always in the context of dyeing multiple small packages. Fabric beam dyeing presents particular problems in view of the size of the package, the practical limitation in many cases to one-way flow and the effect of liquor loss through the sides of the roll.

As dye-liquor is forced outwards through the roll of fabric the pressure and the liquor flow per unit area are reduced(*8*); these effects are counterbalanced by the reduction in velocity of liquor so providing more time for dye exhaustion. It should be noted that the pressure drop is greatest across the innermost layers of fabric when in-to-out flow is used; the pressure is down to half about 36% of the way through the fabric roll. An important factor is that as dye-liquor passes through the roll the concentration of dye is diminished, e.g. a fast-dyeing dye may be absorbed so rapidly by the inner layers of the roll that there is little dye left for the outer layers. To prevent noticeable inside/outside differences in depth of colour it is recommended that, in the early stages of beam dyeing, no more than 2% of dye in the liquor be removed on each passage through the fabric roll. To achieve a commercially acceptable dyeing time the proportion of dye absorbed can be increased as the concentration decreases(*16*). The rate of dyeing is controlled by the rise in temperature and, in the case of acid dyes, pH and use of blocking and/or complexing agents.

Dye exhaustion is followed by dye diffusion into the fibre (temperature controlled) and dye migration. The latter appears to play a very minor role in most beam-dyeing processes – an unlevel beam is very difficult to rectify unless easy-migrating dyes are used.

With the limitation on dye exhaustion per cycle, it is obviously desirable to increase the rate of flow of liquor through the roll of fabric. The rate of flow of liquor is dependent upon (a) the permeability of the fabric, (b) the number of layers of fabric, (c) the pump characteristics and (d) the viscosity of the liquor (which decreases as the temperature is raised).

Measurement of fabric permeability under the conditions used in beam dyeing is not easily carried out and measurement of air porosity has been suggested as a rough guide(*41*), although it has obvious limitations. Many of the problems faced in early beam-dyeing machines, e.g. unlevelness, were overcome when more powerful pumps and motors were fitted to increase the rate of liquor flow. The number of layers of fabric is important, particularly where woven fabrics of only moderate permeability are concerned, e.g. on a 50 cm diameter beam barrel about 600 layers of 100 g/m^2 nylon taffeta can be dyed, i.e. about 1 200 m, giving a total roll diameter of about 80 cm. The dyer learns by practical experience what is the maximum length of a particular construction of fabric that can be dyed uniformly with a particular combination of dyes.

Jet-air Dyeing ('Flotation' Dyeing)

In 1968 a Swedish engineer, Niels Bergholtz at AB Ludwig Svensson, patented (BP 1 225 700) a system for beam dyeing textured polyester fabrics and retaining a much greater degree of the original bulk. The principles were:

(a) Pre-bulking the fabric (for best results)

(b) Winding the fabric onto the roll at very low tension

(c) Reducing the liquor flow during dyeing
(d) Injecting very small bubbles of air at around 10 bar pressure into the liquor feed to the pump at around 100–300 ml per min.

The novelty lies in the principle of injecting air into the pressure system which is contrary to accepted dyeing practices, although air injection into non-pressurized garment dyeing systems has been proposed before (BP 613 068). In the Svensson system it was claimed that the bubbles were so small as to enter between and separate the filaments of the textile material, and it has been suggested(*42*) that, in doing so, the bubbles were deformed and acquired interfacial energy which was subsequently released when they regained their normal shape. As the bubbles pass slowly through the roll some coalesce and the larger bubbles rise to the roof of the beam where they are expelled, along with liquor, into the expansion/addition tank, so establishing an equilibrium between air input and removal from the system.

No entirely satisfactory explanation of the air injection technique has been proposed and, although around two hundred of these units were supplied, including half a dozen in the UK, where many dyers were seeking versatility for their beam dyeing equipment, in most cases the system appears to have been abandoned. The general opinion is that pre-bulking and the use of very low winding tension conferred the major benefits of the technique.

2.4 JET DYEING MACHINES

Until 1967 the machines used for dyeing fabrics in individual dye-lots were the winch, jig and beam. The mid-1960s in the UK saw the introduction of large quantities of textured-filament polyester yarns, knitted into double jersey constructions, and originally these fabrics were dyed in tubular form in the winch, the mechanical action assisting the development of fullness in handle. Even at the full boil and with the addition of carrier, producing medium-dark dyeings was a lengthy and expensive business and all too frequently resulted in barry fabric due to temperature variations in setting the textured yarn. From the experience gained in package dyeing textured-polyester yarn, it was known that the ability to dye at around 125°C would solve many of these problems. Dyeing at this temperature on pressure beams provided a partial answer, but the resulting fabric tended to be flat and thin in handle and, particularly in dark colours, often showed glazing.

In 1965 several dyeing machinery manufacturers started to develop pressure winches comprising the basic essentials of a winch surrounded by a pressure vessel and supplemented by a liquor-circulation system feeding dye-liquor from an external heat exchanger, usually into the back of the vessel, and removing it from the front. To avoid undue disturbance of the plaited ropes of fabric, liquor movement across the vessel was very slow. Pressure winches were very cumbersome, heavy, difficult to load and any fabric entanglement was expensive to rectify. The introduction of the first jet dyeing machines in the UK in 1967 made pressure winches obsolete and they were rapidly replaced.

The Gaston County Machine

The principle of using a jet of dye-liquor to transport fabric in a dyeing machine was first utilized by Victor Fahringer, chief engineer of the Pacific Division of Burlington Industries, and is described in USP 2 978 291. Burlington licensed the invention to Gaston County Dyeing Machine Co., and the first public appearance of a jet machine was in Atlantic City in 1961(*43*). Commercial acceptance of this type of machine was slow until, in 1967, it was found to be almost the ideal machine for dyeing textured polyester double jersey fabrics, then being produced in large quantities in the UK.

This American-designed machine (Figure 2.9) was novel in that it used a high-speed jet of dye-liquor to transport the fabric from the front to the back of the machine at previously unheard-of speeds (up to 400 m/min). The jet plaited the fabric at the back of the machine from where it slid slowly on a perforated plate to the bottom and then started to rise. When lifted from the top of the pile, the fabric passed round driven metering rollers which controlled the rate at which the fabric entered the jet tube. The rollers were smooth and were driven not to provide positive lift of the fabric, but to apply a very small tension prior to the jet. Originally three metering rolls were provided, but these were soon found to be excessive and the number reduced to one. The rolls also assisted in removing some of the partially exhausted dye-liquor from the fabric before the jet saturated it with fresh liquor.

The dye-liquor was withdrawn from the bottom of the machine and passed through a powerful pump and a vertical heat exchanger before being

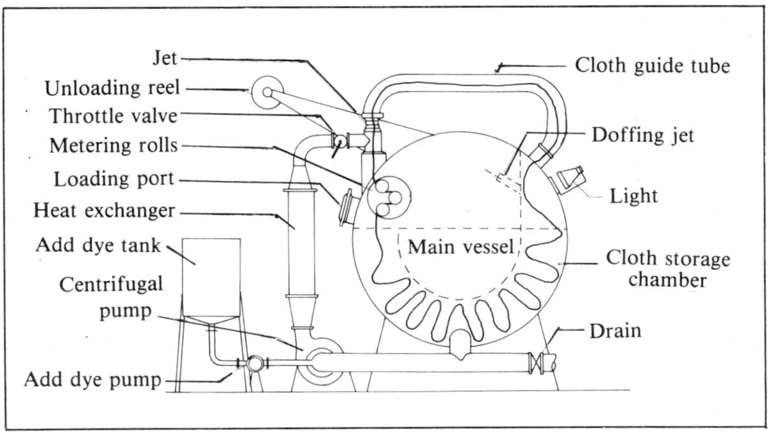

Figure 2.9 – Early Gaston County jet dyeing machine

injected at the top of the machine. The liquor level was kept low and dye additions made via a small secondary pump. The vessel was pressurized by compressed air.

Most jets of this type have two or three tubes mounted side by side within the same pressure vessel, but machines with up to ten tubes have been made. When dyeing double jersey fabric, weighing 600 g/m, loads of up to 150 kg per tube can be accommodated, the entire rope circulating in about 1 min.

Later modifications of the machine incorporated a single driven roll at the front and with the jet tube incorporated inside the pressure vessel.

Compared with the atmospheric-pressure winch for dyeing textured polyester, the Gaston County jet offered the following advantages:

(a) Dyeing at 125°C gave much more rapid dyeing with better penetration, better exhaustion and levelness, less surface dye to be removed during clearing, marginally better colour fastness and dispensed with carrier with its attendant problems.
(b) The lower liquor ratio, around 12:1 or 10:1, provided savings in energy (despite the higher temperature) with more rapid heating and cooling.
(c) Lengthways tension was less and there was slightly greater fullness of handle.
(d) The vigorous circulation of fabric and liquor and greater interchange of liquor on each passage through the jet gave more rapid dyeing.
(e) The frequent movement of fabric round the system minimized creasing.

Disadvantages were:

(a) Greatly increased capital and maintenance costs.
(b) Limited accessibility; the fabric was loaded and unloaded through a small port just below the roller. Four bolts secured the cover over each port. Mechanical cleaning internally was extremely difficult.
(c) Any roughening of inside surfaces that may cause snagging of the fabric was difficult to locate and rectify.
(d) The vigour of the jet discharge encouraged the formation of foam which could completely fill the machine. Chemicals with a predisposition to foam (e.g. many dispersing agents) had to be excluded and/or a de-foaming agent added to the dyebath at the start of operations
(e) The force of the jet was too great for delicate fibres, e.g. triacetate, and sometimes gave an excessively hairy or mussed surface on single jersey acrylic fabrics (removal of excessive quantities of short-staple fibre could also block the filter).
(f) The width of the bottom compartment was designed to accommodate the widest fabric likely to be dyed. Lightweight narrow fabrics did not fill the available space uniformly and sometimes fell over and caused tangling. One answer to this problem was to dye the pieces in parallel, using one complete rope with an equal number of pieces ('flyers') stitched on to successive seams at their leading ends; the flyers were slightly shorter than the corresponding pieces in the rope.

(g) Sampling for colour was difficult as reproduction of the liquor movement and interchange was impossible. The sample had to be taken from the load and one modification was to fit a remote-control 'cigar-cutter' adjacent to a port. The fabric rope was stopped close to a join and the cutter advanced towards the fabric, a sample removed and withdrawn through a pressure lock.

The machine was loaded by means of a tape in each chamber; when unloading, the tape was tied in to the trailing end of the dyed piece so that the machine was always threaded up. If an operative forgot to attach the tape, or a seam broke during dyeing, the machine could be re-threaded with the tape, sometimes painstakingly.

Although lacking the versatility of later jet machines, the Gaston County-type jet remains in the forefront of high-temperature rope-dyeing machines, particularly for dyeing medium to heavy polyester fabrics.

Fully Flooded Jet Machines

When fabric is lifted from the dyebath longitudinal creases are formed and, in order to eliminate this hazard, a new type of jet machine was developed (ca 1969) in which the fabric remained completely immersed throughout the dyeing cycle.

The principle is to pass the fabric through a high-speed jet feeding fresh dye-liquor to the fabric which is carried along a tube to the entrance of a much wider section used as a storage compartment. In this section both liquor and fabric move very slowly until reaching the end when the fabric is drawn off through a narrower tube and fed to the jet. Dye-liquor is drawn off from the system at both ends of the storage compartment and passed through a heat exchanger before being fed back to the jet. By regulating the flow of liquor through the storage compartment, the speed of the fabric in this critical area can be controlled. All commercial machines are capable of operating at temperatures around 130°C.

Several designs of fully flooded machines are shown (Figures 2.10–2.13). That produced by Thies has a very long and comparatively narrow storage compartment; it is a tall machine, economical in floor space, but requiring either a deep pit or a gantry to permit loading and unloading. The Pegg machine is generally similar, but has a much wider and shorter storage section. The machine designed by Then is much lower, which facilitates operations, but occupies more floor space. The Vald Henriksen machine operates in a horizontal plane and is the most compact.

Most production machines have two or three tubes mounted side by side (the Henriksen has horizontal tubes mounted one above the other) with a common pump and heat exchanger. To enable smaller lots to be dyed on a multiple-tube machine, one or more tubes may be isolated from the rest of the system. The maximum fabric load per tube is usually quoted in the range 100–150 kg, the maximum being achieved with heavy cloths. The fabric speed through the jet is usually 100–200 m/min.

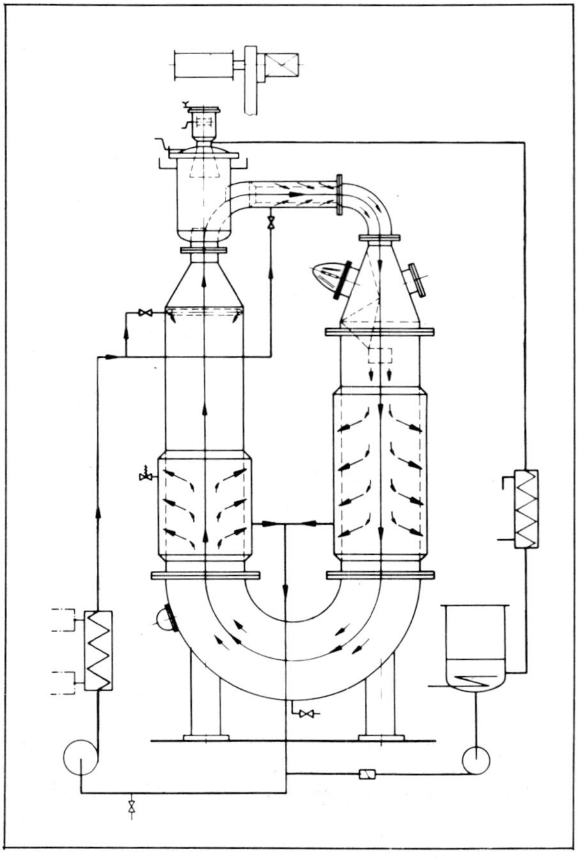

Figure 2.10 – Thies Jet Stream

Lightweight fabrics usually require a smaller diameter jet and early machines were supplied with a range of jets of different sizes; later, adjustable jets became available. Care must be taken to position the jet accurately as any irregularity may impart a spiral motion to the liquor flow which will generate false-twisting in the rope and consequent creasing. It is useful if the jet can be reversed so that, in the event of a blockage due to fabric becoming entangled when leaving the storage compartment (the most common problem), the normal liquor flow can be reversed and the fabric forced back in the hope that the entanglement will clear. One Spanish machine was built symmetrically so that flow in either direction could be used.

Fully flooded machines are particularly useful for delicate fabrics, but are not as easy to operate as later machines. Control of the fabric as it enters the storage section is very limited and, whereas wide heavy fabrics may fill the space uniformly, lightweight narrow fabrics tend either to sink or rise so creating possibilities of tangling. Fabric speed is usually monitored magnetically by fitting a metal strip into the rope, i.e. the time per cycle is measured, some machines being fitted with an alarm to indicate the non-arrival of the strip when a stoppage occurs. Most machines are fitted with one or more port holes with illumination, but, except when producing pale dyeings, fabric movement is difficult to perceive.

It should be noted that with constant settings on the machine, fabric speed will increase considerably as the temperature is raised, e.g. from 100 m/min at 50°C to 150 m/min at 110°C. This increase is due to the reduction in viscosity of dye-liquor with increase in temperature and its effect on pump efficiency and resistance to flow.

As with most high-temperature jets, internal accessibility is extremely difficult, and some makes of machine, which originally appeared as an integral pressure vessel, were later supplied in flanged sections so providing greatly improved access.

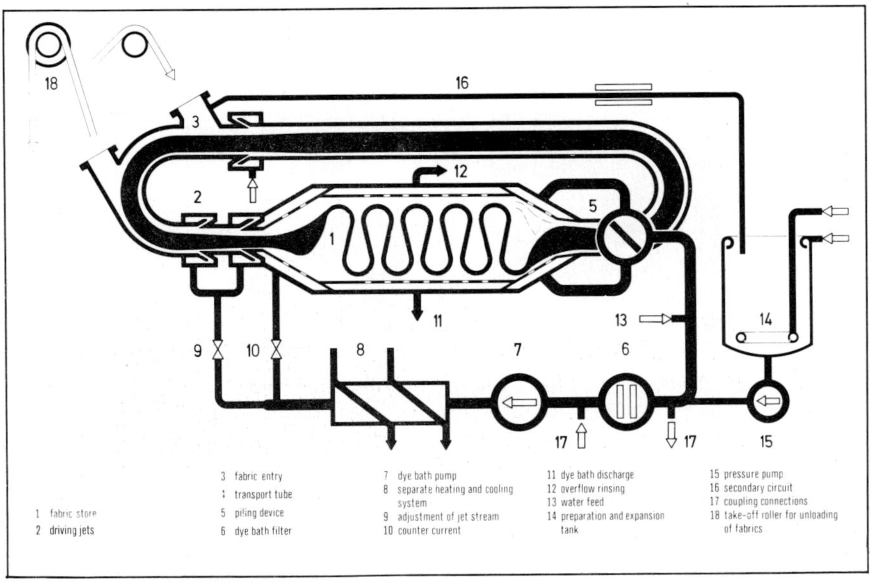

Figure 2.11 – Thenjet machine

The fully flooded machines do not suffer problems of foam formation; liquor ratios employed are normally in the range 12–15:1. As emptying and refilling the machine can sometimes lead to fabric disturbance in the storage chamber, overflow rinsing is often employed.

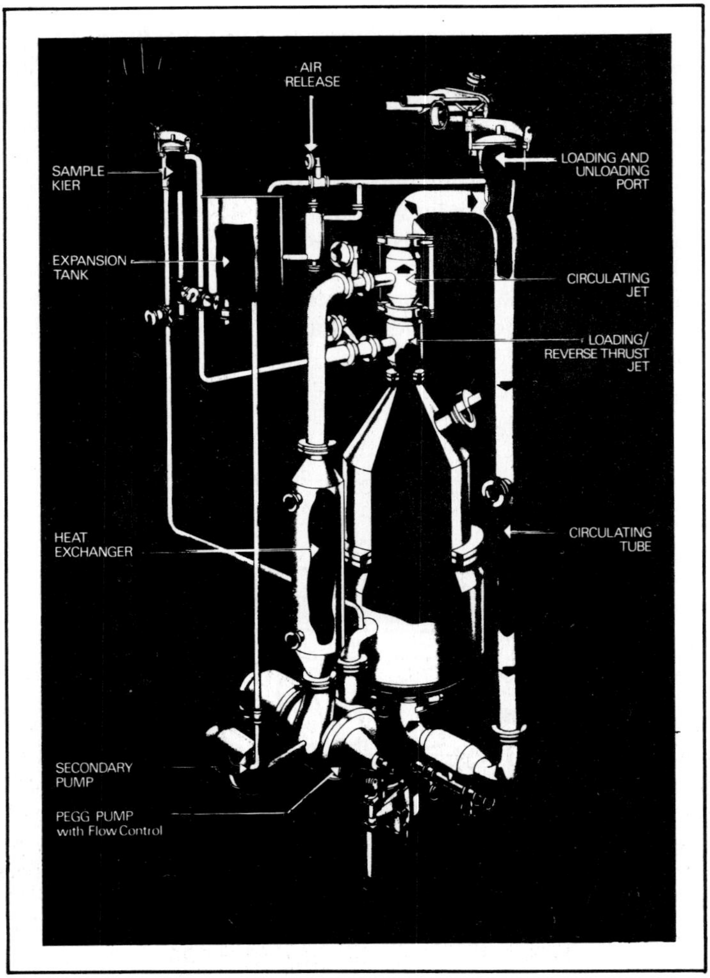

Figure 2.12 – Pegg jet dyeing machine

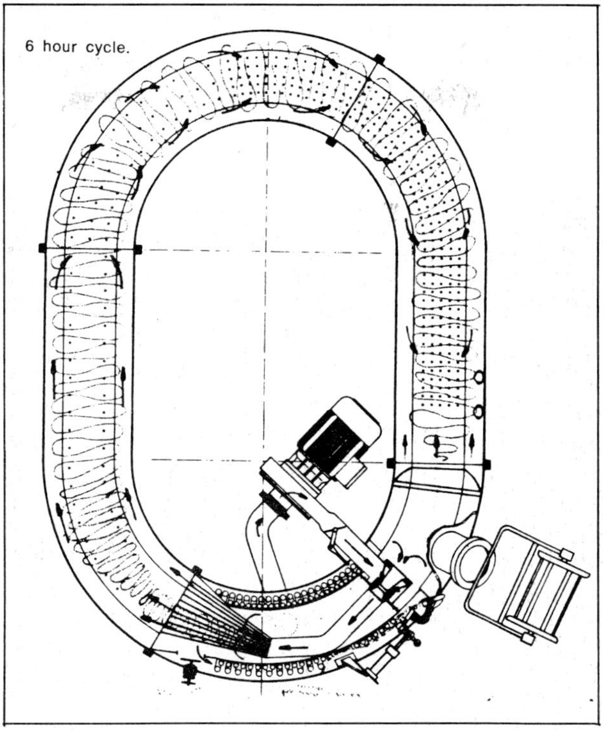

Figure 2.13 – Vald Henriksen jet dyeing machine (plan view)

'Soft-flow' Machines

The first soft-flow machine was produced commercially by Hokuriku in Japan and around 1972 several German manufacturers started building this new type of rope-dyeing machine in which the fabric is lifted briefly from the dyebath by a driven circular reel and then carried to the far end of the machine by a flow of dye-liquor along a transportation tube. This principle is available in atmospheric pressure and high-temperature (pressurized) versions.

Most designs feature the transportation tube outside and above the main vessel (Figures 2.14 and 2.15), but in others the tube is below the main dyebath (Figure 2.16).

In most machines the transportation tube is filled with dye-liquor using an overflow system, and both liquor and fabric flow along the tube under gravity. The reduced speed of the liquor (compared with earlier jets), and consequently gentler action on the fabric, led to this type of fabric movement system being called 'soft-flow'.

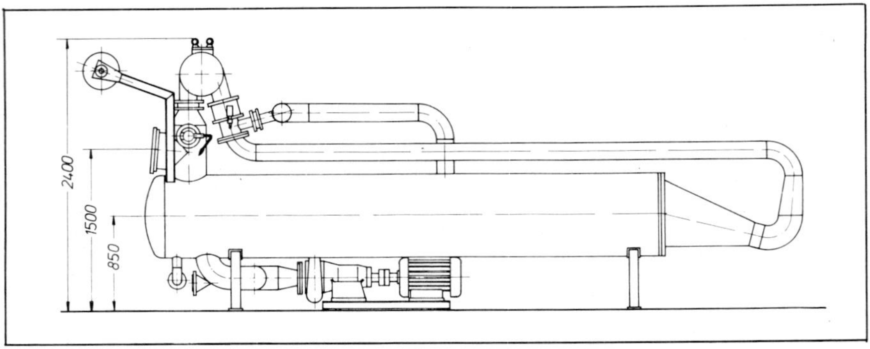

Figure 2.14 – Thies Soft-stream

Some machines use a jet to circulate the liquor and this enables higher speeds to be achieved at some sacrifice of gentle treatment of the fabric.

Using the overflow principle to fill the transportation tube, the pump is required to generate only a low pressure compared with a conventional jet, and consequently it can be driven by a less powerful motor, e.g. 5 kW. In high-temperature versions, a secondary pump is usually available to generate adequate internal pressure. By locating the loading/unloading hatch at the top of the vertical section accessibility is improved.

The use of a driven circular reel provides good control of the fabric speed; any jamming of the fabric increases the load on the driving motor and this can be used to initiate an alarm.

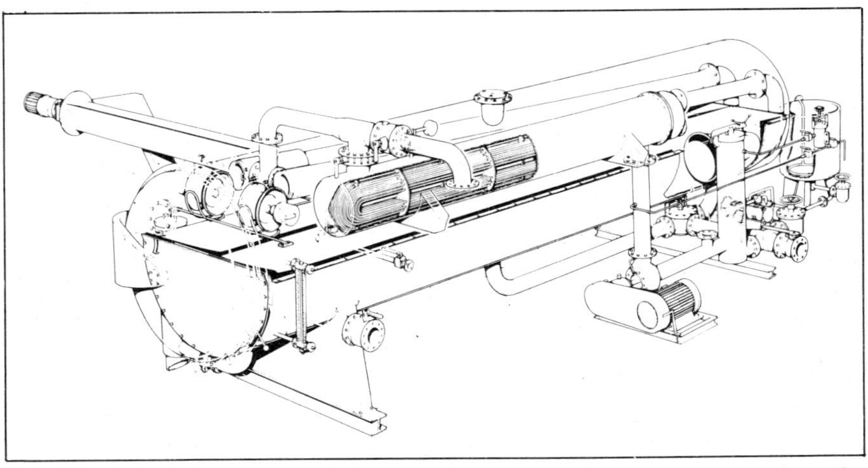

Figure 2.15 – MCS Siluro machine with two reels and tubes within the pressure vessel

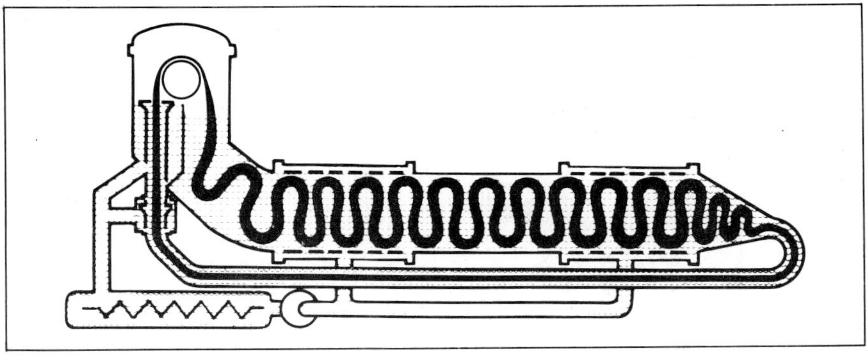

Figure 2.16 – ATYC Rapidsuau machine

'Soft-flow' machines are particularly good at minimizing creasing in sensitive fabrics. The circular reel creates much less strain in the fabric than the conventional winch reel with its elliptical cross-section. The height of the liquor in the vertical section may be raised to within 25 cm of the reel to minimize the extent of lift from the dyebath for delicate fabrics. Allowance should be made for the rise in liquor level due to expansion on heating.

To dye economical weights of lightweight fabrics it is advisable to limit the rope length to 300–400 m maximum and to attach additional lengths of fabric to the main rope as 'flyers'.

'Soft-flow' machines are versatile, generally easier to operate and are more compact than most fully flooded machines. The liquor ratio employed is usually about 15:1 upwards.

Two or three vessels may be located side by side and share one liquor circulation and heating system. Dye additions are made using a secondary pump and, if more than one is required, some liquor may have to be drained off to preserve the desired operating liquor level.

Low Liquor Ratio Reel-jet Machines

The quadrupling of oil prices during 1974–5 emphasized the need to reduce energy requirements during textile operations and the dyeing process was an immediate target for economies in energy use(*44*).

This demand led to the production, in 1975, of comparatively simple dyeing machines capable of operating at liquor ratios down to 5:1, i.e.using less than half the amount of water (and heat) required by the fully flooded and 'soft-flow' machines.

In many respects the essential layout of the machine is similar to the Gaston County Jet (see Figures 2.9 and 2.17). The fabric is plaited down by a jet of dye-liquor and slides on a perforated sheet around the bottom of the dye vessel. At the front the fabric is lifted by a driven roller and fed across the top of the machine to the jet. Fabric movement in this type of machine is

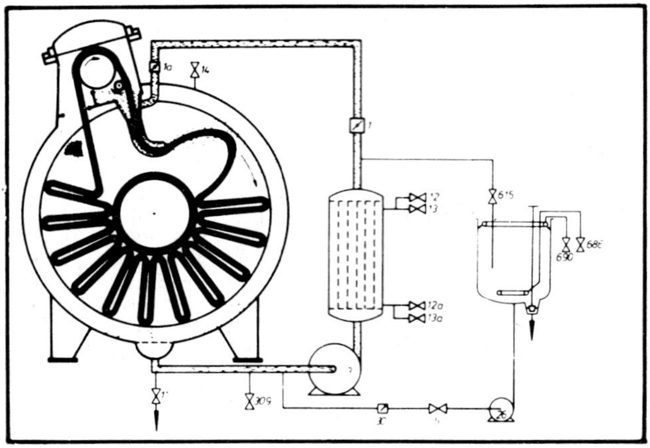

Figure 2.17 – Thies Roto-stream

primarily by the lifting reel supplemented by the jet (in the Gaston County movement was primarily by jet, the metering rolls providing a restraining action). Running speeds up to 400 m/min are employed, the high speeds being particularly useful when clearing excess surface dye.

Both atmospheric and pressurized (high-temperature) models are produced, the former being found to offer particular advantages in the application of reactive dyes to knitted cotton fabrics. The use of less dye-liquor not only gives savings in water consumption, heat and effluent disposal, but also of dyes and chemicals. However, the move towards lower liquor ratios must not be pursued too far otherwise problems of fabric creasing may arise.

The dye-liquor is withdrawn from the bottom of the dye vessel and pumped through a heat exchanger to the jet. As the total volume of dye-liquor is greatly reduced and pumping pressure is low, power consumption is small.

Low liquor ratio machines are usually supplied in multiple units, sharing a common liquor flow and lifting-roller system. In atmospheric pressure machines sampling is no problem, but in pressurized models sampling from the bulk fabric is strongly recommended, either by using a remote-control cutting device or by cooling, depressurizing and opening.

Later developments of this type of machine have tended to be in providing means of movement for the fabric whilst plaiting down; this has been achieved by using a slowly rotating drum (Figure 2.17), a brattice or a circular platform (Figures 2.18 and 2.19). These devices serve as an alternative to the large quantities of dye-liquor used in conventional winches and jet machines for supporting and conveying the fabric slowly from the discharge end of the jet back to the feed.

A further development is the increasing use of rectangular-shaped jets to encourage the opening out of the fabric rope.

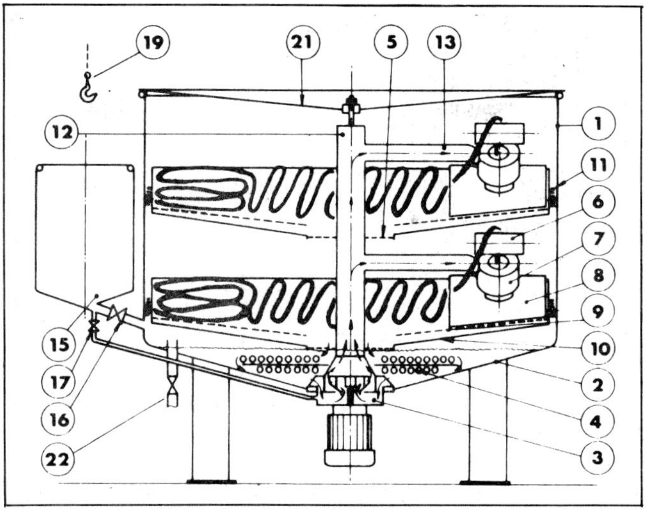

Figure 2.18 – Flow diagram of the VH-Ultimat (elevation)

KEY TO FLOW DIAGRAMS

(1) Cylindrical vessel
(2) Conical bottom – reservoir
(3) Circulating pump
(4) Heating- and cooling coils
(5) Filter plates
(6) Guide roller
(7) Nozzle
(8) Layout funnel
(9) Transport grate
(10) Liquor guide plate
(11) Brake band
(12) Central circulating pipe
(13) Nozzle delivery branch
(14) Fabric front
(15) Addition tank
(16) Ball valve
(17) Ball valve
(18) Throttle valve
(19) Hoist
(20) Inspection hatch
(21) Cover
(22) Overflow pipe

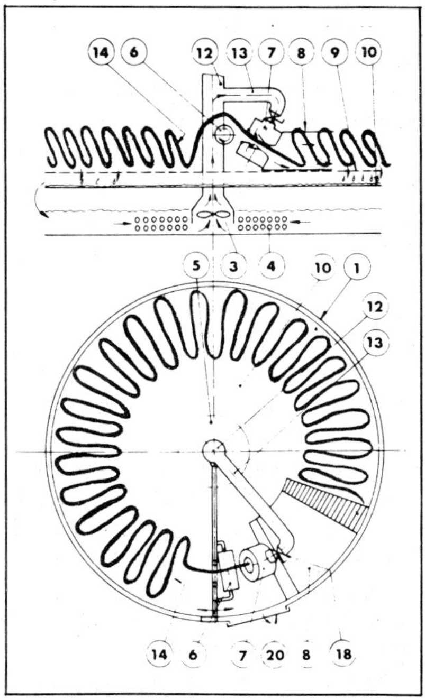

Figure 2.19 – Flow diagram of the VH-Ultimat (plan view)

General

Several comprehensive reviews of jet dyeing machinery have been published(*45–49*). In all the machines described the fabric rope passes through two main phases:

(a) The active phase in which the fabric is moving at high speed, passing through the jet with vigorous pick-up of fresh dye-liquor

(b) The passive phase in which the fabric is moving slowly around the system back to the feed-in to the jet.

The active phase usually occupies about 2% of the total time and the majority of dye absorption from the liquor occurs during this period. During the passive phase diffusion of dye into the fibre takes place.

In the passive phase the fabric is folded and, although the precise conditions are dependent on the fibre, the fabric thickness, and the sharpness of creasing, it is generally recommended that the fabric rope should circulate in about 60 s. This usually implies a maximum rope length of 300–400 m which, although satisfactory for heavy cloths, is uneconomic for lightweight fabrics, which are usually dyed in parallel using the 'flyer' system. Parallel continuous ropes have been tried with very limited success because of the need to have equal lengths of rope (and even if cut equal prior to loading it is unlikely they would contract or extend to the same extent under dyeing conditions).

As the fabric passes through the jet, longitudinal creases are formed, but these are usually eliminated as the fabric balloons out after the jet. Too small a jet diameter for the fabric gives excessive lengthways creasing, too large a jet gives poor fabric movement relative to the volume of liquor.

Heat exchangers on these types of machines must be carefully maintained. Not only are the thermal stresses imposed by high-pressure steam (for heating) and cold water (for cooling) considerable and liable to produce cracking, but dyes and oligomers tend to collect on the heated surface and may subsequently slough off. Regular cleaning is advisable, either with caustic soda and sodium hydrosulphite or, if the exchanger is easily dismantled, by mechanical means.

2.5 STAR FRAMES

A few high-quality velvet fabrics susceptible to crushing when dyed on jig or beam or in rope form, are dyed on a star frame.

The frame, with horizontal arms carrying hooks below, is loaded with fabric by pinning the selvedge and then lowered into the circular dye vessel. A perforated plate is placed on top to assist uniform dye-liquor distribution and the frame slowly raised and lowered in the dyebath either by a lever arm operating from above or by a hydraulic ram frame underneath (Figure 2.20).

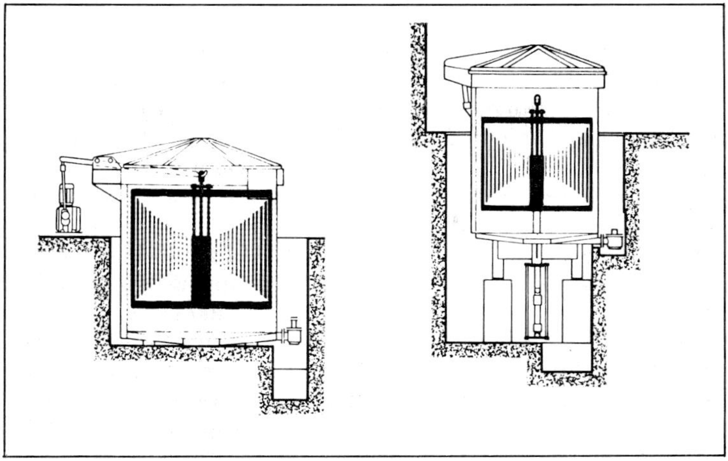

Figure 2.20 – Pesch star frames

Heating is effected by steam coils positioned at the bottom of the dye vessel, and a gentle liquor circulation system is provided. Liquor ratios are generally in the 50:1 area, although low-liquor-level techniques have been introduced using a spray at the top of the machine to reduce the liquor ratio to 15:1.

Star-frame vessels extend up to 2 m in diameter and the hooks are spaced about 10–15 mm apart, providing a maximum load of about 200–250 m of fabric per dye vessel.

The high cost of dyeing by the star-frame technique restricts its use to only exceptionally high-quality fabrics.

3 MACHINES FOR DYEING FABRICS CONTINUOUSLY

3.1 PAD MANGLES

Nearly all processes for dyeing fabric continuously can be divided into two main sections – the application of the dye solution followed by fixation (washing-off surplus dye and auxiliaries may be combined or carried out separately).

The application of dye solution is usually carried out on a pad mangle, although other methods are occasionally used, e.g. Standfast Molten Metal unit (described below), feeding dye-liquor onto the fabric (Polychromatic technique, Kuster TAK method), spraying or the application of foam. Dye fixation may be carried out in steam, in hot air or by chemical treatment, e.g. oxidation.

However, the uniform application of dye-liquor to the fabric is the most critical part of a continuous dyeing process and satisfactory performance of the pad mangle is crucial to success. Pad mangles are also termed dye pads or padders; the expression 'mangle' should be confined to less critical machines for squeezing water from fabric, usually prior to drying (see Section 4.7).

Several excellent papers(*50–54*) have been published on the principles of padding fabrics, but active discussion has continued – indeed, it has been said that 'a dye padder is more a philosophy than a machine'. One reason for the range of opinions is probably the difference in approach between the UK and Europe on the one hand and North America on the other. American market requirements enable large quantities of fabric, say 100 000 m, to be dyed to one colour, and this provides several hours' running at speeds of 120 m/min. In Europe, however, many orders are 10 000 m or less and, for runs of this size, operating speeds of around 50 m/min are adequate. This difference in speed can demand major differences in the pad mangle required.

The padding operation itself consists of two parts: first by the immersion of the fabric in the dye-liquor to achieve thorough impregnation, and second by passing the fabric between two rollers to squeeze air out and force dye-liquor in, and to squeeze surplus dye-liquor back along the fabric. For dyeing most fabrics at speeds of up to 50 m/min a single-dip, single squeeze technique is sufficient; for higher speeds or for heavier fabrics a double-dip, double-squeeze process using a three-bowl mangle is preferable (see Figure 3.1).

The Fabric

The fabric, usually woven, is usually taken from plaited form and, after skying, is fed straight into the pad bath. If run from a large roll, an end cloth should be attached to the tail end, or the fabric run through an accumulator or scray which will continue to deliver fabric to the pad mangle while the end piece is being attached. The fabric must be readily absorbent, as its immersion time in the liquor will be of the order of 0.5 s, and it should be free from loose fly, waxes, oils, etc. Bone-dry cotton fabrics and polyester/cellulosic blend fabrics, which have been heat set immediately prior to padding, are notoriously slow in wetting-out and some dyers arrange for a 're-wetting' agent to be present in the fabric.

The fabric itself must be flat, crease-free and with flat selvedges; knitted fabrics demand special precautions such as low lengthways tension and strategically placed scroll rollers and edge uncurlers. The fabric must be centred accurately on the pad mangle and not be allowed to wander sideways. Woven fabrics should be run under moderate lengthways tension – excessive tension gives poor absorbency and lower liquor pick-up.

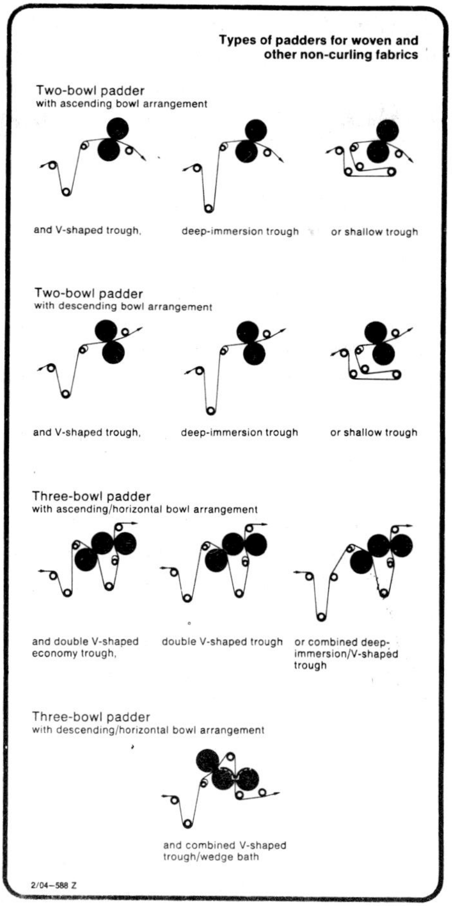

Figure 3.1 – Types of Artos padders for woven and other non-curling fabrics

The Padding Box

Situated below and at the front of the squeeze rollers, the padding box consists of a V-shaped box with a free-running roller at the bottom round which the fabric passes. Dye-liquor is fed in through a perforated pipe running across the back of the bath, the perforations pointing down and away from the fabric. The supply of dye-liquor is controlled automatically to maintain a constant level (where a float switch controls the dye feed care should be taken to ensure that it is not influenced by any froth or foam on the surface).

The cardinal principle is to ensure that liquor in the pad box is absorbed and removed by the fabric quickly. The volume of the pad box should be kept as small as is practicable in order to reduce the amount of dye-liquor wasted at the end of the run and, of even greater importance, to reduce the effect of preferential pick-up of dye or water by the fabric resulting in 'tailing' (see 'Liquor Pick-up'). Ideally the width of the fabric should be only slightly less than that of the box to avoid dead spots at the ends of the box; where there is a large gap a metal filler or blank is inserted. Similarly a displacer is often fitted above the roller to reduce dyebath volume.

For very wide fabrics, a liquor circulation system is often recommended by machinery makers, but it is questionable whether it is necessary. A drain plug is fitted to the bottom of the box to allow rapid drainage at the end of the run (once the end cloth is reached the pad box should be emptied rapidly). The interior of the pad box must be smooth and easy to clean with a hose; dye carry-over from one run to the next is disastrous.

The bottom roller runs in sealed bearings and is usually about 10 cm in diameter. As the fabric starts to run round the roller at the bottom of the bath, the roller exerts a mild pressure to force dye-liquor into the fabric. Similarly, when the fabric leaves the roller, there is a tendency for liquor to be drawn in from outside.

Although most dye-liquors are applied at around 20°C, hot application is sometimes required to achieve rapid and uniform wetting-out and high take-up of dyes into solution(*50*). Some pad boxes are fitted with an external heating jacket circulating hot water or hot oil; the jacket should surround the lower part of the pad box up to within 5 cm of the liquor level (if the jacket goes higher it may cause drying-out of the liquor).

Where two padding boxes are required (Figure 3.1), they are mounted in tandem, the second overflowing into the first (in which is mounted the liquor-level control). Liquor pick-up is much greater from the first than from the second dip.

Pad Bath between Rolls

When fabric is passed downwards between two vertically opposed rolls, the space on the top side of the rolls may be used for containing dye-liquor; plates are fitted to retain the dye-liquor at the ends of the rolls. The advantages of this system are that the dyebath has a very small volume and there is no draining back into the bath with preferential absorption. However, the 'wedge nip', as it is sometimes called, is restricted to padding lightweight fabrics, as thicker fabrics suffer from liquor seepage at the selvedges, difficulties in wetting-out (as the immersion time is very short) and the continuous expulsion of air tends to cause frothing. The setting of the end plates retaining the dye-liquor has to be adjusted precisely in order to provide a seal with the two rollers rotating in opposite directions.

The wedge nip is normally used as the second stage in combination with a conventional dip–squeeze unit.

Nip Rollers

The nip rollers, rolls or bowls are the key to successful pad dyeing. Two bowl mangles are used for lightweight and mediumweight fabrics at moderate speeds – say 50 m/min; three-bowl machines are used for heavier fabrics requiring two immersions and for higher running speeds.

Arrangement of Rollers

Two-bowl mangles can be arranged as vertically opposed in ascending or descending order, or horizontally opposed (Figure 3.1).

With the first three arrangements it is usually necessary to take the fabric over a free-running, smooth roller before entering the nip. This roller should be positioned so that the fabric path leaving the pad box is nearly, but not quite, vertical and so that on leaving the roller the fabric enters straight into the nip. Pad mangles of this type must be fitted with nip guards covering both the entry zone and the sides.

Likewise, on leaving the nip, the fabric should leave both rollers simultaneously otherwise face/back differences may show. It is generally agreed that the fewer points of contact with the squeezed wet fabric the better, and the number of rollers following padding should be cut to a minimum (this also reduces cleaning between successive runs).

The search for a minimum contact with the impregnated fabric led to the development of the horizontally opposed mangle in which the fabric rises vertically from the padding box directly into the nip and straight through into a dryer mounted above the nip. This system has been used widely in the Thermosol process, for dyeing polyester/cotton blends, and has given excellent results despite the difficulty of threading-up through the mangle.

The ascending-bowl arrangement gives better observation of the nip than the descending-bowl system, but the latter has a slight advantage in that the fabric path does not change as much.

The three-bowl arrangements are usually designed to provide maximum impregnation on the first dip–squeeze, and maximum uniformity at the second squeeze. In each case the centre roller is driven in fixed bearings, the two remaining rollers apply pressure and are friction driven.

Dimensions of Nip Rollers

Roller widths extend up to 4 to 5 m for textiles, but for the continuous dyeing of most woven fabrics, mangles of 170–200 cm are used. The rollers should be roughly equal in diameter, usually in the 30–40 cm range, but for continuous, high-speed running (as in the USA) larger diameters are favoured. Increasing the diameter reduces the deflection of the rollers when under pressure and reduces the rev./min for a given fabric speed, but necessitates a significantly heavier roller and larger bearings. Increasing the diameter also increases the area of contact between the two rollers, depending on the pressure applied and deformation of the roller surface. For padding dyes, a minimum contact distance of 1 cm is desirable. The longer the contact distance, the greater the deformation of the roller surface and more rapid the wear. Increasing the area of contact for a given loading reduces the pressure applied to the fabric and this may require increased force to be applied to the ends of the beam when trying to achieve a fairly low pick-up.

Construction of Nip Rollers

Nearly all padding mangle bowls have a steel mandrel which is covered for most of its length by hard rubber with a surface of comparatively soft rubber about 15 mm deep. The hardness of the surface is normally measured by the Shore Durometer, bowls for applying dye-liquors normally registering in the 55–70° region. The softer the rubber the more readily it deforms under pressure to give a greater area of contact with the fabric and to deform around any slubs or thick places in the fabric. With fairly open fabrics, soft rolls will deform readily to reduce pick-up in the fabric interstices. Unfortunately the readiness to deform correlates with the tendency to wear and a compromise must be established. Soft rolls will deform more readily at the selvedge and reduce the amount of dye-liquor carried through at the extreme edge of the fabric. The physical strains set up in the bowl cover can lead to unevenness; this is revealed when wider fabrics are padded. Regrinding to a smooth surface is then essential.

Both rolls in the mangle must be of similar hardness, otherwise there will be a face/back difference on the fabric. Although on old mangles stainless steel, brass or ebonite were used for the driving roller these are now obsolete.

The chemical composition of the rubber surface must be considered carefully as it must resist dye-liquors and the accompanying auxiliaries. In the application of cross-linking finishes to cellulosic fabrics, the rollers should be impervious to solutions containing formaldehyde condensates. Immediately after a run, the rollers of the pad mangle should be hosed down thoroughly, allowing time for all ingredients of the padding liquor to diffuse out of the surface. Soft rolls tend to harden during use, and frequent cleaning is essential to avoid dye becoming engrained.

Pressure is applied to the mandrel at the ends of the roll, the maximum pressure being about 400 N per cm (50 kg/cm) of bowl width, i.e. about 10 tonnes across the total width. The pressure is applied pneumatically to each side of the top roller which is forced, through a fulcrum, to press against the bottom roller. (In three-bowl mangles the centre roll is fixed and the outer two press against it.)

Applying pressure to the ends of a roller naturally causes some deflection of the roller providing lower pressures (and higher pick-up) in the middle of the pad and higher pressures (lower pick-up) at the sides. The deflection of a bowl of known construction can be calculated(*54*) and is dependent on the length of the bowl face and the diameter of the bowl. Alternatively it can be measured using a Shirley Bowl Deflection Indicator.

In order to try and counteract the effect of bowl deflection the surface camber of one or both bowls is modified so that the surface profile nullifies the effect of bowl deflection. This is usually done by grinding down the surface of the roller to taper it slightly towards each end. The amount of camber is correct for only one pressure on the roll.

An alternative method to correct bowl deflection is to skew the axis of the pressure roller relative to the axis of the fixed roller(*54*). This arrangement tends to reduce the pressure between the bowls at the edges, the amount of skewing varying with the deflection. An advantage of this system is that the bowl surfaces are ground parallel and the amount of skewing can be varied to suit the load.

Special Designs of Mangle Bowl

In the UK the Kusters 'swimming roll' mangle has enjoyed an extremely high reputation for many years. The principle of the roll design (Figure 3.2) is to have a stationary mandrel, around which the roller shell rotates in bearings. The space between the mandrel and the shell is divided by pressure seals into two compartments. The smaller compartment, pressurized by oil, is always towards the nip. Oil is constantly pumped into the compartment, some passing the seals and circulating back to the oil reservoir.

Increasing the oil pressure in the compartment tends to raise the surface pressure at the middle of the roll so compensating for the deflection of the mandrel. By controlling the internal oil pressure and the forces exerted hydraulically at the ends of the mandrel, the configuration of the bowl can be adjusted to give a uniform linear pressure at the surface. Swimming rolls are normally used in pairs, but can be used in conjunction with a conventional roll. The maximum internal pressure is about 500 kPa (71 lbf/in^2).

Where the second roller is of conventional design, i.e. with a rotating mandrel, this is the driving roller turning the swimming roller by surface friction. Where two swimming rollers are in combination, the surface shell

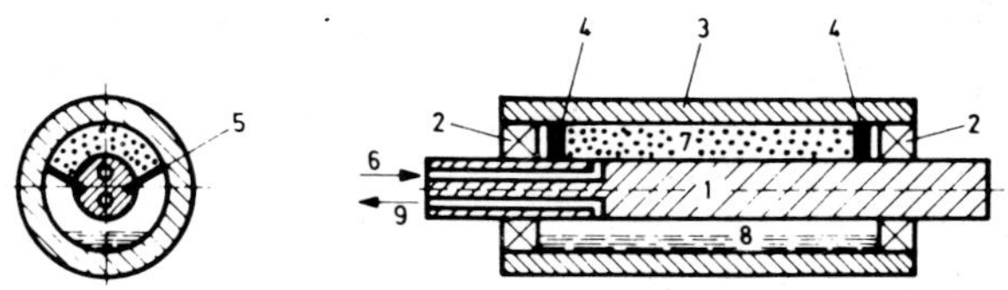

Key

1. Stationary axis
2. Anti-friction bearings
3. Shell
4. Frontal seals
5. Longitudinal seals
6. Oil in
7. Pressurized compartment
8. Oil-collecting compartment
9. Oil out

Figure 3.2 – Principle of the swimming-roll design

of one is driven through a clutch ring at the opposite end from the circulation of oil.

Another type of padding mangle is the Artos stabilized roller system shown in Figure 3.3. The roller consists primarily of three components: at the centre a tube shaft to the ends of which the load is applied, a rubber-covered sleeve on the outside, and an inner sleeve covering only the middle section of the roller and linking the outer sleeve with the inner shaft. Although the inner shaft deflects under applied pressure, the force is transmitted to the outer sleeve via the inner sleeve.

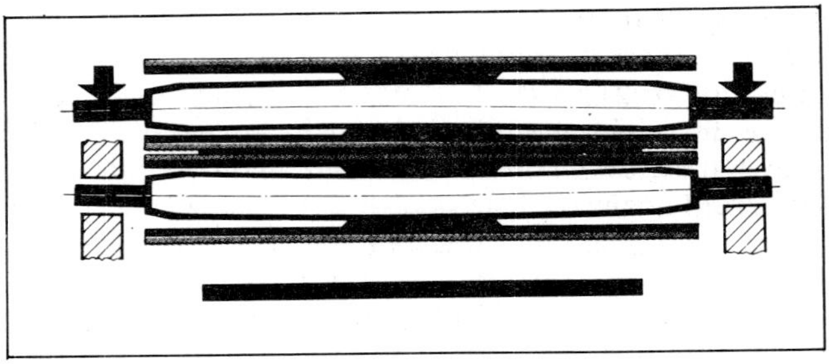

Figure 3.3 – Artos stabilized bowls compensate for deflection of the mandrel under pressure

Care of Rubber Rollers

To preserve the rubber coating in prime condition, rollers should be covered during storage to minimize oxidation of the surface. Excessive heat or cold should be avoided along with dampness or direct sunlight. Rollers should always be stored horizontally on their mandrels, never on end or resting on the rubber surface, which would rapidly develop a flat spot.

After prolonged storage, grinding to remove the outer layer of oxidized rubber is recommended. When installed in the mangle, pressure should not be applied whilst the rollers are stationary. If this should happen and a 'flat' develops, it may be possible to remove it by running for some hours in warm water.

Rollers should always be cleaned immediately after use using plenty of water; organic cleaning liquids are not advised, soap and water and a bristle brush are preferred.

One of the worst disasters to happen to any dyer is for a pad mangle roller to be damaged by a foreign object such as a nail. Any indentation will mark-off onto the fabric, repeating at a distance equal to the circumference of the roller. If the fault cannot be removed by running in warm water, the roller will probably have to be removed from the mangle and re-ground – a very time consuming job. Sometimes small faults can be covered by the careful application of rubber cement.

Application of Pressure to the Rollers

In nearly all two-bowl mangles, the lower bowl is in fixed bearings and the upper bowl is moved to press against the lower. In the separated 'open' position the top roller is only about 2 cm away from the bottom. The bearings of the top roller are mounted on arms pivoted at the side of the mangle and actuated by pneumatic cylinders. The quantity of compressed air required is small and may be obtained either from a 690 kPa (100 lbf/in^2) dyehouse services line or from a small compressor attached to the mangle structure. The compressed-air line is attached to the pneumatic cylinders at each side of the machine so that they operate in conjunction. The compressed air may be switched from one end of the cylinder to the other to give positive control to either the open or closed position.

Most mangles up to 2 m wide operate with a force of up to 5 tonnes on each end of the roll, giving a total weight of 10 tonnes, i.e. 50 kg/cm.

In three-bowl mangles the central bowl is in fixed bearings and the outer two rolls are forced against the centre. In all cases the design must be such that when there is no pressure in the cylinder, e.g. when the mangle is not operating, the top roll will be in the open position. The rolls should be brought together only when the driven roll is in motion.

The Drive

The roller in fixed bearings (whether in a two- or three-bowl machine) is driven from one side of the machine. The electric motor is mounted either at the side of the machine or underneath it. The drive system depends very much on the maximum speed of the mangle, which may range from 25 to 120 m/min.

Provided the fabric wets out rapidly, the liquor pick-up is almost independent of fabric speed over a fairly wide range.

In a continuous dye range, the speed of the pad mangle dictates the speed of the subsequent machines, small differences in speed being controlled by compensator bars.

Foundations

The sides of a pad mangle have to be very robust in order to withstand the forces applied; a typical 2 m wide mangle weighs about 3 tonnes. This weight is borne by the sides whose contact with the floor is about 1 m × 50 cm (each). This gives a floor loading of about 3 $tonnes/m^2$.

To facilitate washing down with a hose, the floor below the mangle should slope to a drain and a small retaining wall should be constructed round the mangle.

The positioning of a pad mangle must be carried out precisely. Laterally the centre of the mangle must be on the centre line of any subsequent operation (such as drying); further, the rollers must be such that they discharge fabric straight down this line – any tendency to throw to one side or the other will tend to give creases. Thirdly the mangle must be level to provide uniform depth of liquor along the padding box.

Liquor Pick-up

In continuous dyeing the objective is to obtain a uniform distribution of a known amount of dye-liquor on the fabric. Liquor pick-up is usually quoted as a percentage relating the weight of dye-liquor to 100 g of dry fabric. It is important to pick up enough liquor to enable the dye to diffuse uniformly across and through the fabric, but not so much liquor as to cause swealing or excessive migration of dye during subsequent drying. For 50:50 and 67:33 polyester/cotton fabrics, pick-ups are usually between 45 and 55%, for all-cotton fabrics around 60–70%, and spun viscose fabrics 80%. The pick-up must first be established for a particular fabric under specific conditions, so that the required concentration of dye in solution can be calculated to provide the correct depth of dyeing.

Whereas determination of liquor pick-up is straightforward, the problems associated with 'tailing' have caused much discussion(*51,52*). Basically the problem derives from differential pick-up of components of the padding liquor leading to variation in composition of the liquor as the run continues. Even mixtures of pigment colours have been known to show tailing on long runs. The most effective answers to the problem are firstly to limit the time for preferential absorption to occur (say 1 s maximum), and secondly to have a small dyebath, the contents of which are replenished rapidly.

Safety

Pad mangles have been the cause of many serious injuries in dyehouses and the in-running nips must always be fitted with guards. Most pad mangles in continuous dye ranges are kept threaded-up with end cloth or with tapes, but where the mangle has to be threaded it *must* be done with the mangle stationary and the pressure off.

Uniformity of Expression

For assessing the uniformity of liquor expression across the width, the classical technique is to use a 15 cm wide strip of carbon paper in a sandwich of white draughtsman's paper(*54*). The sandwich is laid along the top of the lower bowl whilst stationary and then the upper bowl slowly lowered and pressed for a few seconds before release. The mark transmitted to the paper indicates the uniformity of pressure across the width.

3.2 VACUUM IMPREGNATION

One of the main problems in padding thick fabrics is the presence of air occluded in the structure, which prevents the ready absorbtion of dye-liquor. Following the observation that a roll of fabric under vacuum is very readily

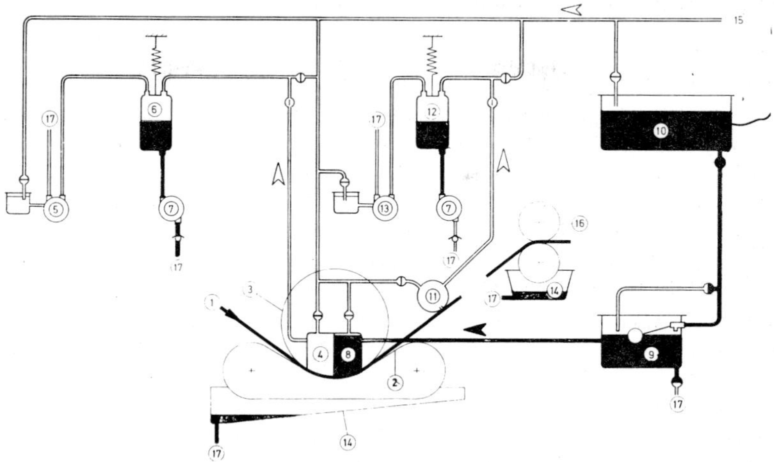

Liquor circuit for Farmer Norton Vacuum Impregnation Machine
(for liquors which are unsuitable for recirculation)

1. Cloth passing through machine
2. Rubber belt
3. Cylindrical screen
4. Vacuum box – impregnation unit
5. High-vacuum pump
6. High-vacuum receiver
7. Liquid extraction pump
8. Liquor box – impregnation unit
9. Liquor level trough
10. Liquor supply tank
11. Suction box – liquor extraction unit
12. Suction receiver
13. Suction pump
14. Drip trough
15. Wash water
16. Squeezer
17. Drains

Figure 3.4 – Liquor circulation in the Farmer Norton vacuum impregnation machine, for liquors unsuitable for recirculation (on most machines a squeeze roll is fitted instead of the suction box (11))

saturated with dye-liquor(*17*), a few machines were constructed to apply vacuum impregnation continuously(*55*).

In the UK Sir James Farmer Norton designed a machine (Figure 3.4). in which the fabric was carried on a rubber belt underneath a perforated rotary screen, in the first half of which suction was applied to the fabric and in the second, dye-liquor was fed at atmospheric pressure. The fabric then passed to a mangle to remove excess liquor which could either be rejected or returned to the dye-feed system.

Trials showed that good impregnation was achieved even on sized fabrics in the loomstate. However, commercial success was limited, probably due to the comparative ease of a two-dip – two-nip system, which is adequate for nearly all fabrics, and the fact that improved penetration on thick, heavy fabrics necessitated increased quantities of dye to obtain a given depth of dyeing.

3.3 DRYING PRIOR TO FIXATION

The purpose of padding is to apply dye-liquor uniformly (and in the right amount) to the fabric; the next stage is to promote diffusion into the fibre(s) and this may be done wet (e.g. for cellulose) or dry (e.g. the Thermosol process for polyester/cellulosic blends). In the latter process it is crucially important, having applied the dye-liquor uniformly, to dry the fabric without permitting migration of the dye.

Conventional drying entails the application of heat to the surfaces of the wet fabric, producing a rise in temperature leading to evaporation of water from the surface and transportation of water from inside the fabric to the surface until the fabric is dry. This movement of water takes place on the surface of all fibres as well as within cellulose, and may also cause the migration of dyes and other non-substantive solutes to the area of evaporation(*56,57*). It is generally agreed that below 30% liquor pick-up on a polyester/cellulosic blend, migration is negligible. After padding to give a liquor pick-up of 50%, it is important to get down to 30% or below quickly.

Accordingly, the accepted optimum arrangement for applying disperse dyes to polyester by the Thermosol technique is to pad the fabric through a horizontally opposed mangle with an infra-red predryer mounted above (see Figure 3.5). The fabric emerges from the nip and, without touching any rollers, is subjected to a powerful blast of radiant heat for a few seconds which raises the temperature of the wet fabric swiftly to around 100°C at which point there is a rapid loss of moisture from the fabric surface. The infra-red heat must be equally balanced on both sides of the fabric or migration will occur.

The infra-red heaters are mounted in banks and may be gas- or electrically-heated. The length of the banks of heaters range from 1.5 to 3 m and, in the case of gas-firing, are estimated to reach a temperature of 800°C. An infra-red radiation of about 3 μm wavelength is required, as this is preferentially absorbed by textile fibres. The heat output is high and difficult to control. The simplest method is to measure the temperature of the air being removed at the top of the machine, which gives a quicker response than a fabric moisture controller at the exit of the machine.

The intensity of heat input from the infra-red heaters causes problems in the event of a fabric stoppage – the fabric would certainly be damaged. Safety systems have been designed so that, if a stoppage occurs, not only is gas or power to the heaters cut off, but either metal shutters drop between the heaters and the fabric or one side of heaters opens away from the fabric.

At the bottom of the heater battery, warm air is blown in and travels upwards with the fabric to assist the removal of steam which is drawn off at the top.

For particularly difficult fabrics, e.g. those with high polyester content and with a large quantity of liquor on the surfaces, a migration inhibitor should be added to the padding liquor.

An interesting, and possibly spectacular, variation on the conventional Thermosol process is to incorporate approximately one-third by volume methanol in the pad bath (the Remaflam process). After passing through the nip the fabric is heated and methanol vapour is ignited. The heat evolved so close to the fabric converts some of the water in the fabric to steam, whilst the fabric remains wet it suffers no damage. Potential problems of flammability and toxicity in the storage and use of substantial quantities of methanol, as well as its relative cost, will determine the progress of this development.

3.4 THE THERMOSOL METHOD

The Thermosol technique for dyeing fabrics containing polyester was first described in 1949 by J W Gibson of Du Pont(*58*). The process was based on the diffusion of disperse dyes into polyester at temperatures of 200–220°C – extremely high for those days. Commercial exploitation of the process for woven polyester/cotton blend fabrics developed firstly in the USA and then, in the 1960s, in Western Europe. In the UK the adoption of the Thermosol technique proceeded much more slowly, probably due to priority being given to weft-knitted textured-filament polyester fabrics (which were all batch dyed), supported by market demands for small orders per colour. It was not until about 1968 that substantial quantities of polyester/cotton fabrics were dyed by Thermosol techniques in the UK.

The process may be divided into four sections as follows:

(a) Padding the dye-liquor onto the fabric
(b) Drying down to a low moisture content to prevent migration
(c) Final drying
(d) Fixing the colour by a treatment at 200–220°C nominally for 30–60 s

The padding operation has been described in Section 3.1 and predrying in Section 3.3. Final drying is usually carried out on steam-heated cylinders and the process may be interrupted at this point to permit fabric examination before proceeding with fixation. However, once confidence is established, the processes can be combined into one operation to save energy, labour and time.

Thermosol Fixation

The fixation stage(*59–62*) may be carried out on a variety of machines capable of uniformly heating the fabric to about 210°C – variation across the width should not exceed ±3.5°C.

Stenters

Originally pin stenters were used for thermofixation and times of about 60 s were found to be necessary for adequate fixation. This led to the installation of long stenters, e.g. with eight or more chambers, necessitating a very long pin chain, much space and high capital and maintenance costs.

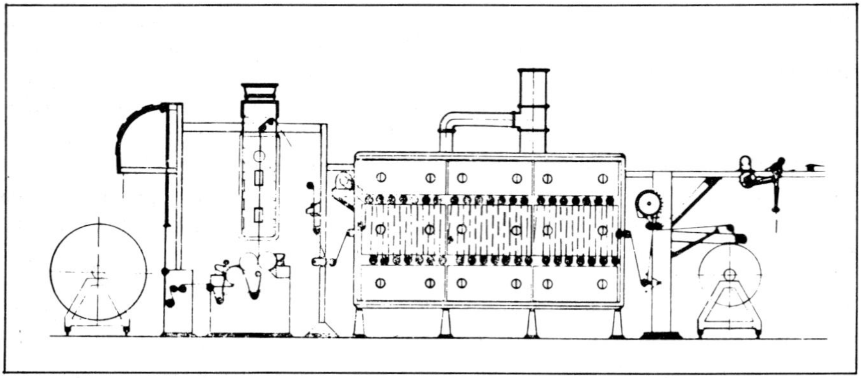

Figure 3.5 – Artos Thermosol dyeing range, showing a double-dip pad mangle, an infra-red dryer mounted above followed by a three-chamber hot flue

Stenters provided excellent widthways control of the fabric, but in some cases the pins and pin plates introduced variation in heating at the selvedges. To eliminate this fault cranked pins were set into the side of the pin plates so that the fabric was held clear of the hot plate.

Stenters are described in more detail in Section 4.8.

Hot Flues

Hot-flue Thermosol units have been widely used in Western Europe, the fabric being led vertically round a series of top and bottom parallel rollers mounted in an oven (Figure 3.5). Air is circulated around heat exchangers at the bottom of the machine up towards the fabric. Machines of this type usually have a capacity of up to 90 m of fabric and occupy comparatively little space.

The top rollers are driven and the bottom are free running, the roller bearings being located outside the machine. In order to avoid fabric creasing, large (18 cm) diameter rollers are used and must be accurately aligned; preferably the distance between top and bottom rollers should not exceed 125 cm. All machines of this type operate with a significant degree of lengthways tension, particularly when the fabric is being lifted; there is no control of the width of the fabric, which will be effectively set at the temperature of thermofixation.

The lengthways tension in the fabric causes it to stretch slightly so that, if the speed of the fabric coming out of the machine equals the speed going in, the fabric loops would extend in length until they drag on the floor of the machine. Consequently the rollers at the exit are set to run very slightly faster than those preceding. As the roller surfaces are smooth there is a significant amount of slippage between the fabric and the rollers, and a fair degree of lengthways tension must be maintained to provide traction. In many machines the drive to the top rollers is divided into two or more sections, the speed of the rollers in each section being controlled by the position of a compensating roller immediately preceding the section. (As with steamers, a more recent development is hydraulic drive to the top rollers.)

The fabric enters and leaves the machine through narrow slots, in order to minimize the amount of cold air taken in with the fabric and to reduce heat loss at the exit. Immediately upon entry the fabric may be subjected to a blast of hot air designed to raise fabric temperature rapidly. The heat exchangers in the floor are often filled with hot oil from a closed circuit. Alternatively, particularly in the USA, gas firing may be used. The inner surfaces of the machine must be well insulated to retain heat. Although hot air is circulated in the machine, the rate of heat transfer is nowhere near as good as that in a stenter and it is important that the fabric is fully dry and hot when it enters the machine. Heat transfer between the rollers and the fabric probably plays an important role. The fabric selvedges tend to heat up more quickly than the body of the fabric, due to the greater exposure to hot air and to radiation from the internal walls.

Access to the inside of the machine is by side doors, but should be required only rarely. During a production run a break-out at a fabric seam, or a roller seizing-up, means several hours' lost production due to the need to cool down before re-threading the fabric; several metres of cloth will inevitably be seriously damaged.

In order to overcome the loss in width during hot-flue thermofixation, Monforts produced a range including a hot-flue unit to provide dye fixation followed immediately by a short pin stenter to pull the fabric out to width (Figure 3.6).

Contact Heating

Although thermofixation systems using hot air as the heating medium were based on times of 30–90 s at temperatures in the range 220–200°C, experience with contact-heating methods indicated that, once the critical temperature range has been reached, sublimation and diffusion of dye into the fibre took place rapidly, probably within 10 s. Whereas with hot-air heating much of the time is taken up with raising the temperature of the substrate to 200+°C, contact heating provides much more rapid heat transfer and reduces overall times down to around 20 s, so reducing machine size and cost for a given rate of production.

Contact heating has been developed and used mainly in the USA, the cylinders or 'cans' being about 60 cm in diameter and mounted in banks. Direct gas firing inside the cans is preferred to hot-oil circulation on account of simplicity, cheapness and ease of maintenance. Fixation on cans tends to give a polished appearance to the fabric which a good salesman will describe as a 'mercerized look'.

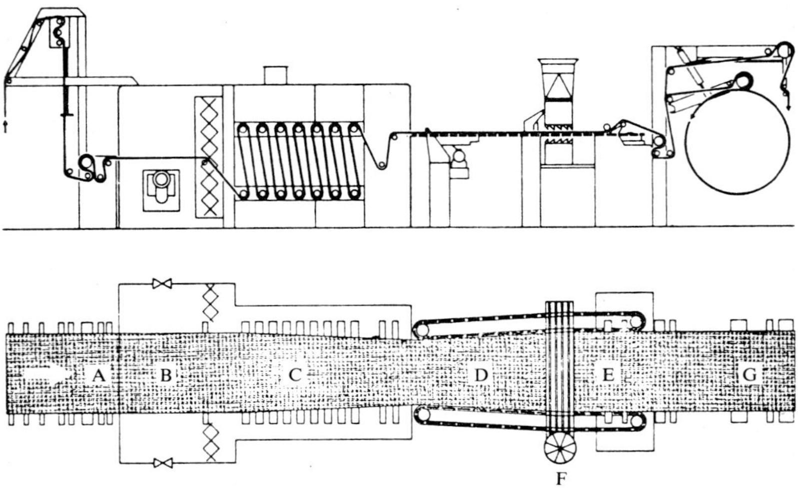

A. Driven draw-in roller
B. Preheater
C. Hot flue
D. Stenter entry for correcting width
E. Parallel zone immediately following cooler
G. Delivery

Figure 3.6 – Monforts Menko range for pre-padded and dried fabric

Perforated Suction Drum

The perforated suction drum technique produced by Fleissner offers a very effective method of heating fabrics to high temperatures. A range (Figure 3.7) may contain four, six or eight drums of 100–140 cm diameter. The drum surface has many perforations and, by fitting an exhaust fan to the drum axis, air is drawn through the fabric as it moves with the rotating drum. The air is heated and recirculated, the air flow through the unused arc of the drum being restricted by internal baffles. The air being drawn through the fabric ensures good contact with the drum surface. A wire mesh cover is usually fitted to diffuse the air flow around the perforations.

The recirculated air is usually heated by a hot-oil or gas-fired heat exchanger. The rate of heat transfer is exceptionally good, combining contact heating with the benefit of hot air being drawn through the fabric in alternate directions rather than (as in a stenter) air being blown at the fabric surface. On leaving the heating chamber, the fabric passes round a cooling cylinder, normally filled with cold water. Dye fixation time is usually in the 15–20 s range with a six-drum machine operating at about 60 m/min.

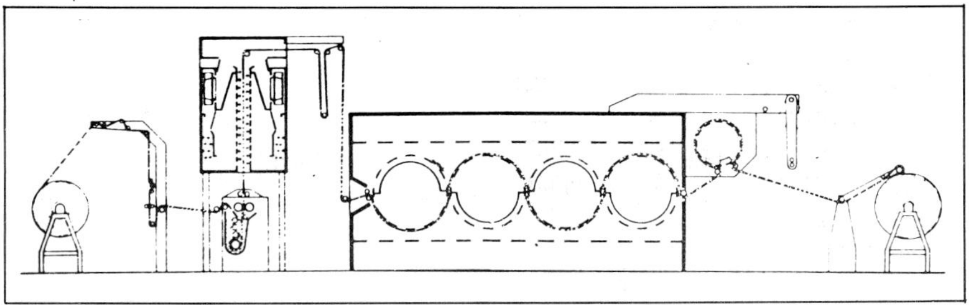

Figure 3.7 – Fleissner perforated suction drum range with predryer; some ranges have a stenter feed section before the drums; the pad bath has multiple rollers

Some Fleissner suction drum ranges were fitted in tandem with padding and pre-drying units and others were supplied with a pin stenter feed in order to pull the fabric out in width immediately prior to contact with the first drum.

It is reported that the action of very hot air passing through the fabric leads to a small amount of dye leaving the fabric and circulating with the hot air stream. To avoid carry-over of dye from one run to the next (particularly after running a deep dyeing) care should be taken to try to keep the dye in the vapour state and to expel contaminated air through the exhaust duct. At the end of the run, circulation and expulsion of hot air should continue for several minutes to remove as much contamination as possible.

3.5 THE STANDFAST MOLTEN METAL MACHINE

This ingenious machine was introduced commercially in 1950 for the application of vat dyes to cotton fabrics(*63–65*).

The essential part of the machine consisted of an iron U-tube (Figure 3.8) having two arms each 140 cm long, 3 cm thick and 180 cm wide for a 150 cm wide cloth. The U-tube was filled with 2 300 kg of a fusible metal alloy containing 50% bismuth, 27% lead, 13% tin and 10% cadmium, with a melting point of 70°C and a density of over 9 g/cm^3. The dyebath floated on the surface of the metal alloy on the inlet side and was about 20 cm deep (dyebath volume was only 5 l).

The reduced vat dye solution was stored at a low temperature until just before addition to the dyebath at around 70°C. The time of immersion of fabric in the dyebath was of the order of 0.4 s (at 27 m/min) down to 0.1 s (at 110 m/min). Liquor was therefore entrained between fibres and yarns rather than in the fibres.

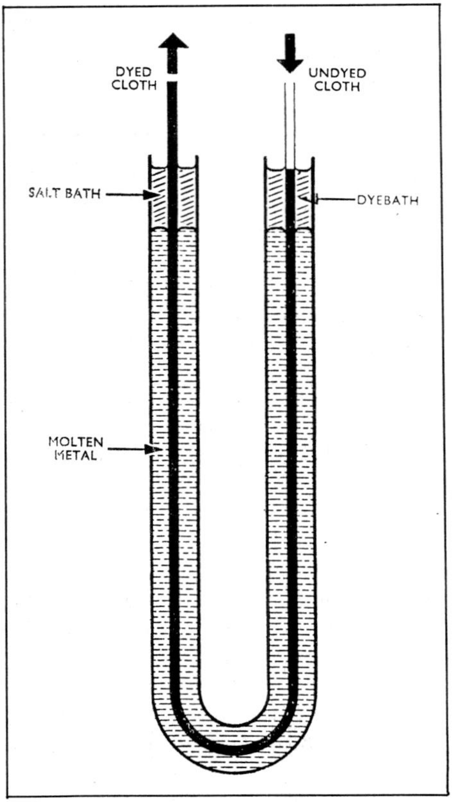

Figure 3.8 – Essential part of Standfast Molten Metal machine

Immersion in the metal, usually kept at 95–100°C, was for 6 to 1.6 s for speeds 27–110 m/min, and during this time the dye-liquor was absorbed by the fabric. The metal exerted a uniform squeezing effect, generating a pressure of around 1.4 kg/cm^2 at the bottom of the machine compared with about 21 kg/cm^2 exerted by a pad mangle. Consequently the liquor pick-up was usually around 140% (on dry weight of cloth) compared with, say, 80% by padding.

On the exit side of the tube the fabric passed through a solution of sodium sulphate to remove metal particles picked up on the surface of the fabric as well as any surface dye. The exit bath was about eight times the volume of the dyebath. The fabric was then washed, oxidized, soaped and finally dried on cylinders.

Later modifications to the machine(*66,67*) extended the maximum width up to 250 cm for sheeting, and provided versions for dyeing narrow warps of yarns; the provision of a steamer attachment between the exit from the molten metal and the entry into the sodium sulphate bath improved the colour yield of many vat dyes. Pre-padding of vat pigment followed by reduction in the machine improved penetration and colour yield.

Features of the Molten Metal machine were the good colour matching and speed in changing colourways (there was very little contamination of the metal bath); it was a compact unit and probably showed good thermal efficiencies (although this was hardly considered in the energy-cheap days of the 1950s!)

Attempts to increase the versatility of the machine included the application of (a) direct dyes to give pale dyeings on cotton and viscose, and (b) both vat and disperse dyes on acetate. Care had to be taken to avoid stretching the fabric lengthways and to minimize creasing.

Regrettably the Molten Metal machine found only limited use in the UK (the most common reason being the shortage of orders suitable for continuous dyeing), but it is still used to apply vat dyes to cotton fabrics in the Middle East and in India.

3.6 THE PAD–ROLL SYSTEM

This method of dyeing was developed in Sweden in 1954(*68*) and initially concerned the application of direct dyes to cellulosic fabrics and acid dyes to nylon. Later it was extended to the application of metal-complex dyes to warp-knitted nylon automotive fabrics.

The technique consisted of padding the fabric with dye-liquor and then heating by infra-red radiation and saturated steam for only a couple of seconds before the fabric was wound onto a large roll. The roll was kept in an atmosphere of saturated steam in a 'cottage' or chamber for several hours for dye fixation to take place. The roll was slowly rotated during this period to prevent the liquor from settling.

The aim was to wind the correctly impregnated fabric, at constant temperature (close to 100°C) and moisture content, onto the roll at low tension without marking the fabric surface with edge guiders. At the same time the heat input had to be carefully regulated to raise the temperature of the wet fabric to near 100°C without drying and without excessive condensation. Very few dyers achieved this objective successfully and consistently and, although the commercial introduction of reactive dyes revived interest in the process, nowadays it is rarely used for coloration. However, it is used successfully for desizing and bleaching cotton and polyester/cotton fabrics. The equipment is relatively simple and cheap – a sophisticated pad mangle is not essential for applying bleaching liquor – and the process is more flexible than fully continuous processes.

3.7 CONTINUOUS STEAMERS FOR DYE FIXATION

Continuous roller steamers are used to provide suitable conditions for the diffusion of dye into the fibre, e.g. vat, sulphur and direct dyes on cotton (particularly in polyester/cotton blends). The principle of the operation is that the uniform distribution of dye on the fabric is undisturbed whilst being subjected to saturated steam at 100°C.

The mechanical requirements are achieved by leading the fabric in open width round alternate rollers in top and bottom horizontal banks in the machine (as in a roller hot-flue dryer). The top rollers are driven, commonly by chains but increasingly by a series of hydraulic motors, and the bottom rollers free-wheel. The rollers must be accurately aligned to avoid the formation of any lengthways creases.

For the diffusion of reduced vat and sulphur dyes times of 20–60 s are usually allowed and capacities are of the order 30–60 m of fabric.

The shell of the machine, particularly the roof, should be well lagged and the roof heated by steam coils or, preferably, steam plates to avoid condensation. A steam supply at 200 kPa is adequate for these plates which feed into a condensate line.

The main steam supply is live steam at around 90 kPa feeding directly into a water sump at the bottom of the machine. The water is rapidly raised to the boil and helps to maintain an equilibrium in the steamer. A narrow sight glass is fitted to show the level of water in the sump and an automatic level control is desirable.

The fabric enters the steamer through a narrow slot, which is heated to prevent droplets of condensation, and against a slight counter-current of steam to minimize the ingress of air with the fabric. In fact, fabrics at temperatures below 100°C heat up very rapidly in saturated steam (in about 1 s) which condenses on the fabric, raising the moisture content and permitting more steam to make contact with the fabric.

It is important to minimize the amount of air in the machine because, in the presence of air, the partial pressure of the steam is reduced below atmospheric pressure and the evaporation temperature of the fabric is reduced. The presence of as little as 10–20% air in the steam is said to reduce the evaporation temperature to 97–94°C respectively(*69*), which can have a significant effect on dye fixation (especially acid dyes on nylon carpets). Furthermore the presence of air reduces the heat transfer rate to the fabric.

Steam, being less dense than air, rises in the steamer and tends to displace the air downwards. Air venting is effected mainly through steam traps, preferably of the thermostatic type (rather than thermodynamic or float type). However, some air will mix with the steam and pockets may persist in the extreme corners of the vessel. Some venting from the roof of the machine is desirable.

The input of live steam into the machine is usually controlled by a thermostat which has to be located away from radiation from the roof heaters (which tend to superheat the steam) and to be in the position of fairly active movement of steam. Thermostat settings of 105–108°C are customary, implying a small amount of superheat and slight liquor loss by the fabric.

When leaving the steamer the fabric passes through an exit leg, often feeding directly into a cold-water seal. Where the next operation is washing-off unfixed dyes and auxiliaries it is advisable to pass the fabric, still full of steam, directly into a hot-wash box in which the steam condenses to provide excellent penetration of the liquor.

Access to the machine is through side doors often fitted with windows for observation.

Specific requirements for a vat dye steamer are that the steam vapour must be dry saturated at 100°C, i.e. with the minimum of superheat; a practical figure would be 102–104°C. The steam vapour must be completely free of air. The wet exit leg must be kept cool by a constant cold-water feed as otherwise it becomes a 'leuco' exhaust dyebath, (see Figure 3.9)

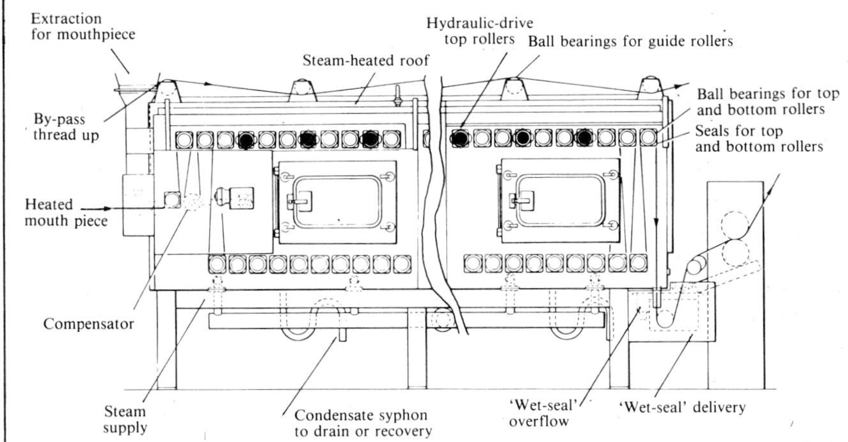

Figure 3.9 – Mather & Platt dye steamer – two section 48 m capacity

4 MACHINES FOR PREPARING AND DRYING FABRICS

The preparation of fabric for dyeing is an essential part of the dyer's craft and epitomized in the expression 'well boiled is half bleached'. Whereas, in batchwise dyeing, the absorption of dye takes place over hours, in continuous dyeing – as practised widely on polyester/cotton woven fabrics – the dye-liquor is applied and the surplus squeezed off in about 1 s. The demand for a consistently high standard of fabric preparation has led to much study and substantial changes in both methods and machinery(*70*).

Usually the initial process given to many cotton and cotton-blend fabrics is a singeing treatment to remove excessively long fibres protruding from the fabric surface (especially projecting polyester and nylon fibres, which, because of their high strength and smooth cross-section, can intertwine and form unsightly 'pills').

The fabric is run at about 100 m/min round a water-cooled small-diameter cylinder, and a gas flame, running the width of the cloth, is played almost tangentially on one side of the cloth to ignite the projecting fibres. The gas/air mixture is regulated to give a soft, colourless flame, the objective being to remove the projecting fibres, leaving the fabric structure unimpaired. The reverse side of the fabric is then treated in similar fashion, before the fabric is padded in an aqueous solution of enzyme (for desizing) or detergent.

In the event of the fabric failing to maintain an adequate speed, the gas is automatically turned off. Singeing machines have been responsible for many fires in dyehouses and fire precautions must be strictly observed. Preferably singeing should be carried out in a building isolated from the rest of the plant with a minimum of beams, joists, etc. for harbouring the inevitable fly from short fibres. Fire extinguishers must be kept ready to hand.

Singeing is a fast, cheap operation especially compared with shearing (cropping), which is carried out at about 5–10 m/min. A limitation of singeing as an initial operation is that subsequent desizing may liberate more long fibre ends; some polyester fibres may form globules at the end, which may take up more dye. In a few cases, therefore, it may be desirable to singe as an intermediate operation.

Another method of singeing is to pass the fabric rapidly over a red-hot plate heated by electricity or gas, but this system now finds little favour in the UK.

4.1 OPEN-WIDTH RELAXING MACHINES

Nearly all methods of producing fabrics entail subjecting the yarns and fibres to stresses and, on initial wetting-out, most yarns tend to relax and the fabric to contract in length. If the fabric (particularly if woven of filament yarn) is creased during the initial wetting, uneven contraction may result in permanent creases. Therefore, with many types of woven fabric, it is highly desirable to wet out the fabric whilst it is in open width and free from creases.

Many dyers and finishers of crêpe fabrics (in which the figure of the crêpe was dependent on the swelling of a highly twisted viscose yarn) built their own relaxing (also called contracting) machines. For lightweight fabrics a brattice conveyor was often used to carry the fabric whilst soap solution was sprayed from above. A variation of this was to feed the fabric down a 3 m

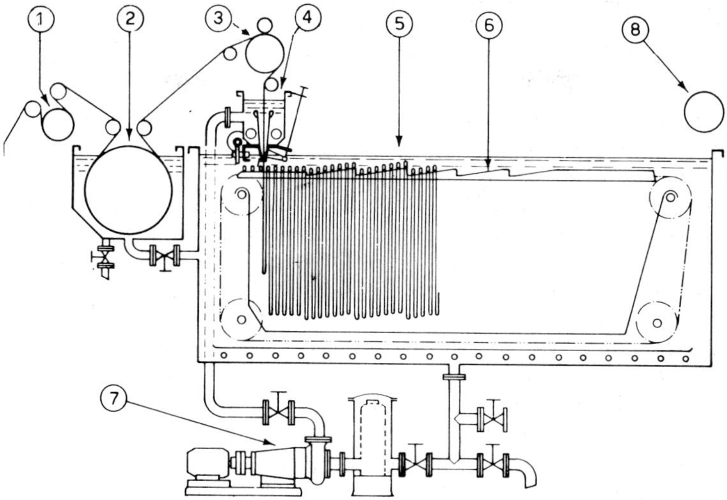

Key:

1. Draw rollers for controlling feed
2. Pre-wetting bath with driven roller
3. Draw rollers
4. Trough with V-slot
5. Main tank
6. Saw-tooth rails
7. Centrifugal pump with filter

Figure 4.1 – Mezzera open-width relaxing/scouring unit

chute, water being sprayed across the full width at the top and the fabric passing into a water trough at the bottom.

Such machines were simply constructed and cheap, but provided only brief dwell times when operating at production speeds. Later, festoon relaxing machines were developed and widely used, the best-known models being Mezzera (Italy) and Hinneken (USA).

The principle of the machine (Figure 4.1) is to suspend the fabric in loops from a series of parallel rods which move very slowly along the length of the machine. The rods are mounted in tracks on either side of the bath and just below the surface of the liquor. The rods rise up a saw-tooth ramp (5 cm) and then drop, so giving the opportunity for the fabric in contact with the rod to be displaced. It is customary for the fabric to be immersed for up to 20 min before being drawn off by a roller and sprayed with warm and then cold water before being plaited down. There is only gentle movement of liquor in order not to disturb the fabric loops. Heating is by steam coils, either in the bottom of the machine or at the sides.

An important feature is the control of fabric as arranged on the rods. The fabric is carefully guided into the centre of the machine and fed over a full-width reel into a V-shaped trough with an open bottom. The trough is kept half-filled with water which pulls the fabric off the reel and carries it down into the bath to form a loop generally about 80 cm long. The bottom of the trough slowly moves forward with the advance of the rod being loaded and then, about every 5–6 s, suddenly rotates back to start loading the next rod. This arrangement also removes air from the fabric before entering the bath.

After carrying the fabric through the machine, the rods move down to the bottom of the bath and return to the feed end.

For delicate filament fabrics the festoon machine provides excellent conditions for relaxing weaving strains; the scouring action is very limited, but fugitive tint and oils can be removed and water-soluble warp sizes softened and dissolved. The maximum speed of operation is dictated by the speed of the water pouring out of the bottom of the V-trough – the gap must be sufficiently wide to permit fabric joins to pass through, but not excessively wide otherwise large quantities of water will be required. Turbulence in the trough should be avoided.

Festoon relaxing machines can be up to 20 m in length and hold large quantities of scouring liquor, e.g. 30 000 l. Good foundations are required for machines of this weight. The amount of heat required at start-up (and the time taken) is considerable and careful production scheduling is required to operate the machine efficiently.

4.2 OPEN-WIDTH CONTINUOUS WASHING MACHINES

The demand for machines that will efficiently scour woven fabrics in a crease-free condition has led to a variety of open-width washing machines being developed(*71*). They are designed to facilitate the removal of readily soluble material from the fabric along with loose soil on the surface, mainly by the interchange of liquor in the fabric augmented by the fabric's rapid movement (about 100 m/min) through relatively stationary liquor and the mechanical effect it experiences passing around the rolls. The basic processes are: squeezing, diluting, and diffusing. Various devices for accelerating liquor/soil removal are described below.

The classic open-width washing machine (Figure 4.2) consists of a series of top and bottom parallel rollers, the lower set being immersed in the scour liquor. The fabric passes alternately between top and bottom rollers, being subjected to a series of immersions in the liquor.

The top rollers are driven usually by chains or V-belts in a bank of five or six, which make up a compartment or unit. The bottom rollers free-wheel and the bearings are sealed and self-lubricating. The bearings of the top rollers are usually mounted outside the wash box to permit ease of maintenance and lubrication. Both top and bottom rollers should be of the largest practicable diameter, say 12–13 cm, and the distance between top

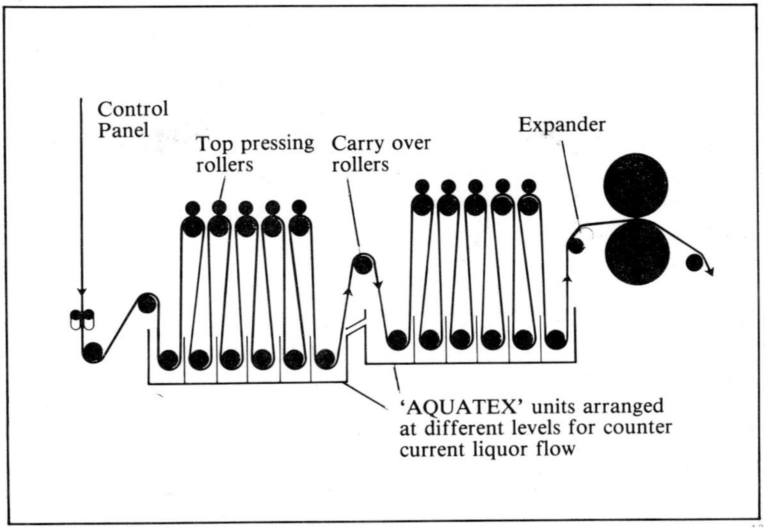

Figure 4.2 – Mather & Platt Aquatex units with final high-expression nip

and bottom rollers kept to a minimum to reduce the tendency for fabric creasing or edge curling.

Where compact machines are required, double threading may be employed in which case additional free-wheeling top and bottom rollers are fitted (Figure 4.3). This arrangement effectively doubles both the fabric capacity per washing unit, the number of immersions and the effective time of washing for a given speed. However, the arrangement is more complicated and costly, and makes threading-up more difficult and reduces accessibility.

The need to improve washing efficiency has led to the adoption of counter-current flow of scouring liquor in the wash box. Vertical metal plates separate the liquor between the bottom rollers; liquor flows in the opposite direction to fabric travel, either by a cascade or serpentine arrangement of the plates. Each successive immersion of the fabric is in cleaner liquor, whereas the liquor becomes increasingly contaminated as it approaches the discharge point close to the fabric entrance.

Where double threading is practised, the depth of the scouring liquor has to be increased to cover both bottom rollers.

The removal of soluble impurities in the fabric increases with the increase in temperature of the scouring liquor due mainly to the lowering of viscosity coupled with an increase in speed of diffusion. For most fabrics, a temperature of about 95°C appears to give the optimum balance between rate of removal of impurities against the cost of heat input. Most machines nowadays are fitted with a lightweight cover to retain both heat and vapour.

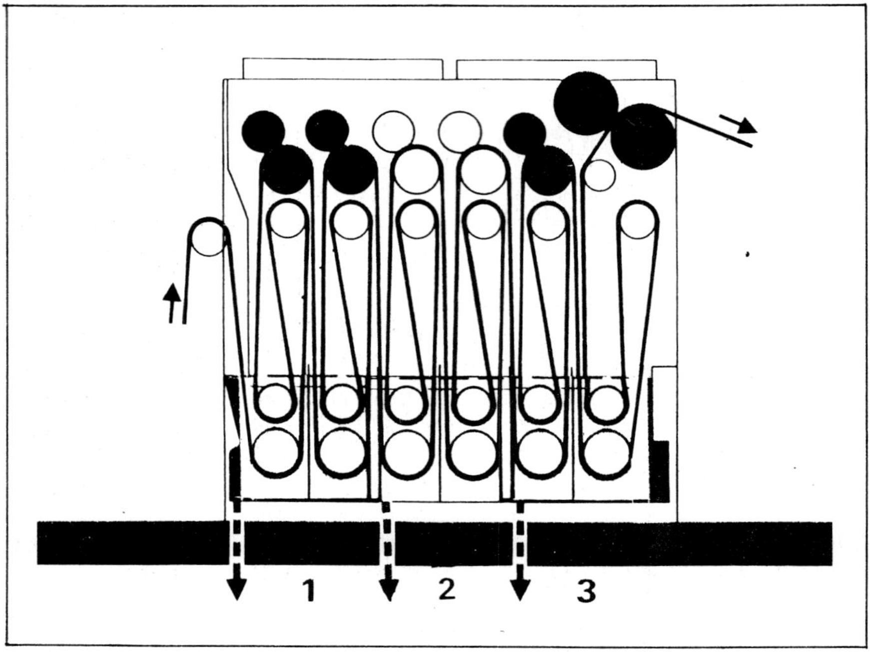

Figure 4.3 – Benninger Becoflex washing compartment

An important factor in open-width washing is the amount of liquor carried over from one immersion to the next. As the fabric is lifted upwards, liquor streams downwards and a small amount is expressed on making contact with the upper roller. Carry-over can be reduced by fitting a pressing roller above the driven rollers.

A continuous washing range is often made up from a number of units, and these may be connected either by a simple carry-over roller, fitted with a tension compensator, or by nip mangles.

With counter-current operation, heating is primarily at the liquor entry to the machine with heating coils in the remaining units to maintain the temperature. In many cases hot water discharged from the washing machine passes through one side of a heat exchanger to raise the temperature of the inlet liquor. For rapid heating direct steam addition through a perforated pipe is used; steam coils are preferred for more controlled heating and economy.

The rate of water inflow should be controlled (outflow is by overflow) to avoid waste of both water and heat. Inflow, coupled to machine speed, ensures that when the machine slows or stops, the flow of water is automatically restricted. A more sophisticated device relates the rate of input to the degree of contamination (measured by opacity) in the outlet liquor.

In recent years major improvements have been made in reducing the amount of water used for washing fabrics, manufacturers quoting reductions from over 10 l/kg of fabric down to 3–6 l/kg range. Comparisons between manufacturer's claims must be viewed with great caution, unless the type of fabric, the initial degree of contamination and the final level of impurity are clearly stated. Where the rate of diffusion is important, the fabric capacity of a wash box has a major influence on the machine performance.

A study of the effects of varying time, temperature and agitation on washing efficiency has been published in a series of papers(*72–74*). One of the first systems to be tried to improve washing efficiency was to install sprays to discharge on to the fabric just below the top rollers. This arrangement is moderately effective and comparatively simple to install. However, if clean water is used, total water and heat consumption rises rapidly; if internal liquor is used, pumps and a filtration system to remove fly are required.

Machine accessibility is important for cleaning purposes and for threading-up, for which a crawl speed on the drive is very useful.

Horizontal Washing Machines

Whereas in the traditional washing machine the fabric moves vertically between top and bottom rollers, recent years have seen the introduction of machines in which rollers are in two vertical banks and the fabric travels mainly horizontally (Figure 4.4). The fabric enters at the bottom and stepwise ascends to the top, the distance between the two banks of rollers being about 70 cm. Water enters at the top and, after being expressed at a roller, is collected in a tray and fed to a lower level. At this point where the fabric meets the rollers a wedge of water is formed on the surface which forces water down through the fabric or over the selvedges. It appears, therefore, that horizontal machines perform better on fabrics which are readily permeable rather than on more solid fabrics where a much higher proportion of liquor spills over the edge and is wasted. Horizontal washing machines are more economical in floor space, but impose higher tension on the fabric; controversy over the merits and limitations of horizontal and vertical machines continues(*75*).

The lower liquor content of the horizontal machine encouraged the development by Kleinwefers of a pressurized continuous washer (called the 100 Plus Washer), constructed to operate at 108°C with a gauge pressure of 50 kPa. Each unit contained six rollers on each side and fabric was fed into a pair of draw rollers at the bottom and then round a swinging compensator roller controlling the speed of one bank of rolls. The total fabric content was about 15 m and access was obtained through a central hatch. The im-

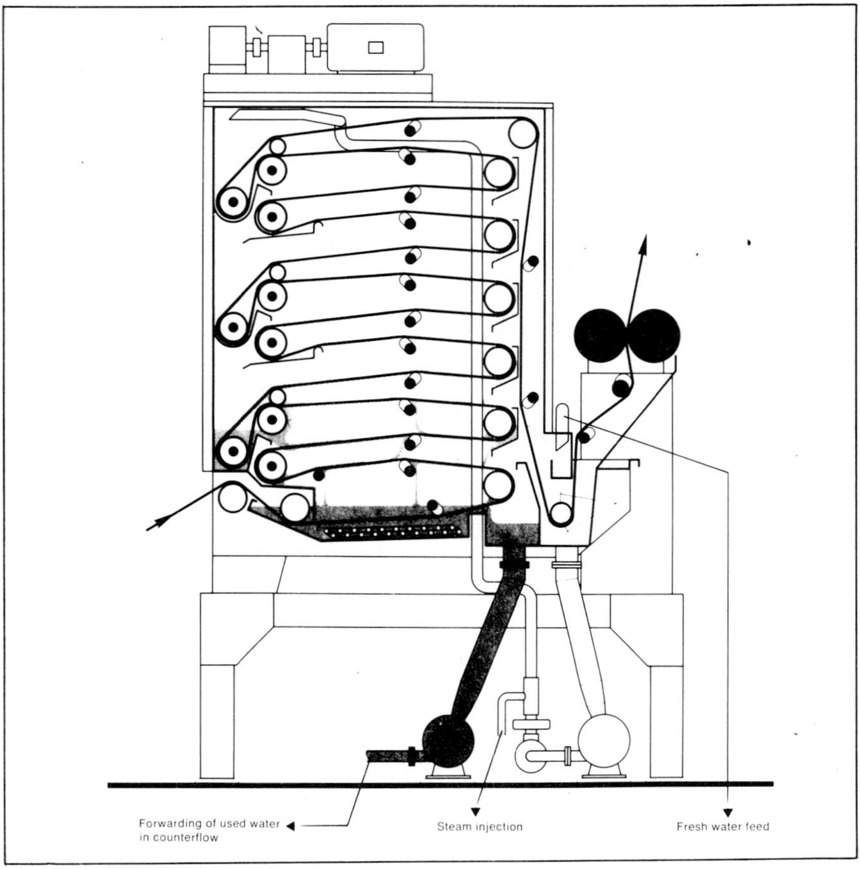

Figure 4.4 – Artos horizontal washer

provement in washing efficiency was difficult to specify and was counterbalanced by increased capital and maintenance costs.

Agitators

Considerable ingenuity has been displayed by machinery manufacturers in devising means by which fabric–liquor interchange can be promoted and mechanical action, e.g. flexing and squeezing, provided to the fabric.

Some successful examples are:

(a) Fluted or wave bottom rollers, which not only agitate the wash liquor in the box but also tend to force liquor through the fabric. These devices are fairly simple to construct and install and are cheap.

(b) Planetary rollers – also fairly simple.

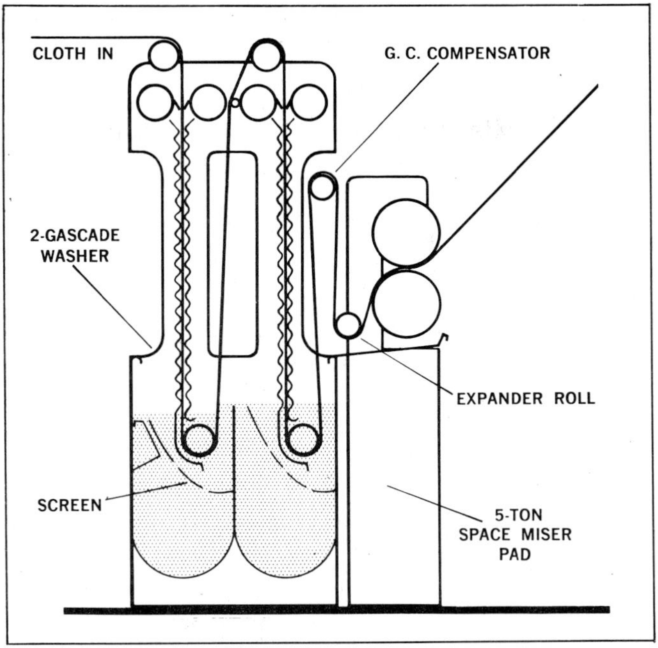

Figure 4.5 – Gaston County alternator washer

(c) Passing the fabric between corrugated formers which distort the liquor movement (and the fabric configuration) alternately. The Gaston County Alternator is a recent example of this (Figure 4.5).
(d) Passing the fabric round a 30 cm diameter perforated cylinder immersed in wash liquor; the axle of the cylinder is vibrated and causes vigorous radial oscillations in the wash box. This type of machine (Kuster Vibrotex) is intended primarily for fairly open fabrics, especially knitted fabrics, which can relax and readjust their structure during this treatment (Figure 4.6).
(e) Passing the fabric round a perforated drum containing an inner fluted roll or a roll fitted with baffles, independently controlled. The rotation of the inner roll produces surges in the liquor which induce a pulsating action on the fabric. The outer roll free-wheels and is taken round by the fabric.

The use of vibratory devices and supersonic oscillators to increase the efficiency of washing has declined in recent years. One specific machine vibrated at a nerve-jarring frequency with such volume that its acceptability on the shop-floor was very short lived. Care should also be taken to ensure that welds and bearings, etc. are capable of withstanding prolonged oscillation.

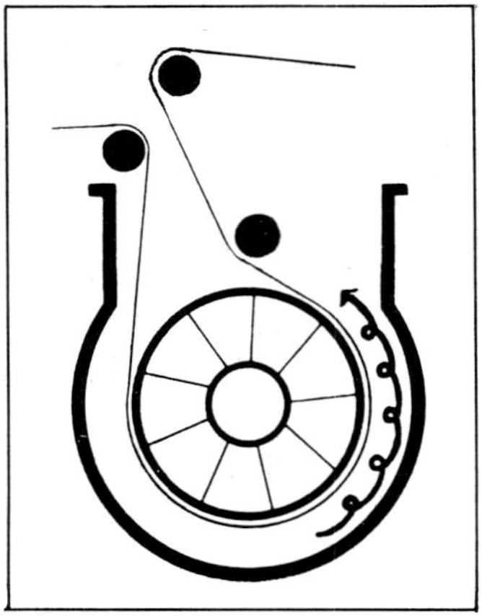

Figure 4.6 – Kuster Vibrotex

Suction-drum Washers

Following the success of suction-drum washers for scouring and dyeing loose stock, wider machines were made for scouring fabrics, particularly porous ones such as knitted fabrics. Scouring liquor is drawn through the fabric interstices, effectively removing water-soluble material; solid dirt, however, may be washed out or washed further into the fabric.

Suction-drum washers usually have two or four bowls driven at the same speed, the fabric covering about 80% of the circumference of the drums. Liquor is pumped out of the end of the drum and re-circulated; if staple fibre fabrics are to be handled, filters are essential. The flow of liquor at the edges of the fabric tends to uncurl them and therefore fabrics should be fed into the machine to curl away from the drum. Edge uncurlers should be fitted before the first drum on the machine. A simple nip may be fitted between successive bowls – this is primarily to advance the fabric rather than to remove surplus liquor (Figure 4.7).

4.3 PRESSURE STEAMERS FOR SCOURING

Best known of these machines is the Mather & Platt Vaporloc, which is essentially a roller conveyor bed located within a pressure vessel(*76*) (Figure 4.8). Two seals are provided, running longitudinally along the top of the vessel, one on each side of a central beam. Each seal consists of a pneumatically inflated flexible 'bag' pressing on a low-friction covering. The pressure inside the seal need be only slightly greater than the internal pressure (170 kPa at 130°C).

Immediately inside the machine positively driven draw rollers feed the fabric to a plaiting mechanism which arranges the fabric on the roller bed

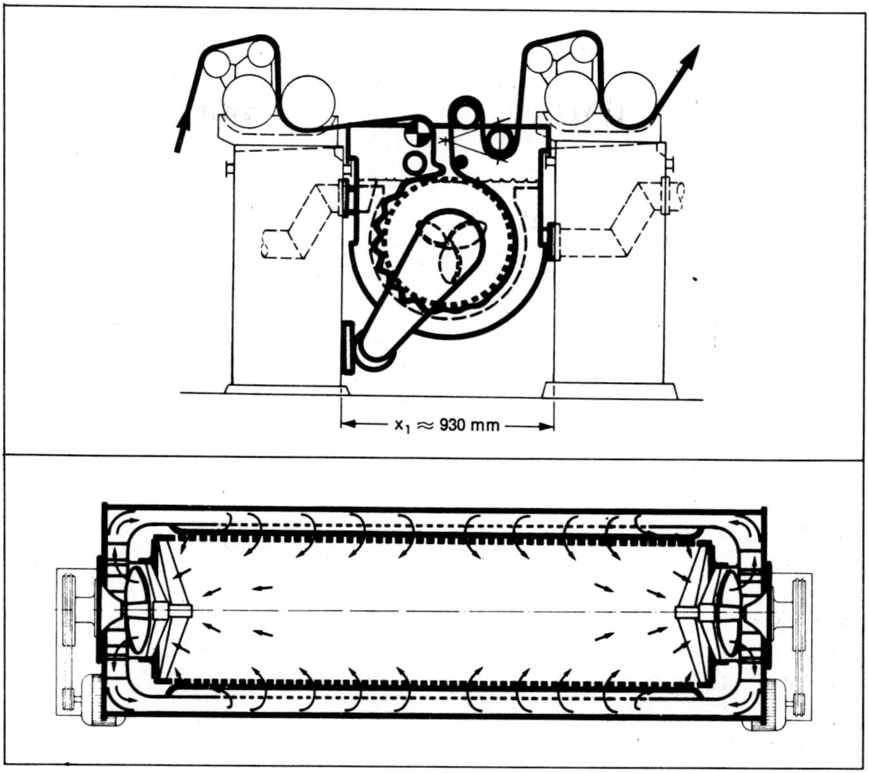

Figure 4.7 – Artos suction drum (elevation and plan view)

conveyor, whose capacity is nominally reckoned to be 200 m. At the delivery end the fabric is lifted from the bed and drawn through the lip seal by delivery rollers outside the vessel.

For maintenance the circular door at the end of the vessel is opened and, after disconnecting from the drive, the roller bed system can be slid out of the pressure vessel. The drive is by shafts running in pressure-sealed bearings to the entry pull rollers and to the roller bed.

Interlocking devices ensure that the door is closed and locked before air can be admitted to the seals and only then can steam be admitted. A modulating valve is fitted to ensure that a steady pressure is maintained. Care must be taken when sewing fabric joins to avoid multiple thicknesses which could damage the seals. In the event of a seal being severely damaged the loss in air pressure cuts off the steam and opens an exhaust valve, while any steam escaping through the damaged seal discharges upwards in comparative safety.

Figure 4.8 – Mather & Platt Vaporloc continuous pressure steamer

Two portholes are provided in the door and internal lighting to permit observation. Externally the machine is fitted with a series of entry rollers to ensure that the fabric is straight and centred when entering the machine. The entry seal has a nip effect and sets an upper limit on the amount of liquor which can be taken in by the fabric.

The continuous pressure steamer is used mainly for scouring cotton and polyester/cotton woven fabrics in dilute caustic soda solution – 2–3 min at about 130°C appears to be adequate.

Continuous pressure steamers have found limited use for bleaching fabrics with hydrogen peroxide, and a consensus of opinion favours a pressure scour followed by an atmospheric bleach. However, it may be noted that, where a dyeworks is situated at a high altitude, say over 2 000 m above sea level, the pressure steamer possesses an unassailable advantage, whether for scouring or for bleaching, in being unaffected by altitude depression of boiling point.

4.4 CONTINUOUS BLEACHING MACHINES

J-boxes

J-boxes are used widely in the United States and to a much lesser extent in the UK for bleaching cotton fabrics in rope form. In the simple Becco unit the fabric is run in at the top of the J, saturated steam blown in the top and the fabric piled down; in the later Du Pont machine a preheating section is fitted.

Most machines are made of stainless steel with a smooth internal finish so that the fabric slides down easily. J-boxes are cheap and compact, with high fabric capacity and no moving parts. However, the weight of wet fabric on the bottom layers is high and not only presses in sharp creases, but can make withdrawal of the fabric difficult.

J-boxes to handle fabrics in open width have been developed, but were more difficult to run and the fabrics (particularly polyester/cotton) showed widthways creasing at the folds. In the USA, where bleaching with hydrogen peroxide is customary, the use of sodium silicate as a stabilizer led to the formation of insoluble salts with any calcium or magnesium ions. These salts not only deposited on the fabric to provide resist marks in subsequent dyeing but also built up as a colourless layer on the inside of the J-box and abraded the fabric. The salts were particularly difficult to remove and the problem was eventually overcome by replacing the silicate with organic stabilizers.

In the UK many bleachers preferred to use sodium chlorite instead of hydrogen peroxide and this rapidly gave problems of corrosion with most grades of stainless steel. Two machines intended for tubular-knitted cotton for underwear were built with the essential J-box liner made of titanium, the capital cost being greatly increased due to the intrinsic price of the metal coupled with the complexity of welding. Nevertheless, the investment was fully repaid with several years production until the liner wore through and had to be replaced.

Later, for technical and economic reasons, the range was changed to bleaching with hydrogen peroxide until it was found to general surprise and dismay that the titanium lining was being corroded at a significant rate.

Roller Conveyors

The growth in UK production of woven polyester/cotton fabric, which required open-width wet processing, led to the development of roller conveyor steamers in which fabric can be given a dwell of about 10 min in a relaxed condition (Figures 4.9 and 4.10).

The fabric enters the machine through an entry leg, designed to minimize the introduction of air, and then passes to a feed system controlling entry to the plaiter. The plaits are formed on a curved bed of parallel rollers which rotate very slowly, edging the fabric forward until it comes to the take-off point when the fabric passes through a scroll roller and centring device before delivery.

In a typical machine about 600 m of fabric can be accommodated on the bed, and the plaiter feed and take-off rollers can be operated independently. As the fabric usually contracts slightly in length, take-off is marginally slower than the plaiter feed. Low-pressure steam is blown into the machine, preferably through a water-sump, to provide saturation.

Useful reviews of American bleaching practices have been published(*77, 78*).

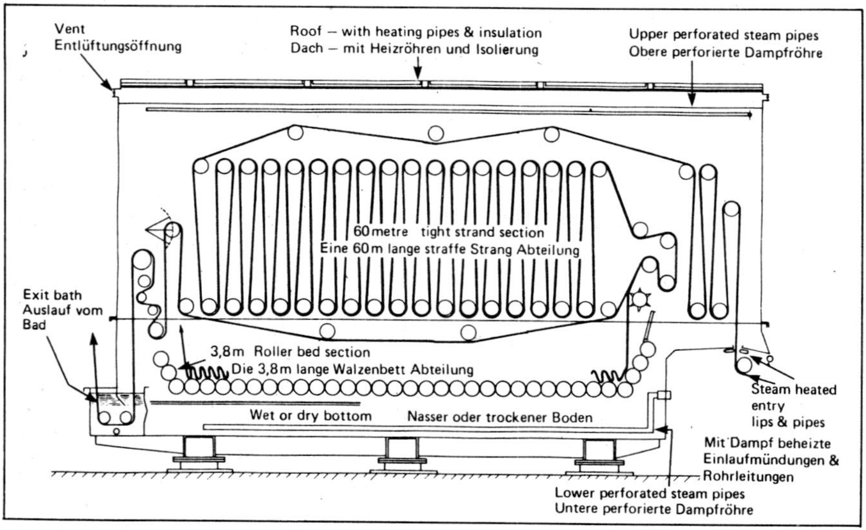

Figure 4.9 – Combined flat-bed and roller-type steamer

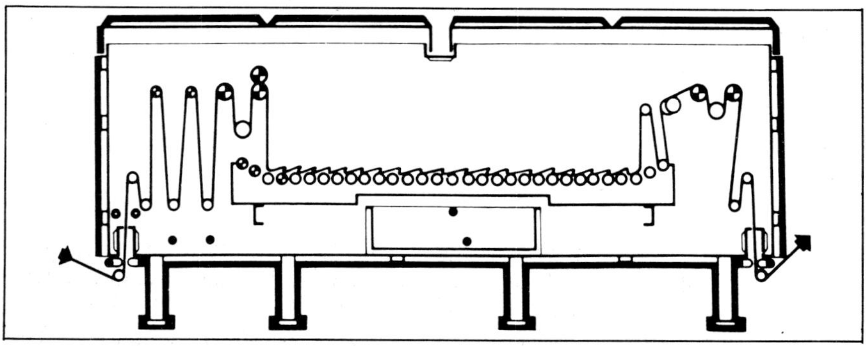

Figure 4.10 – Smith CS200 roller-bed steamer with all rollers driven to transport the fabric in a relaxed condition

4.5 SOLVENT-SCOURING MACHINES

Scouring machines, using non-aqueous solvents, have been developed mainly for two types of processing operation:

(a) The preparation of polyester/cotton woven fabrics prior to continuous dyeing

(b) The scouring of circular-knitted fabric made from textured polyester yarn.

The use of non-aqueous solvents in textile processing has been excellently reviewed by Shipman(*79*).

The advantages of using organic solvents for scouring textiles are:

(a) The ease of removal of oils, lubricants and waxes, particularly from polyester fibres

(b) As solvents are absorbed by fibres to a very limited extent, fibre swelling is very much less than in water and so fabric shrinkage is greatly reduced.

The ease of removal of organic solvents, due to their low latent heat of vaporization and low specific heat, are marginal benefits.

The use of solvents in the textile industry has evolved from commercial dry cleaning and was accelerated by the availability of non-flammable solvents, such as chlorinated hydrocarbons. The important properties of the commonly used solvents are given in Table 4.1(*79*).

TABLE 4.1

Solvent	Threshold limiting value (p.p.m.)	Density (g/ml)	Boiling point (°C)	Specific heat (kJ/kg/°C)	Latent heat of vaporization (kJ/kg)	Surface tension (N/m)	Azeotrope boiling point (°C)	Water, by weight (%)
Trichloroethylene $CCl_2.CH\,Cl$	100	1.46	87	1.0	240	0.0288	72	16
Perchloroethylene $CCl_2.CCl_2$	100	1.62	121	0.9	209	0.0323	88	7
1,1,1-Trichloroethane $CH_3.CCl_3$	350	1.31	74	1.2	227	0.0253	65	4
Water H_2O		1.00	100	4.2	2 256	0.0727		

At the time of writing (mid-1980), the price of perchloroethylene and trichloroethylene is about \$530 per m^3. Obviously scouring machines using solvent must operate with much smaller volumes of liquor which must be recovered and re-used. All commercial machines include equipment for recovering solvent from the exhaust, together with a distillation column and vessels for storing dirty and clean solvent.

Solvent recovery must be of a high order and several manufacturers claim 99% recovery; losses greater than 2% on weight of goods treated are serious in terms of cost. Most machines are fitted with carbon-granule absorbers to take out traces of solvent from the exhaust. The solvent is recovered from the absorber by stripping with steam.

Although the early tumbler machines used white spirit and the first continuous-scouring machines used trichloroethylene, perchloroethylene has largely replaced both these solvents. Both these chlorinated hydrocarbons wet out the fabric readily and are extremely effective in removing oil and waxes. Depending on the time and temperature of treatment, some oligomer is removed from polyester fibre. If the fabric is subsequently to be bleached with sodium chlorite care must be taken to remove any chlorinated hydrocarbon residues from the fabric as they catalyse the decomposition of sodium chlorite, liberating chlorine dioxide.

A problem with most solvent-scouring machines is the disposal of the residues at the bottom of the still. The waste, usually a black evil-smelling tar, is incinerated only with difficulty and then gives off highly corrosive fumes laden with HCl.

Columns for distilling perchloroethylene should be made of mild or stainless steel, but not galvanized steel, as a new zinc surface can generate dichloroacetylene with a TLV of 0.1 p.p.m. After a time the zinc surface becomes deactivated, but the dichloroacetylene can persist in the system for a month.

Although chlorinated hydrocarbons are normally fairly stable compounds they should not be exposed to sources of high temperature. Recirculated solvent vapour can be broken down to produce hydrochloric acid if direct gas-firing is used. In laboratory ovens, chlorinated hydrocarbon vapour, liberated from fabric samples and recirculated over electric heating elements, have been known to decompose to phosgene, with considerable discomfort to the laboratory staff in the area.

Tumbler Solvent-scouring Machines

These machines are identical to those used by commercial dry cleaners and consist of a drum rotating about a horizontal axis. The solvent is sprayed into the drum, which is fitted with baffles or divided into three or four compartments. For fabric scouring, baffles are preferred in order to provide a larger effective capacity. The speed of the drum is variable and is usually programmed to stop and reciprocate after every three cycles, so as to give greater movement of the goods. The fabric is loaded through a glass hatch at the front and care is taken to minimize twisting. Loading and unloading are laborious and time-consuming operations.

After treatment in the solvent the liquor is drained off and the speed of the drum increased to remove more solvent. The drum is then slowed and hot air blown in and exhausted to a condenser and the carbon absorber. The solvent is filtered to remove broken fibres, etc. and then distilled before re-use.

Tumbler solvent machines are used for scouring/bulking fabrics knitted from textured polyester and nylon yarns. Not only does the solvent remove lubricants and waxes, but the mechanical action induces fabric contraction and generates yarn bulk and stretch subsequently regulated by heat setting.

However, the tumbling operation is laborious and too expensive for routine operations, although several dyers keep one tumbler for special lots.

Continuous Solvent-scouring Machines

One of the earliest systems used in the UK for preparing polyester/cotton woven fabrics entailed passing the fabric in open width through a bath of trichloroethylene at a temperature of 60°C(*80*). A treatment of about 10–20 s was sufficient, and was followed by steaming to remove the solvent.

A later development of this principle was the Markal process in which the fabric was wetted out with trichloroethylene containing a surfactant and either hydrogen peroxide or an enzyme(*79,81*). The fabic was then steamed to flash off the solvent, leaving the surfactant finely dispersed throughout the cloth. Steaming also promoted the degradation of the warp size, and, in the presence of hydrogen peroxide, encouraged bleaching. Unfortunately, this neat, multi-purpose operation was short-lived when it was found that the hydrogen peroxide slowly decomposed the solvent, producing HCl which had a highly corrosive effect. The operation involving trichloroethylene – surfactant – enzyme has been used successfully for several years to prepare polyester/cotton fabrics for subsequent dyeing by the Thermosol process.

It should be noted that in the Markal process the solvent was used as a medium to disperse the surfactant and enzyme through the fabric – the waxes, etc, were removed in subsequent hot washing, the solvent being flashed off as an azeotrope and condensed at the bottom of the steaming chamber.

Several manufacturers have produced continuous solvent-scouring machines most of which, in the UK, have been used for weft-knitted fabrics produced from coloured yarns. In many cases continuous machines have been developed to reproduce the effect of solvent scouring in a tumbling machine (Figure 4.11).

The fabric is immersed in a bath of solvent and may be taken round a perforated drum or vibrated to induce fabric relaxation and bulk. On leaving the bath the fabric is squeezed to express solvent and then dried in hot air whilst supported by a perforated drum. The solvent is circulated through the bath and part withdrawn for distillation which is carried out continuously.

In order to reduce solvent loss and to minimize smell most machines of this type operate at slightly sub-atmospheric pressure.

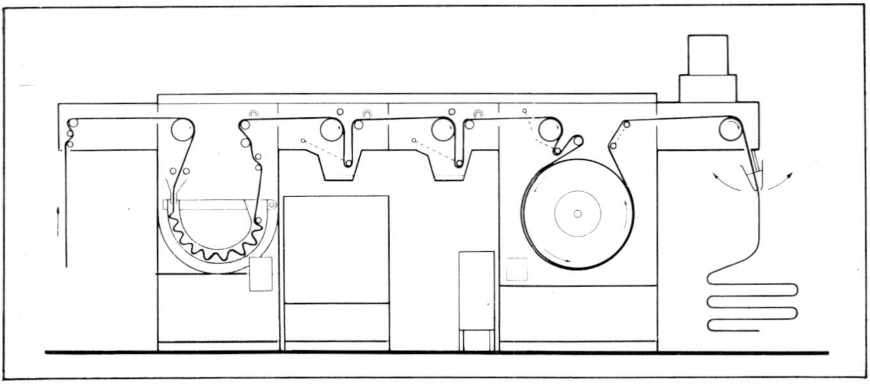

Figure 4.11 – Spencer Libratex continuous solvent-scouring machine

Machines operating on this principle have been adapted to apply fluorescent brightening agents or softeners/resins/water-repellent agents from perchloroethylene, as well as being used for scouring.

4.6 MERCERIZERS

The operation of mercerizing(*82*) entails the impregnation of cotton or linen fabrics with about 30% caustic soda solution which swells and plasticizes the cellulosic fibres releasing many internal strains. As the fibres swell the fabric tends to contract, but this tendency is mechanically resisted by applying tension during dilution of the caustic soda at the subsequent washing stage.

The benefits of mercerization are: a more lustrous and smooth appearance, increased uptake of dye and improved colour value, improved dimensional stability and a smoother, sleeker handle. Depending on the tightness of the weave, some of the convolutions in the cotton fibre are removed and there is evidence to show that subsequent cross-linking treatments can be applied more uniformly.

Although 30% caustic soda is required for full mercerization, semi-mercerization begins around 12% and many operators compromise on about 18% on the grounds of economy and improved penetration of the liquor.

Chain Mercerizers

Chain mercerizers consist of three sections:

(a) Impregnating with caustic soda solution, usually by a saturator and three-bowl pad

(b) A period of at least 40–50 s for the fibres to swell whilst the fabric is taken round a series of metal drums to keep it flat and to maintain lengthways tension
(c) Pulling the fabric out to the original width and then washing out the caustic soda by overhead wash water sprays augmented by suction boxes immediately underneath the fabric. The fabric is pulled out by clips in a strongly built stenter chain up to 25 m long, width extension taking place in the first third. The liquor collected underneath the fabric is pumped up to the preceding spray, i.e. washing is on a counterflow principle.

Finally, the fabric is given a thorough wash off in an open-width scouring machine.

Chain mercerizers provide excellent control of both warp and weft tensions, but are suitable only for woven fabrics. The stenter chains, working under wet conditions, require frequent maintenance.

Chainless Mercerizers

Chainless machines consist of a series of tanks, each with horizontal metal rollers partially immersed in the liquor, with rubber-covered rollers resting above and between the metal rollers, which assist liquor penetration, restrict weft shrinkage and provide fabric traction. Each tank of about twelve rollers has its own d.c. motor drive operating on intermediate pad nips. the first two or three tanks are used as saturation for the caustic soda liquor, giving a dwell time of 45–50 s, the remaining tanks operate as stabilizers during treatment with hot wash water. Thereafter the fabric is further washed on a conventional multi-box open-width washer (Figure 4.12).

Although chainless mercerizers provide only a modicum of weftways control they have advantages of being more gentle to the fabric, requiring less maintenance, as well as offering possibilities of running two fabrics one on top of the other. Fabrics of different width can be run through immediately after each other and, on wide machines, fabrics may be run side by side. With care, selected knitted fabrics can be processed.

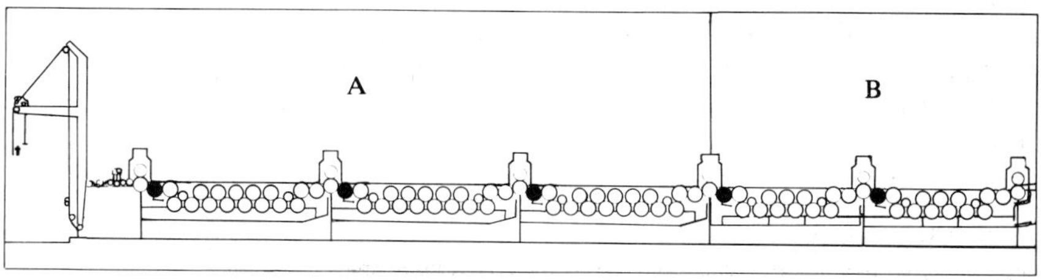

Figure 4.12 – Impregnating (A) and stabilizing (B) zones on a Benninger chainless mercerizing range; each compartment has a tension regulator

In the UK, chainless mercerizers have replaced some of the older chain machines.

A fairly recent development has been the application of near-boiling caustic soda to cotton fabric; this provides greater and more uniform penetration and more rapid action (5 s instead of 45–50), but so far commercial acceptance of hot mercerizing has been limited(*83*).

Semi-mercerization (causticizing) and full mercerization substantially increase the affinity of cellulose for dyes. It follows that, for both chain and chainless types of machine, the design must provide for uniform impregnation with caustic soda solution and uniform dilution during the subsequent washing stages. This is especially important when the dyeing method is continuous.

4.7 WATER EXTRACTION

Textiles require drying after coloration and/or washing. This is carried out in two stages: firstly the mechanical removal of extraneous water followed by the application of heat to complete the process. Using heat to drive off the residual water is expensive and so it is important to maximize the effectiveness of mechanical means which are:
(a) centrifuging, (b) mangling, (c) suction methods and (d)other methods.

Comparative water retention levels of different types of fabric extracted by different methods have been quoted (Table 4.2)(*84,85*).

TABLE 4.2

	Retention (%)		
Fabric	Centrifuging	Mangling	Suction
Cotton	45– 50	45– 70	40–55
Viscose	80–100	60–100	70–80
Sec. cellulose acetate	30	40– 50	27
Nylon 6.6	15– 25	20– 40	14–20
Polyester	12	20– 30	15
Wool	42	58	84

Conditions:

Centrifuging	10–15 min,	700 g centripetal acceleration
Mangling	36 m/min,	0.6 kN/cm^2 nip pressure
Suction slot	8 m/min,	1.2 mm slot width, 94 kN/m^2 vacuum pressure

Comparative power costs in fabric drying (pence per 100 kg of water removed at early-1970s prices);these should be trebled for the early 1980s.

Dewatering		*Thermal Drying*	
Mangling	1.5	Drying cylinders	25
Centrifuging	2.5	Stenter	35
Suction slot	4.5	Radiation	120

Centrifuges

Centrifuges for fabric are very similar to those used for yarn and staple (see Section 2.6), but are usually larger in diameter (150 cm) and revolve at a much lower speed. An inherent problem with centrifuging (called 'swinging' in some dyehouses) fabrics is the tendency to form creases which may be irremovable. For sensitive fabrics, 10 min extraction at 350 rev./min in a 150 cm diameter basket or 400 rev./min in a 120 cm diameter basket is recommended.

Centrifuging is a batchwise process (as opposed to mangling and suction) and the fabric is run in via a pot eye and compressed-air-driven pulley mounted over the axis, the basket being rotated very slowly by hand during loading to achieve a uniform balance.

For safety reasons an interlocking lid is essential on a centrifuge so that the motor cannot be started until the lid is locked, nor the lid raised until the basket is once again stationary.

Mangles

Mangles for water extraction are the same in principle as pad mangles (see Section 4.1), the difference being that the prime requirement for pad mangles for applying dyes and finishes is uniformity, whereas in mangling the efficiency of squeeze is of paramount importance.

The basic principles of mangling woven cloths have been described in an excellent paper(*54*). The factors determining the degree of water retention in a cloth immediately after mangling include:

(a) Pressure applied to the fabric;usually quoted as the total load applied to the ends of the rolls divided by the length of the rolls, i.e. kg/cm.
 An increase in pressure gives lower water retention, the benefit diminishing exponentially depending on the compressibility of the fabric. Generally speaking 20 kg/cm is a low pressure, whereas exceeding 55 kg/cm gives little benefit.

(b) Hardness of the bowl; with most woven fabrics a harder bowl means less deformation, higher pressure and lower water retention. Early mangles had wooden bowls made of sycamore, but these have been supplanted by stainless steel, ebonite, wood fibre, and rubber-covered materials.

(c) Temperature of the water in the fabric; water retention falls as the temperature increases, mainly due to the lower viscosity permitting water to flow back more rapidly from the nip. The change in fibre swelling with temperature also reduces water retention. Padding through hot water is sometimes used to obtain a lower water retention. Moss(*54*) quotes water-retention figures for a particular cotton cloth of 71% at 5°C and 58% at 65°C.

(d) Cloth speed; through the range to 90 m/min increasing the speed has only a small effect, i.e. for a doubling of speed the water retention of cotton increases only 3%, viscose 3½% and nylon 1%.

(e) Bowl diameter; this too has only a small effect. Smaller-diameter bowls have less area of contact and therefore produce higher pressures, but the difference between 50 cm and 15 cm diameter bowls is only 3% in favour of the latter(*54*).

Most mangle bowls are loaded at the ends, which usually produces some deflection and lower pressure at the middle of the bowl. This may be counteracted by cambering one or both bowls so that the diameter at the middle is slightly greater than that at the ends.

An interesting development was the introduction of bowls with super-resilient filling of which the best-known are Roberto bowls. The filling is composed of resilient, rubber-covered fibres about 7 to 10 cm thick under pressure; it appears to be particularly effective on knitted structures or fairly open woven fabrics.

Laboratory trials indicate the effect of altering bowl combinations; the fabrics, in strip form, were wet out in water and passed through a laboratory mangle, weighed, dried in an oven and re-weighed (Table 4.3).

TABLE 4.3

	Water retention on weight of dry fabric (%)							
Bowls	Rubber/ steel		Roberto/ steel		Roberto/ rubber		Roberto/ Roberto	
Pressure (lb/inch)	100	300	100	300	100	300	100	300
Polyester double jersey	102	72	58	44	60	45	35	30
Acrylic double jersey	98	80	68	60	68	62	50	44
Triacetate double jersey	97	88	56	49	56	49	43	35
Acrylic single jersey	112	100	67	60	67	59	49	45
Textured nylon jersey	105	73	60	50	60	50	38	33
40 denier nylon warp-knit	91	80	48	38			32	29

The effectiveness of the super-resilient bowls has been ascribed to three factors, each of which plays a role:

(a) The relative softness of the bowl at the nip enables the surface to deform, so expressing liquor from the holes or 'windows' in the fabric structure.

(b) Whilst in the nip the porous structure of the bowl absorbs liquor from the fabric. This liquor is expressed as the covering enters the nip again.
(c) Similar in concept to (b), but, immediately after the nip, as the pressure decreases and the bowls rapidly regain their original shape, the porous structure absorbs liquor from the surface of the fabric. As in (b), the liquor absorbed by the bowl on the exit side of the nip is expressed at the next entrance.

All mangle bowls, and particularly bowls with porous surfaces, should be well washed after use. Hosing for 5–10 min whilst running under pressure is recommended; bleach liquors and organic solvents should not be used.

Suction Dewatering

Many fabrics which would be badly creased by spinning in a centrifuge or damaged by squeezing between the rollers of a mangle are extracted by passing over a suction slot. The method is particularly suitable for filament woven fabrics and, as the machine takes up little room, it can often be located in front of a stenter or cylinder dryer.

The slot is cut along the top of a horizontal cylinder of about 20 cm diameter, coupled at the bottom to a large-diameter pipe leading to a gauze filter (to trap any loose yarn) protecting a water separator and a vane vacuum pump. Motor sizes run from 10 kW for fabrics to 75 kW for dewatering 4 m wide tufted carpet.

The machine draws air through the fabric into the slot, the inrushing air sweeping out extraneous water from the fabric. Obviously it is important to minimize air leaks and the lengths of slot at either side of the fabric are covered by rubber or plastic sleeves. The fabric must be centred before it passes over the slot and then round a pair of draw rollers.

A study of the principles of suction dewatering, as applied mostly to cellulosic fabrics(*86*), showed that water removal occurred very rapidly – the cloth need be over the slot for only 1–2 ms. For fabric extraction slot widths generally range between 1 and 3 mm and, except for very dense cloths, a vacuum of 40 kPa is adequate.

These results have been confirmed more recently in work on wool fabrics(*87*) that has extended the earlier theory of the opening of void paths through the fabric to include capillary flow along the fabric immediately in front of the slot. Milled fabrics and felts give higher water-retention figures than expected, presumably because the large number of fibre cross-overs restrict both the passage of air and the flow of water to drier areas.

With knitted fabrics, which usually have fairly high porosity, it is important to try to retain a seal at the edges. With fabrics of high porosity it is often advantageous to restrict air flow directly through the fabric and a non-absorbent roller or equivalent device is often fitted immediately over the slot to serve as a baffle and to increase efficiency.

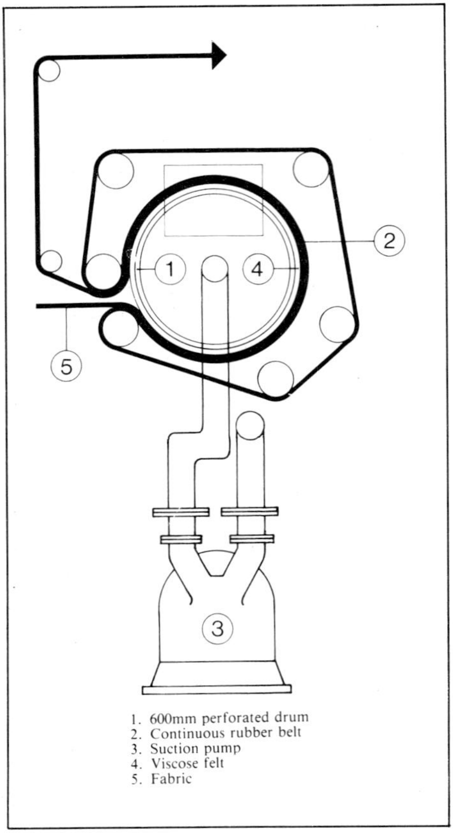

Figure 4.13 – Monforts Matex-Vac

Other Machines

The soaring cost of energy has resulted in several new developments for removing water from textiles.

Machines based on the use of compressed air or high-pressure steam for forcing water out of the fabric have met with little success. More recently a technique used in paper making has been adapted to textiles (Monforts Matex-Vac); the fabric is run with a continuous felt around a perforated drum. Liquor is absorbed by the felt which then separates from the fabric. Suction applied to the inside of the drum then removes water from the felt (Figure 4.13). The effect is probably similar to using a super-resilient bowl, i.e. as the felt and fabric leave the nip the reduction in pressure enables the felt to expand and absorb liquor from the fabric.

4.8 CYLINDER DRYERS

Heating by contact is probably the simplest and cheapest method for the final drying of fabric and is widely used for woven fabrics.

The fabric is passed round a series of cylinders filled with steam under pressure. There are still machines in use with thin walls and using steam at only 70 kPa pressure, but most machines will operate with 210 kPa steam, and some will take up to 600 kPa steam pressure.

The cylinders (low-pressure machines frequently had 32 56 cm diameter cylinders, but higher-pressure machines have fewer cylinders) are arranged staggered in banks to provide maximum contact distance with the fabric. In 32-cylinder machines the banks usually run horizontally, but in smaller machines the banks are in vertical sets. The cylinders are driven from a common drive on one side of the machine; the hollow shaft contains the steam supply to the vessel and a condensate removal pipe, which should extend down to within 5 mm of the bottom of the cylinder.

The installation must be fitted with a safety valve (to guard against excess pressure) and an anti-vacuum valve in each cylinder to permit air to enter when the steam in the cylinder cools and condenses when the machine is no longer in use. In order to expel air rapidly from the cylinders when the steam is turned on, thermostatic air vent units should be fitted to the exhaust side of the machine. When cold these vents are wide open allowing the air to escape, but on approaching the operating temperature the vents close automatically.

Cylinder dryers (sometimes called 'cans') have good thermal efficiency and give an excellent, smooth, lustrous finish to acetate taffetas together with a crisp handle without any pin or clip marks in the selvedges. The contact between the surface of the cylinder and the wet fabric is similar to that of a hot flat iron and the finish is beneficial on some fabrics, but unsuitable where a surface effect or a full handle is required.

The main limitation of a conventional cylinder dryer is that there is no control of fabric width and the finisher often has to give a final treatment on a stenter to ensure that the finished fabric is of consistent and specific width. As adjusting the width of dry fabrics is not easy, in some cases a compromise has been effected by locating a small bank of cylinders immediately prior to a stenter. The aim is to achieve a major proportion of the heating and drying operation on the cylinders and to complete the operation on the stenter.

4.9 STENTERS

Stenters are usually the most expensive and important machines in a dyehouse/finishing works and, by controlling the finished length and width of the fabric, have an effect on the appearance, handle and properties of the finished fabric.

The fabric is held at the edges by two endless chains, which convey it

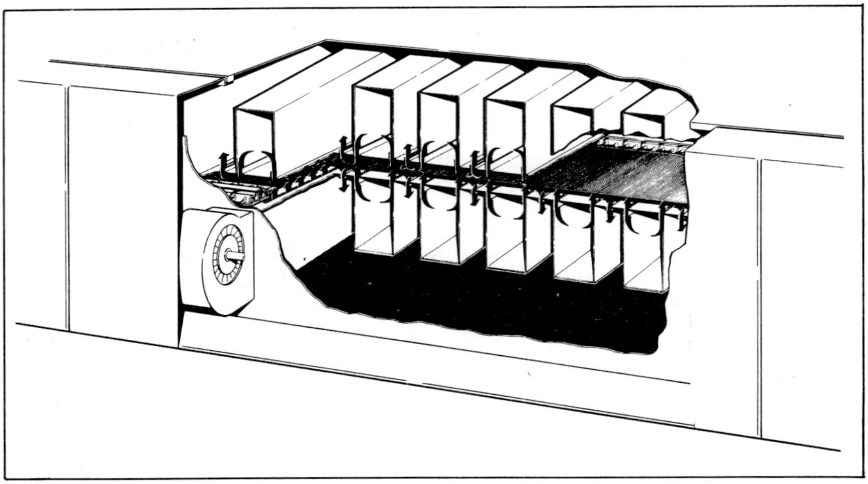

Figure 4.14 – Famatex stenter showing the fabric passing between the tapered air ducts

through a stove in which hot air is blown at the cloth, (Figure 4.14). After emerging from the stove the fabric is disengaged from the chains and either wound into a roll or plaited into a truck. The chains return through the oven back to the starting point.

Pin and Clip Stenters

The fabric is held onto the chains either by pins mounted into a base plate or by clips in which the fabric edge is clamped between two smooth surfaces.

Pin stenters are the more common, the pins being about 5 mm long and mounted in two rows along the base plate, about 24 pins in approximately 7 cm. To prevent the fabric edge coming into contact with the hot base plate the base of the pin may be much thicker than the upper portion.

An alternative design is to have cranked pins projecting from the side of the plate to encourage uniform air flow round the fabric edges.

Whatever the design of stenter pin it is important that any bent or distorted pins be repaired or replaced quickly as continued use leads to further trouble with snagged or torn fabric edges.

In clip stenters the clips are in two sections, upper and lower, the upper being in either the open or closed position. At the fabric feed-in point the upper jaw descends and grips the edge of the fabric; at the delivery point the jaw opens and the fabric is released. Clip stenters require a definite edge, preferably a selvedge, to grip and are usually confined to woven fabrics, particularly furnishings. Clip stenters leave no pin marks in the selvedges, but customer acceptance of pin holes in selvedges has increased in recent years. However, the major reason for the decline in clip stenters is the inherent difficulty of applying overfeed (see below) when entering the stenter; overfeed is necessary to minimize lengthways shrinkage in subsequent washing and dry cleaning.

A few combined pin–clip stenters have been built in which a bracket can be rotated to bring pin plates into operation.

The Stenter Rails

The stenter rails run the length of the machine (20–30 m), and are either horizontal or vertical return depending on plane of rotation. Most stenters in use in the UK are horizontal-return machines.

In vertical-return stenters the chain returns underneath and, at the entry feed appear from below rather than from the sides. At one time this arrangement was claimed to be beneficial for the accurate feeding of warp -knits at speed, but this factor is no longer relevant.

A feature on some stenters is that, after the fabric has been positioned on the pins, the plates are tilted so that the pins are angled about 10° outwards to prevent the fabric coming off the pins during heating/setting. Approaching the delivery point the pin plates resume their vertical position so that the fabric can be stripped off by a take-off roller.

Lubrication of the chain in the rail is important and can be done automatically; the lubricant must be stable to high temperatures and, whenever hot, the chains should be kept moving to prevent any section remaining in the stove for a long time.

The rails are divided into three sections and can be moved laterally about the centre line. The tapering entry section (up to the heating chambers) is usually about 5 m long for woven fabrics and 7 m for warp-knits, which may require to be pulled out more in width before entering the oven. The first 3 m of the pin rails are pivoted so that they can rapidly move in or out. The centre section is as long as the oven (which may have from three to eight heating sections, each about 3 m long) and is usually parallel, unless there is a change in width during drying/setting.

The delivery section is about 4–5 m long to permit sufficient cooling to take place before the fabric is stripped off the pins by a roller and either rolled or plaited. On some machines an air blower is fitted after the heating sections to force cool air onto the fabric.

The rails are moved in or out on motorized cross-screw threads to give the desired width at each stage through the machine.

Fabric Entry

Stenter speeds range from 10 m/min for heavyweight to 100 m/min for lightweight fabrics and the higher the speed, the more vital the need for a good entry system to ensure that the fabric is properly centred and the edges fed onto the pins at the correct speed.

The entry system usually commences with an overhead gantry with free-running rollers; this long run-in helps to align the fabric feed with the stenter and provides an opportunity for creases to drop out. The fabric then descends behind the stenter operator platform and passes underneath to a driven roller just above floor level. It then rises upwards in front of the operator and is contacted first by a scroll roller (each half threaded in opposite directions to take out curl at the edges), then by a smooth idling roll followed by the main feed roller, rubber-covered to ensure good control (Figure 4.15). These three rollers are often in a line inclined at 45°. The fabric is then on the same level as the pins and moves forward to join them (Figure 4.16). Usually, in this distance (40 cm), the sides of the fabric pass through uncurling devices, threaded fingers rotating at a different speed from the cloth. The precise arrangement of rollers and fabric threading-up differs from make to make of stenter.

Finally, the edges of the fabric pass through detectors mounted on the pivoted arms bearing the pin rail, which can move in or out to pick up the fabric edge. Lightly loaded spring feelers are normally used, the fabric edges being presented between rotating uncurlers; some stenters use optical edge detectors.

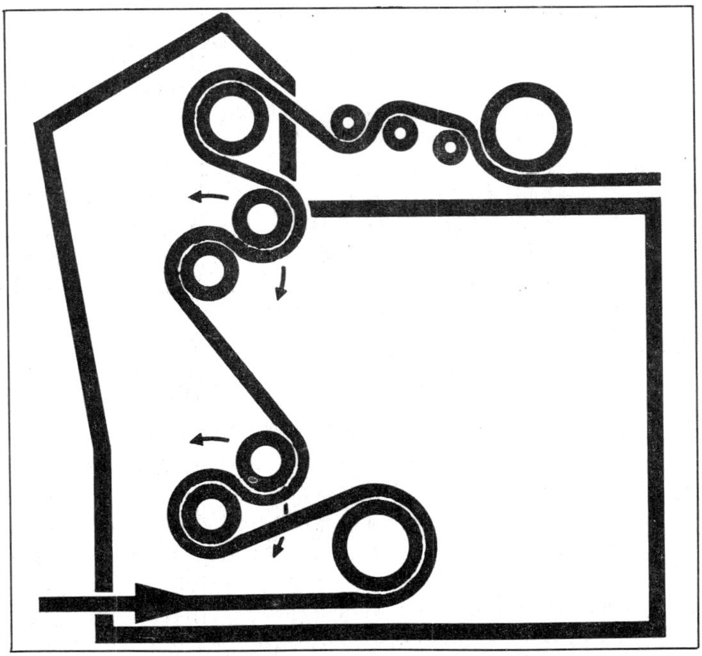

Figure 4.15 – Feeding arrangement for knitted fabrics on a Famatex stenter; two of the scroll rollers can be moved to give greater or less action on the fabric

Figure 4.16 – Feeding a knitted fabric onto a Bruckner stenter; the first uncurler, the pinning wheel and the pinning brush are visible

Pivoted near the top, the feelers hang vertically and maintain contact with the fabric edge. Should the edge move outwards the feeler is displaced and the upper portion operates a relay, actuating a small electric motor to wind out the end of the pin chain together with the edge uncurlers and the feeler. Each feeler operates independently of the other.

Immediately after being fed to the top of the pins the fabric edges pass under a freely rotating circular brush (pinning brush) which presses the fabric onto the pins. The brushes are turned by the movement of the pins.

The speed of the fabric through the machine is controlled by the pin chain, but the speed of fabric entry can be adjusted by the speed of the entry rollers. If slower than the pin chain the fabric is pulled off the rollers and stretched, so gaining length but increasing subsequent lengthways shrinkage on washing. If a fabric is too wide it may be narrowed by applying lengthways tension; fabrics for subsequent raising (brushing), an operation which imposes considerable lengthways tension (and resultant loss in width), may be prepared by stentering with lengthways tension.

However, for many clothing fabrics, good dimensional stability to washing is demanded and this means that lengthways stresses and strains acquired in fabric processing require compensating in the form of feeding fabric slightly more rapidly than the speed of the stenter chain. The amount of overfeed is often a matter of skilled judgement – some stenters claim to impart up to 20% overfeed, but 5% is more frequent. Although the surface speed of the main feed roller may be set 5% faster than the stenter chain, the fabric speed at entry to the chain may be less due to slippage. Excessive overfeed means less final length and, in extreme cases, cockling at the selvedges.

Drive System

The main drive motor for a 25 m long stenter is usually rated at 7.5 kW and is mounted at the exit of the machine. Through a worm reduction gearbox, the motor drives a cross shaft connected to a main chain wheel on each side of the machine. The rotation of the chain wheel pulls the chain along the rail.

The speed of the entry feed rollers, the take-off roller and the winding/plaiting mechanism are linked to the main drive.

Tachometers registering the chain speed have indicating dials at both ends of the dryer housing. Emergency stop buttons are located at both sides of the stenter at both ends. Microphone/loudspeaker systems provide communication between operatives at ends of the machine.

Fabric Delivery

Near the end of the machine the fabric is lifted off the pins by a driven take-off roller and then often passes round a pair of draw rollers which deliver either to an overhead plaiting mechanism or to a take-up roll attached to the stenter or on an A-frame driven by an air or hydraulic motor.

A few machines have means of contact winding fabric round cardboard tubes and delivering single piece lengths wound onto tubes without slowing the machine, but separating fabrics cleanly and consistently whilst travelling at about 70 m/min has proved difficult.

On many warp-knitted fabrics the edges tend to curl and are useless to the customer; on these fabrics it is customary to apply poly(vinyl acetate) in a 5 mm wide continuous line or series of dashes by means of an engraved wheel touching the upper surface of the fabric just inside each set of pins before entry into the drying chamber. After leaving the drying chamber the fabric is cut down the gummed lines by 'bacon-slicer' type knives, and the fabric lifted by the take-off roller; the strips at the edges are removed, usually through a suction line, to a collecting box.

420 cm wide warp-knitted fabrics, in particular, may be sliced into three widths just before leaving the stenter. Four gumming wheels and four circular knives are required and carefully positioned across the stenter.

Stenter Heating

A variety of methods can be used to heat the air circulating in stenters, each having particular advantages.

Steam
High-pressure steam from the main boilers is fed to a heat exchanger; condensate is returned to the boiler house. This method is clean and economical but usually adequate only for drying. A steam pressure of 1 000 kPa at the stenter will provide a maximum air temperature of around 165°C – adequate for drying and curing cross-linking reactants, but inadequate for heat setting or thermofixation.

Circulating Hot Oil
A thermo-stable oil is heated to 250°C + in a separate boiler and circulated through the heat exchanger. The provision of a separate boiler, pipework, pumps, storage/expansion tank, etc., all of which must be well-lagged, increases the capital and maintenance costs.

Oil
Each compartment has its own oil burner discharging into one side of a heat exchanger. Since demand for heat fluctuates, care must be taken to ensure reliable control of the oil spray and ignition.

Oil-fired stenters, in which the combustion products were fed directly into the stenter (as opposed to using a heat exchanger, called indirect heating) were used briefly, but trouble with oil spots and soot due to incomplete combustion showed that this arrangement is unsatisfactory, and all modern oil-heated stenters incorporate heat exchangers.

Compared with gas, oil is more easily transported and can be stored, so guarding against interruption in supply.

Gas
The soaring price of oil and increasing availability of natural gas in the UK has led to a big swing to gas-fired stenters. The burnt gas fumes are fed directly into the stenter chamber and early fears that acetate fabrics dyed with colours sensitive to such fumes would be adversely affected appear to be unjustified. Some spinning oils and monomer fumes given off by unscoured fabric during heat setting are burned up so reducing the amount of smoke exhausted and providing a small amount of additional heat. A recent development has been the invention of a radial gas burner, by Dunlop Engineering, which gives more complete combustion and improved efficiency.

To avoid any build-up of unburned gas in the stenter, the exhaust fans must be operating for 30–60 s before the gas supply is turned on. Should the pilot flame go out the gas supply is shut off automatically.

Indirect firing through a heat exchanger can be used for gas, but the additional capital cost together with lower thermal efficiency (only about 60% compared with direct firing) are major disadvantages.

Electricity
A few stenters have been built with electrical heating (one in a large factory

complex was adjacent to an organic solvent plant where combustion processes were banned), but electrical energy is generally considered to be too expensive.

Stenters for drying and heat setting often employ two forms of heating, e.g. steam in all chambers for drying supplemented by gas or oil firing in the final two or three chambers for setting.

Steam Injection during Heat Setting
Fabrics heat set in the grey state at temperatures of 190–200°C tend to turn yellow or even brown due to the removal of volatiles and oxidation of spin finishes, lubricants, etc. Particularly in the case of warp-knit nylon for lingerie, the yellowness is removable only by chemical bleaching, a lengthy and costly process.

Around 1960 it was discovered that if steam were injected into the stenter the amount of yellowing was greatly reduced. The reduction in yellowing is probably due to steam replacing some of the oxygen and thereby reducing oxidation. Claims were made that the presence of superheated steam produced a softer-handling fabric (true) and enabled setting to be carried out at a lower temperature (unsubstantiated).

The steam should be injected through silencers at the highest practicable temperature, e.g. 165°C at 600 kPa, into those chambers used for heat setting, e.g. the last two chambers in a four-chamber stenter. 300 kg of steam per chamber per h is adequate; any excess is not only wasting steam, but also oil/gas for raising the circulating gases to setting temperature.

The steam injection control valve should be locked off when not in use – especially when maintenance work is being carried out.

Air-flow in Stenters
The hot gases are forced into pairs of tapered ducts extending across the width of the fabric and then discharged at the fabric. In some stenters the ducts have nozzles running the full width to provide uniform air distribution onto the fabric, but other designs feature circular nozzles arranged to strike the fabric surface at an angle. It is most important, particularly when heat setting prior to dyeing, that hot-air distribution is uniform across the full width of the fabric (see Figures 4.17–4.19).

Some stenter designs (often called 'two and two counter type' arrangements) incorporate left- and right-handed ductings and fan and burner locations in alternate chambers so that any widthways velocity variation in one pair will be counterbalanced by the next (see Figure 4.20).

Care must be taken to provide a reasonable balance of distribution between top and bottom ducts. Usually in fabric drying slightly more air is forced into the bottom ducts to support some of the weight of the fabric and because the bottom of the fabric is wetter than the top. Supporting the weight of the wet fabric helps to prevent excessive sagging, particularly in

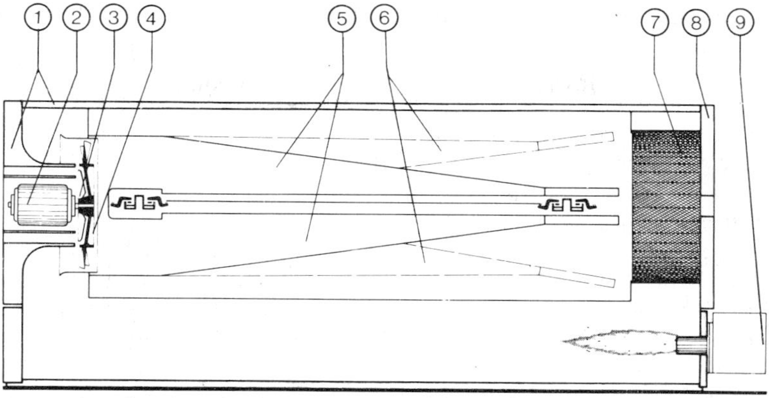

① all-round insulation
② two fan motors . . .
③ . . . with directly coupled motor fan wheel
④ two axial fans
⑤ two upper and two lower nozzle bodies in working position . . .
⑥ . . . and opened
⑦ two circular lint-collecting filters
⑧ two large sliding doors (another sliding door is located on the opposite side of the dryer between the two fan motors)
⑨ a gas or oil burner, or a burner which can be switched as required from gas to oil and from oil to gas

Figure 4.17 – Cross-section of an Artos direct-fired stenter

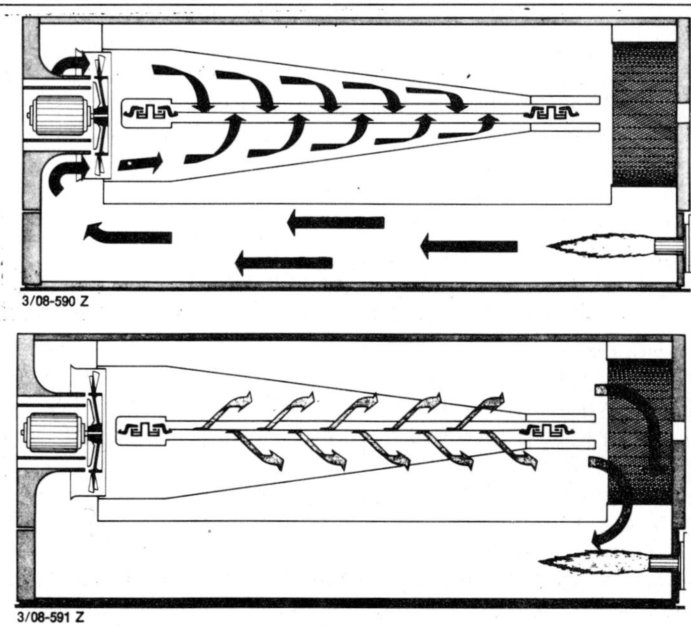

Figure 4.18 – Artos direct-fired stenter showing air flow from burner to fabric (top) and air flow from fabric through the filter to burner (bottom)

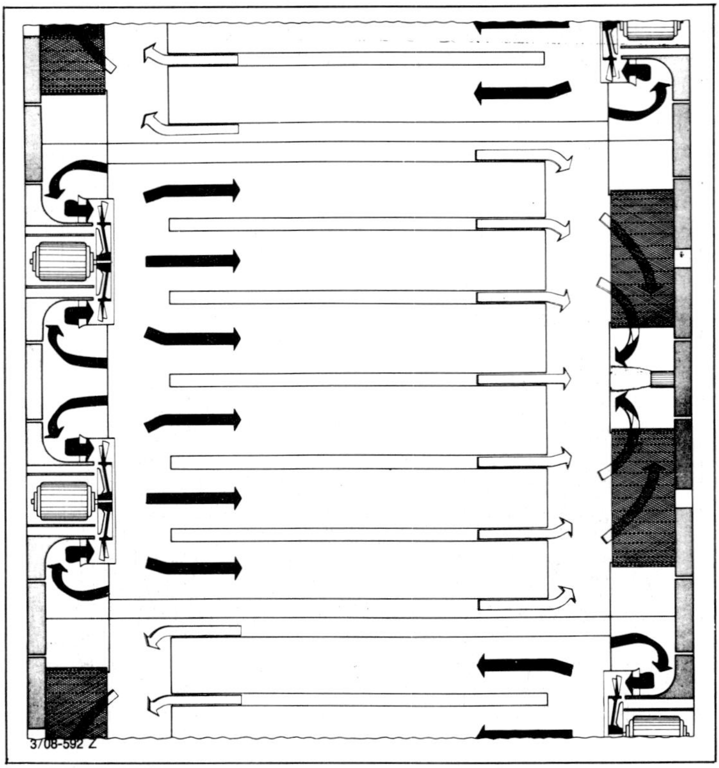

Figure 4.19 – Artos stenter (plan view of air flow)

knitted fabrics, but must not be overdone otherwise the edges of the fabric may be lifted off the pins with disastrous results.

The hot air, after being forced against the fabric surface, makes its way between the ducts to the sides of the stenter where it is collected by one fan or a pair of fans and recirculated. On some stenters the fan motors (which are usually two-speed) are built into the side of the machine, on others they are located on top. A large wire-mesh screen is fitted in the side to remove fly or lint from the system.

Where a pair of fans is used for a 3 m long compartment the fan motors are usually rated at 7.5 kW each, so for a five-compartment stenter the total electrical power requirement will approach 100 kW.

In each compartment a temperature probe is fitted at the inlet to the ducts and regulates the operation of the burner or the flow of air over the heat exchanger.

At the top of the compartment is a damper that regulates the rate at which exhaust gases, including steam, are removed into an exhaust duct. If the damper is wide open steam will be rapidly exhausted, but a large volume of input air will require heating so fuel consumption will be high. If the damper is nearly closed much smaller quantities of air will be demanded and less fuel used, but, as the proportion of water vapour in the air builds up, the rate of drying of the fabric will decrease.

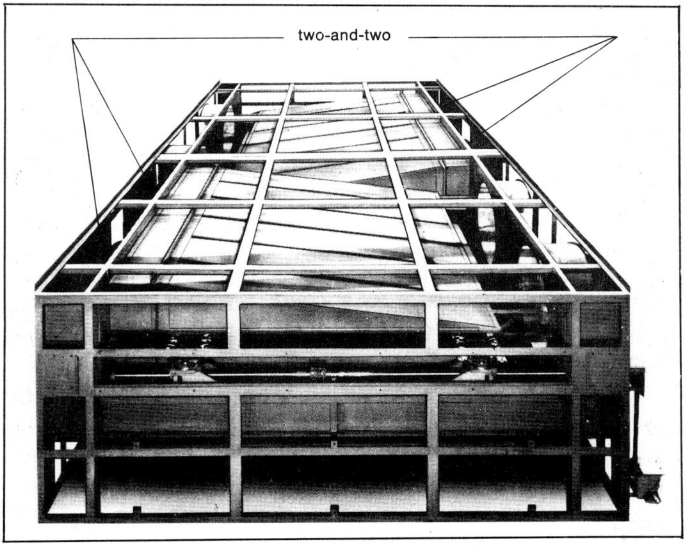

Figure 4.20 – Artos two-and-two counterbalanced arrangement

It is alleged – and is probably true – that on the majority of stenters the setting of the exhaust dampers is never changed from the position adopted when the machine was commissioned. Despite the increased cost of fuel, damper settings are rarely altered as they are not easily accessible and adjustment has little visual effect. It is generally accepted that 10% steam in exhaust air is about the proportion to aim for when drying, but very few stenters are equipped with humidity monitors. If the steam content is down to only 4%, two and half times as much air is being used and is being heated. In an interesting paper(*88*), practical measurements indicated that on an indirect oil-fired four-bay stenter 64% of input heat went out through the exhaust and nearly 29% through the burner exhaust – a total of 93%. The rate of chamber exhaust is quoted in one example as 5 000 ft^3/min at 190°C and in another as 9 500 ft^3/min at 170°C.

Although in theory the amount of exhaustion required when heat setting unscoured fabrics should be small, the amount of volatiles (spinning oils, monomer, etc.) given off provides copious amounts of white smoke which must be removed from the system. On a direct-firing stenter some of these organic materials are burned up when recycled through the burner.

The exhaust duct should be located so that any condensed volatiles do not drip back onto the fabric. Much trouble was caused on heat-setting stenters that, when started after a long break, melted the tarry deposit inside the ducts which subsequently dripped onto the fabric. Removal of exhaust gases at the side of the stenter is preferred and, on a few heat-setting stenters, underfloor ducting with accessibility for cleaning has been installed.

Stenter Casing

The stenter casing usually consists of a double thickness of mild steel sheet about 10 cm apart and filled with rock wool or similar insulation. Despite the large surface area, the amount of heat loss through the casing is usually only about 2%.

On most modern stenters one side of the machine is fitted with wide doors to provide good accessibility. Some stenters have the roof of each compartment hinged along one side so that one side of the roof complete with upper delivery duct can be raised to promote rapid cooling and facilitate cleaning.

With all stenters, and particularly stenters frequently used for heat setting, precautions should be taken to guard against fire, especially the ignition of condensates in the ducting (which used to be a fairly common occurrence). In the event of fire in the stenter the gas or oil supply should be shut off, the fabric input cut and run out, and the inside of the machine drenched with steam. In the absence of a reliable steam supply carbon dioxide cylinders may have to be provided.

Conveyor-band Stenters

Many wet weft-knitted fabrics tend to sag when held by the stenter pins; sagging may be so pronounced as to catch on the lower lip of the entrance to the heating section. Widening the stenter helps to overcome this problem, but tends to stretch the fabric width excessively. To resolve this difficulty a small number of conveyor-band stenters have been built which incorporate a steel-mesh conveyor running just below the stenter chain and supporting the centre of the fabric. Originally the conveyor extended the full length of the stenter, but later models carried the fabric through the entry system up to the oven, in which the weight of the wet cloth could be supported by increased air flow from the bottom nozzles.

Care must be taken to ensure that the wire mesh is free from any snags and that the speeds of the conveyor and stenter chain are identical, otherwise the fabric will develop a bowed position across the width.

Tenters

Tenters (usually in Yorkshire) are multi-layer stenters and are associated with the drying of fairly heavy woollen and worsted cloths. For these fabrics the temperature is usually limited to a maximum of 140°C and, rather than have a very long mono-layer stenter, a tenter makes several passes at different levels through a drying oven. The entry section rises to a height of about 3 m and the chain travels forwards and backwards through the oven descending a step at each end. The chain operates on the vertical-return principle.

The tenter may deliver fabric on the far side of the oven from the entrance (as in a stenter) or may roll up fabric in the space beneath the entry section. This saves space and manpower, as the operatives can see and act at both ends of the machine.

Most tenters are heated by pressurized steam through heat exchangers and, because they handle heavier fabrics, usually run at lower speeds than stenters.

5 GARMENT-DYEING MACHINES

As opposed to dyeing piece lengths of fabric, which are subsequently cut to shape and made up into clothing, a significant section of the dyeing and finishing industry, particularly in the Midlands and Scottish border areas, is involved with the dyeing of knitted garments, e.g. socks, tights, stockings, briefs, jumpers, sweaters, hats, etc. Various types of dyeing machinery have been developed to handle hundreds of individual articles of different shapes and sizes rather than lengths of regular fabric.

5.1 SIDE-PADDLE MACHINES

The side-paddle machine (Figure 5.1) is very versatile, handling most types of knitted garments from fully fashioned knitted outerwear to children's socks. Garments may be dyed loose or in open-mesh bags of polyester or polypropylene. The use of bags reduces the risk of long-sleeved garments becoming entangled so permitting higher machine loadings and, in the case of children's socks, facilitates unloading.

The machine is oval in plan with an annular dye vessel round a central island formed by a perforated sheet. The dye-liquor is circulated round the annulus by a large multi-bladed paddle located on one of the straight sides of the machine.

Production machines usually range from 1 000 up to 5 000 l capacity taking a maximum load of 200 kg. Because of the need for the garments to move freely, 25:1 is about the lowest liquor-to-goods ratio which can be achieved. Garments packed into mesh bags must have room to move to avoid permanent creasing; however, delicate goods should not be packed too loosely as excessive movement may damage the surface of the fabric.

The paddle wheel on a 2 500 l machine is about 1.5 m in diameter with the axis mounted above the highest water level. The paddle rotates at about 5–10 rev./min, each of the five blades penetrating to about half the depth of the liquor, and each blade being slightly swept back primarily to

Figure 5.1 – Pegg side-paddle dyeing machine

minimize the risk of garments catching on it. As each blade strikes the surface of the water any garment or bag near the surface is pushed down into the vessel. The paddle induces an appreciable degree of turbulence into the circulating liquor.

Paddle machines are made of stainless steel, with walls about 92 cm high (for safety) and the annulus about 50–75 cm wide. The central island of perforated sheet contains:

(a) 8 cm diameter water inlet for filling the machine without disturbing the garments unduly.

(b) 12 cm drain valve for emptying (machines of more than 2 500 l capacity usually have two), operated by lift rods with extended handles.

(c) Steam pipe,either discharging low-pressure steam through a sparge pipe into the liquor or a steam coil operating at 400–500 kPa. The former, simpler method is usually employed and gives vigorous diffusion, the latter may require a propeller to supplement the diffusion of hot dyeliquor into the dyeing vessel. However, with increasing size of vessel, heat applied only in the central island becomes more difficult to diffuse uniformly, so steam injection directly into the main vessel is frequently employed. This is achieved by placing three or four slight ramps running across the annulus, each ramp covering a steam pipe. The wall on the downstream side of the ramp has a line of small holes through which the stream is forced, giving both rapid and uniform heating.

(d) With small dyeing machines a facility for concentrated dye-liquor to be poured directly into the central island; in larger machines a dye-feed tank or funnel located outside the machine supplies a dye-feed pipe leading into the island.
(e) Controls for water and steam supply with extended hand wheels to permit safe and easy operation.
(f) Temperature control. At one time, probes were located in the central island in order to keep them away from the garments. However, the temperature in the island is frequently different from that of the main dyebath and, nowadays, the probe is located in a pocket in the outer wall of the vessel close to the bottom, the pocket being angled to avoid garments fouling the probe.
(g) Liquor overflow pipe, generally used for overflow cooling.

The interior surfaces of the machine must be smoothly finished as any burrs may snag the garments. The top of the machine is usually covered in stainless steel with two large hatches hinged to permit good access for loading and unloading.

To expedite cooling after dyeing it is customary to run in cold water at a controlled rate, with excess hot liquor being removed through the overflow pipe. Many machines have a cold-water spray pipe running across and above the dyebath; this enables the rate of addition of cold water to be clearly seen and controlled, ensures that the addition is across the full width of the dyebath and gives maximum separation of cold water in/hot liquor out.

Where garments are packed into bags the loading should be such as to permit good access of dye-liquor and a small amount of movement of goods inside the bag. Bagging is also useful in keeping garments of a particular size separated during dyeing – the bags are marked with coloured discs to denote the garment size.

Unloading a side-paddle machine is a strenuous, wet and messy business. After cooling and rinsing the goods the paddle wheel is set in motion and the operative, usually standing on a temporary platform, leans into the machine and grabs and lifts the bags or garments as they float past; loose garments are retrieved by a perforated paddle, and the liquor level can be lowered as the load is reduced.

This stage of paddle dyeing is crude and laborious, but no satisfactory alternative has been found. Mechanical hoists have been used to lift heavily loaded bags from the liquor, but have found only specialized application.

Side-paddle machines are simple in principle and comparatively cheap to purchase and to operate; they can be used for scouring, bleaching, dyeing and the application of finishes such as softeners, shrink-resisting agents and moth-proofing finishes.

5.2 OVERHEAD-PADDLE MACHINES

Probably the simplest dyeing machine ever built, the overhead-paddle machine consists of a dye vessel with a semi-circular bottom and a four-bladed full-width paddle mounted at the top and back of the machine. The blades dip about 10 cm into the dye-liquor and push the garments and liquor down and round the vessel. A perforated steam pipe, running across the dyebath, is fitted below a perforated false bottom. Filling is by a water pipe mounted just behind the paddle. The drive is usually by a V-belt fitted at the side to a small motor at the base of the machine.

Overhead-paddle machines are used for dyeing stockings (in bags), felts for hats, and, in smaller sizes, for samples.

5.3 ROTARY DRUMS

The principle of the rotary-drum machine is to have the goods in a perforated horizontal cylinder rotating slowly in a vat of slightly greater size. Each internal cylinder or drum is divided into separate quadrants so ensuring that the goods move with the rotation of the drum and each quadrant has its own sliding cover. A standard size of drum is about 116 cm in diameter and 80 cm in width, and it is customary to mount two drums side-by-side on the same revolving shaft. A machine of this type, with a liquor capacity of 1 800 l, has a load capacity of about 90 kg (approx. 375 dozen pair of tights), i.e. a liquor ratio of 20:1.

A few machines have been built with two cylinders mounted back-to-back and sharing the dyebath with heating and other services.

The outer tank holds the dye-liquor and is fitted with upward-sliding doors to give access to the perforated drums within. For a 1800 l machine, an 8 cm diameter water inlet is fitted near the top and a similarly sized overflow pipe at the opposite side. An 8 cm drainage valve is fitted at the same side. Cooling is carried out by overflow rinsing using the water inlet or, as a refinement on large machines, 3 cm cooling pipes with staggered holes fitted near the bottom of the dyebath.

The drive mechanism must also support the weight of the drum and contents, and is often arranged through spur gears and a worm reduction gear and chain to a 1 kW a.c. motor. The rate of rotation of the drum is about 3.3–4 rev./min, with reversal via the a.c. feed about every 20 s.

Heating is by the direct injection of steam by two 3 cm diameter perforated pipes running across the machine, one at the front and one at the back. A temperature gauge is fitted at one side and a dye addition tank located at the top and to one side, feeding by gravity across the width of the dyebath.

Rotary-drum machines are used for dyeing small garments, especially stockings, panty-hose and half-hose. The garments are placed in open mesh non-absorbent bags about 75 by 50 cm. The garments may be put in at random or, where better presentation is required, folded flat and inserted into pockets in the bag. The mechanical action of the machine is very gentle, liquor draining out of the goods as they rise to the top prior to re-immersion. The internal finish of the machine must be of a very high standard to avoid snagging, and the perforations are usually dished.

Sampling for colour is carried out by taking an example from the bulk, paying particular attention to the thorough penetration of any seams in the garments.

The rotary drum is compact, simple to operate and maintain, and relatively cheap to purchase and run.

5.4 TUMBLERS

A few machines, e.g. Dytex (Neil and Spencer Ltd), Milnor (Pellerin Milnor Corp.) and Sancowad (Samuel Pegg Ltd) have been developed using front-loading tumblers for dyeing small garments.

The principle of operation is to load the goods into a perforated drum which rotates about a horizontal shaft at the back of the machine. The drum may be divided into three or four pockets or fitted with baffles to lift the goods, which are usually contained in bags.

In the case of the Dytex(*89*) and Milnor, the height of dye-liquor in the drum may be varied over a wide range, and the maximum load is dependent on the acceptable density of packing. Typical figures for a 1.5 m diameter drum 1 m deep indicate that 90 kg (360 dozen pairs) of panty-hose can be dyed in about 720 l of water so giving a liquor ratio of 8:1.

The speed of rotation is variable, usually in the range 10–30 rev./min for a 1.5 m^3 machine, the lower end of the range being used for dyeing. The speed of rotation of the drum should be such as to get the bags of goods to roll over each other; higher speeds which lift small bags until they drop onto the rest of the load give too vigorous a mechanical action. An alternating motion is normally used in order to rearrange the load frequently.

Heating is usually by direct steam injection outside the drum but some machines are fitted with a closed-coil heat exchanger at the base of the dyebath.

Due to the strong similarity of this type of machine to commercial dry-cleaning machines, the degree of automatic control fitted is usually high, a punched card controlling the conditions of each operation. Likewise, safety interlocks are fitted to ensure that the loading door is locked before the cycle is started and remains locked whilst the drum is in motion.

At the end of the wet-processing cycle, tumbler machines usually incorporate a centrifugal period to remove water from the goods (this action also facilitates the clearing of loose dye and dye-liquor from the goods). The mechanical problems of spinning a load of wet goods in a 1.50 m diameter cage around a horizontal spindle are considerable, and one or more intermediate speeds (e.g. 200 rev./min) are employed before raising to nearly 800 rev./min. At these speeds the load must be evenly balanced to avoid vibration and a self-balancing system has been devised. Any out-of-balance tendency is measured by a commutator at the end of the drive shaft with three segments each corresponding to a tank in each of the three pockets of the drum. If the drive shaft is deflected, contact with the appropriate segment of the commutator opens a valve to inject water into the appropriate tank (or tanks) to balance the drum.

Nevertheless, the mechanical forces involved in centrifuging are considerable, and the machine, weighing 4.5 tonnes, requires a solid foundation. The drive to the drum is through a clutch and V-belts, separate motors being provided for normal and centrifugal drives. A level switch prevents the drum going into the centrifuging condition whilst liquor remains in the dyebath. A wrap-around drum brake is fitted which fails safe in the event of electrical or air-pressure failure.

Dyes are added through a tank and feed pipe into the outer casing. Unloading the centrifuged goods is simple and speedy.

The principle of the Sancowad process developed by Sandoz(*90*) is to spray concentrated dye-liquor on the loose goods contained in the drum to give a pick-up of about 100% on the weight of dry goods. The dye-liquor contains a small amount of foaming agent which helps it to disperse uniformly through the goods, which are tumbled cold for 10 min. It is important that no appreciable colour fixation occurs during this phase. When uniform distribution has been achieved steam is blown into the machine, the temperature rises rapidly and fixation takes place. Rinsing and centrifuging are carried out as in the Dytex and Milnor machines.

All the tumbler machines described can also be used for scouring garments and particularly for milling woollen garments; the speed of rotation of the drum has a marked effect on the rate of milling. Such machines have been called multi-purpose and their range can be extended by adding a hot-air blowing system as used in tumbler dryers, so providing a complete 'dry-to-dry' sequence of operations in one machine. However, the time required for drying is quite extensive, and a dyer may well transfer the centrifuged goods to an existing dryer so enabling the dyeing machine to be used for its intended purpose.

Comparisons of the operating efficiencies of paddle machines and the Dytex, Milnor and Sancowad machines have been carried out by workers at Hatra(*91*). In brief, multi-purpose tumbler dyeing/centrifuging machines are undoubtedly more economical in terms of water and steam and offer slight savings in dyes and auxiliaries. The most important advantage however is in time – multi-purpose machines with their close control of processing conditions and a comparatively short cycle time offer a strong incentive to achieve 'no-add' dyeing techniques, implying comparatively long runs of standard dyeings.

Undoubtedly multi-purpose machines of this type eliminate the practical problems of unloading side-paddle machines and provide a much drier and acceptable working environment.

On the other hand the capital cost of a multi-purpose machine has been estimated to be about four times the cost of a rotary-drum machine of similar capacity, and maintenance costs of the more sophisticated machine are likely to be significantly higher.

In practice it has been found that the differences in mechanical action between tumbler and paddle machines result in a significantly different product. Tumbling promotes shrinkage and bulk in crimped nylon – especially half-hose – giving excessively shrunken and dense garments, causing problems in post-boarding which does not eliminate the differences. Modifying the knitting construction gives an acceptable product and resolves the problem for a dyehouse using *either* paddle *or* tumbler machines; where both types are involved production problems are increased to an unacceptable degree.

5.5 HIGH-TEMPERATURE MACHINES

As in the field of circular-knit fabric dyeing, the introduction of textured polyester yarns into knitted garments in the late 1960s presented the garment dyer with considerable problems, particularly when medium–heavy depths were required. Dyeing at temperatures of 120–125°C was the answer to several of these problems and a limited number of pressurized garment-dyeing machines were developed. The following two examples indicate how very different machines can be devised in response to commercial demands.

Pegg Toroid Machine

This pressurized machine (Figure 5.2) was developed from a long-established circular machine for scouring and milling woollen garments. The principle is to induce a circular flow in a vertical plane around the machine, i.e. to form a torus.

This is achieved by forcing liquor up the centre of the dyebath and then, by using an inverted mushroom, deflecting the liquor outwards and, when approaching the sides of the vessel, upwards. The machine is fitted with an easily removable false bottom of which the outer two-thirds are heavily perforated; the centre pillar to which a lifting eye is fixed forms the stalk of the mushroom. With the pump flow set to maximum there is vigorous agitation of the dye-liquor sufficient to disperse bundles of garments put into the machine. Once the garments are moving freely the rate of flow is reduced to give a more gentle action favouring the movement of garments from the outside towards the core of the torus and so providing more uniform distribution.

Liquor return to the pump is downwards through the perforated false bottom and out of the vessel through a 15 cm diameter pipe; the pump is mounted at the bottom of the machine just outside the dye vessel and is driven by a 45 kW motor (for a 2 m diameter machine) through a belt drive.

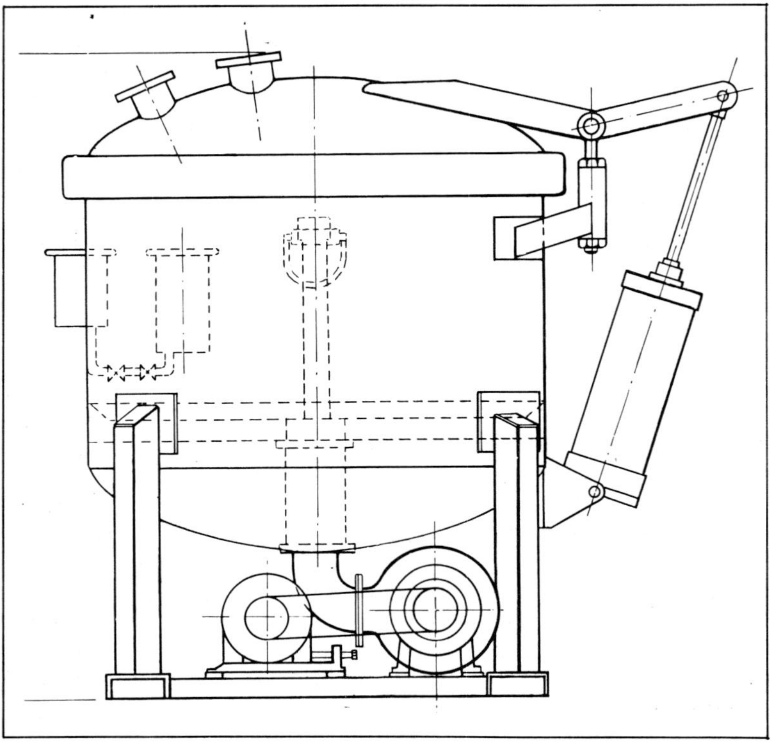

Figure 5.2 – Pegg Toroid machine

Just below the perforated false bottom are the heating and cooling coils arranged horizontally. To allow the pump and motor to be located underneath, the vessel is mounted on four stanchions and either set in a pit or surrounded by a raised working platform. The vessel is fitted with a heavy cover, which is secured by a ring-locking arrangement. Two port holes are provided in the cover together with internal illumination. The lid is lifted hydraulically.

The machine is operated using one-way flow, and is pressurized by heating the dye-liquor as in a domestic pressure cooker. Dye additions are made using a small pump from a small external tank which can be pressurized.

The thermometer probe is usually set in the wall of the machine below the false bottom and out of the way of the garments. This location is entirely satisfactory under normal operating conditions, but must be regarded with great care when the normal liquor flow is disrupted. In the event of failure of electric power, motor or pump, liquor flow will cease and the garments sink to the false bottom so dividing the dyebath into top and bottom sections. Application of cooling water through the coils will effectively reduce the temperature of the lower part – which will be indicated via the temperature probe – but will have little effect on dye-liquor in the upper section, which may still be in a superheated state and lead to a potentially dangerous situation. Expert advice should always be sought before opening pressurized machinery when normal liquor flow has been disrupted.

The Toroid machine is unloaded by lifting the false bottom out of the machine complete with the load of garments using an overhead hoist. A machine with an internal diameter of about 2 m and volume of 3 000 l will take a maximum load of about 100 kg. A sample pot is fitted, but in practice most dyers prefer to take the sample from the bulk goods.

Leemetals Epicyclic Paddle Machine

This machine consists of a circular pressure kier mounted on stanchions, but with an inwards-opening access door mounted in the side. The liquor is moved round the chamber by a two-bladed paddle offset mounted in the roof of the kier. The paddle is also angled so that, when rotating in a clockwise direction, the garments move round the machine in the same direction and are moved from the outside to the centre of the vessel. With anti-clockwise motion the garments are moved to the outside of the vessel, the paddle also imparting a vertical motion to the liquor. Paddle rotation can be varied from 5 to 40 rev./min, 30 rev./min being frequently used for dyeing. To avoid entanglement it is customary to run clockwise for a few minutes, stop for 20 s, and then reverse for 2 min; automatic reversal timing is fitted. A 2 kW motor is mounted on the roof and drives the paddles through V-belts.

Liquor is withdrawn through a grille in the outer wall of the machine and is pumped by a small motor through an external heat exchanger and back into the centre bottom of the machine. Pressurization is by heating the dye-liquor, and dye addition is by pump from an external tank.

A 2 m diameter machine has virtually the same liquor volume (3 000 l) and load (100 kg) as a Toroid, but the power consumption is considerably less. The capital cost is also much lower.

To facilitate unloading a bar can be lowered radially into the dye-liquor to catch the garments as they circulate round the machine and they are removed manually. Nevertheless, unloading necessitates an operative leaning into the machine through the hatch, and is a wet and messy operation.

5.6 DYE-BOARDING MACHINES

The traditional method for dyeing and finishing nylon stockings consisted of pre-steaming, dyeing in a drum and finally setting the stockings in pressure steam whilst mounted individually on vertical flat metal formers called boards (boarding). The final operation was by far the most expensive as it entailed the careful handling of individual delicate garments and was a highly paid job.

Dye-boarders were developed in the 1950s to combine the operations of dyeing and boarding, the precise presentation of the stockings providing a suitable opportunity for the application of dyes(*92*).

The boards were usually made of aluminium alloy with a very smooth finish and mounted upside-down, i.e. with the foot uppermost. The bottom of the board (the widest part) was securely bolted to a trolley which held two rows of 36 boards each about 2.5 cm apart. When loaded with stockings the trolley was moved along rails to a centre section and a pressure chamber (the bell) lowered and locked into position before steam under pressure was applied. Each bell had two trolleys mounted on rails diametrically opposite so that while one trolley was being steamed in the bell the other could be stripped and reloaded. Usually two operatives tended each machine, each working on one side of both trolleys. Each rack of boards could be of different sizes and both care and ample working space were required to ensure that the correct size of stockings were fed to the appropriate boards.

The dye-boarding machine incorporated a spraying system in the roof of the pressure chamber so that dye-liquor could be sprayed onto the feet and then run down the length of the stocking to the welt.

Firstly steam was blown in to the vessel at atmospheric pressure to relax the stockings and to give a small degree of set and to heat the metal formers supporting the stockings. Hot dye-liquor (80–90°C) was then sprayed from the top of the bell onto the feet of the stockings and then flowed down the leg and through the welt. The liquor flow was maintained for about 2 min. The dye-liquor was discharged to drain and saturated pressure steam fed in at 50–100 kPa (gauge) depending on the type of nylon in the stocking (110–115°C for nylon 6 and 115–120°C for nylon 6.6). After 2 min the steam was released to a blow-down pit and hot air blown in to dry the stockings. When dry the bell was raised and the trolley run out for unloading.

Dyeing was limited to the application of easy-levelling disperse dyes (C.I. Disperse Yellow 1, Red 1 and Blue 3 were normally used) and care had to be taken to ensure that the stockings did not bag on the formers or streakiness resulted. Penetration was no problem and it is probable that post-boarding aided dye diffusion into the fibre appreciably.

Finish was included in the dyebath and rinsing was unnecessary. About 110 l of dye-liquor were used for 72 stockings and it is probable that, in the short time permitted, only half the dye and less than half the finish were absorbed by the goods.

Dye-boarding machines were operated to precise time cycles (about 7 min, long enough for two operatives to strip 72 stockings off the formers and replace with 72 new stockings carefully presented). Repeatability of dyeings from cycle to cycle was good and colour matching simple. Dye makers supplied pre-packed mixtures of disperse dyes in appropriate weights for specific colours so no dyeing expertise was required. The operation did not need vast amounts of water and steam and therefore lent itself to knitters who wanted to bypass the commission dyeing field and set up their own colouring and finishing facilities. A Pegg automatic dye-boarder had a rated capacity of 25 dozen pairs of stockings per h and runs of 1 500 dozen pairs per colour were common. Work to extend the dye-boarding principle to heavier goods, e.g. half-hose, met with little success due to poor penetration of the garment seams and the need to dye heavier depths with higher colour fastness to washing.

The introduction of fully contracted stockings (derisively called 'abject objects') demanded relaxed dyeing techniques and the fashion changes towards tights caused a massive swing away from dye-boarding and few machines remain in use today.

6 OTHER FORMS OF DYEING

The dyeing of wool and worsted fabrics (including milling) is covered in another textbook(*93*). Similarly, the machinery required for fabric printing is included in another textbook(*94*).

Details of dyeing and finishing machinery for narrow fabrics have also been published(*95,96*).

REFERENCES

1. Whittaker, J.S.D.C., **77** (1961) 690.
2. Duckworth, Rev. Prog. Coloration, **1** (1969) 3.
3. Fleming and Gaunt, Rev. Prog. Coloration, **8** (1977) 47.
4. Siegrist, Rev. Prog. Coloration, **8** (1977) 24.
5. Horn, J.S.D.C., **81** (1965) 262.
6. Boulton and Crank, J.S.D.C., **68** (1952) 109.
7. Armfield, Boulton and Crank, J.S.D.C., **72** (1956) 278.
8. Denton, J. Textile Inst., **50** (1959) T521.
9. Denton, J. Textile Inst., **54** (1963) T406.
10. Denton, J. Textile Inst., **55** (1964) T228.
11. Denton, Shirley Inst. Memoirs, **36** (1963) 75.
12. Fox, J.S.D.C., **78** (1962) 393.
13. Johnels and Crawshaw, J.S.D.C., **85** (1969) 101.
14. Siegrist and Liddiard, J.S.D.C., **89** (1973) 523.
15. Heane, Hill, Park and Shore, J.S.D.C., **95** (1979) 125.
16. Hoffman, and Mueller, J.S.D.C., **95** (1979) 178.
17. Hadfield and Lemin, J.S.D.C., **77** (1961) 198.
18. Hampson, J.S.D.C., **67** (1951) 369.
19. Boulton, J.S.D.C., **67** (1951) 489.
20. Flanagan, J.S.D.C., **71** (1955) 790.

21. Mason, Park, Shore and Thompson, J.S.D.C., **96** (1980) 246.
22. Meldrum and Ward, J.S.D.C., **74** (1958) 140.
23. Blackburn and Ward, J.S.D.C., **80** (1964) 9.
24. Derbyshire and Lemin, J.S.D.C., **80** (1964) 363.
25. Burley, Rattee and Flower, J.S.D.C., **85** (1969) 187, 193.
26. Burley, Rattee and Flower, J.S.D.C., **87** (1971) 278.
27. Rayment, J.S.D.C., **88** (1972) 122.
28. Dyer, **143** (1 May 1970).
29. Cheetham, J.S.D.C., **76** (1960) 95.
30. Lemin, J.S.D.C., **74** (1958) 746.
31. Nossar, Chaikin and Datyner, J. Textile Inst., **64** (1973) 594, 718; **65** (1974) 300, 464, 581.
32. Dyer, **163** (1980) 470, 475, 477, 478.
33. France, J.S.D.C., **75** (1959) 477.
34. Kilby, J.S.D.C., **76** (1960) 479.
35. Morton, J.S.D.C., **81** (1965) 52, 93, 150.
36. Janas, J.S.D.C., **86** (1970) 73.
37. McFarlane and Wilcock, J.S.D.C., **65** (1949) 148.
38. Ashpole, McFarlane and Wilcock, J.S.D.C., **66** (1950) 17.
39. Fox, J.S.D.C., **84** (1968) 401.
40. United Merchants and Manufacturers Inc., BP 1 088 665.
41. Garratt, Haden and Smith, Dyer, **124** (1960) 405.
42. Beckmann and Langheinrich, Melliand Textileber, **51** (1970) 316, 321.
43. Fortess and Ward, Amer. Dyestuff Rep., **51** (1962) 400.
44. Wyles, Rev. Prog. Coloration, **9** (1978) 35.
45. Paterson, Rev. Prog. Coloration, **4** (1973) 80.
46. J.S.D.C., **88** (1972) 321.
47. Ratcliffe, Rev. Prog. Coloration, **9** (1978) 58.
48. Ratcliffe, J.S.D.C., **96** (1980) 94.
49. Simborowski, J.S.D.C., **96** (1980) 111.
50. Speke, J.S.D.C., **70** (1954) 221.
51. Mann, J.S.D.C., **71** (1955) 7.
52. Marshall, J.S.D.C., **71** (1955) 13.
53. Potter, J.S.D.C., **71** (1955) 20.
54. Moss, J. Textile Inst., **50** (1959) 654.
55. Fox, Marshall and Stewart, J.S.D.C., **83** (1967) 493.
56. Wilson, J.S.D.C., **76** (1960) 84.
57. Wilson, J. Textile Inst., **51** (1960) 590.
58. Gibson, Du Pont Tech. Bull. (1949) 82; USP 2 663 612 and 2 633 613.
59. Bent, Flynn and Sumner, J.S.D.C., **85** (1969) 606.
60. Fox, Glover and Hughes, J.S.D.C., **85** 614.
61. Norris, Textile Inst. & Ind., **8** (1970).
62. Shore, Rev. Prog. Coloration, **10** (1979) 33.

63. Hannay and Kilby, BP 620 584.
64. Boardman, J.S.D.C., **66** (1950) 397.
65. Ardron, Fox and Speke, J.S.D.C., **68** (1952) 249.
66. Ogden, Amer, Dyestuff Rep., No. 13, (27 June 1960) 32.
67. Stone, Dyer, **133** (1965) 605.
68. Erikson, Lundquist and Mellbin, J.S.D.C., **71** (1955) 894.
69. Dawson and Dunkerley, J.S.D.C., **94** (1978) 177.
70. Butcher, Rev. Prog. Coloration, **4** (1973) 90.
71. Shaw and Lewis, Textile Prog., **3** (1972) 4.
72. Parish, J.S.D.C., **78** (1962) 109.
73. Parish, Amer. Dyestuff Rep., **54** (1965) 402.
74. Parish, Textile Recorder, (Feb 1967) 52.
75. Duckworth, Text. Inst. & Ind., **18** (1980) 89.
76. Duckworth, Horsley & Thwaites, J.S.D.C., **88** (1972) 281.
77. Skelly, J.S.D.C., **72** (1956) 12.
78. Easton, Ciba-Geigy Review, (1971) No. 3.
79. Shipman, Rev. Prog. Coloration, **2** (1971) 42.
80. Hollings and Ruddy, Text. Inst. & Ind., **8** (1970) 12.
81. Dyer, **148** (1972) 508.
82. Marsh, 'Mercerising' (London: Chapman & Hall Ltd, 1941).
83. Rusnak, Sarmany and Duckworth, Eighth Shirley International Seminar, Shirley Publn. S24 (1976).
84. Training Course on Mechanical Drying, Shirley Institute, (Manchester 1974).
85. The Wira Textile Data Book – Wira, (Leeds 1974).
86. Parish and Thorp, J. Textile Inst., **56** (1965) T.179.
87. Denton, Holme and Salaman, Textile Inst., Annual Conference, (Edinburgh 1978).
88. Collis, J.S.D.C., **96** (1980) 305.
89. Dyer, **148** (1972) 226.
90. Lister and Turner, Textilveredlung, **6** (1971) 11, 708; Lister, Shirley Institute Seminar (1976).
91. Jaeckel, Knight and Pyle, J.S.D.C., **92** (1976) 157.
92. Rayment, Text. Manuf., (May 1964) 200.
93. Bird, 'Theory and Practice of Wool Dyeing' (Bradford: SDC, 1972).
94. Miles (ed), 'Textile Printing' (Bradford: DCPT, 1981).
95. Partridge, 'Man-Made Textiles' (1963) 58, and (1964) 35.
96. Wyles, Text. Inst. & Ind., **9** (1971) 315.

Chapter 2

Fluid Flow, Heat Transfer and Mass Transfer

W R MURPHY

INTRODUCTION

The intention of this chapter is to provide a self-contained introduction and cultural background to the cognate topics of fluid flow or momentum transfer, heat transfer and mass transfer; it therefore contains no references as such. However, as it is hoped that the reader will be stimulated into seeking additional knowledge on at least some of the subjects, a selected list of standard textbooks has been included at the end.

The textile coloration industry is very diverse, ranging from small, manually controlled batch processes to large computer-controlled ones.

At the design stage, the objectives resulting from information on products, production rates and costs are incorporated into a flow sheet which should include the various unit operations involved in the processes, and the raw materials and services required to operate them. The mass and energy balances associated with the flow sheet should then be optimized to ensure conservation of energy in the various process and service streams. At this stage the specifications of the process equipment, service plant and effluent treatment plant, which may be purchased, should be examined in detail to ascertain how they measure up to the objectives, both as individual items and as part of the integrated whole. Fitting the main items together for optimum efficiency and effectiveness for production and maintenance may require scale models and many calculations.

Considerable experience is required to select the best items of equipment and there are usually so many alternatives that the designer must confine his attention to the most promising in order to keep within the time-scale available. Engineering problems require flexible thinking and imagination; one should not assume that the original designer is always correct – objectives, products, technology and costs do change with time.

At the design stage a balance must be maintained between investment and operating costs, bearing in mind the importance of continuity of production and quality and the fact that technology is available to over-automate processes.

Having selected, installed and commissioned the plant, a system should be instituted to monitor, on a routine basis, the performance of the various process plants against the original or changing objectives. This should certainly include mass and energy balances or audits and an analysis of plant failures, both process and maintenance, to assist management in cost control and decision making.

Metering equipment incorporated to enable effective monitoring to be achieved may take the form of simple items such as pressure gauges, level indicators, temperature indicators, flow meters and ammeters. Even if such items are not to be installed in the original plant it is worthwhile making provision for their inclusion later.

Operating staff should have at least a basic understanding of the design of equipment for fluid flow, heat transfer and mass transfer so that they can appreciate how their system can best be operated as well as how it can be altered to operate under changing objectives and circumstances.

This chapter should assist engineers and non-engineers to communicate better and to tackle common objectives with an awareness of the potentials and limitations of the installed equipment.

1 FLUID FLOW

1.1 INTRODUCTION

Fluid flow is not only an important subject in its own right, it is also involved in the processes of heat and mass transfer. Indeed, the momentum and energy transfer phenomena associated with the flow of fluids are closely linked with the study of heat and mass transfer.

When fluids are conveyed through pipes, ductings, open channels and process equipment, energy is required to overcome the pressure drop caused by friction. To design a plant which will operate efficiently requires a knowledge of the fundamental principles of fluid flow.

Nature and Classification of Fluids

Fluids include both liquids and gases. The distinction between a 'permanent' gas and a vapour is that the latter may be liquified at atmospheric temperature; the critical temperature of a vapour is at or above atmospheric temperature.

Fluids may be classified in two different ways, either according to their behaviour under the action of externally applied pressure or according to the effects produced by shear forces. The former is related to the fluid density, the latter to fluid viscosity.

If the density of a fluid is independent of pressure or temperature it is said to be incompressible. If, however, the density is sensitive to either pressure or temperature it is said to be compressible. All fluids are compressible to some extent, but in normal fluid-flow problems liquids are generally considered to be incompressible. Gases are much more compressible than liquids and appreciable changes in the volume of a gas can occur if its pressure or temperature is altered. For practical purposes a gas may be regarded as incompressible if the percentage change in pressure or absolute temperature is small – typically, less than about 5%.

The relationship between the absolute pressure, p, the absolute temperature, T, and the volume, V, of a real gas is generally complex, but, except at very high pressures, the behaviour of a gas approximates to that of the ideal gas:

$$pV = nRT \tag{1.1}$$

where n = number of moles of gas
R = universal gas constant.

At high pressures, or when pressure changes are large, there may be appreciable deviations from Eqn 1.1 and an approximate equation of state should be used. In such cases Eqn 1.1 may be modified to include a compressibility factor, C:

$$pV = CnRT \tag{1.2}$$

The factor C is a function of pressure and absolute temperature. In general, C is less than one for low pressures and greater than one for high pressures.

When a force acts on a fluid, layers of the fluid slide over one another, thereby setting up stresses which oppose the shearing. The magnitude of the shear stresses depends on the fluid viscosity and the rate of shearing. For a gas, the viscosity is usually low and, even at high rates of shear, the viscous stresses are small. Under such conditions the gas approximates in its behaviour to an inviscid fluid. All real fluids have a finite viscosity and may be called viscous or viscid fluids. However, it is convenient on occasions to assume that a fluid has a zero viscosity. In many problems concerned with the flow of a gas or a liquid, the viscous stresses are important and give rise to appreciable velocity gradients within the fluid; dissipation of energy, generally in the form of heat, occurs as a result of the stresses produced.

Fluids may also be classified as either Newtonian or non-Newtonian fluids.

1.2 PROPERTIES, QUANTITIES, UNITS AND DIMENSIONS

Properties – Density, Viscosity and Surface Tension

In general, the properties of a fluid depend on pressure and temperature. For most practical purposes, the properties of a liquid may be regarded to depend solely on temperature.

Density, ρ, defined as mass per unit volume, is expressed in the SI system of units as kg m^{-3}.

For a unit mass of an ideal gas, having a molecular weight M, Eqn 1.1 becomes:

$$p \cdot \frac{1}{\rho} = \frac{1}{M} \cdot RT \qquad (1.3)$$

$$\text{or } \rho = \frac{pM}{RT} \qquad (1.4)$$

One gram mole of an ideal gas at 0°C and 10^5 N m^{-2} occupies 22.41×10^{-3} m^3.

The specific gravity, or relative density, is the ratio of the density of a substance, at the temperature under consideration, to the density of water at its temperature of maximum density, 4°C. The term, being a ratio, is dimensionless.

In general, liquids become more dense with decrease in temperature. The density of a gas increases with increasing pressure up to its critical point, when the densities of the gas and liquid phases are identical. The densities of mixtures and solutions, in general, are not predictable; a liquid can dissolve considerable quantities of a gas without a significant change in density.

Viscosity is a measure of the internal friction of a fluid; it compares the resistance offered by the fluid to any shearing action tending to distort it. During fluid flow, shear stresses will exist in the fluid, their magnitude depends on the fluid viscosity and rate of shear. A fluid at rest is free from shear stresses; viscosity, therefore, is apparent only when the fluid is in motion.

The dynamic viscosity, absolute viscosity, coefficient of viscosity, or simply, viscosity, μ, is defined as the shear stress, τ, required to move one layer of a fluid past another layer of fluid, unit distance away, with unit velocity:

$$\tau = -\mu \cdot \frac{dv}{dy} \qquad (1.5)$$

where τ = shear stress, N m^{-2} or kg m^{-1} s^{-2}

μ = viscosity N s m^{-2} or kg m^{-1} s^{-1}

dv/dy = velocity gradient, s^{-1}.

For traditional reasons, viscosity is frequently expressed in poises or centipoises. The poise has the c.g.s. units g cm^{-1} s^{-1}, and hence 1 poise is equivalent to 0.1 N s m^{-2} and 1 centipoise is equivalent to 10^{-3} N s m^{-2}.

On occasions, it is better for comparative work to use the kinematic viscosity, ν, which is defined as the ratio of dynamic viscosity, μ, to mass density, ρ:

$$\nu = \frac{\mu}{\rho} \qquad (1.6)$$

(NB Be careful to distinguish v (velocity) from ν (kinematic viscosity))

The units of kinematic viscosity are $m^2\ s^{-1}$, and once again, for traditional reasons, kinematic viscosity is frequently expressed in stokes or centistokes. The stoke has the c.g.s. units $cm^2\ s^{-1}$ and hence 1 stoke is equivalent to $0.1 \times 10^{-3}\ m^2\ s^{-1}$ and 1 centistoke is equivalent to $10^{-6}\ m^2\ s^{-1}$.

The viscosity of a liquid depends primarily on temperature, decreasing with increase in temperature. Poiseuille showed that the relationship between viscosity and temperature for a pure liquid is:

$$\mu_t = \frac{\mu_0}{(1 + at + bt_2)} \tag{1.7}$$

Where μ_t = viscosity at t°C in poises
μ_0 = viscosity at 0°C in poises
a and b = constants which depend on the liquid.

For water μ_0 is 0.0172 poise, a is 0.03337 and b is 0.000221; values for other liquids are available in the literature.

In oil technology, the viscosity index is of more use than a single viscosity at a specified temperature. When plotted on special graph paper (Refutas paper) the relationship between viscosity and temperature is linear; this is useful in problem solving.

The viscosity of a liquid increases slightly with increase in pressure, water is an exception in that its viscosity first decreases and then increases as pressure increases.

In the case of gases, increase in temperature results in increased molecular activity (increased kinetic energy); this increases the magnitude of the intermolecular forces which increases the viscosity. At higher pressures, especially in the neighbourhood of the critical point, the viscosity of a gas increases with increasing pressure.

Surface tension, σ, is defined as force per unit length or energy per unit area, it has the units, $N\ m^{-1}$ or $J\ m^{-2}$.

Within the body of a liquid, a molecule is attracted equally in all directions by the molecules surrounding it. At the interface between the liquid and the vapour or gas the upward and downward attractions are unbalanced. The liquid surface behaves as if it were an elastic membrane under tension. This surface tension is the same at any point on the surface and acts in the plane of the surface, normal to any line in the surface.

Surface tension causes drops of liquid to tend to adopt a spherical shape; it is also responsible for capillary action, which causes a liquid to rise in a fine-bore tube when its lower end is placed in a liquid which wets the tube. If the liquid does not wet the tube, it will be depressed in the fine tube below the outside liquid surface.

If θ is the angle of contact between the liquid and the solid, the upward force due to surface tension, σ, is:

$$\sigma \pi d \cos \theta \tag{1.8}$$

where d = tube diameter, m.

When z is the height to which liquid is raised in the capillary tube, and ρ is the liquid density, then the weight, w, of liquid raised is:

$$w = z\,\rho g \frac{\pi d^2}{4} \tag{1.9}$$

Equating the two forces gives:

$$\sigma \pi d \cos\theta = z\rho g \frac{\pi d^2}{4} \tag{1.10}$$

$$\text{or } z = \frac{4\sigma\cos\theta}{\rho g d} \tag{1.11}$$

Surface tension decreases with increase in temperature. The surface tension of a liquid is sensitive to impurities even in trace quantities.

Quantities – Mass, Weight, Force and Pressure

The units and dimensions of three common physical quantities, mass, weight and force, are important, especially in engineering where forces and pressures are encountered.

Mass is the quantity conserved in chemical reactions. Mass, acceleration and force are related:

$$\text{force} \propto \text{mass} \times \text{acceleration} \tag{1.12}$$

In the SI system of units, a force of 1 newton will give an acceleration of 1 m s^{-2} to a mass of 1 kg, hence the constant of proportionality in Eqn 1.12 is unity:

$$\text{force} = \text{mass} \times \text{acceleration} \tag{1.13}$$

The term weight is frequently, but incorrectly, used as a synonym for mass; weight, unlike mass, varies with location.

Pressure is the force exerted by a fluid on the surfaces with which it is in contact, or by one portion of a fluid on an adjacent portion. The pressure at any point is the force exerted on unit area at that point and is expressed in newton's per square metre, N m^{-2}. The SI unit of pressure is the pascal, Pa; it is equal to one newton per square metre. For ease of comprehension, the unit N m^{-2} has been retained in this chapter, it will be appreciated, however, that to be strictly correct the composite group, N m^{-2}, wherever it appears, should be replaced by Pa. The term bar was introduced to provide a unit of suitable magnitude to replace the unit of pressure in atmospheres; normal atmospheric pressure is 1.013 bar, 1 bar is exactly 10^5 N m^{-2}.

The pressure, p, at a point in a fluid can be expressed in terms of the height, z, of a column of the fluid which causes the pressure, or which would cause an equal pressure if the actual pressure were applied by other means. The pressure is:

$$p = z\rho g \tag{1.14}$$

where p = pressure, N m^{-2} or kg m^{-1} s^{-2}
z = height of fluid above the point, m
ρ = mass density of fluid, kg m^{-3}
g = gravitational acceleration, m s^{-2}.

The pressure head is measured as a 'length' of fluid, the mass of the fluid or its density obviously must be given, since a pressure head of 1 m of water differs from a pressure head of 1 m of air by the ratio of their densities.

The earth is surrounded by an atmosphere of air many miles thick. The pressure at the surface of the earth due to this atmosphere depends upon the head of air above the surface. Atmospheric pressure at sea level is 101.3×10^3 N m^{-2}, equivalent to a head of 10.33 m of water; this pressure decreases with increase in altitude.

Absolute pressure is the pressure measured above the absolute zero; a perfect vacuum. Gauge pressure is the pressure read on a gauge, it is the

TABLE 1.1
Units and Dimensions of Fluid Properties and Quantities

Property or quantity	Symbol	Definition	Dimensions	Units
Mass	m		M	kg
Velocity	v	distance/time	LT^{-1}	m s^{-1}
Acceleration	a	velocity/time	LT^{-2}	m s^{-2}
Force	F	mass × acceleration	MLT^{-2}	N or kg m s^{-2}
Weight	w	force	MLT^{-2}	N or kg m s^{-2}
Pressure	p	force/area	ML^{-1} T^{-2}	N m^{-2} or kg m^{-1} s^{-2}
Work or energy	E	force × distance	ML^2 T^{-2}	J or kg m^2 s^{-2}
Power	P	work/time	ML^2 T^{-3}	W or kg m^2 s^{-3}
Momentum	mv	mass × velocity	MLT^{-1}	kg m s^{-1}
Shear stress	τ	force/area	ML^{-1} T^{-2}	N m^{-2} or kg m^{-1} s^{-2}
Mass density	ρ	mass/volume	ML^{-3}	kg m^{-3}
Dynamic viscosity	μ	stress ÷ velocity gradient	ML^{-1} T^{-1}	N s m^{-2} or kg m^{-1} s^{-1}
Kinematic viscosity	ν	viscosity/density	L^2 T^{-1}	m^2 s^{-1}
Surface tension	σ	force/distance	MT^{-2}	N m^{-1} or kg s^{-2}

pressure measured above atmospheric pressure, or below atmospheric pressure in the case of a vacuum. Hence, absolute pressure is the sum of gauge pressure and atmospheric pressure.

Units and Dimensions

A check list of the definitions, units and dimensions for the common properties and quantities of fluids is given in Table 1.1.

1.3 FLUIDS IN MOTION

This section deals, in general terms, with the flow of a fluid over a solid boundary; the special case of pipe flow is discussed in Section 1.5.

Flow Patterns

The velocity of a fluid particle is a vector quantity having both magnitude and direction, these may vary with time. In general , the velocity of a particle has three coordinates: v_x, v_y and v_z (or u, v, w), one for each space coordinate. Although the path followed by a particle (a streamline or path line) is ordinarily a curve in three dimensions, the individual particles move in the general direction of flow, but not necessarily parallel to one another. In many simple situations, such as flow through a pipe or flow past plates parallel to the flow, all velocity vectors in the field are parallel and only one velocity component is required to describe the flow situation. In this chapter discussions will be limited, in general, to one-dimensional flow.

Constant velocity over a cross-section is shown by equidistant streamlines, and an increase in velocity by closer spacing of the streamlines. Figures 1.1, 1.2 and 1.3 illustrate streamlines in a straight pipe, through a constriction and past an immersed object.

A moving fluid which is not influenced by a solid boundary is not subject to shear, and shear stresses are not present in the fluid. The flow of an incompressible fluid with no shear is called potential flow or undisturbed flow.

When a fluid flows over a solid boundary, the fluid particles in contact with the solid are decelerated to zero and the adjacent particles are retarded by the viscous drag of the fluid. Thus, the velocity in the neighbourhood of the solid will change in a direction at right angles to the flow.

Prandtl, in 1904, stated that, except for fluids moving at very low velocities or possessing high viscosities, the effect of a solid boundary on the flow is confined to a layer immediately adjacent to the boundary. This layer is called the boundary layer, shear and shear stresses are confined to this part of the fluid; outside the boundary layer, potential flow exists. Most flow processes can be analysed by considering the fluid in two parts, the boundary layer and the remaining fluid flowing outside the boundary layer. In some situations the boundary layer may be neglected and in others, such

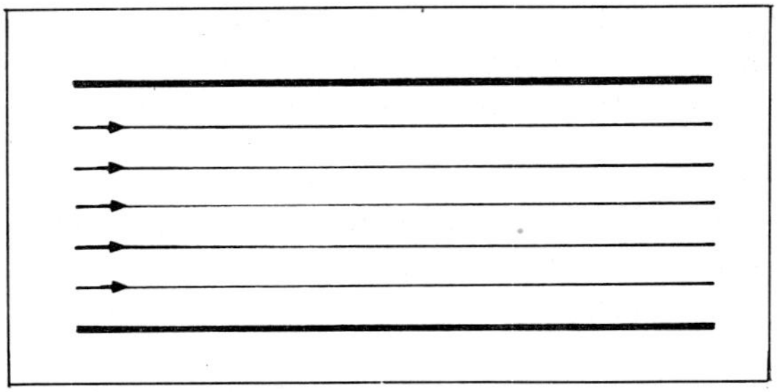

Figure 1.1 – Streamlines in a straight pipe

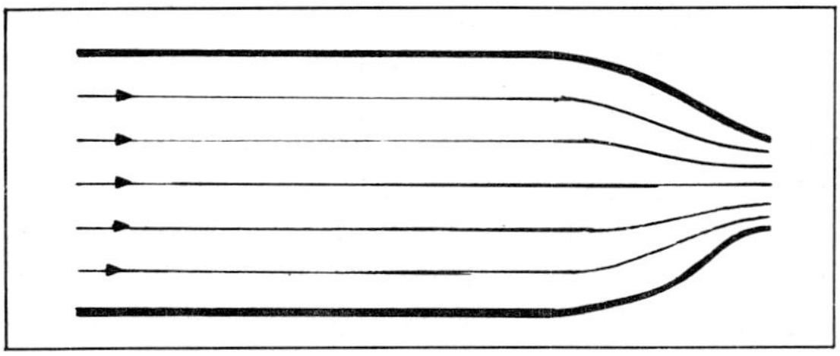

Figure 1.2 – Streamlines at a constriction

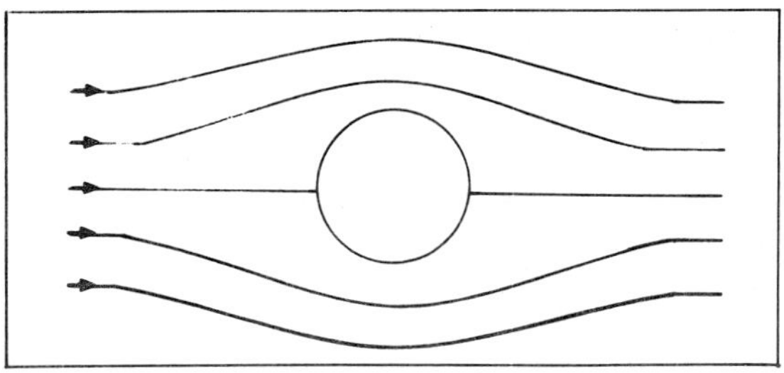

Figure 1.3 – Streamlines past an immersed object

as flow through pipes, the boundary layers from the walls meet at the centre line of the circular pipe and there is no potential flow.

When an incompressible fluid flows under the influence of a solid boundary, four important effects can result:

(a) Formation of velocity gradients and shear stress gradients
(b) Initiation of turbulence
(c) Formation and growth of boundary layers
(d) Separation of boundary layers from the solid boundary.

At low velocities, fluids flow without mixing and adjacent layers slide over each other. There are no cross-currents nor eddies, movement across streamlines occurs as a consequence of diffusion on a molecular scale. This type of flow is called streamline, laminar or viscous.

At higher velocities, the fluid motion becomes erratic and random, the cross-currents produce turbulence and the formation of eddies of various sizes which result in lateral mixing. The degree of mixing increases with increase in fluid velocity. The eddies, which are formed continually, break down into smaller eddies which in turn break down into smaller ones, until finally the smallest eddies, having a diameter of about 1 mm and containing about 10^{16} molecules, are rapidly destroyed by viscous shear. A wide spectrum of eddy sizes exists at any given time; the flow within an eddy is laminar.

The difference between laminar and turbulent flow was first demonstrated by Osborne Reynolds in 1883. He introduced a fine stream of coloured water into a glass pipe through which water was flowing and observed that at low flow rates the stream of coloured water remained as a thin thread with no lateral mixing. He deduced that the water was flowing in parallel straight lines and that the flow was laminar. When the flow rate of water was gradually increased, the thin thread of coloured water was disturbed to an increasing extent until at a critical velocity the thread of colour was mixed uniformly throughout the entire cross-section of the pipe. Reynolds then deduced that the water was moving in the form of cross-currents and eddies, and that the flow was turbulent.

The factors which determine the transition from laminar flow are related by the dimensionless Reynolds group:

$$\mathrm{Re} = \frac{L v \rho}{\mu} \tag{1.15}$$

where L = a characteristic linear dimension, for pipes this is the internal diameter, d, m

v = fluid velocity, m s^{-1}

ρ = fluid density, kg m^{-3}

μ = fluid viscosity, N s m^{-2} or kg m^{-1} s^{-1}

Re = m. m s^{-1}. kg m^{-3} ÷ kg m^{-1} s^{-1}, i.e. it is dimensionless.

Laminar Flow

Figure 1.4 illustrates the velocity profile for the one-dimensional laminar flow in an incompressible fluid flowing over a plane solid surface. The distances, y, are measured at right angles to the surface and direction of flow. The velocity, v, is a function only of the distance from the solid surface, y; its value ranges from zero at $y=0$ to the undisturbed or free stream velocity, v_{fs}, of the main fluid stream at $y=y_m$.

The velocity increases with distance from the solid at a decreasing rate, it approaches the main stream velocity asymptotically. The velocity gradient is the reciprocal of the slope of the velocity profile. At position y_1, the velocity gradient, dv/dy, is the limit of $\Delta v/\Delta y$ as Δy approaches zero. It will be observed that the velocity gradient is a maximum at the solid surface and approaches zero asymptotically with distance from the solid surface. The velocity gradient may be considered as the time rate of shear for the system.

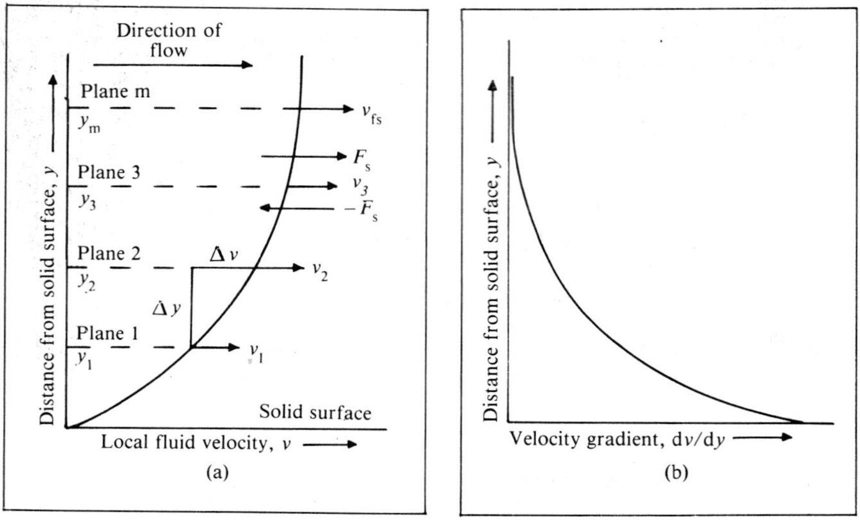

Figure 1.4 – (a) Velocity profile in laminar flow
(b) Velocity gradient profile in laminar flow

An actual fluid resists shear and hence, when there is a rate of shear, a shear force must exist to have created it. In one-dimensional flow, the shear force acts parallel to the plane of the shear. In Figure 1.4, at plane 3 the shear force F_s acts in the direction of flow. This force is exerted by the fluid beyond plane 3 on the fluid between plane 3 and the solid. By Newton's third law of motion (to every force there is an equal and opposite force) an equal and opposite force, $-F_s$, acts on the fluid beyond plane 3, from the fluid between plane 3 and the solid. The shear stress, τ, is then defined in terms of the total force, F_s, and the area of the shearing plane, A_s:

$$\tau = \frac{F_s}{A_s} \tag{1.16}$$

where τ = shear stress, N m^{-2} or kg m^{-1} s^{-2}.

Newton's Law of Viscosity

The viscosity of a fluid can be determined in terms of the force required to overcome its resistance to flow. Figure 1.5 illustrates the laminar flow of a fluid sandwiched between two flat plates, each of area A, separated by a distance D. The lower plate is stationary and the upper plate is moving with a constant velocity, v_0. For a Newtonian fluid the velocity profile is linear and the velocity, v, is zero at $y=0$, and v_0 at $y=D$. The velocity gradient, $-\mathrm{d}v/\mathrm{d}y$, is a constant and equals v_0/D; the negative sign implies that the velocity decreases in the direction of the stationary plate.

The viscous frictional effects in the fluid resist the motion of the top plate, and a force, F, is required to maintain its velocity at v_0. This shear force, which acts over the surface area, A, is related to the shear stress, τ, by Eqn 1.16.

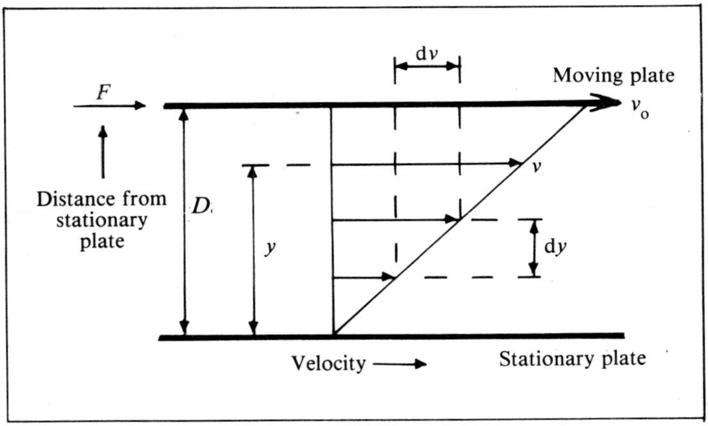

Figure 1.5 – Shear between two plates

Newton's law of viscosity states that the shear stress, τ, is proportional to the velocity gradient, $-\mathrm{d}v/\mathrm{d}y$, in the fluid. The constant of proportionality is known as the coefficient of dynamic viscosity, μ. Newton's law of viscosity can then be written:

$$\tau = -\mu . \frac{\mathrm{d}v}{\mathrm{d}y} \tag{1.17}$$

The ratio of the shear stress to the rate of shear is constant for gases and most non-colloidal liquids; these fluids are called Newtonian fluids. Liquids which do not obey Newton's law of viscosity are called non-Newtonian fluids and their 'apparent viscosity' is a function of the rate of shear.

For Newtonian fluids in laminar flow, the velocity gradient, $-\mathrm{d}v/\mathrm{d}y$, is also the rate of shear, $\mathrm{d}\gamma/\mathrm{d}t$ or $\dot{\gamma}$. The viscosity law can then be written to show that dynamic viscosity, μ, equals the shear stress divided by the shear-rate:

$$\mu = \frac{\tau}{\dot{\gamma}} \tag{1.18}$$

A graph of shear rate (abscissa) against shear stress (ordinate) for a Newtonian fluid is a straight line through the origin; the numerical value of the slope of this line is the viscosity of the fluid.

Example 1.1
A liquid fills the space between two parallel plates 0.025 m apart. If a force of 0.300 N is required to move one plate, having a surface area of 0.100 m^2, with a uniform velocity of 0.050 m s^{-1} while the other plate remains stationary, calculate the viscosity of the liquid.

Velocity gradient $= -dv/dy = v/y = 0.050\ m\ s^{-1}/0.025\ m = 2.0\ s^{-1}$

Shear stress, $\tau = F_s/A_s = 0.300\ N/0.100\ m^2 = 3.00\ N\ m^{-2}$

Viscosity, $\mu = \tau/(dv/dy) = 3.00\ N\ m^{-2}/2.0\ s^{-1} = 1.5\ N\ s\ m^{-2}$

Problem 1.1
A liquid, having a kinematic viscosity of 5.00×10^{-3} $m^2\ s^{-1}$ and a density of 900 kg m^{-3}, fills the space between two parallel plates 5.00×10^{-3} m apart. Calculate the force required to move one plate past the other stationary plate with a velocity of 2.00×10^{-3} m s^{-1}, when the moving plate has an area of 1.00 m^2.

(*Answer*: 1.8 N)

Problem 1.2
A shaft, diameter 80.00×10^{-3} m rotates in a bearing, diameter 80.12×10^{-3} m and length 95×10^{-3} m. The annular space between the shaft and the bearing is filled with oil having a dynamic viscosity of 0.096 N s m^{-2}. Calculate the power required to overcome viscous resistance in the bearing when the shaft rotates at 25.0 rev./s. Power equals viscous force times peripheral velocity.

(*Answer*: 1 510 W)

Non-Newtonian Fluids

In the previous section it was mentioned that Newtonian fluids obey Newton's law of viscosity, which states that the shear stress, τ, is directly proportional to the velocity gradient, dv/dy, or the rate of shear, $\dot{\gamma}$, the constant of proportionality being the viscosity of the fluid.

Figure 1.6 shows a series of typical rheograms at constant temperature and pressure. Curve 1, which has a constant slope and passes through the origin, is typical of a Newtonian fluid.

There are three classes of fluids which do not obey Newton's law of viscosity, non-Newtonian fluids:

(a) Shear dependent, time independent
(b) Time dependent
(c) Viscoelastic

For these non-Newtonian fluids, the term viscosity has no meaning unless it is related to a particular rate of shear, $\dot{\gamma}$. An apparent viscosity may be defined:

$$\mu_a = \frac{\tau}{\dot{\gamma}} \tag{1.19}$$

(a) Curves 2, 3 and 4 in Figure 1.6 are rheograms of shear-dependent fluids. Curve 2 represents a Bingham or ideal plastic; the fluid will not flow until a certain minimum or threshold shear stress, τ_0, has been provided; when τ_0 has been exceeded the fluid viscosity is a constant. Typical examples of Bingham plastics are sewage sludge and clay suspensions.
Curve 3 represents a pseudo-plastic fluid; the curve passes through the origin, is concave downward at low rates of shear and becomes linear at high rates of shear. Typical examples of 'shear thinning' fluids are solutions of high polymers, paper pulp, rubber latex and suspensions of polystyrene in organic liquids.
Curve 4 represents a dilatant fluid; the curve is convex downward at low shear rates and becomes linear at high shear rates. Typical examples of 'shear thickening' fluids are starch solutions and sand-filled emulsions.

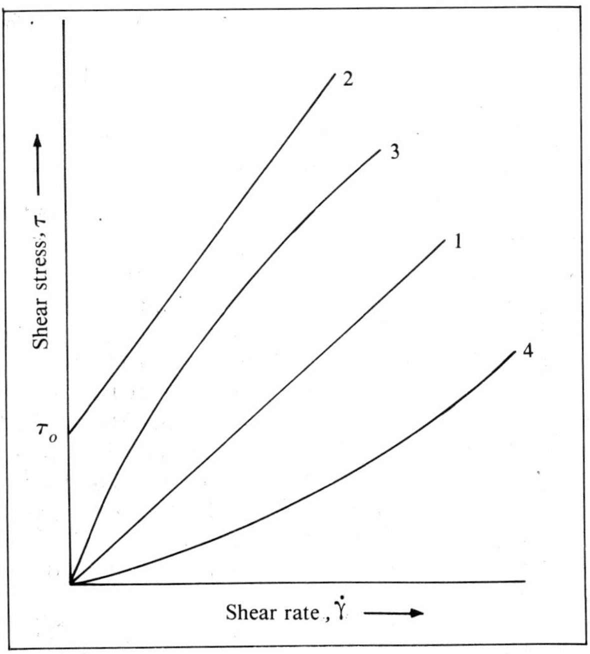

Figure 1.6 – Rheograms for Newtonian and non-Newtonian fluids

(b) Time-dependent non-Newtonian fluids are common. The additional variable of time complicates the analysis. In the loop technique of analysis, the fluid is subjected to an increasing shear rate and then a decreasing shear rate, returning to a shear rate of zero. If no time dependence exists, the two curves will coincide. If, however, the apparent viscosity changes with time a hysteresis effect is observed and two separate curves in the form of a loop are generated. Rheopectic fluids, such as certain sols, show an increase in apparent viscosity with shear time. Thixotropic fluids, such as paints, show a decrease in apparent viscosity with shear time.

(c) Viscoelastic fluids exhibit both viscous and elastic properties. When viscoelastic fluids, such as molten polymers, are extruded through a die or spineret, the cross-section of the extruded material is greater than the orifice through which it has been extruded.

Turbulence

Turbulence can result from contact of the flowing fluid with solid surfaces or from contact between two layers of fluid moving at different velocities. The former is called wall turbulence, the latter free turbulence. Wall turbulence can occur when the fluid flows through open or closed channels or past solid objects immersed in the fluid. Free turbulence can occur when a boundary layer separates from a solid surface and flows through the bulk liquid.

The shear forces existing in turbulent flow are much larger than those occurring in laminar flow. Turbulent shear stresses are called Reynolds stresses, τ_t. The total shear stress in a turbulent fluid is the sum of the viscous stresses and the turbulent stresses.

The Boundary Layer and Its Significance

It has already been mentioned that fluid in motion relative to a solid surface develops a velocity gradient in a plane at right angles to the solid. If the solid is immersed in an infinite medium, then the velocity gradient is a maximum at the solid surface and decreases asymptotically with distance from the surface to a point in the main stream fluid where its magnitude is reduced to a negligible value. The layer of fluid within the velocity gradient is called the boundary layer. If two or more immersed solid surfaces are in close proximity, then their individual boundary layers may interact with one another. A typical example of this is flow in a circular pipe; the boundary layers progressively thicken until they meet at the axis of the pipe. The thickness of the fully developed boundary layer is thus the radius of the pipe.

The concept of the boundary layer not only greatly simplifies the analysis of fluid-flow problems, it is also useful in convective heat transfer and mass transfer studies.

As the velocity gradient disappears asymptotically and not abruptly into the main stream, the boundary of the gradient is more readily defined as the thickness, δ, of the layer extending to a point where the velocity is 99% (or 99.9%) of the free stream velocity. This region is referred to as the hydrodynamic boundary in momentum transfer. A similar type of boundary layer, the thermal boundary layer, exists in heat transfer. Under certain circumstances in convective heat transfer the hydrodynamic and thermal boundaries coincide.

The boundary layer does not form immediately, as fluid approaches a solid boundary it builds up gradually along the solid surface, from zero thickness at the leading edge, to a maximum, some distance from the leading edge; this distance is known as the entry length, L_e.

Figure 1.7 illustrates the formation of a boundary layer for fluid flowing parallel to a thin flat plate in an infinite medium. Only one boundary layer is shown, a similar boundary layer would form on the underside of the plate. Upstream of the plate the fluid is assumed to be undisturbed by the plate, and everywhere has the free stream velocity v_{fs}; the local or point velocity is represented by v. The arrows indicate the direction and magnitude of the fluid velocity.

The velocity of the fluid increases from zero at the solid–fluid interface to the free stream velocity at some distance from the plate. Each of the velocity profiles shown in the figure correspond to a definite distance, L,

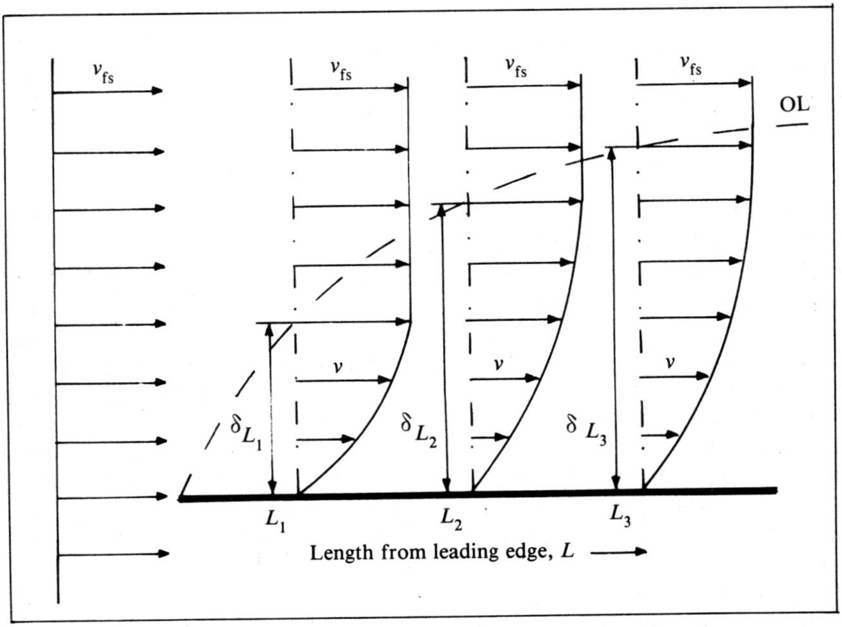

Figure 1.7 – Formation of a boundary layer on a flat plate

from the leading edge of the plate. The dashed line, OL, passes through all points where the velocity is 99% (or 99.9%) of the free stream velocity. Hence the line OL is the outer limit of the boundary layer and separates the fluid stream into two parts: one in which the fluid velocity is constant and equal to the free stream velocity and the other in which the velocity varies from zero, at the solid surface, to a velocity 99% (or 99.9%) of the free stream velocity.

The symbols δ_{L1}, δ_{L2} and δ_{L3} represent the boundary layer thickness at positions L_1, L_2 and L_3 from the leading edge of the plate.

In the initial stages the velocity distribution in the boundary layer is continually changing, but after some distance from the leading edge the velocity distribution no longer changes with distance, the flow then is said to be 'fully developed'.

Very near the solid boundary the velocities are small and the flow in that part of the boundary layer is laminar. Further away from the solid surface, but still in the boundary layer, the velocities become greater and flow in that part may be turbulent, depending on the local Reynolds number; it is perfectly possible for a boundary layer to be entirely laminar. Between a laminar layer and a turbulent layer is a transitional or buffer layer of intermediate character. A turbulent boundary consists of three zones or regimes: the laminar sublayer, the transitional zone and the turbulent zone. In the laminar sublayer viscous forces predominate, and in the turbulent zone viscous forces are insignificant.

When the flow in the boundary layer is completely laminar, the thickness, δ, of the layer increases with $L^{0.5}$, where L is the distance from the leading edge. For turbulent boundary layers the thickness of the fully developed layer increases with $L^{0.8}$. Both these relationships have important consequences, not only in fluid flow but also in convective heat and mass transfer.

Figure 1.8 illustrates the initiation of turbulence, for one side only of a thin flat plate; it is characterized by a sudden increase in the thickness of the boundary layer. Near the leading edge, due to deceleration, the fluid velocity is low and the boundary layer is thin and entirely laminar. As the layer thickens, at distances further from the leading edge, a point is reached where turbulence begins. For fully developed turbulent flow, the laminar sublayer may be in the order of 0.2 mm, it decreases with increasing fluid velocity. In the transitional zone the flow fluctuates between laminar and turbulent flow.

For flow parallel to a flat plate, turbulent flow first appears at a Reynolds number somewhere in the range 0.1×10^6 to 3×10^6. A local Reynolds number less than 0.1×10^6 indicates that the boundary layer is entirely laminar. The change to turbulent flow occurs at 0.1×10^6 for rough surfaces and 3×10^6 for smooth surfaces.

The initial development of the boundary layer depends on the nature of the leading edge of the plate. If it is neither thin nor sharp then the situation can be more complex than illustrated in Figure 1.8.

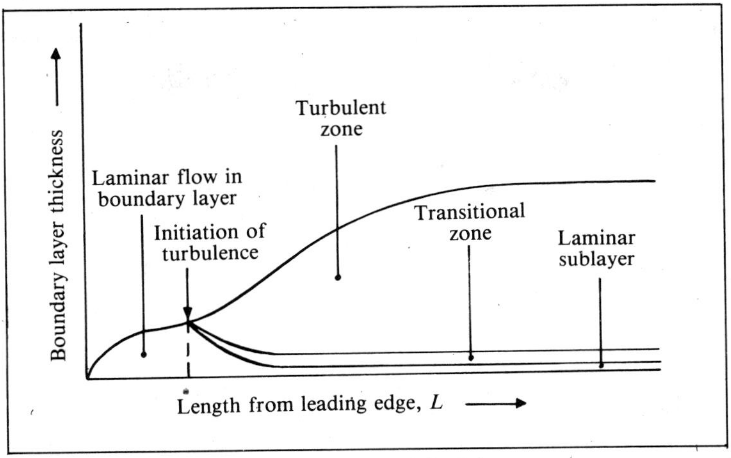

Figure 1.8 – Development of a turbulent boundary layer on a flat plate

Boundary Layer Separation and Wake Formation

The growth of a boundary layer at the leading edge of a solid surface has been discussed, but what happens at the trailing edge? If the trailing edge is parallel to the direction of flow, then, in the case of the flat plate, the boundary layers on the two sides of the plate have grown to a maximum thickness. For a short distance downstream of the plate the boundary layers and velocity profiles persist. The velocity gradients gradually fade, the boundary layers intermingle and disappear. The fluid once again flows with a uniform undisturbed velocity, provided that it does not encounter any other solid surfaces.

Separation of a boundary layer from a solid surface is encountered whenever a change in the direction or magnitude of the fluid velocity is too large for the boundary layer to remain in contact with the surface. Typical examples are a sudden enlargement, a sudden contraction, a sharp bend, or when fluid flows around an object.

Figure 1.9 illustrates a cross-section of fluid flowing at right angles to a cylinder of sufficient length so that the end effects do not interfere with the flow patterns. The area available for flow changes as the fluid flows around the cylinder; it gradually decreases from A towards B (and D) and gradually increases from B (and D) towards C. For continuity reasons, the fluid velocity gradually increases to a maximum at B (and D), the positions of minimum flow area, and gradually decreases downstream of B (and D). It is a consequence of the energy equation (see Section 1.4) that, as velocity decreases, pressure increases; there is thus a pressure variation around the cylinder. A pressure variation of this type is called an adverse pressure gradient.

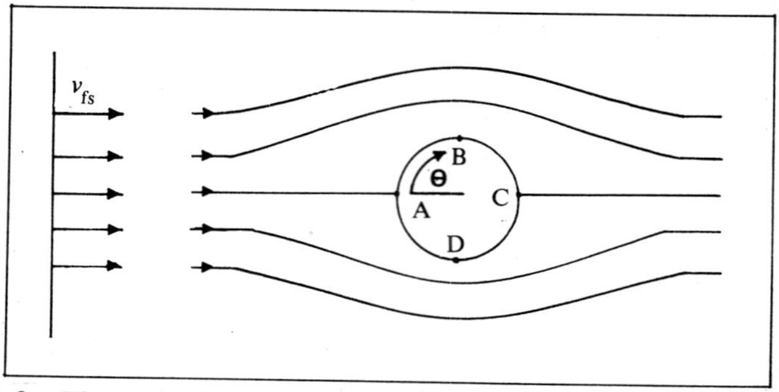

Figure 1.9 – Fluid flowing around a long cylinder

A slowly moving boundary layer is very sensitive to the retarding forces produced by the increasing pressure downstream of B (and D), and may eventually decelerate to zero or even be reversed in its direction. The point on the solid surface where this first occurs is called the separation point; it marks the separation line between the boundary layer and the relatively stationary fluid below it.

Figure 1.10 illustrates the separation of the boundary layer on a curved surface due to a flow reversal. Point B is the separation point. Its position in actual flow depends on whether the boundary layer is wholly laminar or has a turbulent (and buffer) region. Flow separation depends on the dissipation of kinetic energy, $v^2/2$. The greater the kinetic energy content of the boundary layer the longer will it persist in contact with the solid surface. A laminar boundary layer will therefore separate more readily than a turbulent one. For flow around a long cylinder the angle θ in Figure 1.9 would

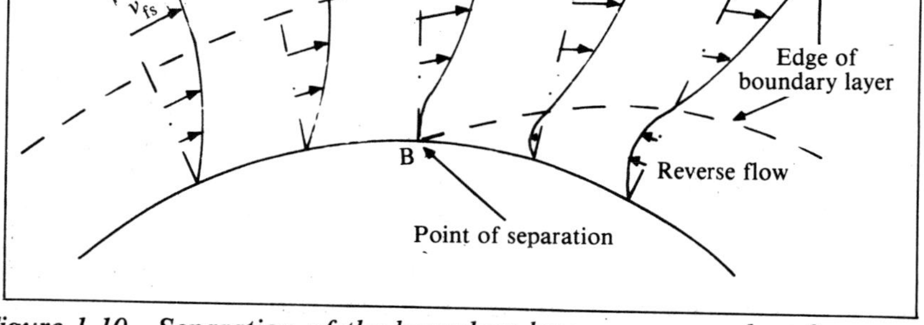

Figure 1.10 – Separation of the boundary layer on a curved surface

be about 85° and 140°, respectively, for laminar and turbulent boundary layers.

Figure 1.11 illustrates fluid flowing past three differently shaped bodies, all presenting the same projected area at right angles to the flow. The flow is assumed to be perfectly smooth around the streamlined body (a); there are no sudden changes in the magnitude of the velocity; friction losses are due entirely to skin or wall friction or drag. Flow around the sphere (b) and disc (c) is assumed to be such that there are sudden changes in both the direction and magnitude of the velocity. The bodies are then non-streamlined and are called 'bluff' bodies. To an extent, whether a body is streamlined or bluff depends not only on its shape, but also on the Reynolds number of the system. At a sufficiently high Reynolds number, boundary layer separation could occur with body (a).

When boundary layer separation does occur there is a zone of highly decelerated fluid at the rear of the body. In this wake, large eddies, called vortices, are formed. These vortices consume considerable energy and may lead to a large pressure loss in the fluid. Immersed bodies should, where

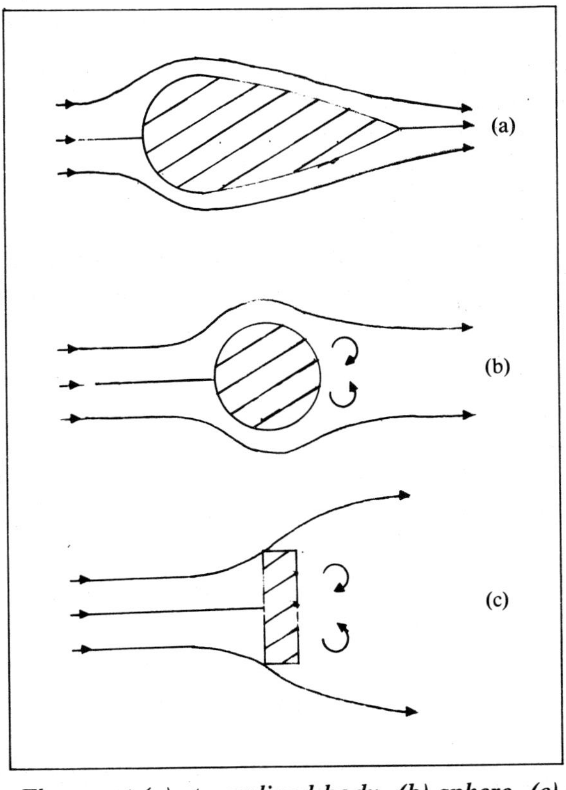

Figure 1.11 – Flow past (a) streamlined body, (b) sphere, (c) circular disc

possible, be streamlined to minimize energy losses. There are, however, certain situations, for example, the mixing of liquids and convective heat transfer, where boundary layer separation is encouraged. The additional frictional loss due to wake formation is called 'form-friction' or 'form-drag', the name being derived from the form or geometry of the immersed body. Neighbouring bodies can exert an influence on the development, shape and magnitude of the wake produced behind a particular body.

In any flow situations dealing with immersed bodies, the total energy loss due to frictional effects is the sum of skin-drag and form-drag. Their relative contributions depend on the Reynolds number of the situation.

1.4 THE MASS BALANCE, MOMENTUM AND ENERGY EQUATIONS

Mass Balance Equation

The mass balance, or continuity, equation states that, for steady state conditions, the rate of mass entering a system equals that leaving it. Based on the law of conservation of mass, the continuity equation may be applied to any number of sections in a system. The continuity equation, between points 1 and 2 along a pipe then takes the form of a simple mass balance:

$$Q = v_1 A_1 = v_2 A_2 \tag{1.20}$$

or

$$m = \rho_1 v_1 A_1 = \rho_2 v_2 A_2 \tag{1.21}$$

where Q = volumetric flow rate, $m^3\ s^{-1}$
v = fluid velocity, $m\ s^{-1}$
A = cross-sectional area occupied by fluid, m^2
m = mass flow rate, $kg\ s^{-1}$
ρ = fluid density, $kg\ m^{-3}$.

In many problems, the mass flow rate per unit area, or mass velocity, G, is the important quantity:

$$G = \dot{m}/A = \rho v \tag{1.22}$$

where G = mass velocity, $kg\ m^{-2}\ s^{-1}$.

It will be appreciated that unlike density, ρ, the mass velocity, ρv, is independent of temperature.

Problem 1.3

A liquid flows through a pipeline which gradually converges from 0.250 m diameter at position 1 to 0.200 m diameter at position 2, and then branches, one branch being 0.100 m diameter and discharging at position 3, the other branch, diameter 0.150 m, discharging at position 4. If the velocity at position 1 is 1.50 m s^{-1} and the velocity at position 4 is 2.50 m s^{-1}, calculate the discharge rate at position 3 and the velocities at positions 2 and 3.

(*Answers*: 0.0294 m^3 s^{-1}; 2.30 m s^{-1} and 3.74 m s^{-1})

Momentum Equation

The rate of change of momentum for a fluid flowing through a changing cross section is:

$$F = \dot{m}\,(v_2 - v_1) = \rho Q(v_2 - v_1) \qquad (1.23)$$

where F = force or rate of change of momentum, N or kg m s^{-2}

Energy Equation

The total mechanical energy of a fluid in motion is the sum of the three mechanical energy components:

(a) Potential energy
(b) Pressure energy
(c) Kinetic energy.

A fluid may also have non-mechanical energy contributions, such as internal energy and heat energy. In practice some of the mechanical energy of a moving fluid is converted into heat energy through frictional forces.

Consider now the three mechanical energy contributions of a fluid in motion, each in terms of unit mass of fluid.

(a) The potential energy is the energy that a fluid has because of its position in the earth's gravitational field. The work required to raise unit mass of fluid to a height, z, above a datum level against gravitational acceleration, g, is zg. This work is equal to the potential energy of a unit mass of fluid at a height, z, above the datum level.
(b) The pressure energy is due to the weight of the fluid or the pressure it exerts and the energy or work required to introduce the fluid into the system without a change in volume. If p is the pressure and V the volume of a mass, m, of fluid, then pV/m is the pressure energy per unit mass of fluid. The ratio, m/V, is the fluid density, thus the pressure energy per unit mass of fluid may be written as p/ρ.
(c) Kinetic energy is due to fluid motion. The kinetic energy of unit mass of fluid is $v^2/2$, where v is the linear velocity of the fluid, relative to some fixed point.

To summarize:

(a) Potential energy $= zg$ (1.24)

(b) Pressure energy $= p/\rho$ (1.25)

(c) Kinetic energy $= v^2/2$ (1.26)

It will be observed that each of the energy terms has the dimensions of energy per unit mass: $ML^2\,T^{-2}.\,M^{-1}$ or $L^2\,T^{-2}$, with the units of $m^2\,s^{-2}$.

The total mechanical energy, E, of unit mass of a fluid in motion is:

$$E = zg + \frac{p}{\rho} + \frac{v^2}{2} \tag{1.27}$$

Frequently it is convenient to consider fluid flow problems in terms of head rather than energy. The potential head, pressure head and velocity head, each having the dimensions of length are, respectively: z, $p/\rho g$ and $v^2/2g$.

Figure 1.12 illustrates an inviscid fluid flowing down a pipeline from a reservoir, the level of which is maintained constant by an inflow of fluid. Two manometers are installed in the pipeline at positions 1 and 2.

From the law of conservation of energy, it follows that for the situation illustrated in Figure 1.12 the total change in mechanical energy must be zero:

$$E = z_0 g = z_1 g + \frac{p_1}{\rho_1} + \frac{v_1^2}{2} = z_2 g + \frac{p_2}{\rho_2} + \frac{v_2^2}{2} \tag{1.28}$$

This is known as the Bernoulli equation, which is stated more generally:

$$zg + \frac{p}{\rho} + \frac{v^2}{2} = \text{constant energy, } E \tag{1.29}$$

The Bernoulli equation written in terms of head is:

$$z + \frac{p}{\rho g} + \frac{v^2}{2g} = \text{constant head, } h \tag{1.30}$$

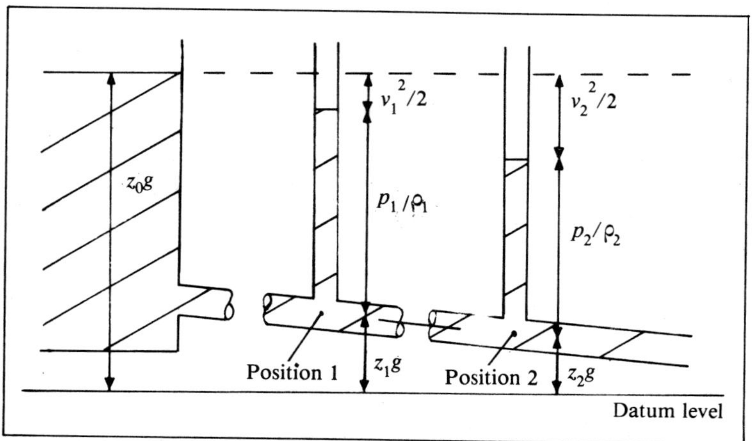

Figure 1.12 – The mechanical energy contributions to the energy of an inviscid fluid in motion

Strictly speaking, the Bernoulli equations, as written above, only apply to the flow of a fluid along a streamline. For the steady state flow of a fluid in a pipe, the Bernoulli equations above can be written in a modified form:

$$E = z_0 g = z_1 g + \frac{p_1}{\rho_1} + \frac{v_1^2}{2\alpha_1} = z_2 g + \frac{p_2}{\rho_2} + \frac{v_2^2}{2\alpha_2} \tag{1.31}$$

where v is the mean or volumetric velocity in the pipe and α is the dimensionless velocity distribution factor, which accounts for the velocity distribution across the pipe. It is shown in Section 1.5 that α equals 0.5 and approximately 1.0 for laminar and turbulent flow respectively.

In practice, a real fluid requires to do a positive amount of work to overcome viscous frictional forces. The total mechanical energy is therefore not constant along a pipe, but always decreases in the direction of flow. To preserve the conservation of energy equation, heat, equivalent to the loss in mechanical energy, is generated.

The Bernoulli equation for real incompressible fluids is corrected by adding an additional term, W_f, the energy lost per unit mass when a fluid does work to overcome friction; or, in terms of head, h_f, the head loss due to friction. The Bernoulli equation can then be written in terms of energy for a real incompressible fluid flowing from position 1 to position 2, in Figure 1.12:

$$\left(z_2 g + \frac{p_2}{\rho_2} + \frac{v_2^2}{2\alpha_2} \right) - \left(z_1 g + \frac{p_1}{\rho_1} + \frac{v_1^2}{2\alpha_1} \right) = -W_f \tag{1.32}$$

or in terms of head:

$$\left(z_2 + \frac{p_2}{\rho_2 g} + \frac{v_2^2}{2\alpha_2 g} \right) - \left(z_1 + \frac{p_1}{\rho_1 g} + \frac{v_1^2}{2\alpha_1 g} \right) = -h_f \tag{1.33}$$

If a pump were installed between positions 1 and 2 in Figure 1.13 to impart work or energy, ΔW, or head, Δh, to a unit mass of the fluid, then the corresponding modified Bernoulli equations would be:

$$\left(z_2 g + \frac{p_2}{\rho_2} + \frac{v_2^2}{2\alpha_2} \right) - \left(z_1 g + \frac{p_1}{\rho_1} + \frac{v_1^2}{2\alpha_1} \right) = \Delta W - W_f \tag{1.34}$$

and

$$\left(z_2 + \frac{p_2}{\rho_2 g} + \frac{v_2^2}{2\alpha_2 g} \right) - \left(z_1 + \frac{p_1}{\rho_1 g} + \frac{v_1^2}{2\alpha_1 g} \right) = \Delta h - h_f \tag{1.35}$$

where the units of ΔW and Δh are $m^2\ s^{-2}$ and m respectively.

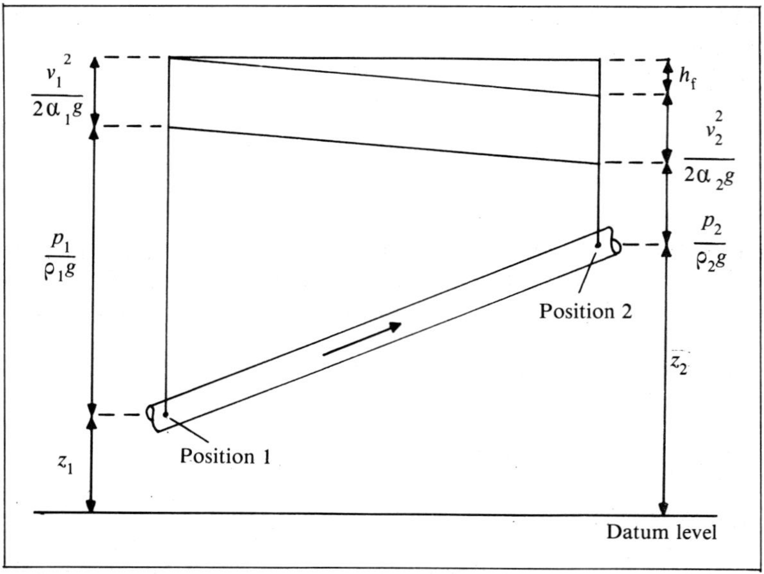

Figure 1.13 – Representation of the Bernoulli equation for the flow of a real fluid

Eqns 1.34 and 1.35 apply to incompressible fluids; however, they can be used in a gas flow system where the density change through the system is less than about 5%. When the density change is greater than about 5%, as in the case with high-velocity flow, or when large temperature differences occur, the gas must be treated as a compressible fluid.

The Bernoulli equation can be modified to include all situations in which the conservation of energy applies. The modified form, which includes the effects of heat, external work and energy losses in the flow of a real fluid, is called the flow equation:

$$zg + \frac{p}{\rho} + \frac{v^2}{2\alpha} + \Delta q + \Delta W + W_f = \text{constant} \tag{1.36}$$

where q, ΔW and W_f represent, heat, work and energy losses per unit mass of fluid respectively. Heat, Δq, and work, ΔW, may be either added to or removed from the system and as such may have either positive or negative values. The energy loss, W_f, must always have a positive sign on the output side of the flow equation.

Example 1.2

Water, at a height of 20.0 m above a datum line, flows in turbulent flow with a velocity of 12.0 m s^{-1} and a pressure of 225×10^3 N m^{-2}. Calculate the total energy per unit mass of water above the datum line, when the gravitational acceleration is 9.807 m s^{-2} and the density of water is 1 000 kg m^{-3}.

Total energy per unit mass is: $zg + p/\rho + v^2/2$

$$= (20.0 \times 9.807) + (225 \times 10^3/1\,000) + (12.0^2/2)$$
$$= 196 + 225 + 72$$
$$= 493 \text{ m}^2 \text{ s}^{-2}$$

Example 1.3

Calculate the velocity head, in metres of water, and the corresponding pressure, when water flowing along a pipe has a velocity of 2.50 m s^{-1} and a density of 1 000 kg m^{-3}. Assume the gravitational acceleration is 9.807 m s^{-2}.

$$\text{Velocity head} = v^2/2g = 2.50^2/(2 \times 9.807)$$
$$= 0.139 \text{ m of water}$$
$$\text{Pressure} = z\rho g = 0.139 \times 1\,000 \times 9.807$$
$$= 3\,120 \text{ N m}^{-2}$$

Problem 1.4

Gas pressure is used to transport 13.3×10^{-3} m^3 s^{-1} of a process liquid having a density of 1 250 kg m^{-3}, over an effective vertical distance of 14.0 m through a 0.075 m diameter pipeline from a large batch tank to a head tank operating at an absolute pressure of 100×10^3 N m^{-2}. Ignoring frictional effects and the liquid velocity in the batch tank, calculate the absolute gas pressure required when gravitation acceleration is 9.807 m s^{-2}.

(*Answer*: 277×10^3 N m^{-2})

Problem 1.5

A liquid having a density of 950 kg m^{-3} is conveyed up a converging pipe at a rate of 90.0×10^{-3} m^3 s^{-1}. The pipe is 0.150 m diameter at the lower end and 0.300 m diameter at the upper end; the two ends are separated by a vertical height of 3.00 m. When the pressure at the lower end is 345×10^3 N m^{-2}, calculate, assuming no friction losses, the pressure at the upper end.

(*Answer*: 329×10^3 N m^{-2})

Applications of the Energy Equation

I Discharge through a Small Orifice

An opening in the side of a vessel can be called a small orifice when the velocity through the orifice does not vary appreciably from top to bottom. By this definition an opening in the bottom of a vessel is a small orifice.

The general Bernoulli equation for the turbulent flow of a fluid between positions 1 and 2, in the terms of head is:

$$\left(z_2 + \frac{p_2}{\rho_2 g} + \frac{v_2^2}{2g} \right) - \left(z_1 + \frac{p_1}{\rho_1 g} + \frac{v_1^2}{2g} \right) = 0 \qquad (1.37)$$

For the case of an incompressible liquid of density ρ, flowing through an orifice, the change in potential head is essentially zero and Eqn 1.37 simplifies:

$$\frac{v_2^2 - v_1^2}{2g} = \frac{p_1 - p_2}{\rho g} \tag{1.38}$$

This equation relates the kinetic energy, actually kinetic head, of the liquid to the pressure loss.

Figure 1.14 illustrates the discharge of a liquid through a small orifice from a vessel, operating at atmospheric pressure, of such a size that the velocity, v_1, can be considered negligible. Substituting the values of velocity and pressure for positions 1 and 2 into Eqn 1.38 gives:

$$\frac{v_2^2 - 0}{2g} = \frac{p_{at} + h\rho g - p_{at}}{\rho g}$$

$$\text{or } \frac{v_2^2}{2g} = h$$

$$\text{or } v_2 = (2gh)^{0.5} \tag{1.39}$$

The velocity v_2 is the theoretical discharge velocity and will be designated v_t. The theoretical discharge rate, Q_t, is the product of the theoretical discharge velocity and the cross-sectional area of the orifice, A:

$$Q_t = A(2gh)^{0.5} \tag{1.40}$$

In fact, the actual discharge rate is less than the theoretical discharge rate for two reasons. The fluid streamlines converge at the orifice, forming a *vena contracta*, the area of which is less than the area of the orifice; the actual area of the *vena contracta* is C_cA, where C_c is the dimensionless coefficient of contraction, and is the ratio of the areas of the *vena contracta* and

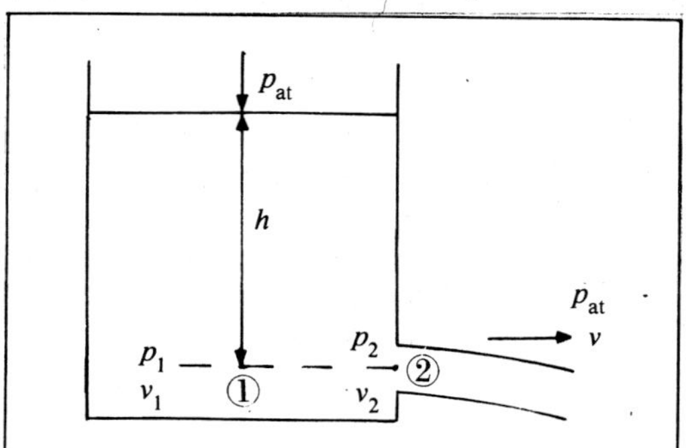

Figure 1.14 – Discharge of a liquid through a small orifice

the orifice. The liquid experiences frictional resistance as it flows through the orifice, the velocity is therefore reduced. The actual velocity of the discharge is:

$$C_v v_t = C_c(2gh)^{0.5} \tag{1.41}$$

where C_v is the dimensionless coefficient of velocity.

The actual discharge rate is:

$$Q = C_c C_v A(2gh)^{0.5} \tag{1.42}$$

The coefficents of contraction and velocity are generally combined in the dimensionless coefficient of discharge, C_d, and Eqn 1.42 written:

$$Q = C_d A(2gh)^{0.5} \tag{1.43}$$

Eqn 1.43 gives the theoretical flow rate in terms of the liquid head, h. If, however, when the fluid flows through the orifice there is a head loss, h_f, the actual net head available for this flow is $h - h_f$, and the actual flow rate is:

$$Q = A[2g(h - h_f)]^{0.5} \tag{1.44}$$

Combining this equation with Eqn 1.43 gives a relationship for the fluid head, the head loss and the coefficient of discharge:

$$h_f = h(1 - C_d^2) \tag{1.45}$$

The coefficient of discharge is usually determined experimentally, but it can be evaluated from values of C_c and C_v. The former can be measured, and the latter is determined indirectly from experimental data. Figure 1.15 gives approximate values for the coefficient of discharge for some typical orifices.

Example 1.4
A process liquid, density 1 050 kg m^{-3}, is discharged to the atmosphere through a 0.050 m orifice in the bottom of a large tank. Assuming the orifice coefficient of discharge to be 0.62, calculate the actual flow rate when the liquid level in the tank is 2.5 m and there is a gas pressure of 135×10^3 N m^{-2} above atmospheric pressure above the liquid. The gravitational acceleration may be taken as 9.807 m s^{-2}.

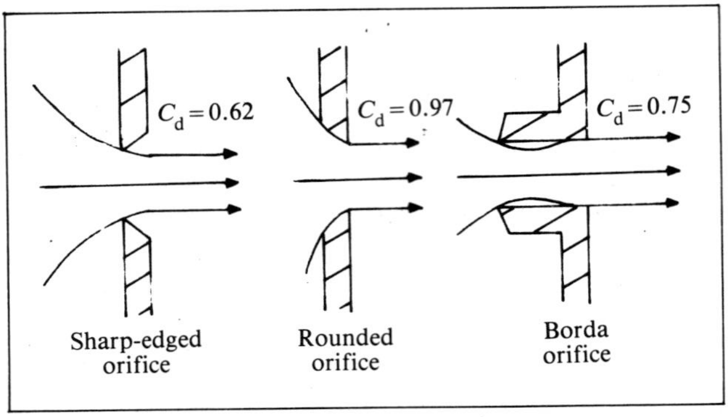

Figure 1.15 – Coefficients of discharge for typical orifices

$$\frac{v_t^2}{2g} = \frac{p_{at} + 135 \times 10^3 + h\rho g - p_{at}}{\rho g}$$

$$v_t = \left[\frac{2(135 \times 10^3 + h\rho g)}{\rho}\right]^{0.5}$$

$$= \left[\frac{2(135 \times 10^3 + 2.5 \times 1050 \times 9.807)}{1050}\right]^{0.5} = 17.5 \text{ m s}^{-1}$$

$$A = \pi d^2/4 = \pi(0.050)^2/4 = 1.96 \times 10^{-3} \text{ m}^2$$

$$Q = C_d A v_t = 0.62 \times 1.96 \times 10^{-3} \times 17.5 = 0.0213 \text{ m}^3 \text{ s}^{-1}$$

Problem 1.6

Water is discharged through a 0.040 m diameter orifice in the bottom of a tank. The discharge rate is 0.0144 m^3 s^{-1} when the water level above the orifice is 5.50 m and the gas pressure above the water is 115×10^3 N m^{-2} gauge. When the *vena contracta* diameter is 0.032 m, calculate the coefficients of contraction, discharge and velocity, and the head loss in metres of water. The density of water is 1 000 kg m^{-3} and the gravitational acceleration is 9.807 m s^{-2}.

(*Answers*: 0.64; 0.62; 0.97; 10.6 m water)

Problem 1.7

A large gas holder, storing gas of density 1.25 kg m^{-3}, at 2 450 N m^{-2} gauge pressure, develops a leak to atmosphere through a 0.010 m diameter hole. Calculate the leakage rate (kg s^{-1}) assuming the coefficient of discharge to be 0.62. The gravitational acceleration is 9.807 m s^{-2}.

(*Answer*: 3.81×10^{-3} kg s^{-1}

II Pumping

Pumps supply energy, ΔW, or head, Δh, to a fluid in motion to overcome frictional losses and to raise the fluid to a higher level. When a pump is installed in a pipeline between positions 1 and 2, the relationship for steady state flow is given by Eqn 1.35, where α_1 and α_2 depend on the pipe Reynolds number. In most industrial applications the fluid would be in turbulent flow, and hence α_1 and α_2 would be unity. This modified relationship is examined in detail in Section 1.8.

1.5 FLOW IN PIPES AND DUCTS

Figure 1.16 illustrates the development of boundary layer flow in a pipe which is submerged in liquid in a large tank. The pipe projects through the end of the tank and has a valve to change the mean volumetric velocity of liquid through the pipe as required. The steady state flow condition is preserved by maintaining a constant liquid level in the tank. The rounded entrance to the straight section of pipe ensures that boundary layer separation does not occur. A sharp entrance of a pipe, without a rounded inlet, could cause boundary layer separation and could also lead to the formation of a *vena contracta*.

When fluid flows through the pipe the layers of fluid adjacent to the walls are decelerated and a boundary layer, similar to that described in Section 1.3, begins to form at the entrance of the pipe. As the fluid flows along the pipe the boundary layers gradually increase in thickness until at some distance downstream from the pipe entrance they meet at the axis of the pipe. At this point the velocity distribution in the pipe has achieved its final form and it will remain unchanged until it encounters a disturbance such as the outlet valve. Downstream of the point where the velocity distribution has reached its final form, the fluid is said to be in 'fully developed flow'.

Up to that point the boundary layers occupy a gradually increasing portion of the cross-section of the pipe; the total fluid stream consists of a core of fluid flowing in a piston-like manner at constant velocity and an annular boundary layer zone between the core and the pipe walls . When the

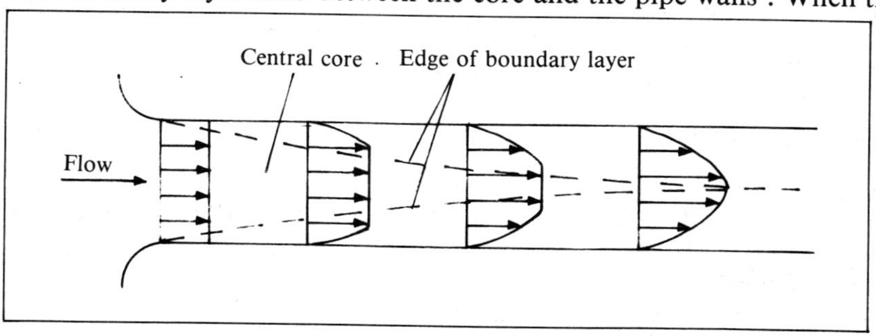

Fig 1.16 – Development of boundary-layer flow in pipe

boundary layers converge, the central core disappears. In Figure 1.16 the length of the solid-headed arrows is proportional to the velocity.

If the flow rate of liquid through the pipe is such that the boundary layers are entirely laminar, then when they meet at the pipe axis, laminar flow will persist in the fully developed flow region. If the flow rate is such that transition to turbulent flow has already occurred before the boundary layers meet, then turbulent flow will persist in the downstream region. As is the case for a flat plate, the turbulent boundary layer has a laminar sublayer adjacent to the solid surface and a transitional region between the laminar and turbulent zones. The velocity in the laminar sublayer is very much less than the mean volumetric velocity and this has important consequences in convective-heat and mass transfer. The laminar sublayer and indeed the transitional zone can be affected by pipe roughness, if it projects into the flow regions.

Reynolds Number for Pipe Flow

It has already been mentioned in Section 1.3 (Flow Patterns) that Osborne Reynolds in 1883 demonstrated that the type of flow which occurs in a pipe depends on four factors. These four parameters define the Reynolds group, Re:

$$\mathrm{Re} = \frac{dv\rho}{\mu} \qquad (1.46)$$

where d = pipe diameter, m
v = fluid velocity, m s^{-1}
ρ = fluid density, kg m^{-3}
μ = fluid viscosity, N s m^{-2} or kg m^{-1} s^{-1}.

The Reynolds group may also be defined:

$$\mathrm{Re} = \frac{dG}{\mu} \qquad (1.47)$$

or $$\mathrm{Re} = \frac{dv}{\nu} \qquad (1.48)$$

where G = mass velocity, ρv, kg m^{-2} s^{-1}
ν = kinematic viscosity, μ/ρ, m^2 s^{-1}.

The Reynolds group can be regarded as a measure of the relative magnitude of the inertial force, ρv^2, to the viscous force or shear stress, $\mu . v/d$, from Newton's law of viscosity, Eqn 1.17. The Reynolds group may now be written:

$$\mathrm{Re} = \frac{\rho v^2}{\mu v/d} = \frac{dv\rho}{\mu} \qquad (1.49)$$

The larger the Reynolds number, the greater will be the relative contribution to inertial forces leading to turbulence. The smaller the Reynolds number the greater will be the relative magnitude of viscous forces.

In general, for pipe flow, when the Reynolds number is less than 2 100 the flow is streamline, viscous or laminar, between 2 100 and 4 000 the flow is likely to be transitional, and above 4 000 the flow is likely to be turbulent. Since Reynolds' original work, it has been shown that both the 2 100 and 4 000 limits are affected by factors such as the type of pipe entry and exit, initial disturbances, vibration and surface roughness. It is possible, under rather special conditions, to have laminar flow maintained at Reynolds numbers up to about 10 000 for distances up to 100 diameters downstream of the pipe entrance. For this reason, it is usual to arrange for the pipe Reynolds number to exceed about 10 000 if turbulent flow is required.

In coiled pipes, a secondary circulation of fluid, the Dean effect, occurs in a plane at right angles to the main flow. This circulation stabilizes laminar flow, thus increasing the critical Reynolds number:

$$\mathrm{Re}_{\mathrm{critical}} = 20 \times 10^3 \left(\frac{d_\mathrm{i}}{d_\mathrm{c}} \right)^{0.32} \tag{1.50}$$

where d_i = pipe internal diameter, m
d_c = mean coil diameter, m.

If, for example, the ratio of the pipe diameter to coil diameter were 1:10, the critical Reynolds number, below which laminar flow exists, becomes about 9 600 instead of 2 100.

It is not possible to maintain transitional or turbulent flow below a Reynolds number of about 2 100, and hence this value is generally regarded as an absolute limit below which, for pipe flow, laminar flow will exist.

In Figure 1.16 the length of pipe required to establish fully developed flow is called the entry length, L_e. An approximate experimental correlation for the entry length is:

$$\frac{L_\mathrm{e}}{d} = 0.057\ \mathrm{Re} \tag{1.51}$$

where L_e = entry length, m
d = pipe diameter, m.

For a pipe Reynolds number of 2 100 and a diameter of 0.050 m, the entry length from Eqn 1.51 is 6.0 m. The entry length for water flowing at a typical Reynolds number of 0.1×10^6 in a 0.025 m diameter pipe is 142 m.

Fortunately, in most industrial equipment, the pipe entry is not smooth, and in any case turbulence is caused by many factors, such as bends, valves, and pumps. The entry length is then almost independent of the Reynolds number and is about 25 to 40 pipe diameter; for a 0.100 m diameter pipeline this would be 2.5 to 4.0 m. This length is often short when compared with the pipe length having fully developed flow and, in engineering calculations, it is usually assumed that there is no entry length.

For non-circular pipes or ducts the characteristic linear dimension, the equivalent diameter, d_e, replaces the pipe diameter, d, in the Reynolds group. This equivalent diameter is equal to 4 r_H, where r_H, is the hydraulic mean radius and is defined:

$$r_H = \frac{\text{cross-sectional area filled with fluid}}{\text{length of wetted perimeter}}$$

The wetted perimeter equals the actual perimeter only when pipes run full, i.e. always for gases.

Table 1.2 gives the equivalent diameter to be used in the Reynolds group for a number of common cross-sections running full of fluid.

TABLE 1.2

Section	r_H	$d_e = 4r_H$
Square duct, side a	$\frac{a^2}{4a}$	a
Rectangular duct, sides a and b	$\frac{ab}{2(a+b)}$	$\frac{2ab}{(a+b)}$
Annulus, diameters D and d	$\frac{\pi(D^2-d^2)}{4\pi(D+d)}$	$(D-d)$

If the core of the annulus is not concentric with the larger pipe substantial deviations from the equivalent diameter shown in Table 1.2 may occur.

Example 1.5
Calculate the Reynolds number when water, having a density of 1 000 kg m^{-3} and a viscosity of 1.01×10^{-3} N s m^{-2}, flows through a 0.075 m diameter pipeline at a velocity of 1.50 m s^{-1}.

$$\text{Re} = \frac{dv\rho}{\mu}$$

$$= \frac{0.075 \text{ m} \times 1.50 \text{ m s}^{-1} \times 1\,000 \text{ kg m}^{-3}}{1.01 \times 10^{-3} \text{N s m}^{-2}}$$

$$= 111 \times 10^3$$

Problem 1.8
Calculate the Reynolds number when steam, having a viscosity of 13.2×10^{-6} N s m^{-2} and a specific volume of 0.669 m^3 kg^{-1}, flows through a 0.150 m diameter pipeline with a velocity of 30.0 m s^{-1}.

(*Answer*: 0.510×10^6)

Problem 1.9
Calculate the Reynolds number when oil, having a kinematic viscosity of 10.5×10^{-6} m^2 s^{-1} and a specific gravity of 0.850, flows at 10.0 kg s^{-1} in a 0.100 m diameter pipeline.

(*Answer*: 14 300)

Problem 1.10
Water, having a viscosity of 1.14×10^{-3} N s m^{-2} and a density of 1 000 kg m^{-3}, flows through a 0.100 m diameter pipeline. Calculate the velocity corresponding to the Reynolds numbers:

(a) 2 100

(b) 4 000

(c) 10 000

(*Answers*: (a) 0.0239 m s^{-1}; (b) 0.0456 m s^{-1}; (c) 0.114 m s^{-1})

Problem 1.11
Calculate the Reynolds number when oil, having a kinematic viscosity of 12.5×10^{-6} m^2 s^{-1}, flows at 1.50 m s^{-1} through:

(a) A rectangular duct, sides 0.100 m and 0.150 m

(b) A concentric annulus, diameters 0.200 m and 0.160 m.

(*Answers*: (a) 14 400; (b) 4 800)

Velocity Distribution in Pipes
The velocity, v, used in the equations for the pipe Reynolds number is the mean or average volumetric velocity, obtained by dividing the volumetric flow rate, Q, by the cross-sectional area of the pipe, A. This mean velocity is not the actual velocity at a particular point, the point velocity. The fluid velocity at the pipe wall is zero, it increases to a maximum at the centre line of a circular pipe. The velocity at any point is a function of the space coordinates at the point.

Figure 1.17 illustrates typical velocity profiles for fully developed laminar and turbulent flow.

Figure 1.18 illustrates the relationship between the ratio of the average velocity, v, to the maximum velocity, v_{max}, at the centre line of the pipe in terms of the pipe Reynolds number. It will be observed that for laminar flow, Reynolds number less than 2 100, the ratio, v/v_{max}, is independent of the Reynolds number and equal to 0.5. In other words, the average velocity is exactly half the maximum velocity at the axis of a circular pipe. The relationship between the point velocity, v_x, at a radial distance y from the centre line of a circular pipe of radius r, and the average volumetric velocity, v, is:

$$v_x = 2v\left[1 - \left(\frac{y}{r}\right)^2\right] \tag{1.52}$$

which is the equation of a parabola.

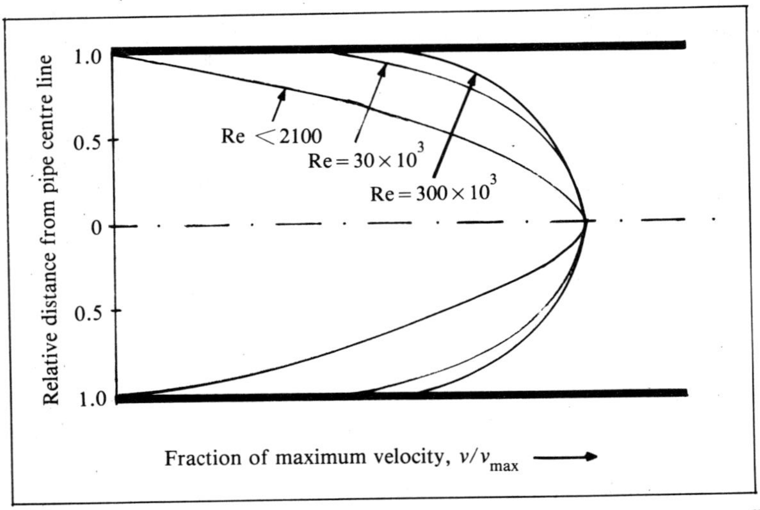

Figure 1.17 – Velocity distribution across a pipe for fully developed flow

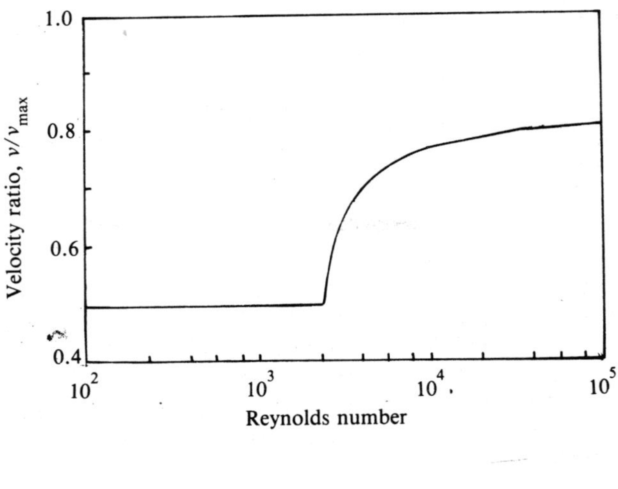

At the pipe axis $y=0$ and v_x has its maximum value; Eqn. 1.52 becomes:

$$v_x = v_{max} = 2v \tag{1.53}$$

For turbulent flow, the velocity distribution in Figure 1.17 is more flat than for laminar flow in the region of the pipe axis, and the difference between the average velocity and the maximum velocity is considerably less than for laminar flow. For turbulent flow in a circular pipe, the ratio v/v_{max} is usually taken to be 0.8, compared with 0.5 for laminar flow. The actual relationship of v/v_{max} to the Reynolds number is illustrated in Figure 1.18.

Considerable study, both experimental and theoretical, has been directed to determining the velocity profile in turbulent flow. Although the results from theoretical calculation agree reasonably with experimental data, the problem has not been successfully solved. However, the work on boundary layer analysis and pipe friction has resulted in a number of equations for the velocity distribution in turbulent flow.

The work of Blasius can be approximated to give a relationship for the turbulent velocity profile:

$$\frac{v_x}{v_{max}} = \left(1 - \frac{y}{r}\right)^n \tag{1.54}$$

where y = distance from centre line of pipe, m
r = pipe radius, m
n = a dimensionless index whose numerical value depends on the Reynolds number, it varies from 1/6 at a Reynolds number of 4 000 to 1/10 at a Reynolds number of 3×10^6.

In engineering, a very commonly used form of Eqn 1.54 is the seventh-root equation, where n has the value 1/7. In spite of the approximations involved in arriving at this equation, it fits the experimentally determined velocity distribution data with reasonable accuracy.

The ratio v/v_{max} is the dimensionless velocity distibution factor , introduced in Section 1.4. The limiting condition of $\alpha = 1$ is never actually achieved, although it can be approached by fast-moving gases in plug flow. It is usual to assume that α is unity in turbulent flow calculations, and the velocity head in the Bernoulli equation is then $v^2/2g$.

Resistance to Flow in Pipes

For real fluids the Bernoulli equation (1.33) includes the term h_f to represent the frictional head loss in the system. The calculation or determination of this head, for any fluid, from its condition of flow and physical properties is a very important aspect of fluid flow problems. As the phenomena associated with laminar and turbulent flow are different, it is to be expected that different laws of frictional resistance apply to the two types of flow; this indeed is the case.

A considerable number of people have studied the relationship between the pressure drop or head loss due to friction for fluids of different density and viscosity flowing in pipes of different diameters and surface roughness. When these results for long straight pipes are plotted, on a log–log basis, in a dimensionless form, $2\tau_w/\rho v^2$ or f against the pipe Reynolds number, a series of interesting curves, similar to those in Figure 1.19, is produced.

It will be observed in Figure 1.19 that, for a particular surface roughness, a single curve expresses the results for all fluids, pipe diameters and fluid velocities. This curve has three parts for smooth and relatively smooth pipes and four parts for rough pipes.

(a) For Reynolds numbers less than 2 100, corresponding to laminar flow, a straight line represents all data, irrespective of surface roughness. The friction factor decreases with increase in the Reynolds number.

(b) For Reynolds numbers in the range 2 100 to 4 000, corresponding to transitional flow, reproducible values of the friction factor, f, cannot be obtained; however, the value of the friction factor is greater than it would be if the straight line for laminar flow were extended. The friction factor increases with increase in the Reynolds number.

(c) For Reynolds numbers greater than 4 000 the friction factor is a function of both the Reynolds number and the dimensionless relative roughness term. The relative roughness, e/d, is defined as the size of

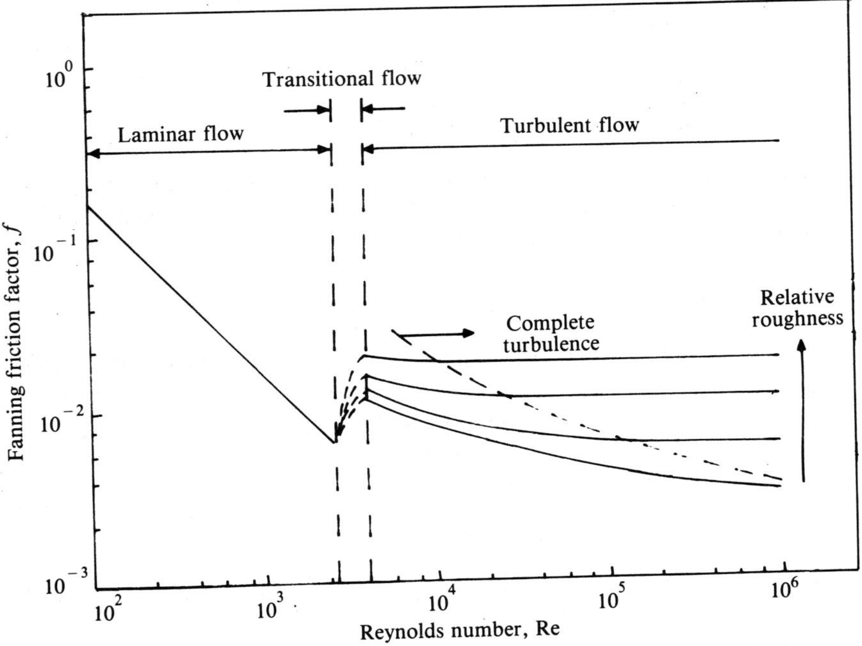

Figure 1.19 – Fanning friction factor for long straight pipes

surface imperfections or equivalent roughness, e, in metres, divided by the pipe diameter, d, in metres. For a particular Reynolds number, the greater the relative roughness, the greater is the friction factor.

(d) For each pipe, except those which are very smooth, the friction factor plot has a region where the friction factor is independent of the Reynolds number. This can be regarded as a region of complete turbulence where the surface imperfections project into and disturb the laminar sublayer, and perhaps the transitional zone, and increase the resistance to flow. The lowest Reynolds number at which this occurs depends on the relative roughness. The complete turbulence curve is included in Figure 1.19.

Friction Factor

It should be noted that there are two pipe friction factors commonly used in the literature, both being represented by the factor f. Unfortunately, one is four times the magnitude of the other. Care, therefore, should be taken to determine which one is being used. In this chapter the symbol f will always refer to the Fanning friction factor (f_F), which is four times less than the Blasius friction factor (f_B).

The Fanning friction factor, f, may be defined as the ratio of the wall shear stress, τ_w, to the product of kinetic energy, $v^2/2$, and density, ρ:

$$f = \frac{\tau_w}{\frac{\rho v^2}{2}} = \frac{2\tau_w}{\rho v^2} = 2j_f \quad (1.55)$$

where j_f is the basic friction factor.

In the laminar region of the friction factor plot, the slope of the straight line is -1, and the intercept on the f-axis is 16, when the Reynolds number is unity. The equation of the line is therefore:

$$f = 16\,\mathrm{Re}^{-1} \quad (1.56)$$

In the turbulent regime, formulae have been developed to fit specific regions of the smooth pipe curve. For Reynolds numbers in the range 4×10^3–100×10^3, Blasius proposed the equation:

$$f = 0.079\,\mathrm{Re}^{-0.25} \quad (1.57)$$

For Reynolds numbers in the range 4×10^3–3×10^6, Koo proposed the equation:

$$f = 0.00140 + 0.125\,\mathrm{Re}^{-0.32} \quad (1.58)$$

For reasons which will become clear later, when the relationships between momentum, heat and mass transfer are discussed, a very useful equation for Reynolds numbers in the range 5×10^3–0.2×10^6 is:

$$f=0.046\ \mathrm{Re}^{-0.20} \tag{1.59}$$

The main problem in the investigation of friction losses in rough pipes is to establish a reliable index of roughness which can be used to analyse the experimental data. A number of people have used smooth surfaces artificially roughened with a coating of uniformly sized grains of sand of known size. For Reynolds numbers greater than 10×10^3, Nikuradse proposed the equation:

$$f^{-0.5}=4.0\log\left(\frac{d}{e}\right)+2.16 \tag{1.60}$$

A commercial pipe does not have a single scale of roughness; it will have a spectrum of roughness. Therefore, the data from experimental work using artificially roughened pipes are not precisely correct for commercial pipes. Cokebrook evaluated the friction factors from a series of tests on untreated commercial pipes and obtained a relationship which can be used for all pipes. This relationship is rather tedious to solve manually and it is usually presented in the form of a graph, similar to Figure 1.19. Values for the equivalent roughness, e, for some commercial pipes are given in Table 1.3.

TABLE 1.3

Type of commercial pipe	Size of surface imperfections or equivalent roughness, e (m)
Drawn tubing	1.5×10^{-6}
Steel or wrought iron	45×10^{-6}
Asphalted cast iron	120×10^{-6}
Galvanized iron	150×10^{-6}
Cast iron	260×10^{-6}
Concrete	300×10^{-6}–$3\,000\times10^{-6}$

Energy Loss in a Straight Pipe

The relationship between pressure loss, Δp, and wall shear, τ_w, for fully developed flow of an incompressible fluid is:

$$\Delta p = \frac{4\tau_w L}{d} \tag{1.61}$$

Substituting for τ_w from Eqn 1.55 gives:

$$\Delta p = 4f \cdot \frac{L}{d} \cdot \frac{\rho v^2}{2} \tag{1.62}$$

This is the Fanning or Darcy equation where f is the Fanning friction factor. In terms of head loss due to friction in the Bernoulli equation, Eqn 1.62 becomes:

$$h_f = \frac{\Delta p}{\rho g} = 4f \cdot \frac{L}{d} \cdot \frac{v^2}{2g} \tag{1.63}$$

It is frequently convenient to have the pressure drop in terms of the volumetric flow rate, Q, rather than in terms of fluid velocity, v. Substituting in the Fanning equation for $v = 4Q/\pi d^2$ gives:

$$\Delta p = 4f \cdot \frac{L}{d} \cdot \frac{\rho}{2} \left(\frac{16Q^2}{\pi^2 d^4} \right) \tag{1.64}$$

$$= \frac{32 f L \rho Q^2}{\pi^2 d^5} \tag{1.65}$$

$$\text{or } h_f = \frac{32 f L Q^2}{\pi^2 d^5 g} \tag{1.66}$$

To evaluate the pressure drop or head loss along a straight pipe for a fluid in fully developed turbulent flow, one would calculate the pipe Reynolds number and then determine the friction factor from a friction factor chart, or calculate it from one of the many correlations available. The pressure drop, or head loss, is then calculated by substituting values into Eqn 1.62 or 1.63.

It will be observed that, in the Fanning equation, the pressure drop is proportional to the friction factor and the square of the fluid velocity. The friction factor can be regarded as being proportional to the velocity:

$$f = 0.079 \left(\frac{dv\rho}{\mu} \right)^{-0.25} \tag{1.57}$$

$$\text{or } f = 0.046 \left(\frac{dv\rho}{\mu} \right)^{-0.20} \tag{1.59}$$

The pressure drop can then be regarded as being proportional to the velocity:

$$\Delta p \propto v^n \tag{1.67}$$

where n = dimensionless index in the general range 1.75 to 1.80.

For Reynolds numbers less than 2 100, streamline flow exists in the pipe. In this situation the value of the friction factor is 16 Re^{-1} and the Fanning equation becomes:

$$\Delta p = 4 \cdot \frac{16\mu}{dv\,\rho} \cdot \frac{L}{d} \cdot \frac{\rho v^2}{2} = \frac{32\mu\, Lv}{d^2} \qquad (1.68)$$

This is the Poiseuille equation, which can be written in terms of head loss:

$$h_f = \frac{\Delta p}{\rho g} = \frac{32\mu\, Lv}{d^2\, \rho g} \qquad (1.69)$$

It will be observed that for laminar flow the pressure loss is directly proportional to the fluid viscosity and fluid velocity. Written in terms of the volumetric flow rate, the Poiseuille equation is:

$$\Delta p = \frac{128\mu\, LQ}{\pi d^4} \qquad (1.70)$$

In practice, additional pressure losses may occur due to (a) sudden contractions or enlargements and (b) pipe fittings, such as bends, valves and metering equipment. Additional pressure losses due to non-isothermal flow may occur. These additional losses are considered below.

Example 1.6

Calculate the frictional pressure loss per 100 m of 0.150 m diameter pipeline when oil, having a density of 900 kg m^{-3} and a viscosity of 19.2×10^{-3} N s m^{-2}, flows at a rate of 25.0 kg s^{-1}. Assume a Fanning friction factor of 0.0080.

$$\text{Flow rate, } Q = \frac{25.0 \text{ kg s}^{-1}}{900 \text{ kg m}^{-3}} = 0.0278 \text{ m}^3 \text{ s}^{-1}$$

$$\text{Fluid velocity, } v = \frac{4Q}{\pi d^2} = \frac{4 \times 0.0278 \text{ m}^3 \text{ s}^{-1}}{\pi (0.150 \text{ m})^2} = 1.57 \text{ m s}^{-1}$$

$$\text{Reynolds number, Re} = \frac{dv\rho}{\mu} = \frac{0.150 \text{ m} \times 1.57 \text{ m s}^{-1} \times 900 \text{ kg m}^{-3}}{19.2 \times 10^{-3} \text{ N s m}^{-2}}$$

$$= 11\ 000$$

As the oil is in turbulent flow, the pressure loss is given by the Fanning equation:

$$\Delta p = \frac{4 \times 0.0080 \times 100 \text{ m} \times 900 \text{ kg m}^{-3} (1.57 \text{ m s}^{-1})^2}{0.150 \text{ m} \times 2}$$

$$= 23.7 \times 10^3 \text{ N m}^{-2} \text{ (or kg m}^{-1} \text{ s}^{-2})$$

Problem 1.12

An oil, having a density of 950 kg m^{-3} and a viscosity of 0.105 N s m^{-2}, flows through a 0.100 m diameter, horizontal pipe at a velocity of 1.05 m s^{-1}. Calculate the pressure loss per 100 m length of pipeline.

(*Answer*: 35.3×10^3 N m^{-2})

Problem 1.13

Oil, having a kinematic viscosity of 15.45×10^{-6} m^2 s^{-1} and a specific gravity of 0.85 is pumped through a 0.150 m diameter pipeline at a rate of 15.5 kg s^{-1}. Calculate the frictional head loss, h_f per 100 m length of pipeline when the Fanning friction factor is given: $f=0.079\ \mathrm{Re}^{-0.25}$.

(*Answer*: 1.14 m of oil)

Problem 1.14

A liquid, having a viscosity of 0.0105 N s m^{-2} and a density of 1 200 kg m^{-3} flows at a rate of 10.2×10^{-3} m^3 s^{-1} through a commercial steel pipe, having a diameter, *d*, of 0.050 m and a pipe roughness, *e*, of 100×10^{-6} m. Using the Nikuradse relationship, Eqn 1.60, calculate the frictional pressure drop per 100 m of pipeline.

(*Answer*: 770×10^3 N m^{-2})

Energy Losses due to Suddenly Changing Cross-sections and Pipe Fittings

I Suddenly Changing Cross-sections

The pressure loss that occurs at a sudden enlargement, such as an open-ended pipe or a connection to a vessel, is due to the loss of kinetic energy associated with the change in fluid viscosity. In Figure 1.20(a) the fluid passes from a pipe of cross-section A_1 into a pipe of larger cross-section, A_2. For continuity reasons, the fluid velocity decreases from v_1 to v_2. The fluid cannot follow the suddenly changing shape, the boundary layer separates, and eddies are trapped at the 'corners' of the larger diameter pipe. These eddies consume a lot of energy from the main fluid stream.

The pressure loss for a sudden enlargement is:

$$\Delta p = \frac{\rho(v_1 - v_2)^2}{2} \tag{1.71}$$

In the special case, where a pipe conveying fluid, velocity v, discharges into a large reservoir through a sharp exit, conditions are equivalent to a sudden enlargement; the fluid velocity in the large reservoir may be taken as zero and the pressure loss is:

$$\Delta p = \frac{\rho v}{2} \tag{1.72}$$

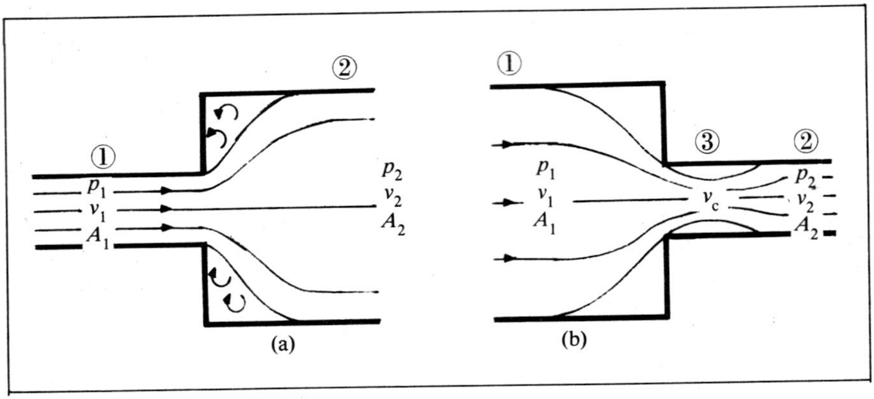

Figure 1.20 – (a) Sudden enlargement
(b) Sudden contraction

The loss in kinetic energy due to the sudden enlargement is frequently referred to as the 'exit loss'.

The pressure loss which occurs with a sudden contraction, such as the connection from a vessel to a pipe, is due to the additional turbulence in the form of eddies, produced in the initial length of the smaller-diameter pipe. This pressure loss is frequently referred to as the 'entrance loss'.

The sudden contraction situation is complicated by the formation of a *vena contracta* – position 3 in Figure 1.20(b). The pressure loss for the sudden contraction in terms of the fluid velocity at the *vena contracta*, v_c, and the downstream velocity, v_2, is:

$$\Delta p = \frac{\rho(v_c - v_2)^2}{2} \tag{1.73}$$

For continuity reasons the flow rate, Q, velocity, v, and cross-sectional area, A, are related:

$$Q = v_2 A_2 = v_c A_c \tag{1.74}$$

The coefficient of contraction, C_c, is defined as A_c/A_2, and the velocity at the *vena contracta*, v_c, is therefore:

$$v_c = v_2 \cdot \frac{A_2}{A_c} = \frac{v_2}{C_c} \tag{1.75}$$

Substituting for v_c in Eqn 1.73 gives:

$$\Delta p = \frac{\rho\left(\frac{v_2}{C_c} - v_2\right)^2}{2} \tag{1.76}$$

$$= \frac{\rho v_2^2 \left(\frac{1}{C_c} - 1 \right)^2}{2} \quad (1.77)$$

$$= K_c \cdot \frac{\rho v_2^2}{2} \quad (1.78)$$

where K_c is the dimensionless contraction loss coefficient.

For laminar flow, K_c is less than 0.1 and the entrance loss may be neglected. For turbulent flow, K_c may be obtained either from experimentally determined data in the form of tables or from correlations, typified by:

$$K_c = 0.5 \left[1 - \frac{A_2}{A_1} \right] \quad (1.79)$$

where $A_1$and A_2 are the upstream and downstream cross-sectional areas respectively.

If specific information is not available to calculate K_c (actually C_c in Eqn 1.76), the value of K_c is usually taken to be 0.5.

II Pipe Fittings

When a fluid flows round bends or through fittings of varying cross-sectional area, the normal flow lines are disturbed and energy losses occur which are not recovered. For laminar flow, the pressure loss round bends and fittings is small and is usually neglected. For turbulent flow, as might be expected, losses have been found to be proportional to the velocity head of the flowing fluid, $v^2/2g$. The loss is difficult to express mathematically and the common approach is to regard the bend(s) or fitting(s) as a number of velocity heads, $nv^2/2g$, or as an equivalent length of pipe, L', which is then added to the actual length of straight pipe for the calculation of pressure loss or head loss, using the modified Fanning equation:

$$\Delta p = 4f \cdot \frac{L + L'}{d} \cdot \frac{\rho v^2}{2} \quad (1.80)$$

$$\text{or } h_f = 4f \cdot \frac{L + L'}{d} \cdot \frac{v^2}{2g} \quad (1.81)$$

where L' is the equivalent length of straight pipe and equals the sum of the equivalent pipe diameters multiplied by the pipe diameter:

$$L' = d \Sigma N \quad (1.82)$$

Table 1.4 lists some average values of the number of equivalent diameters for bends and pipe fittings.

TABLE 1.4
Equivalent Diameters for Bends and Pipe Fittings

Fitting	Pipe diameters, N
45 degree bend	15
90 degree standard bend	30
90 degree long radius bend	20
180 degree close radius bend	70
Entry from branch of T-piece	60
Entry to branch of T-piece	90
Unions and fittings	negligible
Globe valve – fully open	300
Gate valve – fully open	7.5
– three-quarters open	50
– half open	200
– one-quarter open	800

The experimental values in Table 1.4 are based on a single fitting between two lengths of straight pipe. If two or more fittings are situated, such that normal flow is not re-established between them, then the values given will be too low and the pressure loss is likely to be underestimated.

III Coils
When fluid flows through a coil, the secondary circulation mentioned above (Velocity Distribution in Pipes) causes a pressure loss in excess of that which would occur for the same length of straight pipe. A number of equations are available to calculate the pressure loss when a fluid flows through a helical coil. For turbulent flow, in the Reynolds number range 15×10^3–100×10^3, one such equation, in terms of the Fanning friction factor, is:

$$f = 0.079\ \mathrm{Re}^{-0.25} + 0.012 \left(\frac{d_i}{d_c}\right)^{0.5} \tag{1.83}$$

where Re = Reynolds number based on the pipe internal diameter
d_i = pipe internal diameter, m
d_c = mean coil diameter, m.

Eqn 1.83 could be regarded as a modified Blasius equation.

Example 1.7
A horizontal piping system, which connects two large-diameter vessels, has

the components:
(a) 100 m of 0.150 m straight pipe
(b) Four 90° long radius bends
(c) One fully opened globe valve
(d) One half-opened gate valve.

When a fluid, having a density of 1 150 kg m^{-3}, flows with a velocity of 2.00 m s^{-1} in the straight pipe, calculate the total pressure loss for the system. Assume a friction factor of 0.0080, and ignore the entrance and exit losses.

$$L' = d\,\Sigma N = 0.150\ (4 \times 20 + 300 + 200) = 87\ \text{m}$$

$$\Delta p = 4f \cdot \frac{L + L'}{d} \cdot \frac{\rho v^2}{2}$$

$$= \frac{4 \times 0.008\ (100 + 87)\ 1150 \times 2.00^2}{0.150 \times 2}$$

$$= 91.8 \times 10^3\ \text{N m}^{-2}$$

Energy Loss Due to Non-isothermal Flow

When a fluid is being heated or cooled, the temperature gradient will cause a corresponding change in the physical properties along the direction of the temperature gradient. For a liquid, viscosity is the physical property which is most sensitive to temperature change and therefore appreciable viscosity changes may occur as a result of the temperature gradient.

Since the friction factor is proportional to the Reynolds number, and therefore the fluid viscosity, as the temperature increases the viscosity of the liquid will decrease, causing a corresponding decrease in the Reynolds number. The friction factor will increase if the liquid is being cooled.

Sieder and Tate suggested the use of a modified friction factor, f/ψ, instead of the simple friction factor, f, for non-isothermal flow. For laminar flow they proposed the correlation:

$$\psi = 1.1 \left(\frac{\mu_b}{\mu_w} \right)^{0.25} \qquad (1.84)$$

and for the turbulent flow:

$$\psi = 1.0 \left(\frac{\mu_b}{\mu_w} \right)^{0.14} \qquad (1.85)$$

where μ_b = fluid viscosity, evaluated at the arithmetic mean bulk temperature

μ_w = fluid viscosity, evaluated at the arithmetic mean wall temperature.

If water flows in turbulent flow through a pipe such that its arithmetic mean bulk temperature is 40°C, and the arithmetic mean wall temperature is 100°C, then $\mu_b = 0.651 \times 10^{-3}$ N s m^{-2} and $\mu_w = 0.279 \times 10^{-3}$ N s m^{-2}. The value of the Sieder–Tate correction factor, ψ, is:

$$\psi = 1.0 \left(\frac{0.651 \times 10^{-3}}{0.279 \times 10^{-3}} \right)^{0.14} = 1.13$$

This is equivalent to an error of 13% if the temperature effects are ignored.

Economical Pipe Diameter

The investment for piping and pipe fittings can be a substantial part of the total investment of a process plant. It can be as high as 80% of the installed equipment cost or 20% of the fixed capital investment. It is therefore necessary to choose pipe diameters which optimize the pumping and capital charges. Nomograms giving the estimated cost and installation time for pipes and piping systems are given in the literature.

The labour costs involved in piping consist of cutting, fitting, welding or threading and assembly on site. The ratio of labour costs to material costs depends on the material of construction.

The economics of a pipe system hinges on a balance between capital and operating costs. For a given flow rate, and assuming that the velocity is not important, an increase in pipe diameter reduces pressure losses and hence pumping costs, but increases the capital cost. The most economical pipe diameter may be obtained by optimizing the capital and operating costs; methods for doing this are available in the literature.

Many short-cut methods have been proposed for estimating the optimum pipe diameter; some general empirical relationships are shown in Tables 1.5 and 1.6.

TABLE 1.5

Recommended Fluid Velocities in Pipes

Fluid (turbulent flow)	Velocity range (m s^{-1})
Water or fluid of similar density and viscosity	1–3
Low-pressure steam, up to 300×10^3 N m^{-2}	15–30
High-pressure steam, over 700×10^3 N m^{-2}	30–60
Compressed air, up to 500×10^3 N m^{-2}	15–30

TABLE 1.6

Recommended Fluid Velocities for Viscous Liquids

Pipe diameter (m)	Recommended velocities ($m\ s^{-1}$) for viscosities (in $N\ s\ m^{-2}$) of 0.05	0.10	1.00
0.025	0.5–0.9	0.3–0.6	0.1–0.2
0.050	0.8–1.1	0.5–0.8	0.2–0.3
0.100	1.1–1.5	0.8–1.1	0.3–0.4

1.6 FLOW MEASUREMENT

It is frequently necessary, for the purposes of plant operation, process control and standard costing, to use a flowmeter to measure the flow rate of a fluid in a pipe, duct or open channel.

Flowmeters have a primary element which detects some property or change in property of the fluid and a secondary element which indicates or records this property or change in property.

There are essentially two types of flowmeters: quantity and inferential. The quantity or displacement flowmeters measure fluid flow rate directly, in terms of quantity per unit time, by the cyclic displacement of one or more mechanical elements which have a reciprocating or rotary motion. Inferential flowmeters do not measure fluid flow rate directly; they measure fluid velocity and from this infer the flow rate. The inferential flow meters most commonly used in process plant are:

(a) Pitot tube
(b) Orifice meter
(c) Venturi meter
(d) Nozzle meter
(e) Weirs
(f) Rotameters.

Hydrodynamic or Head Meters

The primary element of a head meter has a constriction placed in the fluid flow path. This constriction increases the velocity and hence the kinetic energy of the fluid and produces a corresponding pressure or head loss. The pressure difference is measured, and then indicated or recorded, by the secondary element which may be a manometer, differential pressure (DP) cell or pressure transducer. The Bernoulli equation is used to calculate the flow rate from the pressure difference across the primary element.

Having selected the appropriate type and size of hydrodynamic flow meter in a suitable material of construction, its successful operation

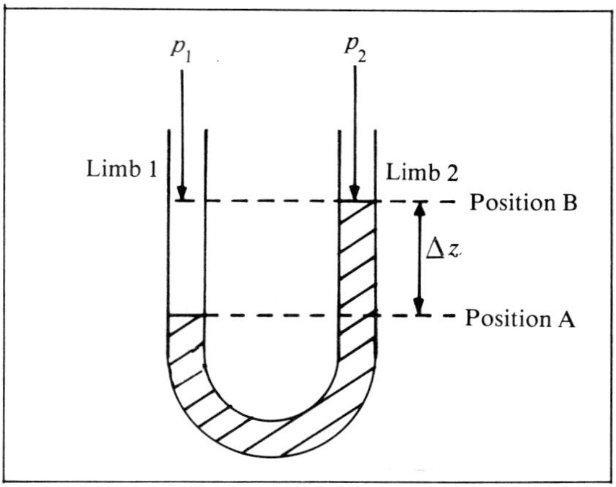

Figure 1.21 – U-tube manometer

depends on routine maintenance and accuracy in measuring the differential pressure which is produced across the primary element. The simplest, easy-to-use, pressure-measuring device is the U-tube manometer.

Figure 1.21 illustrates a simple U-tube manometer consisting of a transparent tube, containing the fluid whose pressure is to be measured, and a manometric fluid. Limb 1 is connected to the higher pressure tapping point and limb 2 to the lower pressure tapping point. The manometric fluid is different to, immiscible with and generally more dense than the flowing

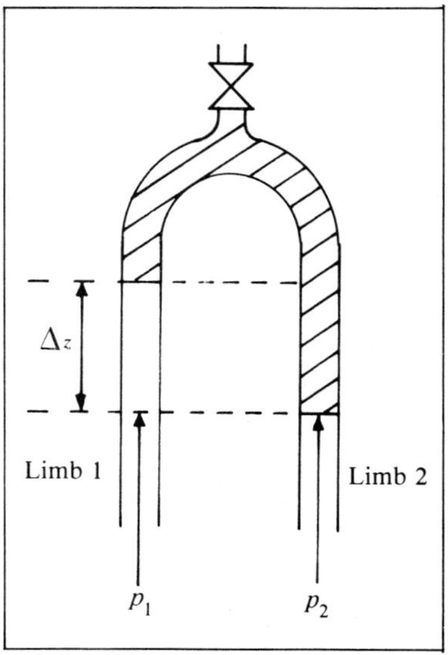

Figure 1.22 – Inverted U-tube manometer

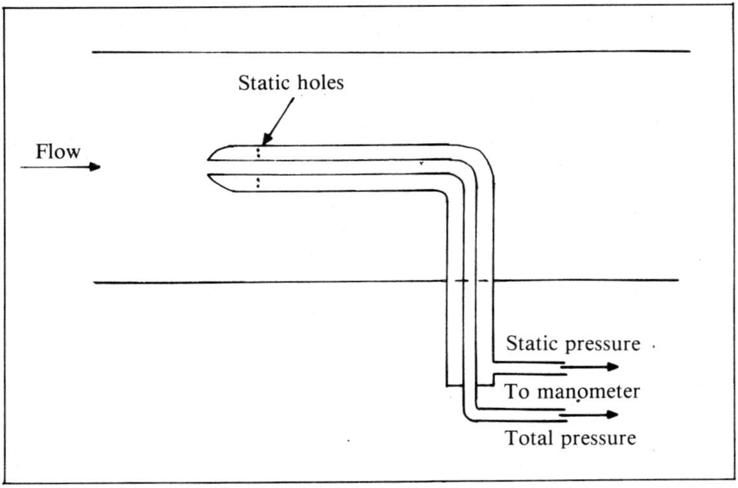

Figure 1.23 – Pitot-static tube

fluid; it is usually, but not always, a liquid, such as water, kerosene, carbon tetrachloride or mercury. It is most unfortunate that a liquid with a specific gravity of about six does not exist. If it is desirable to use a manometric fluid which is less dense than the flowing fluid, an inverted U-tube manometer, similar to that shown in Figure 1.22, can be used.

(a) Pitot Tube

A pitot tube is a simple device which measures a point velocity; it depends on a pressure difference resulting solely from kinetic energy changes. A small quantity of the flowing liquid is decelerated to zero at an orifice situated at right angles to the direction of flow. The local or point fluid velocity is obtained by measuring the difference between the total pressure and the static pressure.

The pitot-static tube, illustrated in Figure 1.23, consists of two concentric tubes arranged to be parallel to the flow of fluid. The outer tube is perforated with a ring of small (0.5 to 1.0 mm) holes which are at right angles to the flow and lead into the annular space. This annular space is otherwise sealed, except for the manometer connection. The entrance of the inner tube is at right angles to the flow, the inner tube is connected to the other limb of the manometer. The leading edges of the inner and outer tubes are connected in a manner which limits form drag.

The relationship between the point velocity, v_x, the manometer reading, Δz, the density of the flowing fluid, ρ, and the manometric fluid, ρ_m, and the gravitational acceleration, g, may be derived from the Bernoulli equation:

$$v_x = \left[\frac{2\Delta z(\rho_m - \rho)g}{\rho} \right]^{0.5} \tag{1.86}$$

A perfect pilot tube would obey Eqn 1.86, but all actual pitot tubes should be calibrated and a discharge coefficient, C_d, usually between 0.98 and 1.00, allocated.

Because of the velocity gradient that occurs across a pipe, one of two procedures must be adopted to obtain the true average flow velocity and hence the average flow rate. The pitot tube may be inserted at the centre line of the pipe and the average velocity calculated from the maximum velocity by the relationship $v = \alpha v_{max}$, where α is obtained from a graph similar to the one illustrated in Figure 1.18. The alternative procedure is to to perform a pitot tube traverse and calculate the mean velocity by graphical integration.

When used for gases at velocities in excess of about 60 m s^{-1}, compressibility becomes important, and Eqn 1.86 must be modified.

Provided that the pitot-static tube is well designed, has a diameter less than one fiftieth the diameter of the pipe, and is perfectly aligned in the pipe, it is possible to have errors less than 1% over a wide range of Reynolds numbers.

(b) Orifice Meter

An orifice meter is a simple, cheap, reliable instrument which can measure the mean volumetric flow rate to an error of about 2%.

The primary element is a thin flat plate, with an orifice, clamped between a pair of flanges in a pipeline. The orifice is usually, but not necessarily, circular and concentric with the pipe axis. It acts as a constriction in the flow line, thus causing the fluid to accelerate. The corresponding pressure drop is measured by the secondary element. Figure 1.24 illustrates a typical orifice meter with *vena contracta* pressure tapping points. The orifice should have sharp edges and its larger opening must face downstream. The diameter

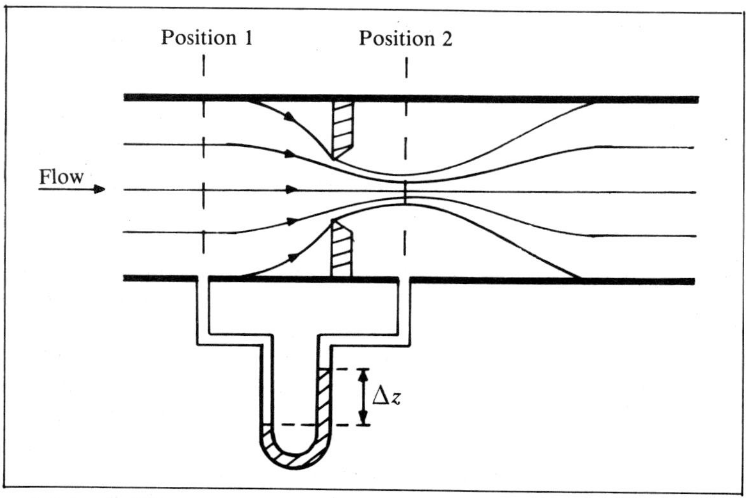

Figure 1.24 – Orifice meter

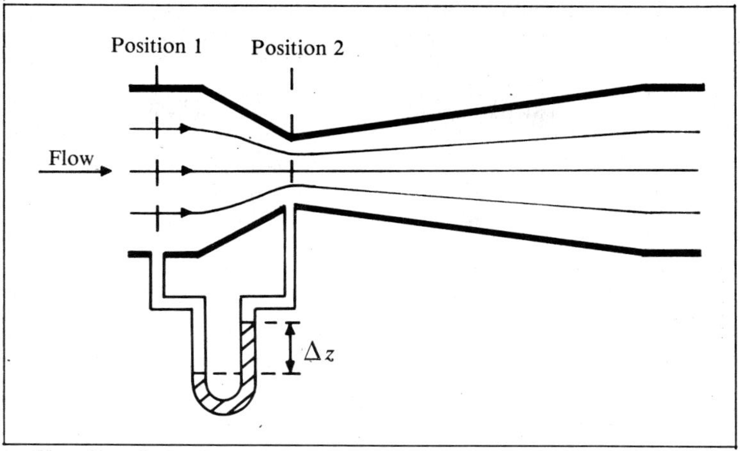

Figure 1.25 – Venturi meter

should be chosen to give a convenient pressure drop across the meter. British Standard 1042 specifies the dimensions of an orifice plate, and the dimensions and positions of the various combinations of pressure taps which may be used.

The Bernoulli equation (Eqn 1.33) can be written for the energy changes between positions 1 and 2 (the *vena contracta*) in Figure 1.24 to give a theoretical equation for the flow rate, Q, in terms of the pressure loss, Δp, the fluid density, ρ, the pipe area, A_1, and the area of the *vena contracta*, A_2. In practice, there is a substantial loss due to form drag resulting from separation of the boundary layer on the downstream side of the orifice and hence the actual flow rate is less than that given by the equation.

(c) Venturi Meter

The principal disadvantage of the orifice meter is the permanent pressure loss, typically about 60%, due to the sudden contraction. The venturi meter is designed to change the fluid velocity gradually so that there is no loss of energy owing to form drag.

The primary element of a venturi meter, as illustrated in Figure 1.25, is a tube which gradually decreases from the pipe diameter to a throat and then gradually increases to the pipe diameter. The taper on the upstream side of the throat is between 20° and 30° and on the downstream side of the throat is no more than 7°. The tapers are sufficiently smooth and gradual to substantially eliminate boundary layer separation and hence energy losses due to form drag. A properly designed venturi meter has a permanent pressure loss of about 10% compared with about 60% for a typical orifice meter. British Standard 1042 gives the exact dimensions of the standard venturi tube.

The pressure difference is measured across the throat and a point in the

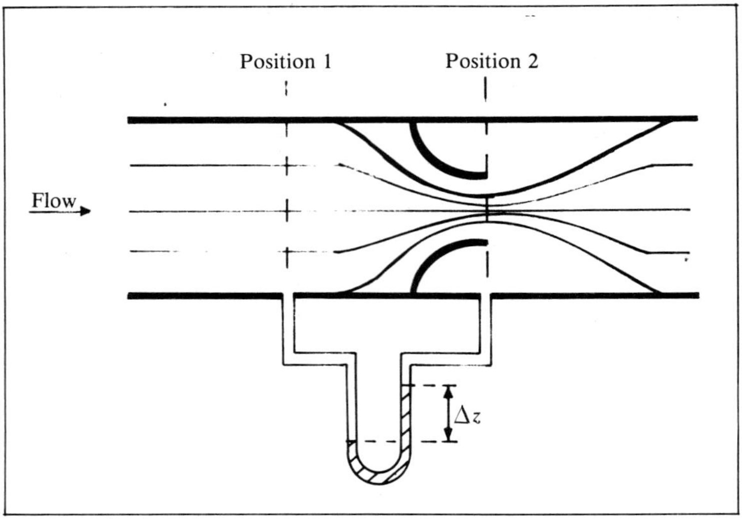

Figure 1.26 – Nozzle meter

pipe, upstream of the venturi meter. In normal practice the flow rate can be measured to an accuracy of about 2%.

Venturi meters are bulky and relatively expensive, but, for a given pipe diameter and pressure difference, a venturi meter can pass up to 60% more flow than an orifice meter.

(d) Nozzle Meter

The nozzle meter, illustrated in Figure 1.26, is a typical example of a group of relatively inexpensive devices which have been introduced to preserve the low permanent pressure loss associated with the venturi meter. It is virtually a venturi tube without the downstream cone. The outlet from the throat corresponds to the *vena contracta* in an orifice meter. The coefficient of discharge is typically 0.98 and the permanent pressure loss typically 40%.

(e) Weirs

Weirs are commonly used to measure the volumetric flow rate of liquids, especially water and liquid effluents, in open channels. Although the rectangular notch and the V-notch weirs are the two most frequently used, notches with special characteristics are used in particular situations.

When flowing over a weir or through a notch in the weir, the pressure energy of the fluid is converted into kinetic energy and hence the Bernoulli equation can be used to establish a relationship between flow rate, head and notch characteristics.

Details of notch construction and refined equations for particular situations are given in the literature.

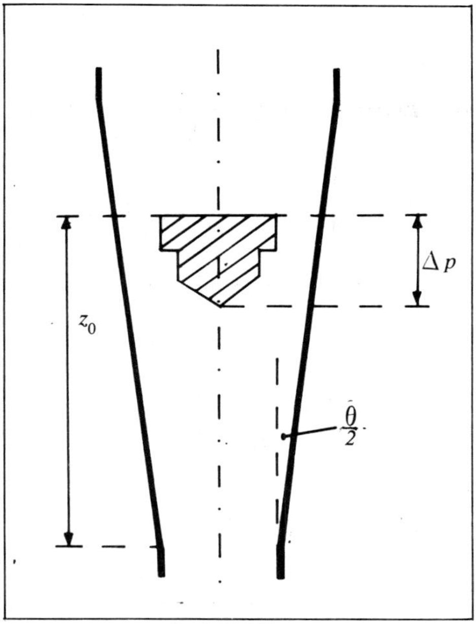

Figure 1.27 – Rotameter

(f) Rotameters

The flow meters which have been considered thus far are variable headmeters. They have no moving parts and are based on the common principle of measuring a variable pressure difference across a fixed constriction in the path of the flowing fluid.

The rotameter is an example of a variable-area meter in which the flow area is varied to produce a constant differential pressure. The rotameter, illustrated in Figure 1.27, consists of a gently tapering vertical tube, usually glass, containing a float which is free to move inside. The rotameter has a flow rate scale marked along its length. The fluid enters the bottom of the tube and, as it flows upwards, it exerts a force on the bottom of the float. The float moves to a position in the tube where this upward force is equal to the downward-acting gravitational force. The fluid flows through the annular space between the float and the tube; its flow rate is measured by noting the reading on the scale corresponding to the upper rim of the float.

Rotameters are usually designed such that, for a particular tube and float combination, the maximum flow rate is ten times the minimum flow rate. A single tube in combination with floats of common geometry, but different densities, can be used for a flow rate range of about 200:1. Rotameters have the advantage that they do not require a calming section either before or after them.

The Bernoulli equation also can be used to derive the equation for a rotameter.

Miscellaneous Techniques for Metering Flow
Thermal and dilution methods are frequently employed to measure the flow rates of fluids.

When heat is added at a steady rate to a fluid, the flow rate can be calculated from the upstream and downstream temperatures and a simple heat balance. An alternative technique, used in hot-wire anemometry, is to immerse an electrically heated wire in the fluid, the rate of heat loss from the wire being a function of the flow rate.

In the dilution technique, a second fluid is added at a known flow rate and concentration to the fluid whose flow rate is to be measured. A sample of the mixture is withdrawn for analysis of the second fluid at a point sufficiently downstream from the addition point to ensure complete mixing. The flow rate is calculated from a simple mass balance.

Liquid flow rates can be measured by the addition of a salt solution, such as sodium chloride or lithium chloride, which can be determined easily at low concentrations. The method is applicable to gases, in which case a second gas or vapour, such as carbon dioxide or ammonia, may be used.

1.7 VALVES

Valves are devices used to control the flow of a fluid by placing an obstruction in the fluid's path. The cross-sectional area of the valve opening, which may be closed by the obstruction, in relation to the cross-sectional area of the pipe in which the valve is situated, depends on the primary purpose of the valve. The selection of a valve for a specific duty depends initially on whether it will be used as a stop valve or a throttling valve.

A stop valve is used for 'on-or-off' service and, as such, it is either fully open or fully closed. When fully open, the resistance to flow of the fluid should be small to minimize the pressure loss through the valve. This means that the valve opening should be as large as possible, ideally having the same cross-sectional area as the pipe in which it is situated.

A throttling valve is used to regulate the flow of fluid. This is usually achieved at the expense of a relatively large pressure loss through the valve. Thus, even when the valve is fully open, the cross-sectional area available for flow is considerably reduced relative to the pipe area.

The most common types of valves are:
(a) Cocks
(b) Ball valves
(c) Globe valves
(d) Gate valves
(e) Diaphragm valves
(f) Pinch valves

(g) Butterfly valves
(h) Check valves
(i) Safety and relief valves
(j) Pressure control valves
(k) Automatic process control valves.

(a) Cocks

Cocks are the simplest method for controlling the flow of fluids; they have been in use since about 200 B.C. A typical cock or plug valve consists of a body casing in which fits a conical plug with a cylindrical passage through it (Figure 1.28). Rotating the plug one quarter of a turn changes the valve from fully open to fully closed. The flow passage changes rapidly with a slight amount of rotation when the valve is just opened and changes little when the valve is nearly fully open. It is difficult, therefore, to regulate the flow rate, especially when fractional openings are required. When trapezoidal section port cocks are used the area of the passage is nearly proportional to the angle through which the plug is turned. Cocks tend to be used for on/off service in small-diameter lines.

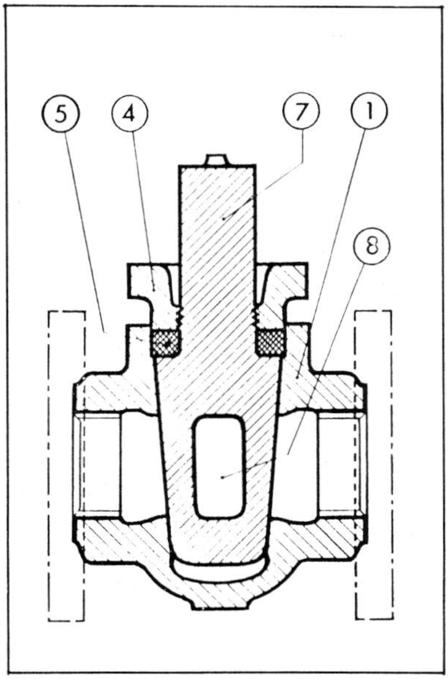

Figure 1.28 – Cock

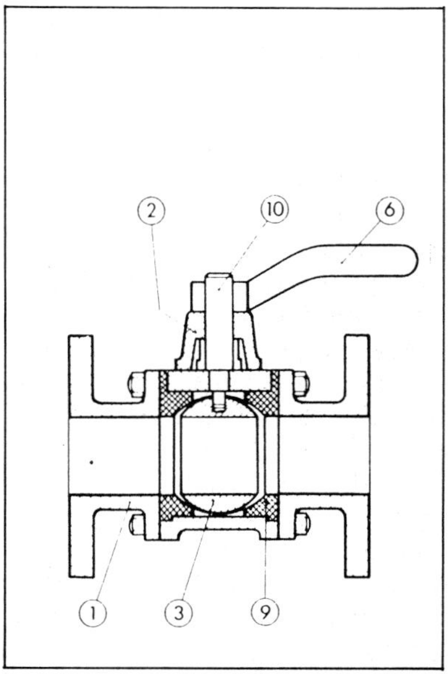

Figure 1.29 – Ball valve

1. Body
2. Bonnet
3. Spherical plug
4. Gland
5. Gland packing
6. Handle
7. Plug
8. Plug port
9. Seat
10. Spindle

(Courtesy: British Valve Manufacturers' Association)

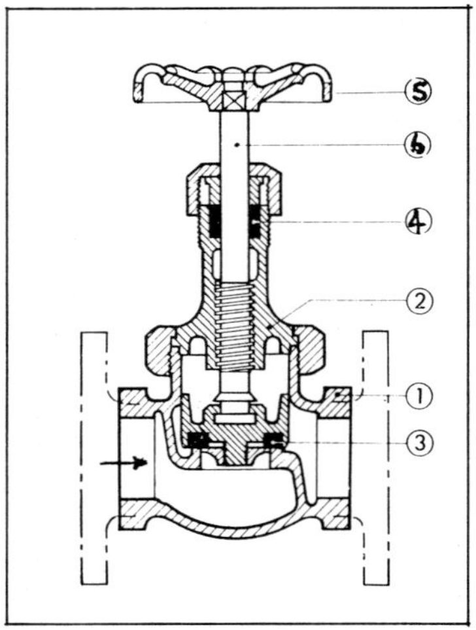

Figure 1.30 – Globe valve

1. Body
2. Bonnet
3. Disc
4. Gland packing
5. Handwheel
6. Stem

(Courtesy: British Valve Manufacturers' Association)

(b) Ball Valves

A ball valve is a modern variation of the cock. The essential difference is that the ball valve has spherical plug which rotates on a resilient seating, usually PTFE (Figure 1.29).

When fully opened, the area of the port is approximately equal to the cross-sectional area of the pipe in which the valve is situated; the pressure loss through the valve is therefore minimal.

Unlike many types of valves, the full port ball valve is not subject to debris, such as flock or lint, lodging upstream of the port and eventually blocking it.

(c) Globe Valves

The essential features of a globe valve are illustrated by Figure 1.30. The horizontal partition in the globular body contains a circular passage in which is inserted a renewable annular ring or seat. When fully closed the valve disc, which is attached to the stem, sits on the valve seat. When the valve is opened the fluid passes up through the annular opening between the valve disc and the seat. In so doing the fluid changes direction several times and the pressure loss through the valve is relatively large.

When the globe valve is closed the valve seat and disc should mate completely, otherwise leakage through the valve can occur. This can happen if the valve seat is damaged or if debris lodges in the annular opening.

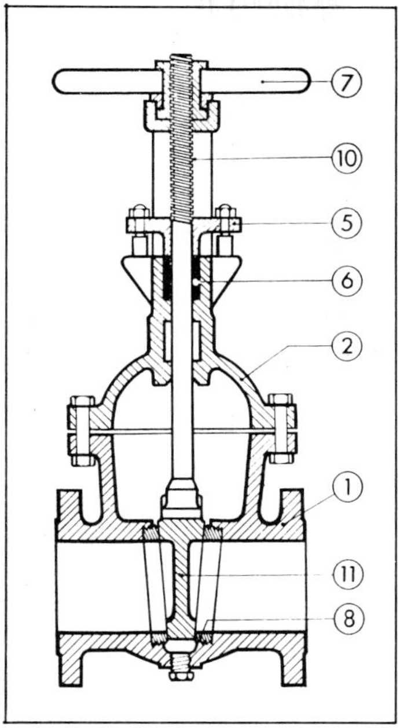

Figure 1.31 – Solid-wedge gate valve

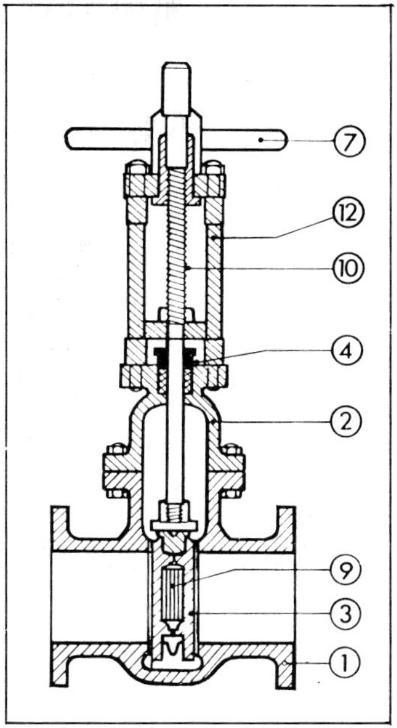

Figure 1.32 – Parallel-slide gate valve

1. Body
2. Bonnet
3. Disc
4. Gland
5. Gland flange
6. Gland packing
7. Handwheel
8. Seat ring
9. Spring
10. Stem
11. Wedge
12. Yoke

(Courtesy: British Valve Manufacturers' Association)

Globe valves are used for regulating the flow of fluids. The valve seat material should be selected to be compatible with the fluid being handled.

(d) Gate Valves

Since 1839, when they were invented, gate valves have been widely used for on/off duty. The faces of the gate or plug may be parallel or wedge shaped, corresponding to the angles of the seating faces in the valve body (Figures 1.31 and 1.32). The gate itself may be solid, split or spring loaded, depending on the security of closure required.

As the valve is opened, the gate rises into the bonnet. When the valve is fully open the passage through the valve is essentially full bore and the pressure loss is relatively small.

Even though a gate valve has two sealing faces, leakage can occur if the gate or seats are damaged or if debris lodges between the faces. In some makes of gate valve a drain plug is provided to remove any accumulated build-up of solid material.

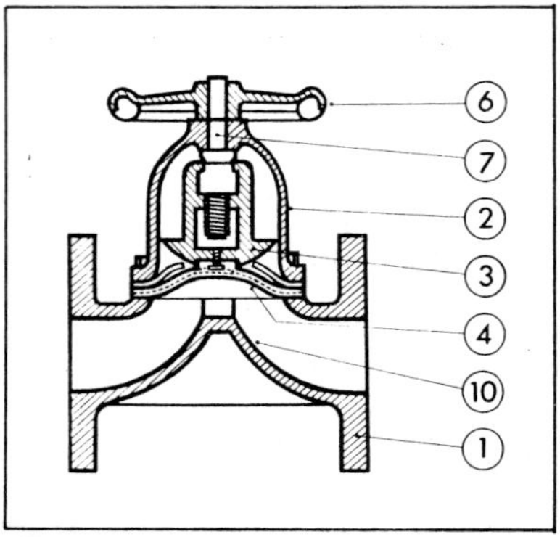

Figure 1.33 – Diaphragm valve

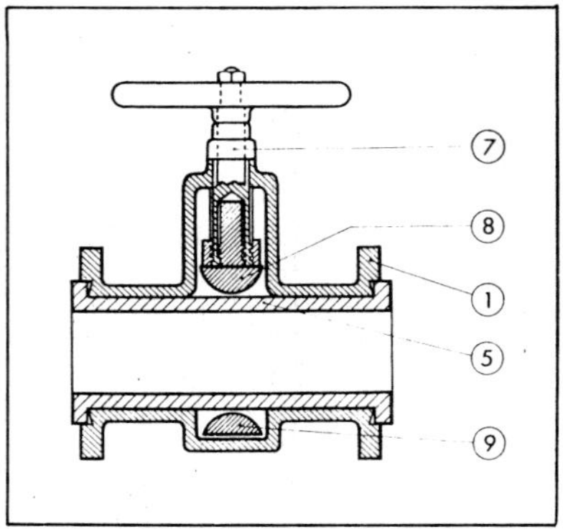

Figure 1.34 – Pinch valve

1. Body
2. Bonnet
3. Compressor
4. Diaphragm
5. Flexible tube
6. Handwheel
7. Spindle
8. Top pinch bar
9. Lower pinch bar
10. Weir

(Courtesy: British Valve Manufacturers' Association)

(e) Diaphragm Valves

Diaphragm valves were developed for compressed-air systems, but now find increasing application as regulating and on/off valves for a very wide range of gases, liquids, suspensions and slurries. A diaphragm physically isolates the bonnet assembly from the fluid being handled (Figure 1.33). This obviates the necessity of a packed gland.

The valve can be designed with a low weir to suit special applications.

(f) Pinch Valves

In the pinch valve, illustrated by Figure 1.34, the fluid flow rate is regulated by compressing a flexible tube. This flexible tube isolates the moving parts of the valve from the fluid being handled. When fully open, the pressure loss through the valve is minimal.

(g) Butterfly Valves

Butterfly valves, as illustrated by Figure 1.35, are ideal for regulating the flow rate of fluids, especially gases. As the valve disc is always in a state of near balance, a small force is required to vary its position. The pressure loss through the valve is relatively low.

Depending on the seating arrangement, butterfly valves may be used for tight shut-off duty, low leakage rate duty or regulating duty or a combination of these.

(h) Check Valves

Check valves or non-return valves are typified by the swing check and piston lift check valves (Figures 1.36 and 1.37). They are used when it is im-

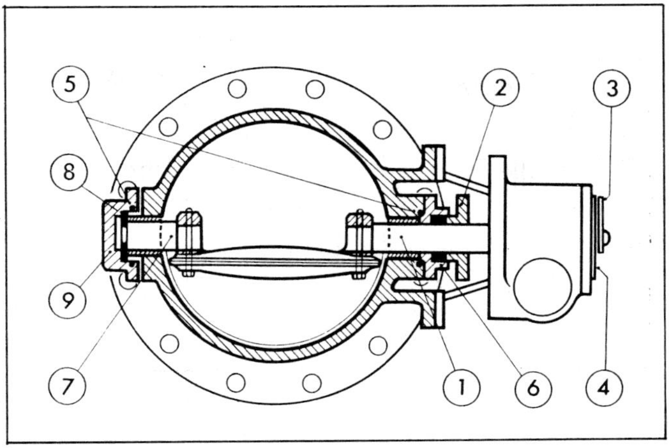

Figure 1.35 – Butterfly valve

1. Drive shaft
2. Gland
3. Indicator
4. Manual operator
5. O-ring
6. Packing
7. Stub shaft
8. Thrust collar
9. Trunnion cover

(Courtesy: British Valve Manufacturers' Association)

portant that one-directional flow is maintained. The valve is opened by the fluid flowing in the desired direction. If the flow stops, or tends to reverse, the valve automatically closes either by gravity or by a spring. The pressure loss is greater through the piston lift type of check valve.

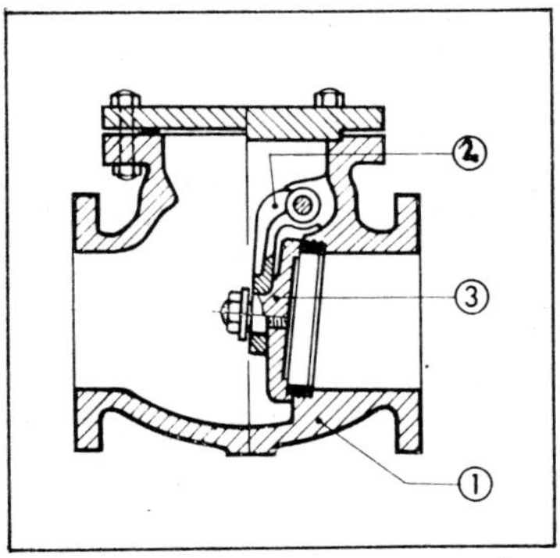

Figure 1.36 – Horizontal swing pattern check valve

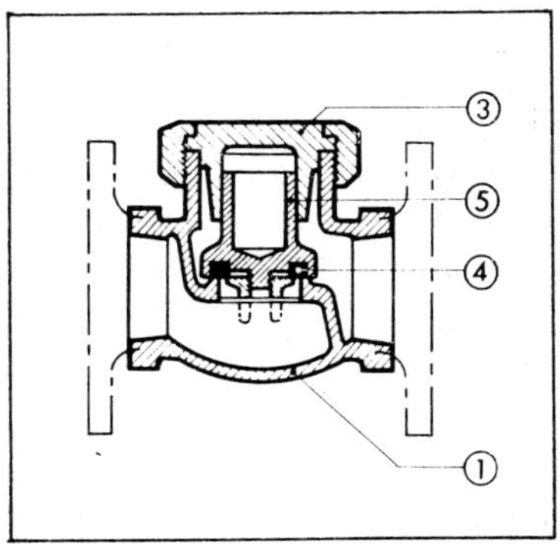

Figure 1.37 – Piston lift check valve

1. Body
2. Hinge
3. Cover
4. Disc
5. Disc holder

(Courtesy: British Valve Manufacturers' Association)

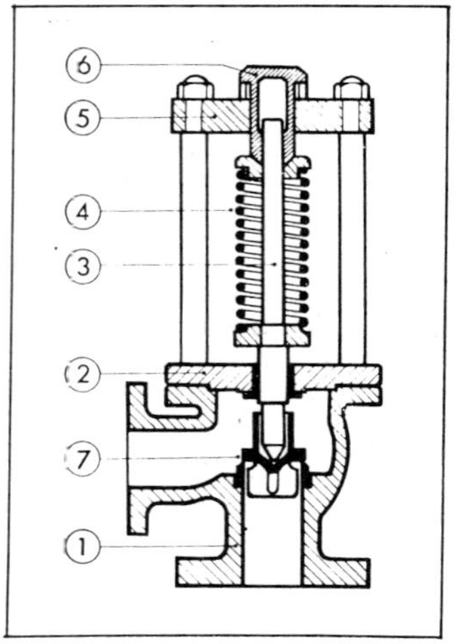

Figure 1.38 – Safety valve

1. Body
2. Cover
3. Spindle
4. Spring
5. Spring cover
6. Adjusting screw
7. Valve head

(Courtesy: British Valve Manufacturers' Association)

(i) Safety and Relief Valves

Although, in fact, the valve illustrated by Figure 1.38 is a safety valve, in principle it could operate as a relief valve. Safety valves and relief valves automatically discharge fluid to prevent a predetermined safe pressure being exceeded. The tension on the spring is adjusted to just balance the predetermined pressure; when this pressure is exceeded the valve opens.

Safety valves are used for compressible fluids, such as steam and compressed gases, which require a quick release of fluid to reduce the pressure. Relief valves are used for incompressible liquids, when immediate full-bore discharge is not required, since a relatively small discharge of liquid will reduce the pressure. Both types of valve automatically close when the excess pressure has been vented.

(j) Pressure Control Valves

Pressure control valves are available to maintain a constant downstream pressure (pressure reducing) or a constant upstream pressure (pressure retaining). When no external power source is required, the valves are classed as 'self-operating'. When an external power source is used to actuate the valve it is classed as a specific form of automatic process control valve. Figure 1.39 illustrates a typical pressure-reducing valve. Once the adjusting screw has been set, the valve will produce continuously and automatically a constant downstream pressure, provided that the upstream pressure is in ex-

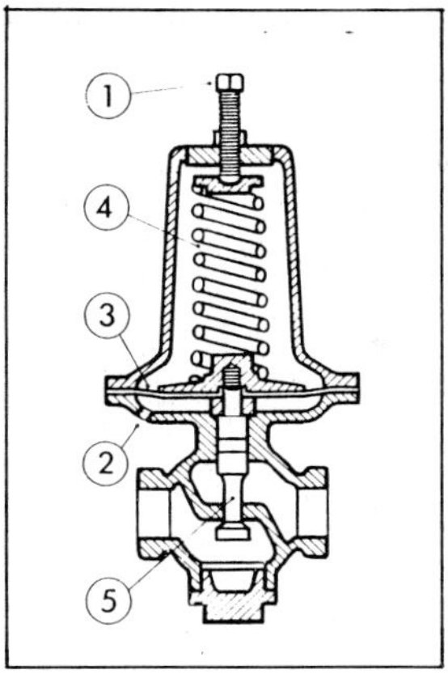

Figure 1.39 – Pressure-control valve

1. Adjusting screw
2. Control-pressure connection
3. Diaphragm
4. Spring
5. Plug

(Courtesy: British Valve Manufacturers' Association)

cess of the required downstream pressure and that the valve plug and seats are not damaged.

(k) Automatic Process Control Valves

A typical automatic process control valve is illustrated by Figure 1.40. It may be considered to consist of four separate but integrated components: the actuator, the packing gland, the body and the valve plug.

Normally the sensing element upstream of the valve measures a particular process parameter (flow, temperature, pressure, level, pH). If this value deviates from the desired value, the gas (air) pressure on the diaphragm in the pressure casing is altered automatically to deflect the diaphragm to move the value stem and plug, thereby altering the flow area in the valve to change the fluid flow rate.

Selection of the appropriate type and size of valve and valve plug to suit a particular application is a specialized job.

Although the majority of automatic process control valves are globe valves, diaphragm valves are used, especially when abrasive or solids containing liquids are to be handled.

1.8 PUMPING OF LIQUIDS

In process plants, liquids have to be transported through pipelines and through equipment such as heat exchangers, filters and beds of particulate solids. Wherever possible an engineer makes use of gravity, but when this is

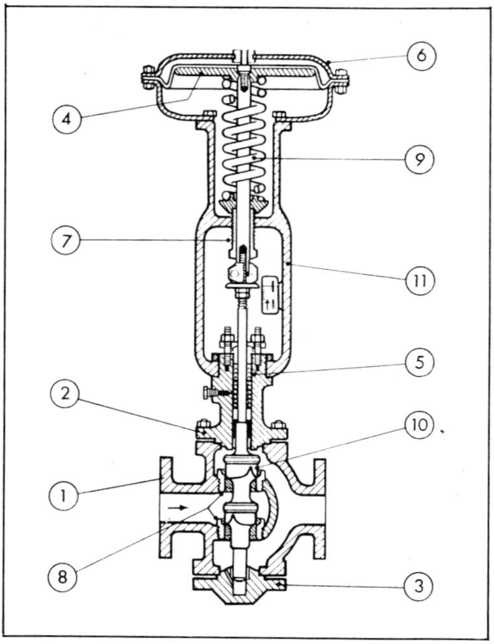

Figure 1.40 – Automatic process-control valve

1. Body
2. Bonnet
3. Bottom flange
4. Diaphragm
5. Packing gland
6. Pressure casing
7. Adjusting screw
8. Seats
9. Spring
10. Plug
11. Yoke

(Courtesy: British Valve Manufacturers' Association)

not practicable some mechanical device has to be employed. Pumps, which may be electrically, steam, air or water driven, convert mechanical energy into fluid energy, the main objective being to produce fluid flow and pressure. In some instances the purpose is simply to overcome the frictional pressure loss along a pipeline system, in others it is to raise the liquid to a higher level.

The liquids used in the process industries vary considerably in both physical and chemical properties and it has been necessary, therefore, to develop a wide range of pumps. Although pumps differ in their principle of operation and mechanical construction, they may, in general, be considered either as impeller pumps or positive displacement pumps.

Suction, Discharge, Total and Net Positive Suction Heads

The energy which a pump requires for a particular duty depends on the height through which the liquid is raised, the lengths and diameters of the pipeline system, the liquid flow rate and the physical properties of the liquid, especially its density and viscosity.

When a pump, situated between positions 1 and 2 in a pipeline system, imparts a head, Δh, to a liquid in turbulent flow, the Bernoulli equation for steady state conditions is:

$$\left(z_2 + \frac{p_2}{\rho g} + \frac{v_2^2}{2g} \right) - \left(z_1 + \frac{p_1}{\rho g} + \frac{v_2^2}{2g} \right) = \Delta h - h_f \qquad (1.35)$$

where the terms z, $p/\rho g$ and $v^2/2g$ are the potential (or static), pressure and velocity heads, respectively, and h_f is the frictional loss.

Figure 1.41 illustrates several typical pumping systems. In each case the datum level is the centre line of the pump, and z_s, z_d, Δz, p_s, p_d are the static suction head, static discharge head, differential static head and gas pressure above the liquid on the suction and discharge sides of the pump respectively. The suction and discharge pressures in the vessels could be above, below or at atmospheric pressure. In the event that none of the three systems illustrated in Figure 1.41 is operated under steady state conditions, then, as fluid is transferred from one vessel to the other, the differential static head, Δz, will change with time. For batch operation, $(z_s)_a$ and $(z_s)_b$ will decrease, $(z_d)_a$ will increase, $(z_s)_c$ will increase (but it is negative relative to the datum level), $(z_d)_b$ and $(z_d)_c$ will remain unaltered.

The important heads to consider in a pumping system are:

(a) Suction head, h_s
(b) Discharge head, h_d
(c) Total of differential head, Δh
(d) Net positive suction head (NPSH).

These heads are now defined with reference to Figure 1.41, where conventionally the datum level is the centre line of the pump.

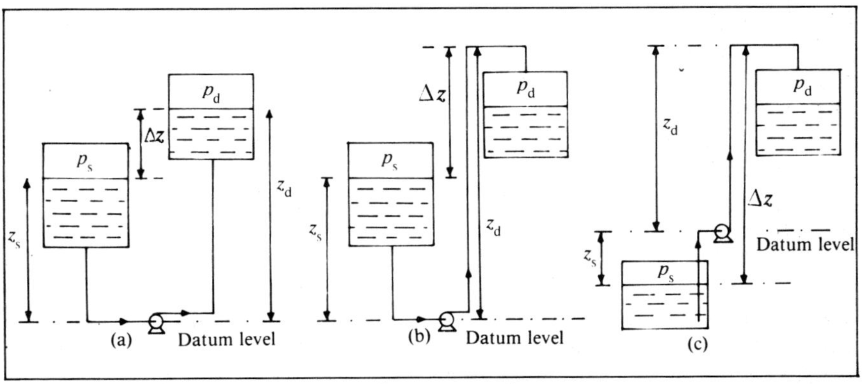

Figure 1.41 – Pumping systems showing suction, discharge and total heads

(a) The *suction head* is:

$$h_s = z_s + \frac{p_s}{\rho g} - h_{fs} \tag{1.87}$$

where h_{fs} is the frictional head loss on the suction side of the pump. It was mentioned in Section 1.5 that the frictional head loss due to bends, fittings, contractions and enlargements may be expressed in terms of a number of velocity heads or pipe diameters. In terms of pipe diameters, h_{fs} may be expressed:

$$h_{fs} = 4f \cdot \frac{L + L'}{d} \cdot \frac{v_s^2}{2g} \tag{1.88}$$

where the subscripts refer to the suction side and L' refers to the equivalent length of straight pipe obtained from, for example, Table 1.4.

If the liquid level on the suction side were below the level of the pump, z_s would be negative and the pump has a negative suction (Figure 1.41(c)). This does not necessarily mean that if the liquid level were below the pump level the suction head must be negative, since $p_s/\rho g$ could have a numerical value greater than z_s.

(b) The *discharge head* is:

$$h_d = z_d + \frac{p_d}{\rho g} + h_{fd} \tag{1.89}$$

where h_{fd} is the frictional head loss on the discharge side.

In terms of equivalent diameters, h_{fd} may be expressed:

$$h_{fd} = 4f \cdot \frac{L + L'}{d} \cdot \frac{v_d^2}{2g} \tag{1.90}$$

where the subscripts refer to the discharge side and L' refers to the equivalent length of straight pipe.

It will be observed that both h_{fs} and h_{fd} in Eqns 1.88 and 1.90 increase with increasing fluid velocity.

(c) The *total head*, Δh, which the pump supplies to the liquid, is the difference between the discharge head and the suction head, $h_d - h_s$:

$$\Delta h = \left(z_d + \frac{p_d}{\rho g} + h_{fd}\right) - \left(z_s + \frac{p_s}{\rho g} - h_{fs}\right) \tag{1.91}$$

$$\text{or } \Delta h = \left(z_d - z_s\right) + \left(\frac{p_d - p_s}{\rho g}\right) + \left(h_{fd} + h_{fs}\right) \tag{1.92}$$

(d) A liquid has a *vapour pressure*, p_{vp}, which must be taken into account when considering a pumping system, since it will decrease the suction head by a value equal to the vapour pressure (expressed in metres head of fluid) corresponding to the liquid temperature in the suction line. The suction head available to move the liquid through the suction piping into the pump is called the net positive suction head, NPSH, and is defined:

$$\text{NPSH} = z_s + \left(\frac{p_s - p_{vp}}{\rho g} \right) - h_{fs} \tag{1.93}$$

or, from the definition of suction head in Eqn 1.87:

$$\text{NPSH} = h_s - \frac{p_{vp}}{\rho g} \tag{1.94}$$

Clearly, the NPSH of a system must always be positive; in other words the suction head must be greater than the vapour pressure, expressed in metres, $p_{vp}/\rho g$.

The frictional head loss, h_{fs}, increases with increasing flow rate and hence the available NPSH decreases with increasing flow rate. At the boiling temperature of a liquid the suction pressure, p_s, and the vapour pressure, p_{vp}, are equal and, from Eqn 1.93, the NPSH becomes $z_s - h_{fs}$. In this event no suction lift is possible since z_s must be positive.

Liquids can be lifted from below the centre line of the pump, in which case z_s is negative, if the term $p_s - p_{vp}$ in Eqn 1.93 is sufficiently large.

Reducing the pressure in a pump suction line below the vapour pressure of the liquid being handled can cause some of the liquid to vaporize in the pump inlet and bubbles of the vapour to be carried into the pump impeller. These bubbles will collapse violently at some point of higher pressure downstream of the pump inlet, usually in the impeller. This can produce sharp crackling noises and may damage the metal surfaces in the immediate vicinity. The phenomenon is known as cavitation. When cavitation occurs in the suction line no liquid can be drawn into the pump; the pump is said to be vapour locked. It is perhaps interesting to note that, for example, water at 50°C increases its volume about 58 000 times when it vaporizes.

The *available* NPSH is a function of the system and depends on the suction head, friction losses and the vapour pressure of the liquid at the operating temperature. The available NPSH can be altered by physically changing the pipe work, the flow rate or the operating temperature.

The *required* NPSH is a function of the pump design, specifically pump speed and pump capacity, and cannot be altered by the user.

Centrifugal Pumps

The main element of a centrifugal pump is its impeller, which is essentially a flat disc with curved blades attached. When the pump is in operation, the

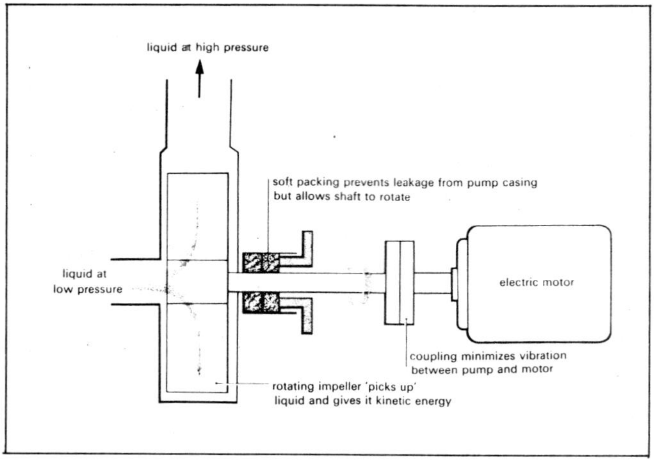

Figure 1.42 – Centrifugal pump

impeller rotates continuously at high speed, typically 25 rev. s^{-1} which, for a 0.150 m diameter impeller, corresponds to a peripheral velocity of about 12 m s^{-1}. The main function of the impeller is to generate kinetic energy by the action of centrifugal force and then to convert this energy to pressure by gradually reducing the fluid velocity.

Centrifugal pumps are used extensively because of their relative simplicity of design, low capital cost, low maintenance and flexibility of application. They span the capacity range 75×10^{-6}–45 m^3 s^{-1} and the head range 1–1 000 m of water. They can be 'throttled' by a valve on the delivery side for control purposes, or even operate against a closed valve. In the process industries they are used for capacities up to about 0.075 m^3 s^{-1} and heads

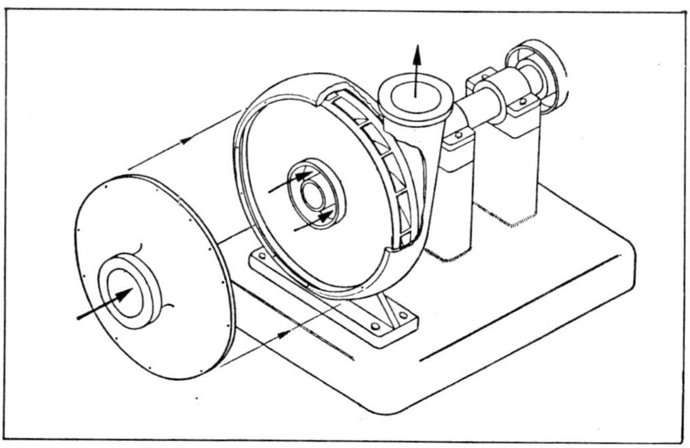

Figure 1.43 – Centrifugal pump

up to about 100 m of flowing fluid. They are available in a wide range of materials of construction including: mild steel, stainless steels, plastics, glass, glass-lined metal, rubber-lined metal.

Centrifugal pumps may be built up as single- or multi-stage units. The maximum head that is practicable to generate with a single-stage unit is about 100 m of water (about 10 atmospheres). If greater heads than this are required, multi-stage units are used. In these a number of impellers or rotors are mounted on a single shaft. The impellers operate in series, the discharge from one feeding the suction of the next so that the head increases. A 0.050 m inlet, six-stage centrifugal pump is available to deliver against a head of 500 m of water.

Axial flow or propeller pumps are frequently used in closed-loop circulation systems, such as jet dyeing machines, where the pump casing is merely an elbow in the pipeline.

In general, centrifugal pumps are employed to deliver relatively large volumes of liquid against medium heads. Axial flow pumps are used to deliver large volumes against relatively low heads. Mixed flow pumps are available to cover the intermediate duty range.

Centrifugal pumps include those with vertical and horizontal shafts, with and without glands, and with different types of impellers to suit the differing requirements for pumping various types of liquids. Figures 1.42 and 1.43 illustrate typical centrifugal pumps.

The prime mover, which is usually, but not necessarily, an electric motor, drives a shaft, generally through a coupling. This shaft carries an impeller which rotates inside a casing. The impeller usually has radially curved blades or vanes, shaped to ensure a smooth flow of liquid between the vanes. The open impeller has the blades fixed to a central hub; in the closed impeller the blades are held between two supporting plates. The blades are usually curved backwards at a continously increasing slope relative to the direction of rotation and the liquid generally approaches the blades radially.

The head developed by a pump is determined mainly by the angle of the blades and the peripheral speed of the impeller. The discharge volume, however, is determined by the cross-section of the passages. The actual impeller design depends on the nature of the fluid being handled, its viscosity and the nature of any solids in suspension.

In the typical centrifugal pump, illustrated in Figure 1.44, the liquid enters the pump axially into the centre or eye of the rotating impeller and is discharged tangentially from the pump casing. As the liquid moves progressively from the eye to the periphery of the impeller, its velocity, and hence kinetic energy, is continuously increased by centrifugal forces.

The liquid leaving the periphery of the impeller is constrained by the pump casing. The discharge pressure is produced by gradually converting the velocity head of the liquid, according to the Bernoulli equation, into pressure head as the liquid flows around the continuously increasing cross-section of the stationary eccentric space or volute.

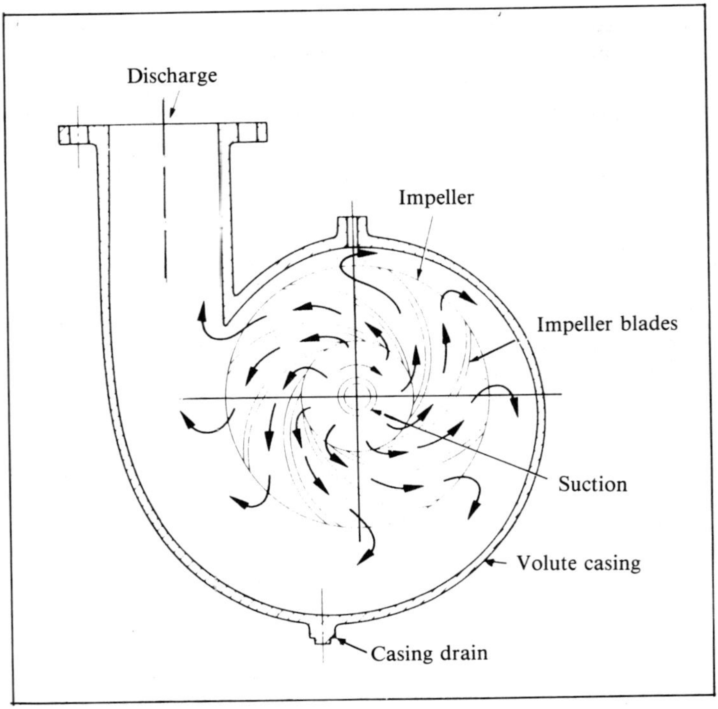

Figure 1.44 – Centrifugal pump

The volumetric flow rate, Q, and the total head, Δh, developed by the pump are related to the rotational speed of the impeller, N, and the impeller diameter, d:

$$Q \propto N d^3 \tag{1.95}$$

$$\Delta h \propto N^2 d^2 \tag{1.96}$$

The actual performance of a centrifugal pump is usually expressed in the form of characteristic curves, similar to those illustrated in Figure 1.45, which shows the interrelation of total head, quantity, efficiency and brake power for a pump at a particular speed. The characteristic curves supplied by manufacturers for a particular pump are actual calibration curves and, therefore, generally refer to water, which is the normal calibration liquid. The size of a centrifugal pump generally refers to its discharge connection; it is usually rated according to the capacity and total head at the point of maximum efficiency. For the pump illustrated by Figure 1.45 the maximum efficiency would be 80% for a capacity of 0.060 $m^3\ s^{-1}$ at a total head of 140 m when consuming 25×10^3 W.

Positive Displacement Pumps

The essential features of a positive displacement pump are as follows.

(a) It will deliver a definite and predetermined volume of liquid at every stroke or revolution of the prime mover, regardless of the head against which the pump has to operate.

(b) The capacity is directly proportional to the operating speed.

(c) It can generate a high pressure and therefore, unlike a centrifugal pump, cannot operate against a closed valve. The pump should be protected by a relief valve installed in a by-pass loop round the pump. If a valve downstream of the pump is closed or if a blockage occurs, the relief valve will lift and liquid will be circulated around the by-pass loop from the discharge to the suction side of the pump.

(d) It can pump air and is therefore self priming.

(e) The initial cost and maintenance costs tend to be higher than those for a comparable capacity centrifugal pump.

Positive displacement pumps are very useful as metering pumps, especially for the proportional feeding of several liquids into a system, as in the case of reactive dyeing.

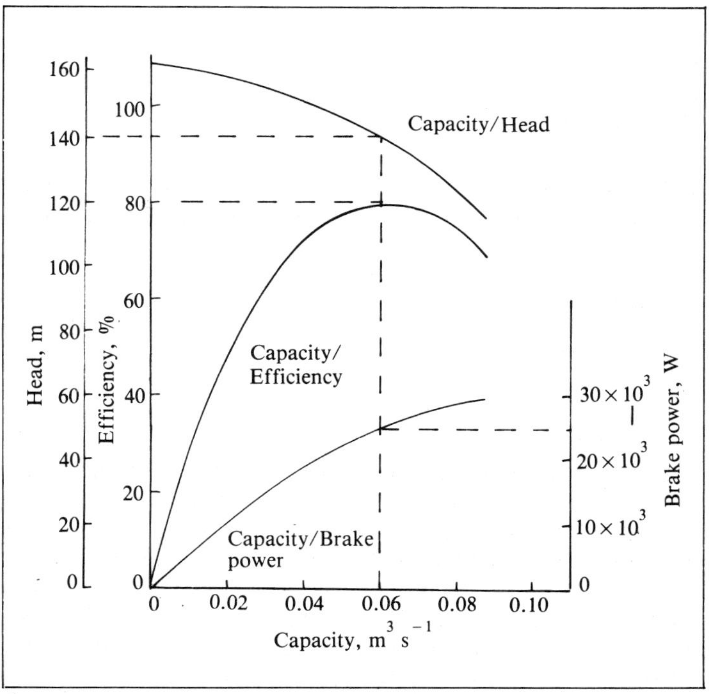

Figure 1.45 – Characteristic curves for a centrifugal pump

Positive displacement pumps may be classified:
(a) Reciprocating pumps
(b) Rotary pumps
(c) Miscellaneous pumps.

(a) *Reciprocating pumps* are typified by the piston or ram pump. They transfer liquid by continuously changing their internal volume. The pumping rate is varied by altering either the frequency or the length of the stroke. Non-return valves are required on both the suction and discharge sides of the pump. The piston is reciprocated in the cylinder by the prime mover. When the piston is retracted the pressure in the cylinder decreases, the discharge valve closes and the suction valve opens, admitting liquid to the pump. When the piston is advanced the pressure in the cylinder increases, the suction valve closes and the discharge valve opens, discharging liquid to the exit line.

The pulsed discharge from a reciprocating pump can be decreased by using a double-action pump or increasing the number of cylinders, both of which increase the capital and maintenance costs.

The diaphragm pump has been developed for handling corrosive liquids and liquids containing suspensions of abrasive solids that could damage the piston and cylinder of a ram pump.

(b) *Rotary pumps* are typified by the gear pump. The two rotating gear wheels of equal size mesh within a fixed casing. As the rotating gears continually unmesh they create a partial vacuum, which causes liquid from the suction line to be drawn into the pump. The continual meshing of the gears generates an increase in pressure which forces the liquid into the discharge line.

Gear pumps can handle liquids of high viscosity, but they should not be used for liquids containing solids.

(c) A number of positive displacement pumps are available, which have some of the characteristics of both reciprocating and rotary pumps. These are typified by the eccentric cam pumps and peristaltic pumps.

1.9 PUMPING OF GASES

In the process industries a considerable number of different gases and vapours have to be conveyed, sometimes over long distances, at a variety of pressures, both above and below atmospheric pressure. The engineer often has to cope with corrosive gases, which may carry erosive solids in suspension.

In general terms, fans, blowers and compressors are used to convey gases or vapours at low, medium and high positive pressures respectively.

For the relatively low pressure change across a fan it can be considered that the density of the gas being conveyed is unchanged, and hence the incompressible flow equations may be assumed. For blowers, and especially compressors, the gas is compressed either adiabatically or isothermally and the simple Bernoulli equation is not adequate to describe the process.

Gas-handling equipment is usually rated in terms of the 'standard cubic metre'. The volume of a gas in standard cubic metres is that measured at a specific temperature and pressure, irrespective of the actual temperature and pressure of operation. Various standards are used in different countries and industries. The SI standard is based on 100×10^3 N m^{-2} and 15°C, which corresponds to a molal volume of 0.024 m^3 mol^{-1}.

Fans

Fans are used for conveying gases or vapours at low positive pressures, frequently for pressures less than 5×10^3 N m^{-2} gauge (0.5 m water gauge).

Fans may be divided into two main classes: (a) centrifugal and (b) axial flow. Centrifugal fans may have impeller blades of one of a number of different shapes:

(i) Straight blades
(ii) Forward-curved blades
(iii) Backward-curved blades
(iv) Radial blades.

(a) Centrifugal Fans

The operating principle of centrifugal fans is the same as that of centrifugal pumps and the same equations can be derived and applied. The impellers are mounted in relatively thin sheet-metal casings. The clearances between the impeller and casing are large and the discharge pressure heads relatively low. Operating efficiencies are in the range 40–70%. They can be used to handle either clean or dust-laden gases depending in their design.

(i) The impeller of a straight-blade centrifugal fan is relatively large and rotates at relatively low speeds. A typical impeller will have between five and twelve straight blades, resembling a paddle wheel, Figure 1.46.

(ii)–(iv) The forward-curved, backward-curved and radial fans generally have between thirty and sixty blades and are frequently referred to as muti-blade fans. Compared with straight-blade fans they have smaller-diameter impellers, operate at higher speeds, have higher efficiencies and deliver larger volumes for a particular size of casing.

(b) Axial-flow Fans

These have a comparatively simple flow pattern, parallel to the fan shaft. They can be either disc or propeller type and are used for general gas, usually air, circulation.

Blowers

Cycloidal or gear pumps may be used as blowers; they are simple and deliver large volumes against pressures which are too high for fans to handle. The cycloidal or 'Rootes' blower, illustrated in Figure 1.47, is used for pressures from 3×10^3 to 100×10^3 N m^{-2} gauge.

Compressors

Rotary or centrifugal compressors are more compact, less expensive and have higher efficiencies than positive displacement compressors. Gas compression requires a substantial expenditure of power. The compression of a gas causes an increase in temperature and, since more power is required to compress a hot gas than a cold one, compressors are generally provided with cooling units; this is certainly true for multi-stage compressors.

Compressors have small clearances between the moving and static parts and are unsuitable for gases containing solids; these are fitted, therefore, with filters on the suction side.

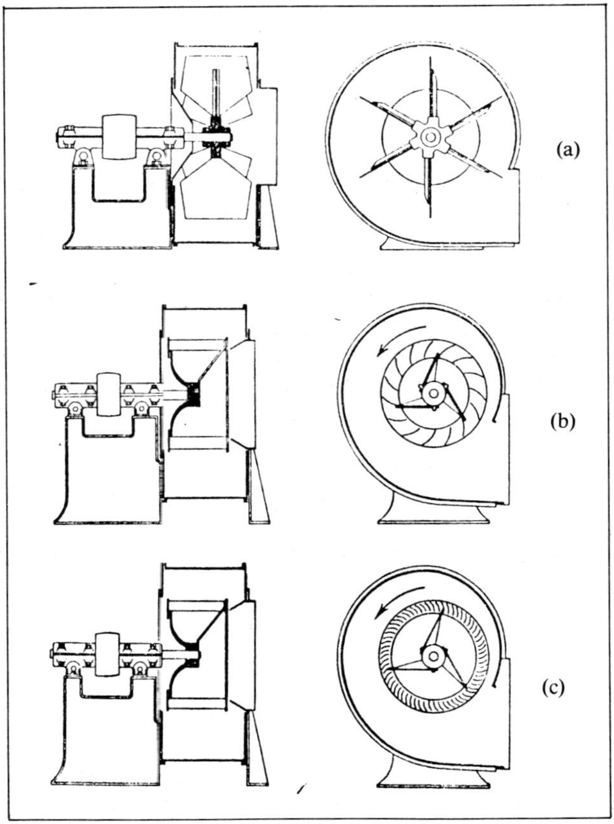

Figure 1.46—(a) Straight-blade fan
(b) Backward-blade fan
(c) Forward-blade fan

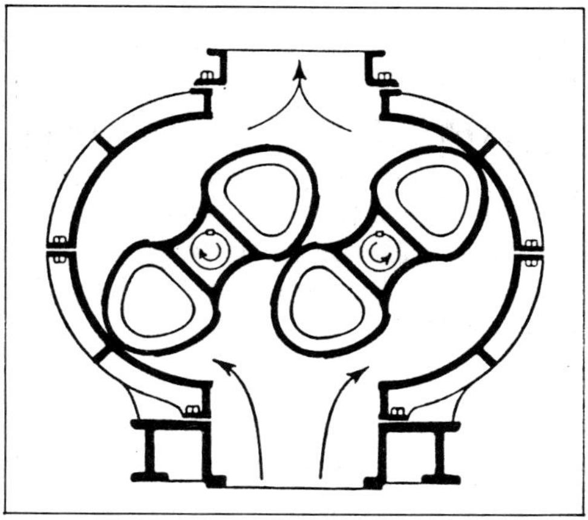

Figure 1.47 – Cycloidal blower

1.10 LIQUID MIXING

Mixing may be required for dissolution, chemical reaction, physical dispersion, heat transfer and mass transfer. It is still an art rather than a science, and equipment is designed from experience and quantitative experiments rather than from agreed fundamental theory. Equations, based on dimensional analysis and standard tank configurations, are available, however, to calculate the power required for various types of agitators in certain situations. Optimizing mixing times, mean residence times and power consumption for a particular process tends to require experimentation.

Types of Impeller

The circulation currents which are developed in a vessel by a mixer depend on the vessel shape, baffles, the physical properties of the liquid and the design, location and rotational speed of the impeller.

The multitude of impeller configurations may be grouped into four main categories:

(a) Propellers
(b) Turbines
(c) Paddles
(d) High-shear impellers.

(a) Propellers

These are essentially high-speed, axial-flow type impellers, the liquid discharge is parallel to the agitator shaft; liquid circulation rates are high.

Propellers generally operate at speeds between about 10 and 25 rev. s^{-1}. The three-bladed marine propeller, illustrated in Figure 1.48(a), is the one most commonly used; it rarely exceeds 0.500 m in diameter. It is not uncommon to have more than one propeller on the same shaft.

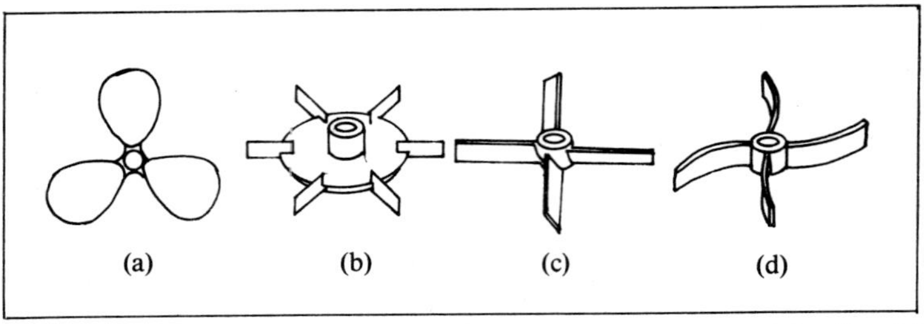

Figure 1.48 – (a) Propellor, (b) flat-blade turbine, (c) inclined-blade turbine, (d) curved-blade turbine

(b) Turbines

The term turbine defines an impeller which has its blades set at a constant angle to the vertical plane over their entire length. The blades are either vertical or inclined, they may be straight or curved; there are numerous variations. The straight-blade radial-discharge, the pitched-blade or inclined-blade axial-flow and the curved-blade turbine are illustrated in Figure 1.48. Turbines generally operate in the speed range 1–3 rev./s and have diameters typically 30–50% of the vessel diameter.

(c) Paddles

Paddles have been used in industrial liquid mixing for a considerable time. The basic flat-blade paddle, illustrated in Figure 1.49(a), is not suitable for high-viscosity liquids and, for this reason, numerous other configurations have evolved from it; a selection of these are illustrated in Figure 1.49. Paddles operate in the speed range 0.3–3 rev. s^{-1}, diameters are typically 50–80% of the vessel diameter.

(d) High-shear Impellers

The high-shear impellers, illustrated in Figure 1.50, have a relatively small diameter, but operate at high speeds. They are essentially low-flow, high-velocity radially discharging impellers.

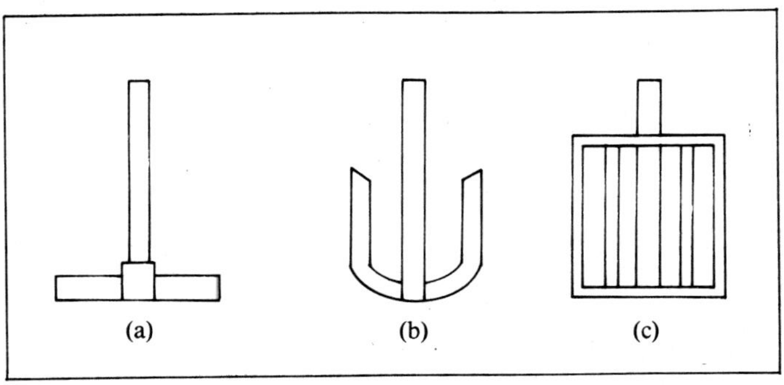

Figure 1.49 – (a) Flat paddle, (b) anchor paddle, (c) gate paddle

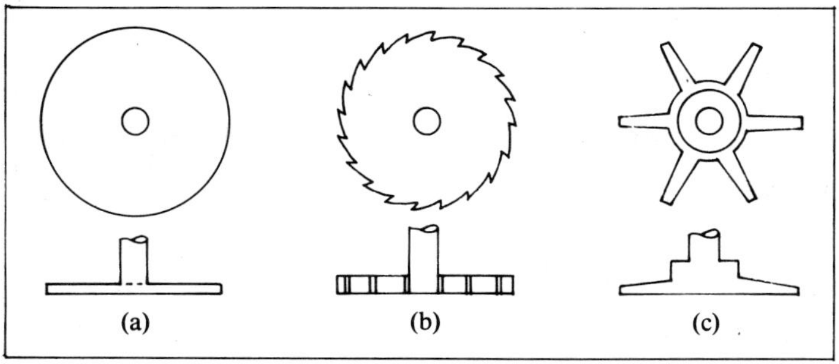

Figure 1.50 – High shear impellers, (a) disc, (b) modified disc, (c) modified turbine

Flow Patterns in an Agitated Vessel

When an impeller is mounted centrally in an unbaffled vessel there is a tendency for a swirling flow pattern to develop, regardless of the type of impeller used. The vortex produced tends to stratify the liquid at the various levels, leading to little or no relative motion between liquid particles. Vortex formation limits the power consumption of an impeller; once the vortex reaches the impeller, severe air entrainment may occur and oscillating surges of liquid can be generated. Either of these can damage the impeller, shaft, gearbox or motor.

To reduce or eliminate the swirling motion, and hence achieve vertical mixing, entails creating an unbalanced flow pattern. This can be achieved by off-centre mounting, angled mounting, side-entry mounting or baffling.

The actual effect of the baffles depends on the type of flow – the Reynolds number. Typical baffles are flat vertical strips of metal set radially from the vessel wall. A vessel which has sufficient baffles, such that the addition of one more does not significantly alter the mixing characteristics, is said to be 'fully baffled'. Even in large vessels, four equi-spaced baffles are usually sufficient to provide full baffling. Standard baffles are between 8% and 10% of the vessel diameter. When mixing suspensions, it is usual to locate the baffles at a small distance from the vessel wall to prevent accumulation of solids at the baffles.

Power Consumption

The power requirement of an impeller cannot be calculated. It can be determined, however, by a combination of dimensional analysis and quantitative experiment. The Reynolds group for mixing is defined by:

$$\mathrm{Re} = \frac{d^2 N \rho}{\mu} \tag{1.97}$$

where d = impeller diameter, m
N = impeller speed, rev. s^{-1}
ρ = fluid density, kg m^{-3}
μ = fluid viscosity, N s m^{-2}.

2 HEAT TRANSFER

2.1 INTRODUCTION

In most chemical processes, whether batchwise or continuous, heat is either required or generated. The transfer of this heat from one region to another at the correct rate is an important aspect of plant design and operation. It is also important to ensure that heat losses from pipes, vessels, buildings, hot effluent streams and hot exhaust gases are kept to an economic minimum. Here we are concerned, in the main, with steady state heat transfer.

Modes of Heat Transfer

A temperature difference is the driving force which causes heat to be transferred in the direction of the lower temperature in an attempt to establish thermal equilibrium. When the temperature gradient does not alter with time, the process is said to be steady state. Unsteady state or transient heat transfer processes, where the temperature gradient changes with time, are encountered in batch processes and during start-ups and shut-downs of continuous processes.

The three mechanisms by which heat may be transferred are:

(a) Conduction
(b) Convection
(c) Radiation.

Even though the three mechanisms of heat transfer are inherently different, and hence may be studied separately, they often operate in combination; the type of system and temperature levels dictate their relative contributions. In some situations it is possible to determine the three contributions individually, in others it is possible only to determine the overall heat transferred.

(a) *Conduction* is the process by which heat is transferred by direct physical contact on a molecular scale; molecules possessing greater kinetic energy transmit part of their energy to those possessing less energy. In solids, there is little or no relative movement of the molecules, heat is transferred by a transfer of vibrational energy. In fluids, the heat is transferred in the form of kinetic energy during molecular collisions. Conduction is the only mechanism by which heat can be transferred within solids; whilst within fluids, conduction is usually combined with convection and, in certain circumstances, with radiation.

(b) *Convection* is the process by which heat is transferred by the actual bulk mixing of molecules on a macroscopic scale. Natural or free convection occurs when the fluid motion results from density differences caused by temperature gradients. In forced convection, the fluid motion is due to some mechanical means, such as pump, fan or agitator. Naturally, convective heat transfer is closely associated with the laws of fluid motion; it is particularly important in the transfer of heat between a solid surface and a fluid. Convection is usually accompanied by conduction, and the contribution due to radiation depends on the temperature levels.

(c) *Radiation* is the process by which heat is transferred through space from a body at some temperature, without the aid of a heat transfer medium. Unlike conduction and convection, radiation can occur through a vacuum. All bodies emit radiant energy. When this radiation falls on a second body it may be partially absorbed, reflected or transmitted; it is only in the first case in which the second body is heated. Heat transfer by radiation becomes increasingly important as the temperature of the emitting body increases.

2.2 PROPERTIES, QUANTITIES, UNITS AND DIMENSIONS

A number of properties and quantities of fluids have been discussed in Section 1.2, only the additional ones pertinent to heat transfer are discussed here.

Properties – Specific Heat Capacity, Specific Latent Heat, Thermal Conductivity and Thermal Diffusivity

The specific heat capacity is the heat required to increase the temperature of one kilogram of a substance by one kelvin; it is expressed as $J\ kg^{-1}\ K^{-1}$. The word specific means per unit mass. Thermodynamically, the increase in heat content of a substance may be stored in a number of ways; thus the actual definition of specific heat capacity depends upon the way in which the heat is added to the substance. The two commonly used specific heat capacities are c_p and c_v, which are measured at constant pressure and constant volume respectively. A gas heated at constant pressure will expand and thermal energy equivalent to the work done in the expansion must be applied to the system. Hence, for a gas, c_p is greater than c_v. The ratio of the specific heat capacities, $c_p/c_v = \gamma$, varies from 1.66, for monatomic gases, to just over 1.00 for polyatomic gases.

Except at extremely high pressures, the specific heat capacity of a solid or liquid is independent of pressure. It is, however, temperature dependent, especially over large temperature changes. Gases and vapours show a more pronounced temperature dependence than is the case for liquids and solids.

The specific latent heat of a substance, λ, is the heat required to effect a phase change at constant temperature; it is expressed as $J.kg^{-1}$. The latent heats of fusion, vaporization and sublimation are associated with the phase changes: solid to liquid, liquid to vapour and solid to vapour respectively. The latent heat of vaporization of a substance decreases with increasing pressure up to the critical pressure of the substance when the latent heat is zero.

The thermal conductivity of a substance, K, is defined as the quantity of heat which will flow across unit surface area in unit time when the temperature difference across a unit thickness of the material is one kelvin; it is expressed as W m^{-2} K^{-1} per metre thickness, or W m^{-1} K^{-1}.

The thermal conductivities of materials range from about 0.001 to 0.01 W m^{-1} K^{-1} for gases, 0.01 to 1.0 W m^{-1} K^{-1} for liquids and 1 to 400 W m^{-1} K^{-1} for solids. Conductors have a high thermal conductivity, insulators have a low thermal conductivity. A change in the internal structure of a metal can affect the thermal conductivity significantly. The slightest trace of arsenic in copper reduces the thermal conductivity to about one-third that of pure copper. The thermal conductivity of an alloy may be less than the value of the thermal conductivity of either constituent.

The thermal conductivity of a substance varies with temperature. The variation may be neglected if the temperature range under consideration is not too large. For many materials, the variation of thermal conductivity with temperature can be represented by a linear relationship:

$$K_t = K_0(1 + \beta t) \tag{2.1}$$

where K_t = thermal conductivity at t°C
K_0 = thermal conductivity at 0°C
β = temperature coefficient of thermal conductivity.

The value of β may be negative or positive. It is generally positive for gases and negative for liquids and solids. There are many exceptions to this general rule; indeed the thermal conductivity of some substances passes through a maximum or minimum with increase in temperature.

There is a relationship between thermal and electrical conductivity. For metallic solids, both properties have high values, for non-metallic solids, insulators, both properties have low values. For gases, there is a relationship between thermal conductivity and viscosity.

Thermal diffusivity, α, is a physical property of a material, associated with all types of unsteady state heat transfer problems involving conduction. The rate of temperature change in conductive heat transfer depends on the numerical value of the thermal conductivity of the material undergoing heat transfer. Thermal diffusivity is analogous to the momentum diffusivity or kinetic viscosity in momentum transfer. It is defined:

$$\alpha = \frac{K}{c_p \rho} \tag{2.2}$$

where α = thermal diffusivity, m^2 s^{-1}
K = thermal conductivity, W m^{-1} K^{-1}
c_p = specific heat capacity at constant pressure, J kg^{-1} K^{-1}
ρ = density of material, kg m^{-3}.

Typical values of thermal diffusivity together with the corresponding typical values of thermal conductivity are:

	K(W m^{-1} K^{-1})	α (m^2 s^{-1})
Gases	0.01	10×10^{-6}
Liquids	0.1	0.05×10^{-6}
Non-metallic solids	2	1×10^{-6}
Metallic solids	100	25×10^{-6}

Quantities – Heat, Heat Content, Rate of Heat Transfer, Heat Transfer Coefficients

Heat is the energy possessed by a substance in the form of the kinetic energy associated with molecular motion. It is expressed in joules, J. The main observable physical effect of a change in heat content of a body may include a change in the temperature (sensible heat) or a change in phase (latent heat).

The heat content of a body, H, at a temperature t°C above the arbitrarily chosen datum of 0°C, assuming no phase change occurs between 0°C and t°C, is:

$$H = m\,c\,t \tag{2.3}$$

where H = heat content of the body, J
m = mass of body, kg
c = specific heat capacity of body, J kg^{-1} K^{-1}
t = temperature *difference* between t°C and datum of 0°C, K.

If the body undergoes a complete phase change at t°C, then its heat content at t°C may be taken to be:

$$H = m(ct + \lambda) \tag{2.4}$$

where λ = specific latent heat of phase change, J kg^{-1}

If the body undergoes a complete phase change at a temperature between 0°C and t°C, say at t_{PC}, then the heat content at t°C may be taken to be:

$$H = m[c_{P1}t_{PC} + \lambda + c_{P2}(t - t_{PC})] \tag{2.5}$$

where c_{P1} and c_{P2} = specific heat capacities of phases 1 and 2, respectively, J kg^{-1} K^{-1}.

If the phase change is vaporization, the terms within the square brackets represent, respectively, the sensible heat, the latent heat and the superheat.

As the specific heat capacities of liquids, vapours and gases vary with temperature, it is frequently more convenient to present the specific heat contents or specific enthalpies, J kg^{-1}, in a tabular or graphical form as a function of temperature. This is especially true for steam and other heat transfer fluids, which, in process use, often undergo both a temperature and pressure change.

The rate of heat transfer, Q, is the heat transferred per unit time, J s^{-1} or W; it may be defined:

$$Q = m\ c\ (t_1 - t_2) \tag{2.6}$$

where Q = rate of heat transfer, W
m = mass flow rate, kg s^{-1}
c = specific heat capacity, J kg^{-1} K^{-1}
$(t_1 - t_2)$ = temperature difference between final and initial temperatures of material, K.

As the specific heat capacity varies with temperature, the value of c in Eqn 2.6 should be the mean value evaluated at the arithmetic mean temperature, $(t_1 + t_2)/2$. Alternatively, if values are available, Eqn 2.6 could be written in terms of specific enthalpies at t_1 and t_2, h_{t1} and h_{t2}, respectively:

$$Q = m(h_{t1} - h_{t2}) \tag{2.7}$$

For a phase change, Eqn 2.6 would normally be written:

$$Q = \dot{m}\,\lambda \tag{2.8}$$

The heat flux is the heat transfer rate per unit surface area, Q/A, and it is expressed as W m^{-2}.

Whether the rate of heat transfer given by Eqns 2.6 to 2.8 can be achieved in a heat exchanger depends on the overall heat transfer coefficient for the process, U, the surface area of the heat exchanger, A, and the overall temperature driving force, ΔT:

$$Q = UA\ \Delta T \tag{2.9}$$

where Q = rate of heat transfer, W
U = overall heat transfer coefficient, OHTC, W m^{-2} K^{-1}
A = effective surface area of heat exchanger, m^2
ΔT = temperature driving force, K.

The overall heat transfer coefficient depends on the properties of the heat transfer surface, the properties and flow conditions of the process and heat transfer fluids.

Units and Dimensions

TABLE 2.1
Units and Dimensions in Heat Transfer

Property or quantity	Symbol	Dimensions	Units
Specific heat capacity	c, c_p, c_v	$L^2\ T^{-2}\ \theta^{-1}$	$J\ kg^{-1}\ K^{-1}$
Specific latent heat	λ	$L^2\ T^{-2}$	$J\ kg^{-1}$
Thermal conductivity	K	$MLT^{-3}\ \theta^{-1}$	$W\ m^{-1}\ K^{-1}$
Thermal diffusivity	α	$L^2\ T^{-1}$	$m^2\ s^{-1}$
Heat content	H	$ML^2\ T^{-2}$	J
Rate of heat transfer	Q	$ML^2\ T^{-3}$	W
Mass flow rate	$\dot{m}$	MT^{-1}	$kg\ s^{-1}$
Specific enthalpy	h	$L^2\ T^{-2}$	$J\ kg^{-1}$
Overall heat transfer coefficient	U	$MT^{-3}\ \theta^{-1}$	$W\ m^{-2}\ K^{-1}$

2.3 CONDUCTION

The theory of conductive heat transfer was established by Fourier in 1822. The Fourier rate equation, for the steady state undirectional flow of heat through a homogeneous material, may be written:

$$Q = -KA \cdot \frac{dt}{dy} \tag{2.10}$$

where Q = rate of heat transfer in the y-direction only, $J\ s^{-1}$ or W
K = thermal conductivity of material, $W\ m^{-1}\ K^{-1}$
A = area of material at right angles to the flow of heat, m^2
dt/dy = temperature gradient through the material in the y-direction only, $K\ m^{-1}$.

The negative sign in the temperature gradient is the result of conventionally defining a positive heat flow in the direction of a negative temperature gradient.

Conduction through a Single Plane Wall

Figure 2.1 illustrates the unidirectional steady state conductive heat transfer through a cross-sectional area, A, of a single plane wall of homogeneous material, having a constant thickness, L, and a constant thermal conductivity, K, due to a temperature difference $(t_1 - t_2)$ or ΔT.

The rate of heat transfer is:

$$Q = \frac{KA\Delta T}{L} \tag{2.11}$$

$$\text{or } Q = \frac{\Delta T}{\dfrac{L}{KA}} \tag{2.12}$$

Example 2.1

Calculate the rate of heat transfer across a flat surface, 0.150 m thick, when the temperatures of the cold and hot faces are 20°C and 200°C. The mean thermal conductivity may be taken as 0.043 W m^{-1} K^{-1}.

$$Q = \frac{KA\Delta T}{L}$$

$$= \frac{0.043 \times A(200-20)}{0.150}$$

$$= 51.6 \text{ W m}^{-2}$$

Eqn 2.12 can be written in terms of the rate of heat transfer:

$$\text{rate} = \frac{\text{driving force}}{\text{resistance}} \tag{2.13}$$

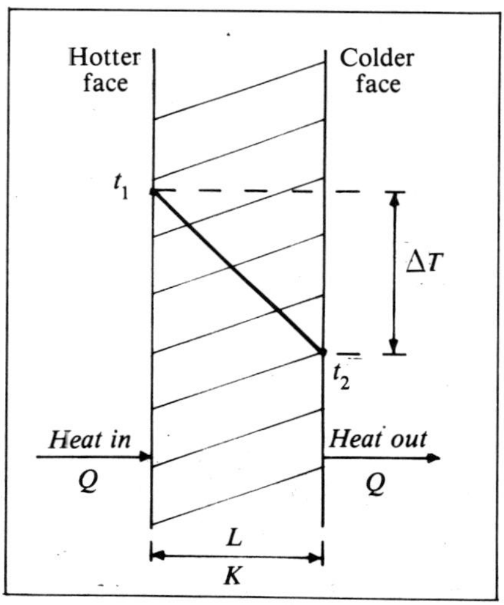

Figure 2.1 – Steady-state conduction through a plane wall

This is analogous to Ohm's law describing the flow of electricity:

$$\text{current density} = \frac{\text{potential difference}}{\text{resistance}} \tag{2.14}$$

This electrical analogy is useful in analysing complex conduction problems.

Conduction through a Composite Plane Wall

Figure 2.2 illustrates the steady state conductive heat transfer through a composite wall made up of three layers of material, in complete thermal contact, having thicknesses L_1, L_2 and L_3 and thermal conductivities K_1, K_2 and K_3.

For steady state, the rate of heat transfer through each layer must be equal and hence: $Q = Q_1 = Q_2 = Q_3$. For an area, A, at right angles to the direction of heat flow, the heat transfer rate equations may be written:

$$Q = \frac{K_1 A \Delta T_1}{L_1} \quad \text{or} \quad \Delta T_1 = Q \cdot \frac{L_1}{K_1 A} \tag{2.15}$$

$$Q = \frac{K_2 A \Delta T_2}{L_2} \quad \text{or} \quad \Delta T_2 = Q \cdot \frac{L_2}{K_2 A} \tag{2.16}$$

$$Q = \frac{K_3 A \Delta T_3}{L_3} \quad \text{or} \quad \Delta T_3 = Q \cdot \frac{L_3}{K_3 A} \tag{2.17}$$

The overall temperature difference, ΔT_o, is $\Delta T_1 + \Delta T_2 + \Delta T_3$ and, from Eqns 2.15 to 2.17, may be written:

$$\Delta T_o = Q \left(\frac{L_1}{K_1 A} + \frac{L_2}{K_2 A} + \frac{L_3}{K_3 A} \right) = Q \sum \frac{L}{KA} \tag{2.18}$$

Hence the rate of heat transfer in terms of the overall process is:

$$Q = \frac{\Delta T_o}{\sum \frac{L}{KA}} = \frac{\Delta T_o}{\Sigma R} \tag{2.19}$$

where ΣR = sum of resistance in series.

It will be appreciated that the ratio of any individual temperature difference to the overall temperature difference is numerically the same as the ratio of the corresponding individual resistance to the overall resistance. Thus:

$$\Delta T_1 : \Delta T_o = \frac{L_1}{K_1 A} : \sum \frac{L}{KA} \tag{2.20}$$

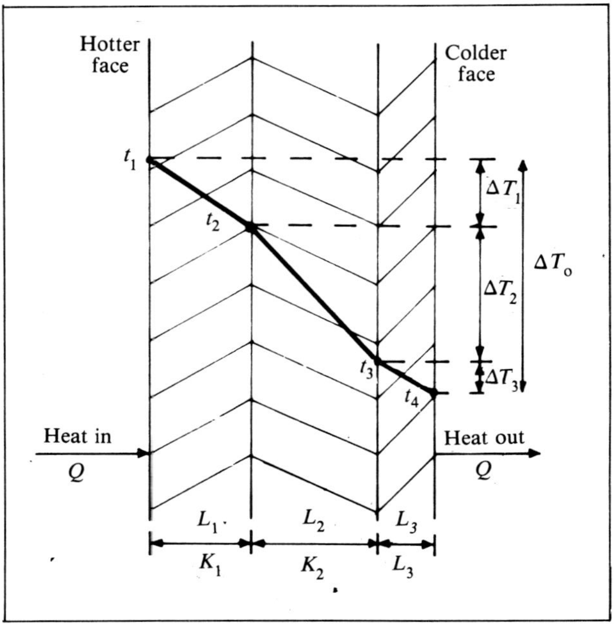

Figure 2.2 – Steady-state conduction through a composite wall

Problem 2.1
A composite plane wall of a furnace consists of three layers: 0.100 m of firebrick on the inside, 0.150 m of insulating brick in the middle and 0.100 m of building brick on the outside. The thermal conductivities are respectively: 0.140, 0.420 and 0.780 W m^{-1} K^{-1}. Calculate the heat loss through the wall and the temperatures at the two interfaces when the inside face temperature is 650°C and the outside face temperature is 30°C.

(*Answers*: 517 W m^{-2}; 281°C and 96°C)

Conduction through a Thick-walled Cylindrical Pipe
Figure 2.3 illustrates the steady state conductive heat transfer from the inner wall of a thick-walled cylindrical pipe. The surface area, A, increases continuously in the direction of the heat flow. The inside and outside radii of the pipe are r_1 and r_2 respectively.

Consider a very thin cylinder, of radius r, concentric with the pipe. The thickness of the wall of this cylinder is dr and, if dr is small compared with r, the lines of heat transfer may be considered to be parallel. For unit length of the cylinder, the surface area at right angles to the flow of heat is: $2\pi r$, and the Fourier equation, from Eqn 2.10 is:

$$Q = K \cdot 2\pi r \cdot \frac{dt}{dr} \qquad (2.21)$$

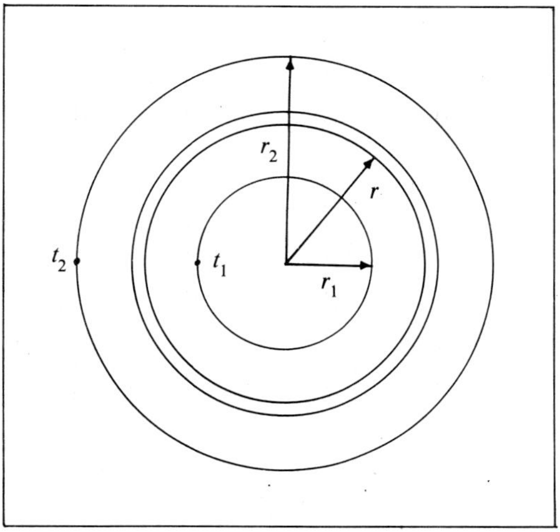

Figure 2.3 – Steady-state conduction through a thick-walled cylinder

Integrating this equation between the limits of r_1 and r_2, gives:

$$Q = \frac{2\pi K(t_1 - t_2)}{\ln(r_2/r_1)} \tag{2.22}$$

The mean surface area, A_m, across which heat is transferred, is the area of a cylinder of unit length and mean radius r_m. The logarithmic mean radius is defined in terms of the outside and inside radii of the cylinder, r_2 and r_1:

$$r_m = \frac{r_2 - r_1}{\ln(r_2/r_1)} \tag{2.23}$$

Hence, per unit length of cylinder, the mean area, A_m, is:

$$A_m = 2\pi \frac{r_2 - r_1}{\ln(r_2/r_1)} \tag{2.24}$$

Equation 2.22 can be rewritten in terms of the mean area:

$$Q = \frac{\Delta T}{\dfrac{L}{KA_m}} \tag{2.25}$$

where ΔT = temperature difference $(t_1 - t_2)$, K
L = metal thickness, $r_2 - r_1$, m.

When dealing with pipes where the ratio r_2/r_1 is less than 2.0, the arithmetic mean area can be used instead of the logarithmic area term, and the resultant error will be less than 4%. When the ratio r_2/r_1 is less than 1.5, using the arithmetic mean area produces an error less than 1.5%.

Problem 2.2
A 0.150 m inside-diameter, 0.175 m outside-diameter metal pipe has an outer surface temperature of 100°C when the inside surface temperature is 105°C. If the mean thermal conductivity of the metal is 45.0 W m^{-1} K^{-1}, calculate the heat loss per square metre of outside surface area.

(*Answer*: 16.7×10^3 W m^{-2} outside area)

Conduction through a Composite Cylindrical Pipe
Figure 2.4 illustrates the steady state heat transfer from the inner wall of a thick-walled cylindrical pipe, through the pipe wall and through the insulation around the pipe. The surface area, A, increases continuously in the direction of the heat flow. The inside and outside radii of the pipe are r_1 and r_2 respectively, the outside radius of the insulation is r_3.

The mathematical treatment for the composite cylindrical pipe is similar to that of the composite plane wall. Thus, the heat transfer rate through the composite pipe, per unit length of pipe, is:

$$Q = \frac{\Delta T_o}{\dfrac{L_1}{K_1 A_{m1}} + \dfrac{L_2}{K_2 A_{m2}}} \qquad (2.26)$$

where T_o = overall temperature difference, $(t_1 - t_3)$, K
L = thickness of metal or insulation, $(r_2 - r_1)$ and $(r_3 - r_2)$, respectively, m
K = thermal conductivity of metal or insulation, K_1 and K_2, respectively, W m^{-1} K^{-1}
A_m = logarithmic mean surface area of metal or insulation, A_{m1} and A_{m2}, respectively, m^2.

Eqn 2.26 for multilayer pipes is:

$$Q = \frac{\Delta T_o}{\sum \dfrac{L}{K A_m}} \qquad (2.27)$$

Problem 2.3
A 0.150 m inside-diameter, 0.175 m outside-diameter metal pipe, having a thermal conductivity of 45.0 W m^{-1} K^{-1}, is covered with a 0.025 m thick concentric layer of insulation, having a thermal conductivity of

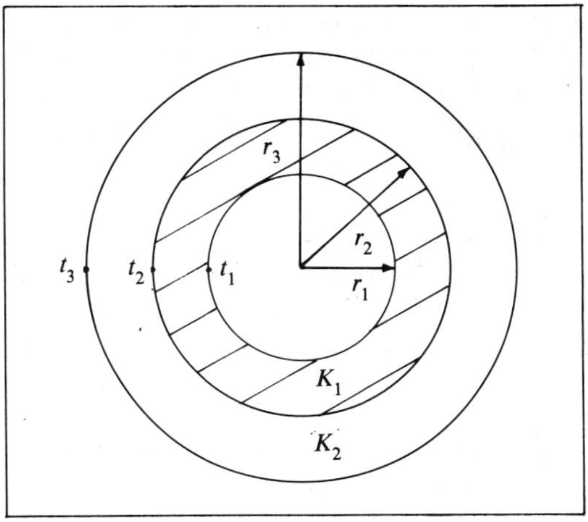

Figure 2.4 – Steady-state conduction through a composite cylinder

0.0680 W m^{-1} K^{-1}. The inside surface temperature is 105°C and the outside insulation surface temperature is 20°C. Calculate the heat loss per square metre of outside surface area of the insulation and the outside pipe surface temperature.

(*Answers*: 204 W m^{-2} outside area; 104.9°C)

Problem 2.4

When conveying a process fluid, a stainless steel pipe, 0.085 m outside diameter, 0.075 m inside diameter, has an inside surface temperature of 125°C and a mean thermal conductivity of 15.6 W m^{-1} K^{-1}. Calculate the thickness of insulation, having a mean thermal conductivity of 0.068 W m^{-1} K^{-1}, required to limit the outside insulation temperature to 25°C, thereby limiting the heat loss from the insulated pipe to 150 W per metre run of pipe.

(*Answer*: 0.014 m)

Thermal Insulation

The steady state situation of heat loss from an insulated pipe is preserved by the heat which arrives at the outer surface of the insulation being removed by natural convection. The air in the immediate vicinity of the pipe is heated, thereby reducing its density. The warm air rises and is replaced continually by cold air. In this way the outer surface is maintained at a constant temperature.

Adding insulation to a hot pipe decreases the outer surface temperature but also increases the surface area available for heat loss; it is therefore possible that an optimum insulation thickness exists.

Figure 2.5 illustrates the heat loss from a pipe by combined conduction and convection. It is assumed that the thermal conductivity of the insulation is much lower than that of the pipe. Hence the temperatures of the inside and outside surfaces of the pipe are essentially the same and independent of the insulation thickness.

The film heat transfer coefficient at the outer limit of the insulation is conventionally represented by h, which has the units W m^{-2} K^{-1}. It is anologous to the reciprocal of conductive resistance K/L, which also has the units W m^{-2} K^{-1}. For air, the value of h may be considered to be constant and independent of the insulation thickness and surface temperature.

The overall temperature gradient causing the heat transfer is $(t_1 - t_a)$, where t_a is the ambient air temperature.

The equation for the heat loss per unit length of pipe is:

$$Q = \frac{(t_1 - t_a)}{\dfrac{1}{2\pi r_3 h} + \dfrac{\ln(r_3/r_2)}{2\pi K}} \tag{2.28}$$

The heat loss, Q, will be a maximum when the resistance is a minimum. Differentiating the resistance with respect to insulation radius, r_3, and equating to zero for a minimum gives the critical radius, r_c, of the insulation:

$$r_c = \frac{K}{h} \tag{2.29}$$

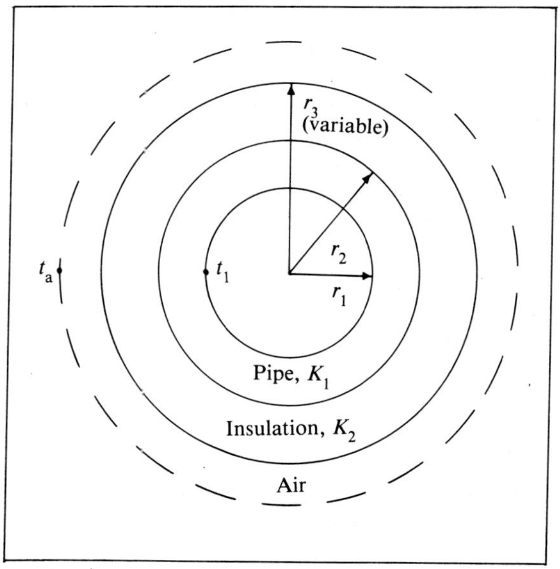

Figure 2.5 – Heat loss from an insulated pipe – critical thickness of insulation

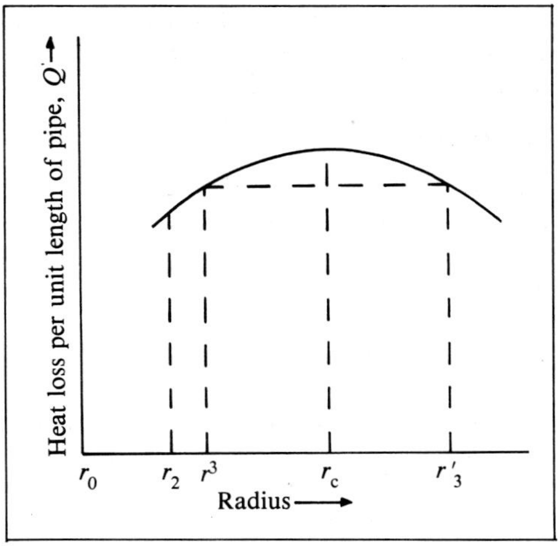

Figure 2.6 – Variation of heat loss with insulation thickness

The value of the film heat transfer coefficient, h, depends on air velocity and the general geometry of the pipe location, and to an extent is outside the designer's control. The critical radius is minimized, therefore, by selecting an insulation material with a low thermal conductivity.

In Figure 2.6 it will be observed that if the insulation radius, r_3, is less than the critical radius, r_c, the heat loss will increase with increasing insulation thickness, and that an increase in insulation thickness of $(r_3' - r_3)$ is necessary before the heat loss begins to decrease. This is a common occurrence with electrical heating elements and small-diameter pipes.

In most situations in process plants, the pipe radius will be greater than the critical radius and hence any insulation added will decrease the heat loss. The optimum thickness of lagging is determined by a cost analysis, including the capital and installation charges of the insulation, and the cost of fuel or power.

Insulation materials owe their insulation property to a structure consisting of minute pores filled with air or a gas which has a low thermal conductivity. If the pore volume is reduced by water absorption, compression or fusion, the thermal conductivity value will increase.

2.4 CONVECTION

Convective heat transfer occurs by the bulk mixing of hot and cold portions of the same fluid on a macroscopic scale. Portions of the cold fluid adjacent to a hot surface receive heat by conduction from the surface and then move along the temperature gradient to the colder regions of the fluid. On most occasions the rate of heat transfer is limited by conduction.

The fluid motion in convective heat transfer may be due to bouyancy resulting from the fluid's thermal expansion which causes density differences, or it may be due to mechanical means. The former is free or natural convection, the latter is forced convection.

As convection is generally concerned with heat transfer between a fluid and the surface of some solid body, it can be considered under four broad headings:

(a) Forced convection in turbulent flow
(b) Forced convection in transitional flow
(c) Forced convection in laminar flow
(d) Free convection.

Since convective heat transfer in the transitional flow region is uncommon in industrial practice, it will not be considered in any detail.

Although fluid velocities in free convection are generally lower than in forced convection, it is not unusual to have turbulent conditions in free convection.

Thermal Boundary Layer

Figure 2.7 illustrates heat transfer between a flat plate and a fluid flowing parallel to the plate. The fluid approaching the plate has a uniform temperature, t_{fs}, which is greater than the uniform temperature of the plate, t_w. The fluid adjacent to the plate cools and creates a temperature gradient; the fluid within the temperature gradient $t_{fs} - t_w$ is the thermal boundary layer; OL is the outer limit of this boundary layer. In the special case of the Prandtl number for the fluid being unity, the hydrodynamic and thermal boundary layers are coincident.

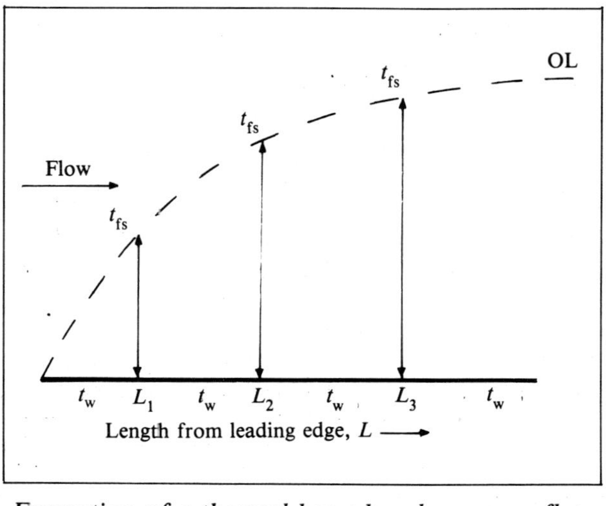

Figure 2.7 – Formation of a thermal boundary layer on a flat plate

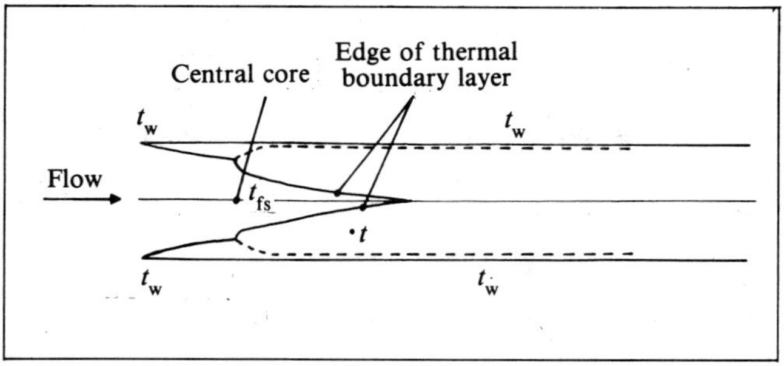

Figure 2.8 – Development of thermal boundary layer in a pipe

Figure 2.8 illustrates the development of the thermal boundary layer caused by heat transfer between the wall of a pipe and the fluid flowing through it. The fluid enters the pipe at a uniform temperature, t_{fs}, which is greater than the uniform temperature of the pipe wall, t_w. The thermal boundary layer increases in thickness with increasing distance from the pipe entrance and finally meets at the centre line of the pipe at a distance downstream of the pipe entrance – the thermal entry length. Downstream of this point the temperature profile becomes flatter and, if the pipe were sufficiently long, it would become completely flat at a temperature t_w. If this happens the temperature gradient disappears and heat transfer ceases.

The temperature profile across a pipe is influenced by the velocity profile, itself, in turn, influenced by the temperature profile, especially when the fluid viscosity is temperature sensitive. In this event the velocity profile and energy loss for isothermal and non-isothermal flow may be quite different.

Film Concept and Heat Transfer Coefficients

The film concept developed from the fact that the major part of the temperature gradient occurs within the relatively thin laminar and transitional regions of the thermal boundary layer. The resistance to heat transfer may therefore be regarded as a 'stagnant film' which covers the solid surface. The thickness and hence resistance to heat transfer of this film is reduced as the average fluid velocity increases. Normally no attempt is made to measure the film thickness, which depends on the Reynolds number and the geometry of the system.

The flow of heat by conduction through the film, which is assumed to adhere to the solid surface in contact with the fluid, could be expressed:

$$Q = \frac{KA\Delta T}{L} \tag{2.11}$$

where L is the effective thickness of the film. Since this is generally not known, the heat transfer equation is generally written:

$$Q = hA(t_1 - t_2) \tag{2.30}$$

where h = film heat transfer coefficient, W m^{-2} K^{-1}
$(t_1 - t_2)$ = temperature difference across the film, K.

The film coefficient is analogous to K/L in conductive heat transfer, hence $1/(hA)$ is a thermal resistance and Eqn 2.30 can be rewritten:

$$Q = \frac{t_1 - t_2}{\dfrac{1}{hA}} \tag{2.31}$$

Even in the idealized situation of heat transfer between a pipe wall at constant temperature and a fluid flowing through the pipe, the fluid temperature varies both along and across the pipe. Hence, in normal engineering problems of convective heat transfer, assumptions must be made. One such assumption is the average film heat transfer coefficient. It will be appreciated that there are also point and maximum heat transfer coefficients. Another general assumption is that the fluid properties may be evaluated at the arithmetic mean temperature of the fluids entering and leaving the system.

Figure 2.9 illustrates the temperature gradients for the case where heat is transferred from a hot fluid through a solid surface to a colder fluid. The dashed lines F_1F_1 and F_2F_2 on either side of the solid represent the boundaries of the 'stagnant film'. Outside these boundaries the fluids are assumed to be in turbulent flow.

The temperature gradient from the bulk of the hot fluid to the solid surface is represented by the line $t_a t_b t_c$, where t_a is the maximum temperature of the hot fluid, t_b is the temperature of the boundary between the turbulent and 'stagnant' regions, and t_c is the temperature at the solid surface. The line $t_d t_e t_f$ has a similar significance for the cold fluid.

In heat transfer calculations it is customary and also convenient not to use t_a, t_b or t_c, but the 'average fluid temperature' or bulk fluid temperature – that temperature which could result from complete mixing of the fluid – as represented by the lines MM and NN for the hot and cold fluids respectively.

The steady state heat transfer equations for the system illustrated by Figure 2.9 are:

$$Q = h_H A_H \, \Delta T_H = K A_W \, \Delta T_W / L = h_C A_C \;\; T_C \tag{2.32}$$

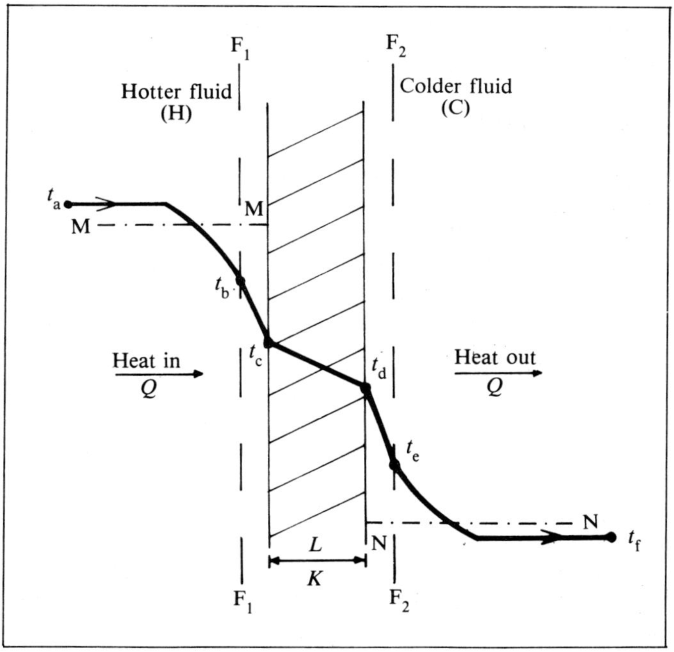

Figure 2.9 – Temperature gradients for convective heat transfer

In terms of the overall temperature difference for the process, ΔT_o, Eqn 2.32 may be rewritten:

$$Q = \frac{\Delta T_o}{\dfrac{1}{h_H A_H} + \dfrac{L}{K A_W} + \dfrac{1}{h_C A_C}} = \frac{\Delta T_o}{\Sigma R} \tag{2.33}$$

where the letters H, W, and C, refer to the hot fluid, solid surface or wall, and cold fluid respectively. The term R represents the total thermal resistance.

When the heat transfer is occurring across a plane surface the area terms A_H, A_W and A_C are all the same, and Eqn 2.33 may be written:

$$Q = UA\ \Delta T_o \tag{2.34}$$

where U = overall heat transfer coefficient, W m^{-2} K^{-1}.

The reciprocal of the overall heat transfer coefficient is a resistance, R. The summation of resistances in series is: $R = R_1 + R_2 + R_3 +$ etc. Since U,

h_H, K/L and h_C are conductances, then:

$$\frac{1}{U} = \frac{1}{h_H} + \frac{L}{K} + \frac{1}{h_C} \tag{2.35}$$

If the hot and cold surface areas are not equal, as is the case for a cylindrical pipe, then the heat transfer formulae should contain modified area terms and not the simple area, A. The overall heat transfer coefficient, U, for a cylindrical pipe may be based on the area of the hot fluid, A_H, the area of the cold fluid, A_C, or the mean area of the pipe, A_m. Therefore, before an overall heat transfer coefficient is established, a definite area must be selected and the coefficient so determined is automatically based on this area. If in Eqn 2.33 the area selected is that corresponding to the hot fluid, then the overall heat transfer coefficient based on the hot fluid area is:

$$\frac{1}{U_H} = \frac{1}{h_H} + \frac{L}{K}\left(\frac{A_H}{A_m}\right) + \frac{1}{h_C}\left(\frac{A_H}{A_C}\right) \tag{2.36}$$

Since diameter d is proportional to area A, Eqn 2.36 may be written in a more convenient form:

$$\frac{1}{U_H} = \frac{1}{h_H} + \frac{L}{K}\left(\frac{d_H}{d_m}\right) + \frac{1}{h_C}\left(\frac{d_H}{d_C}\right) \tag{2.37}$$

It will be appreciated that the products $U_H A_H$, $U_m A_m$ and $U_C A_C$ are identical, hence the larger the area term the lower will be the numerical value of the corresponding overall heat transfer coefficient.

The individual film heat transfer coefficients can be associated with:

(a) Heat transfer to or from fluids across a solid surface without phase change
(b) Heat transfer from condensing vapours
(c) Heat transfer to boiling liquids
(d) Fouling or dirt factors.

The value of the film heat transfer coefficient varies considerably depending on the process. A selection of typical values is given in Table 2.2.

TABLE 2.2
Values of Film Heat Transfer Coefficient

Fluid	h(W m^{-2} K^{-1})
Liquid metals – forced convection	5×10^3–100×10^3
Steam – dropwise condensation	30×10^3–100×10^3
Steam – filmwise condensation	6×10^3–20×10^3
Organic vapours – condensing	1×10^3–2×10^3
Water – boiling	2×10^3–50×10^3
Water – forced convection	300–20×10^3
Oils – forced convection	60–2×10^3
Steam – superheating	30–100
Air – forced convection	10–200
Air – free convection	2–10
Gas flow across tube banks	10–60
Liquid flow across tube banks	60–600

Typical values of the overall heat transfer coefficient for various processes are given in Table 2.3.

TABLE 2.3
Typical Values of the Overall Heat Transfer Coefficient

Process (forced convection)	U(W m^{-2} K^{-1})
Hot gas to water	10–300
Steam to gas	30–300
Steam to medium-viscosity oils	300–600
Steam to water	1 000–4 000
Hot water to water	1 500–3 000

In practice, allowance should be made for the thermal resistance of the deposits which accumulate on the surfaces in contact with the hot and cold side fluids. This fouling builds up with use. The degree of fouling for a particular system tends to increase with increase in temperature level, but decreases with increase in fluid velocity.

The overall heat transfer coefficient, which includes the fouling factors, is called the design or dirty coefficient. Based on the hot fluid, the design overall heat transfer coefficient is:

$$\frac{1}{U_H} = \frac{1}{h_{Hf}} + \frac{1}{h_H} + \frac{L}{K}\left(\frac{d_H}{d_m}\right) + \frac{1}{h_C}\left(\frac{d_H}{d_C}\right) + \frac{1}{h_{Cf}}\left(\frac{d_H}{d_C}\right) \quad (2.38)$$

where h_{Hf} and h_{Cf} are the fouling heat transfer coefficients for the hot and cold fluids respectively.

Clearly, heat transfer surfaces require periodic cleaning to limit the build-up of solid deposits which increase the thermal resistance and decrease the value of the overall heat transfer coefficient. Numerical values of fouling factors usually represent the thermal resistance which is likely to accumulate over a period of one year. Typical values of fouling factors are given in Table 2.4.

TABLE 2.4

Typical Values of Heat Transfer Fouling Factor on the Basis of One Year's Service

Fluid	h_f(W m^{-2} K^{-1})
Distilled water	12 000
Hard water	2 000
Organic vapours	12 000
Organic liquids	6 000
Fuel oil	1 000

Problem 2.5

Steam condenses on the outside of a clean 0.025 m outside-diameter brass pipe which has a wall thickness of 1.0×10^{-3} m. The heat is transferred to water flowing through the pipe. If the steam side and water side film heat transfer coefficients are 11×10^3 and 10×10^3 W m^{-2} K^{-1} respectively, and the thermal conductivity of the brass is 110 W m^{-1} K^{-1}, calculate the overall heat transfer coefficient based on outside area.

After a period of duty the fouling heat transfer coefficient on each side is 10×10^3 W m^{-2} K^{-1}, calculate the new overall heat transfer coefficient based on outside area.

(*Answers*: 4 780 W m^{-2} K^{-1}; 2 390 W m^{-2} K^{-1})

Overall Temperature Difference

When heat is transferred from one fluid to another across a pipe surface, the temperature of both fluids may vary over the entire length of the pipe as illustrated by Figure 2.10.

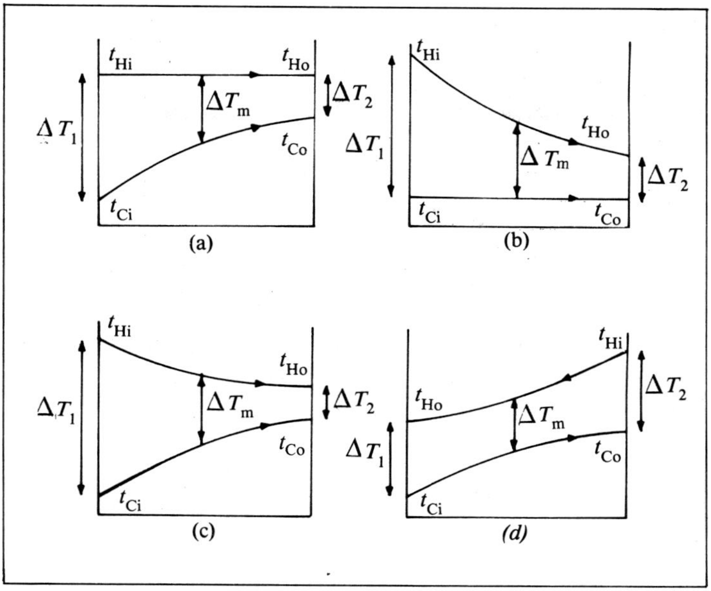

Figure 2.10 – Temperature profiles for: (a) boiling liquid; (b) condensing vapour; (c) co-current flow; (d) counter-current flow

Figure 2.10(a) represents heat transfer to a cold fluid from a vapour condensing at constant temperature. Figure 2.10(b) represents the vaporization of a liquid at constant temperature by heat transfer from a hot fluid. Figures 2.10(c) and (d) represent heat transfer without phase change for co-current and counter-current fluid flow respectively. In Figure 2.10 t_{Hi} and t_{Ho} refer to the inlet and outlet temperatures of the hotter fluid; t_{Ci} and t_{Co} have the same significance for the colder fluid. The maximum, mean and minimum temperature differences between the two fluids are respectively ΔT_1, ΔT_m and ΔT_2.

The equation for heat transfer from, or to, a fluid flowing inside a pipe, as illustrated by Figure 2.10, is:

$$Q = UA\ \Delta T_m \tag{2.39}$$

where ΔT_m = logarithmic mean temperature difference, LMTD, K.

The logarithmic mean temperature difference is defined:

$$\text{LMTD} = \frac{\Delta T_1 - \Delta T_2}{\ln\left(\dfrac{\Delta T_1}{\Delta T_2}\right)} \tag{2.40}$$

where ΔT_1 = larger terminal temperature difference, K
ΔT_2 = smaller terminal temperature difference, K.

Figure 2.10 shows that the actual definitions of ΔT_1 and ΔT_2 depend upon whether the two fluids are flowing co-currently or counter-currently.

Problem 2.6
A liquid is cooled counter-currently in a heat exchanger from 70°C to 30°C at a rate of 1.24 kg s^{-1} by 2.75 kg s^{-1} of water which has an inlet temperature of 20°C. When the mean surface area of the heat exchanger is 7.50 m^2 and the mean specific heat capacities of the liquid and water are 3.25×10^3 and 4.18×10^3 J kg^{-1} K^{-1} respectively, calculate the exit water temperature and the overall heat transfer coefficient.
(*Answers*: 34°C; 1 060 W m^{-2} K^{-1})

Dimensionless Groups
The five main techniques which have contributed to the present state of understanding of convective heat transfer are:
(a) Dimensional analysis combined with the experimental determination of the numerical constants
(b) Empirical formulae from an experimental data
(c) Pilot-plant studies and scale-up factors
(d) Mathematical analysis of momentum transfer, translated into thermal units by anologies between momentum and heat transfer
(e) Exact mathematical solutions of the boundary layer equations.

Dimensional analysis combined with experimentation is the technique most commonly used in determining equations for predicting film heat transfer coefficients for process calculations. Dimensional analysis reveals the dimensionless groups which are involved in a particular heat transfer process, but it does not give the values of the dimensionless constants which appear in the correlations. These must be obtained for investigations which study the results of varying the dimensionless groups.

The parameters which can influence the thermal conductivity and thickness of the 'stagnant film' and hence the film heat transfer coefficient are given in Table 2.5.

TABLE 2.5
Parameters which Influence Film Heat Transfer Coefficient

Parameter	Symbol	Dimensions
Size and shape of surface, expressed in terms of 'diameter'	d	L
Thermal conductivity of fluid	K	$MLT^{-3}\theta^{-1}$
Specific heat capacity of fluid	c_p	$L^2T^{-2}\theta^{-1}$
Fluid viscosity	μ	$ML^{-1}T^{-1}$
Fluid density	ρ	ML^{-3}
Fluid velocity	v	LT^{-1}
Temperature difference across film	ΔT	θ
Coefficient of thermal expansion	β	θ^{-1}
Gravitational acceleration	g	LT^{-2}

The mathematical relationship between the film heat transfer, h, and the variables in Table 2.5 is:

$$h = f(d, K, c_p, \mu, \rho, v, \Delta T, \beta, g) \tag{2.41}$$

Applying dimensional analysis and grouping the variables, it can be shown that the heat transfer process can be represented:

$$\mathrm{Nu} = C\,\mathrm{Re}^x\,\mathrm{Pr}^y\,\mathrm{Gr}^z \tag{2.42}$$

where Nu = Nusselt group, hd/K
Re = Reynolds group, $dv\rho/\mu$
Pr = Prandtl group, $c_p\mu/K$
Gr = Grashof group, $d^3\rho^2 g\beta\Delta T/\mu^2$.

In forced convection, the Grashof number is negligible compared with the Reynolds number and hence Eqn 2.42 may be written:

$$\mathrm{Nu} = C\,\mathrm{Re}^x\,\mathrm{Pr}^y \tag{2.43}$$

In free convection, the Reynolds number is negligible compared with the Grashof number, hence in free convection the equation for the heat transfer may be written:

$$\mathrm{Nu} = C\,\mathrm{Pr}^y\,\mathrm{Gr}^z \tag{2.44}$$

The constants in Eqns 2.42 to 2.44 differ for various heat transfer situations and hence must be determined experimentally.

Nusselt Number

$$\mathrm{Nu} = hd/K \tag{2.45}$$

The Nusselt number, which is sometimes referred to as the dimensionless heat transfer coefficient, is a measure of heat transfer across a 'film'. The film heat transfer coefficient, h, is the average coefficient; there are also point or local values of the film heat transfer coefficient, h_x, and the Nusselt group, Nu_x.

Reynolds Number

$$\mathrm{Re} = dv\rho/\mu \tag{2.46}$$

As the Reynolds number increases, the degree of turbulence increases, and therefore it is to be expected that in equations involving turbulent flow the Reynolds number will be important. As the Reynolds number increases the temperature distribution and hence the fluid viscosity alters. This means that one heat transfer correlation does not hold over the entire Reynolds number range.

Prandtl Number

$$\mathrm{Pr} = c_p\,\mu/K \tag{2.47}$$

The Prandtl group, which contains only the properties of a fluid, provides a useful index to the thermal properties of a fluid and may be used to divide fluids into four main classes.

Fluid	*Prandtl number range*
(a) Liquid metals	0.004–0.04
(b) Gases	0.6–0.9
(c) Low-viscosity liquids	1–10
(d) High-viscosity liquids	up to about 1 000

The low Prandtl numbers associated with liquid metals is due mainly to their high thermal conductivities.

When experimental data are not available for gases, their Prandtl numbers may be estimated from the correlation:

$$\mathrm{Pr} = \frac{4}{9 - \dfrac{5}{\gamma}} \tag{2.48}$$

where γ is the ratio of the specific heat capacities of the gas, c_p/c_v. The value of γ for monatomic, diatomic and polyatomic gases may be taken as 1.66, 1.40 and 1.28 respectively.

The shape of the hydrodynamic boundary layer depends on the value of the momentum diffusivity or kinematic viscosity of the fluid, ν or μ/ρ. The shape of the thermal boundary layer depends on the thermal diffusivity of the fluid, α or $K/(\rho c_p)$.

The Prandtl group may be written as the ratio of the momentum and thermal diffusivities:

$$\mathrm{Pr} = \frac{c_p\mu}{K} = \frac{c_p\mu\,\rho}{k\,\rho} = \frac{\mu}{\rho}\cdot\frac{\rho c_p}{K} = \frac{\nu}{\alpha} \tag{2.49}$$

Hence the Prandtl number relates the temperature profile to the velocity profile.

If, in Figures 2.7 and 2.8, the temperature, t, at any point in the boundary layer is replaced by the dimensionless variable $(t-t_w)/(t_o-t_w)$, or $\Delta T_{actual}/\Delta T_{maximum}$, then this temperature ratio is analogous to the velocity ratio, v/v_{max}, associated with the hydrodynamic boundary layer. For fully developed flow in a pipe, t_o is the temperature of the fluid at the centre line of the pipe.

Figure 2.11 illustrates the relationships between the temperature and velocity profiles for fully developed turbulent flow in a pipe. When the Prandtl number is unity, the temperature and velocity profiles are coincident. For Prandtl numbers less than unity, the temperature gradient near the pipe wall is less steep than the velocity gradient, vice versa at Prandtl numbers greater than unity; the majority of the temperature gradient occurs in the laminar sublayer.

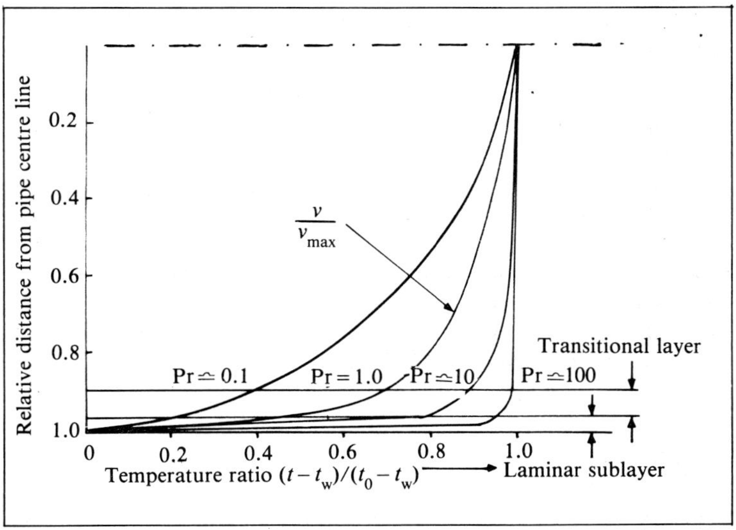

Figure 2.11 – Effect of Prandtl number on the temperature profile for fully developed turbulent flow

Grashof Number

$$\mathrm{Gr} = d^3 \rho^2 g \beta \Delta T / \mu^2 \tag{2.50}$$

The Grashof group applies to free convective problems; it will be noted that a velocity term, as such, does not appear in the group. Since the motion is ultimately due to gravity acting upon the density variation, gravitational acceleration, g, must be included. As the Grashof number increases, the degree of free convection increases. For free convection to be appreciable, the Grashof number should exceed about 0.1×10^9. For free convection with turbulence, the Grashof number should exceed about 10×10^9.

Stanton and Peclet Groups

Two derived dimensionless groups associated with convective heat transfer are the Stanton group, St, and the Peclet group, Pe; they are defined:

$$\mathrm{St} = \mathrm{Nu}\ \mathrm{Re}^{-1}\ \mathrm{Pr}^{-1} = \frac{h}{c_p \rho v} \tag{2.51}$$

$$\mathrm{Pe} = \mathrm{Re}\ \mathrm{Pr} = \frac{d v \rho c_p}{K} \tag{2.52}$$

The physical properties of a fluid that are temperature dependent will change as the fluid undergoes heat transfer. Generally, the numerical values of the properties used to calculate the dimensionless numbers associated with heat transfer are those corresponding to the arithmetic mean of the inlet and outlet fluid temperatures. For large temperature differences, specific temperature functions are used.

Problem 2.7

Using the data below, calculate the corresponding Nusselt, Reynolds, Prandtl, Grashof, Stanton and Peclet numbers.

Data:

Pipe diameter = 50×10^{-3} m
Fluid velocity = 25.0 m s^{-1}
Fluid density = 1.10 kg m^{-3}
Fluid viscosity = 20.0×10^{-6} N s m^{-2}
Specific heat of fluid = 1 000 J kg^{-1} K^{-1}
Thermal conductivity of fluid = 28.0×10^{-3} W m^{-1} K^{-1}
Film heat transfer coefficient = 45.0 W m^{-2} K^{-1}
Coefficient of thermal expansion = 3.20×10^{-3} K^{-1}
Gravitational acceleration = 9.807 m s^{-2}
Temperature difference = 25.0 K

(*Answers*: 80.4; 68.8×10^3; 0.71; 297×10^3; 1.64×10^{-3}; 49.1×10^3)

Forced Convection, Turbulent Flow

I Inside Film Heat Transfer Coefficients

For heat transfer to or from a fluid flowing in turbulent flow inside a clean circular pipe, the general dimensionless Eqn 2.43 reduces to the Dittus–Boelter equation:

$$\mathrm{Nu} = 0.023\ \mathrm{Re}^{0.8}\ \mathrm{Pr}^{n} \tag{2.53}$$

where $n = 0.3$ when the fluid inside the pipe is being cooled
$n = 0.4$ when the fluid inside the pipe is being heated.

The difference in the value of the exponent on the Prandtl group is due to the temperature effects on fluid viscosity, which affects the thickness of the laminar sublayer. The Dittus–Boelter equation has been found to give reasonable agreement with experimental results over the Reynolds number range 10×3^3 to 0.5×10^6 and Prandtl number range 0.7 to 100.

For viscous fluids, there is a marked difference, at any cross-section, between the viscosity of the fluid adjacent to the pipe wall and the viscosity of the bulk fluid. For Prandtl numbers up to about 10×10^3, the Sieder–Tate equation is recommended:

$$\mathrm{Nu} = 0.027\ \mathrm{Re}^{0.8}\ \mathrm{Pr}^{0.33}\ (\mu_b/\mu_w)^{0.14} \tag{2.54}$$

where μ_b = fluid viscosity evaluated at the arithmetic mean bulk temperature of the fluid, N s m^{-2}
μ_w = fluid viscosity evaluated at the pipe wall temperature, N s m^{-2}.

An equation which is useful in momentum, heat and mass transfer analogy studies involves the Colburn *j*-factor. If the Prandtl exponent in the Dittus–Boelter equation is taken as 0.33, the equation can be rearranged by dividing by Re $\mathrm{Pr}^{1/3}$:

$$\mathrm{St}\ \mathrm{Pr}^{2/3} = 0.023\ \mathrm{Re}^{-0.2} \tag{2.55}$$

The term St $\mathrm{Pr}^{2/3}$ is the Colburn *j*-factor for heat transfer, j_h.

For fluids in turbulent flow inside a circular pipe the Fanning friction factor may be expressed:

$$f = 0.046\ \mathrm{Re}^{-0.2} \tag{1.59}$$

hence $$f/2 = 0.023\ \mathrm{Re}^{-0.2} \tag{2.56}$$

and hence $$j_h = f/2 = j_f \tag{2.57}$$

where j_f is the basic friction factor or the *j*-factor for turbulent fluid flow in Eqn 1.55.

Example 2.2
Using the Dittus–Boelter equation, calculate the film heat transfer coefficient when water at a velocity of 1.12 m s^{-1} is heated as it flows through a 50×10^{-3} m diameter pipe. The mean properties of the water are: density 1 000 kg m^{-3}; viscosity 0.544×10^{-3} N s m^{-2}; specific heat capacity 4.18×10^{3} J kg^{-1} K^{-1}; thermal conductivity 0.643 W m^{-1} K^{-1}.

$$\mathrm{Re} = \frac{dv\rho}{\mu} = \frac{50 \times 10^{-3} \times 1.12 \times 1000}{0.544 \times 10^{-3}} = 103 \times 10^{3}$$

$$\mathrm{Re}^{0.8} = 10.2 \times 10^{3}$$

$$\mathrm{Pr} = \frac{c_p\mu}{K} = \frac{4.18 \times 10^{3} \times 0.544 \times 10^{-3}}{0.643} = 3.54$$

$$\mathrm{Pr}^{0.4} = 1.66$$

$$\mathrm{Nu} = 0.023\ \mathrm{Re}^{0.8}\ \mathrm{Pr}^{0.4}$$

$$= 0.023 \times 10.2 \times 10^{3} \times 1.66 = 389$$

$$h = \frac{389\,K}{d} = \frac{389 \times 0.643}{50 \times 10^{-3}} = 5\ 000\ \mathrm{W\ m^{-2}\ K^{-1}}$$

Problem 2.8
A liquid, flowing at a velocity of 2.44 m s^{-1} inside a pipe, 26.7×10^{-3} m inside diameter, 31.7×10^{-3} m outside diameter, is heated by steam condensing on the outside surface of the pipe. Calculate the overall heat transfer coefficient, based on the outside surface area of the pipe, when the steam-side film heat transfer coefficient is 8 500 W m^{-2} K^{-1}, based on outside area, and the liquid-side film heat transfer coefficient is given by the Dittus–Boelter equation: $\mathrm{Nu} = 0.023\ \mathrm{Re}^{0.8}\ \mathrm{Pr}^{0.4}$.

The thermal conductivity of the metal is 97.8 W m^{-1} K^{-1} and the mean properties of the liquid at the operating conditions are: thermal conductivity 0.145 W m^{-1} K^{-1}; density 801 kg m^{-3}; specific heat capacity 1.88×10^{3} J kg^{-1} K^{-1}; viscosity 2.10×10^{-3} N s m^{-2}.

(*Answer*: 1 090 W m^{-2} K^{-1} outside surface area)

II Outside Film Heat Transfer Coefficients
For heat transfer to or from a fluid in turbulent flow at right angles to circular pipes, the average film heat transfer coefficient can be determined from equations of this type:

$$\mathrm{Nu_o} = C\,\mathrm{Re_o}^{x}\,\mathrm{Pr}^{y} \qquad (2.58)$$

where the subscript o refers to the values of the dimensionless groups based on the outside diameter of the pipe. The value of the constant y is generally taken to be 0.33. The values of C and x depend on the value of the Reynolds number, the number of rows of pipes and the configuration of the pipes. Values of C and x are tabulated in the literature.

Convection with Phase Change

I Condensation

When a saturated pure vapour, such as steam, condenses by transferring its latent heat to a solid surface, the condensation may occur by one of two distinct mechanisms:

(a) Filmwise condensation
(b) Dropwise condensation.

In filmwise condensation, the angle of contact between the liquid and the solid surface is small and the liquid 'wets' the surface to form a continuous film of condensate. This film may flow in laminar or turbulent conditions, depending upon the rate of condensation, the path length of the film, the fluid properties and the geometry of the system.

In dropwise condensation, the angle of contact between the liquid and the solid surface is greater than about one hundred degrees. The liquid does not wet the surface, but forms small droplets at microscopic nucleation sites, such as scratches, pits and specks of dirt. The droplets grow and coalesce with their neighbours to form visible fine drops. These drops continue to grow and finally run down the surface in rivulets, leaving an apparently bare surface exposed to the vapour. The resistance to heat transfer at these areas is low, and hence the heat transfer rates are high compared with filmwise condensation. The heat transfer coefficients for the filmwise and dropwise condensation of steam under comparable process and system conditions are typically: 10×10^3 W m^{-2} K^{-1} and 100×10^3 W m^{-2} K^{-1} respectively.

The factors that cause condensation to occur by one or other of the two mechanisms are largely unknown. A system may change rapidly, entirely or partially, from one mechanism to the other. In general, smooth, clean, non-wettable surfaces tend towards dropwise. In the laboratory it is possible, using chemical dosing, to promote a stable dropwise condensate of steam for short periods of time.

In the case of dropwise condensation there is no method available for predicting heat transfer coefficients.

For filmwise condensation, from a vapour free of non-condensable gases, Nusselt derived equations to predict the heat transfer coefficients. For horizontal pipes the equation is:

$$h_H = 0.725 \left(\frac{K^3 \rho^2 g \lambda}{d \mu \Delta T} \right)^{0.25} \tag{2.59}$$

and for vertical pipes:

$$h_V = 0.943 \left(\frac{K^3 \rho^2 g \lambda}{L \mu \Delta T} \right)^{0.25} \tag{2.60}$$

where h_H and h_V = film heat transfer coefficients for horizontal and vertical pipes, W m^{-2} K^{-1}
K = thermal conductivity of condensate, W m^{-1} K^{-1}
ρ = density of condensate, kg m^{-3}
g = gravitational acceleration, m s^{-2}
λ = specific latent heat of vaporization, J kg^{-1}
d = outside diameter of pipe, m
μ = viscosity of condensate, N s m^{-2}
ΔT = temperature difference between vapour and pipe wall, K
L = active length of pipe, m.

In Eqns 2.59 and 2.60 the physical properties of the condensate are evaluated at the arithmetic mean temperature of the saturated vapour, t_{SV}, and of the pipe wall, t_W.

It will be observed from the form of the Nusselt equations that the film heat transfer coefficient decreases as the temperature difference, ΔT, increases.

Any non-condensable gas mixed with the vapour will act as thermal insulation and reduce the rate of heat transfer. One percent of a non-condensable gas can reduce the heat transfer rate by 50%.

A selection of typical values of film heat transfer coefficients for filmwise condensation is given in Table 2.6.

TABLE 2.6

Typical Values of Film Heat Transfer Coefficients for Filmwise Condensation

Condensing vapour	h(W m^{-2} K^{-1})
Steam	10 000
Benzene	1 800
Ammonia	6 000
Freon – 12	1 800
Ethanol	2 000

When the conditions of operation of condensing equipment cannot be predicted with certainty, it is usual to assume a film heat transfer coefficient of 10 000 W m^{-2} K^{-1} for condensing steam.

Problem 2.9

Using the Nusselt equations for filmwise condensation and the data given below, calculate the average outside film heat transfer coefficients for a 25×10^{-3} m outside-diameter, 1.50 m long pipe in the horizontal and vertical positions when the temperature difference, ΔT, is 10°C.

Data:

Gravitational acceleration = 9.807 m s^{-2}
Specific latent heat = 2 260 × 10^3 J kg^{-1}
Thermal conductivity of condensate = 0.680 W m^{-1} K^{-1}
Density of condensate = 960 kg m^{-3}
Viscosity of condensate = 290 × 10^{-6} N s m^{-2}

(*Answer*: 12 500 W m^{-2} K^{-1} and 5 850 W m^{-2} K^{-1})

Problem 2.10

Dry saturated steam, having a specific latent heat of 2 230 × 10^3 J kg^{-1} undergoes filmwise condensation on the outside of a horizontal pipe. The heat is transferred to water flowing through the pipe at 2.45 m s^{-1}. When the inside and outside diameters of the pipe are 25×10^{-3} and 30×10^{-3} m and the temperature difference, ΔT, is 20°C, calculate the overall heat transfer coefficient based on outside area.

The thermal conductivity of the pipe metal is 45.0 W m^{-1} K^{-1}, and the properties of the condensate and water at the appropriate mean operating conditions are:

	Condensate	*Water*
Thermal conductivity, W m^{-1} K^{-1}	0.684	0.658
Density, kg m^{-3}	950	980
Viscosity, N s m^{-2}	252×10^{-6}	430×10^{-6}
Specific heat capacity, J kg^{-1} K^{-1}		4.19×10^3

(*Answer:* 3 870 W m^{-2} K^{-1})

II Boiling

Boiling is a convective heat transfer process consisting of the formation and growth of bubbles of vapour within the bulk of the liquid. It cannot occur until the temperature close to the heater has increased sufficiently for the vapour pressure to exceed the pressure on the liquid, i.e. superheating must occur.

Heat transfer to boiling liquids is very important in numerous operations, such as steam generation, evaporation and refrigeration.

The different regimes of boiling have radically different mechanisms. These regimes have been identified and the values of the film heat transfer coefficients associated with them measured. The film heat transfer coefficients depend on many parameters, including the temperature difference, ΔT, between the heating surface and the boiling liquid. The relationships between ΔT and the film heat transfer coefficient are illustrated by Figure 2.12.

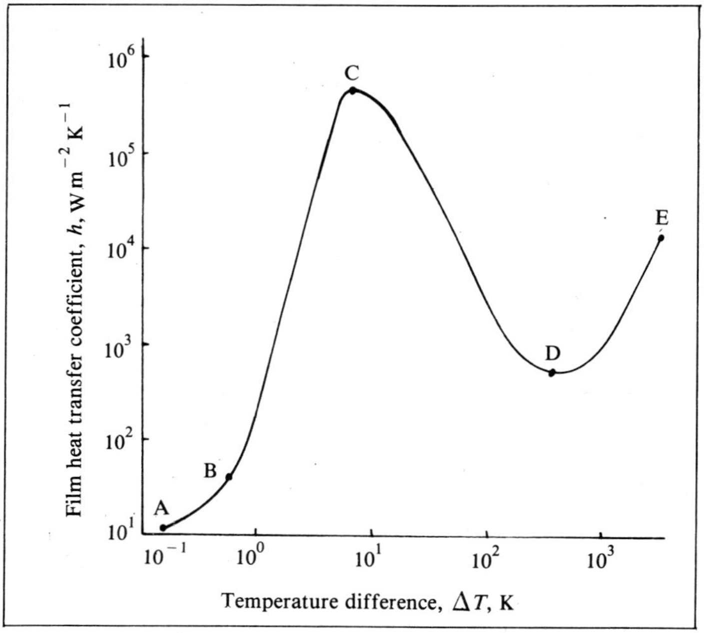

Figure 2.12 – Relationship between ΔT and film heat-transfer coefficient for boiling at a submerged surface

In region AB, heat is transferred through the bulk liquid by superheated liquid rising to the liquid–vapour interface where vaporization occurs.

In the region BC, more bubbles are created from more nucleation sites as a consequence of the larger ΔT values. The concentration of bubbles reaches a maximum at point C. When this occurs the vapour bubbles form a complete vapour blanket at the heater surface preventing liquid from reaching it. The temperature difference at point C is called the 'critical excess temperature'.

Nucleate boiling is the most important regime in industrial boiling. Its effectiveness depends primarily on the ease with which bubbles form and detach themselves from the heating surfaces.

In the unstable region CD, the heat has to be transferred by conduction through the vapour blanket which covers the heater surface. The film heat transfer coefficient decreases with increasing temperature difference from C to D.

Further increase in ΔT in region DE results in film boiling and an increase in the film heat transfer coefficient. In this region ΔT is about 300°C and the heater surface is glowing. The heat is transferred through the vapour film by a combination of conduction, convection and radiation, radiation becoming more important as ΔT increases.

Point E corresponds to the melting temperature of the heater material and, if achieved, the condition is known as 'burn-out'.

2.5 RADIATION

Radiative heat transfer usually occurs simultaneously with conductive and convective heat transfer; its relative importance increases with increasing temperature.

Any body at a temperature above absolute zero radiates energy continuously, in all directions, in the form of electromagnetic waves. These electromagnetic waves travel in straight lines and a second body can intercept the energy only if it can 'see' the radiating body. The energy falling on the second body may be partially absorbed, partially reflected and partially transmitted through the body as illustrated by Figure 2.13. The fractions that are absorbed, reflected and transmitted are called respectively: the adsorptivity, α, the reflectivity, ρ, and the transmissivity, τ. The sum of the three fractions is unity:

$$\alpha + \rho + \tau = 1.0 \tag{2.61}$$

If a body absorbs all the incident radiation, then $\alpha = 1$ and $\rho + \tau = 0$. A surface which absorbs all the visible light falling on it appears black. By analogy, a surface which absorbs all the thermal radiation it receives is called black, irrespective of its actual colour in visible light.

Electromagnetic waves cover a very wide range of wavelengths, from cosmic rays, having a wavelength of about 10^{-14} m, to radio waves, having wavelengths of about 10×10^3 m. The wavelengths involved in radiative

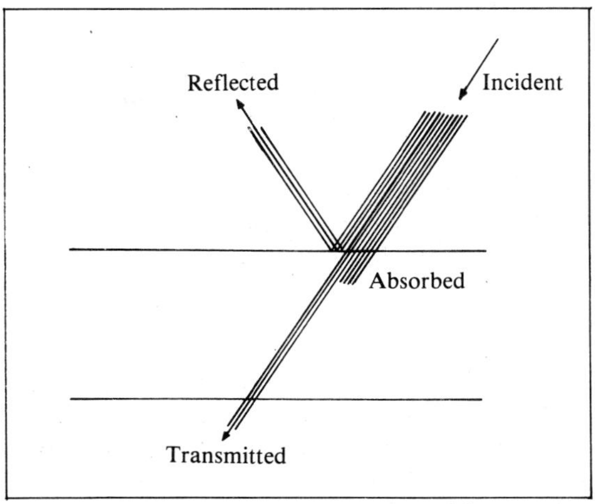

Figure 2.13 – Absorption, reflection and transmission of radiant energy

heat transfer are generally considered to cover the range 0.5×10^{-6} to 50×10^{-6} m. Visible light rays cover the range 0.4×10^{-6} to 0.8×10^{-6} m.

Fundamentals of Black Body Radiation

Stefan–Boltzmann Law

The total emissive power of a black body is proportional to the fourth power of its absolute temperature. The resulting relationsip is known as the Stefan–Boltzmann law:

$$Q_b = \sigma A T^4 \tag{2.62}$$

where Q_b = rate at which energy is radiated from unit area of the black body at an absolute temperature T, W
σ = Stefan–Boltzmann constant, 56.69×10^{-9} W m^{-2} K^{-4}
A = area of radiating surface, m^2
T = absolute temperature, K.

For convenience in calculations, Eqn 2.62 is often written:

$$Q_b = 5.669A \left(\frac{T}{100} \right)^4 \tag{2.63}$$

Planck's Distribution Law

A black body at any particular temperature will emit radiation at various wavelengths. Figure 2.14 illustrates the spectral energy distribution for a black body at two temperatures. The relationship represented by a curve of emissive power against wavelength is known as Planck's distribution law:

$$Q_{b\lambda} = \frac{2 \pi h c^2 \lambda^{-5}}{e^{(ch/k\lambda T)} - 1} \tag{2.64}$$

where $Q_{b\lambda}$ = monochromatic emissive power at wavelength λ, W m^{-2}
h = Planck's constant, 662.5×10^{-36} J s
c = velocity of light, 2.998×10^8 m s^{-1}
λ = wavelength, 10^{-6} m
k = Boltzmann's constant, 13.81×10^{-24} J K^{-1}
T = absolute temperature of black body, K.

In the energy distribution diagram, Figure 2.14, it will be observed that an increase in temperature causes a decrease in the wavelength, λ_{max}, at which maximum emissive power, Q_{max}, occurs. Increase in temperature causes a rapid increase in energy emission at any particular wavelength.

The total rate of energy emission at any temperature is given by the area under the curve for that temperature.

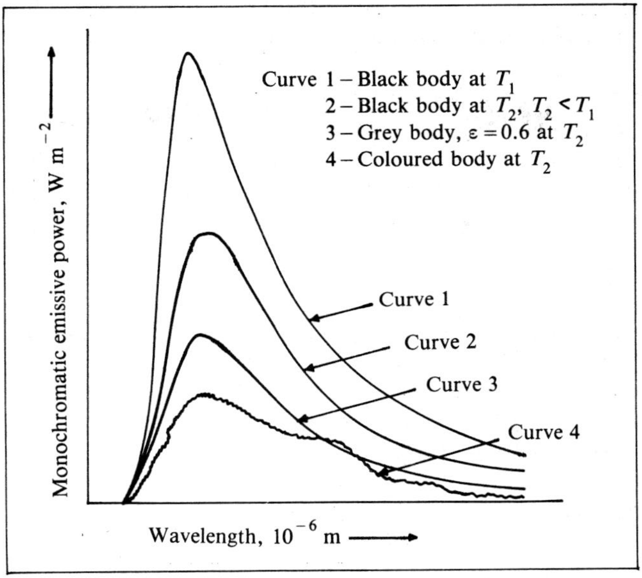

Figure 2.14 – Spectral energy distribution of different bodies

Wien's Displacement Law
Wien's displacement law states that at any particular temperature the maximum wavelength for maximum monochromatic emissive power varies inversely with absolute temperature:

$$\lambda_{max} T = \text{constant} \tag{2.65}$$

Kirchoff's Law
Kirchoff's law states that at thermal equilibrium the emissivity, ε, and the absorptivity, α, of a body are the same. Emissivity, ε, is defined as the ratio of the emissive power of an actual surface to that of a black body.

Inverse Square Law
The intensity of radiation from a body decreases with the square of the distance from the body.

Black and Non-black Bodies
All materials have absorptivities and reflectivities between zero and unity. A black body has an absorptivity of unity and a reflectivity of zero; it absorbs all the incident energy it receives and reflects none. No real surface is truly 'black' and hence all real surfaces have absorptivities and emissivities less than unity.

For a black body, the total radiation emitted is the maximum possible at a particular temperature, so too is the amount emitted at each wavelength. At a particular temperature, the energy distribution curves for real bodies will lie below that of a black body, see Figure 2.14.

The ratio of the emissive power of a real body at any particular wavelength, λ, to that of a black body at the same wavelength and temperature is called the monochromatic emissivity, ε_λ.

A 'grey' body is defined as a material whose monochromatic emissivity is independent, or essentially independent, of wavelength.

Materials which exhibit considerable variation in emissivity with wavelength are called 'coloured' bodies or selective emitters.

In general, the emissivities of materials fall into several groups:

(a) Polished materials $\varepsilon = 0.03–0.08$
(b) Oxidized metals $\varepsilon = 0.60–0.85$
(c) Non-metals $\varepsilon = 0.65–0.95$.

The emissivity of any material varies widely with surface condition and tends to increase with increase in temperature and with time, owing to dirt and oxidation affecting the surface condition.

The lower the value of the emissivity the lower will be the heat loss from a hot surface. If, under black body conditions, the situation is such that the heat loss from a black body is 100 W m^{-2}, then the heat loss from stainless steel, $\varepsilon = 0.65$, and polished copper, $\varepsilon = 0.04$, would be 65 W m^{-2} and 4 W m^{-2} respectively.

Rate of Heat Transfer

The rate of radiant heat transfer from a black body at an absolute temperature, T_1, to black body surroundings (which reflect back none of the radiation) at a lower temperature, T_2, is given, from Eqn 2.63:

$$Q_b = 5.669A\left[\left(\frac{T_1}{100}\right)^4 - \left(\frac{T_2}{100}\right)^4\right] \tag{2.66}$$

The corresponding radiant heat transfer from a grey body of constant emissivity, ε, is:

$$Q = 5.669\varepsilon A\left[\left(\frac{T_1}{100}\right)^4 - \left(\frac{T_2}{100}\right)^4\right] \tag{2.67}$$

The heat transfer from a coloured body, which has emissivities of ε_{T1} at T_1 and ε_{T2} at T_2 is:

$$Q = 5.669A\left[\varepsilon_{T1}\left(\frac{T_1}{100}\right)^4 - \varepsilon_{T2}\left(\frac{T_2}{100}\right)^4\right] \tag{2.68}$$

In terms of the radiative heat transfer coefficient, h_r, the rate of radiant heat transfer from a grey body at T_1 to black body surroundings at T_2 may be written:

$$Q = h_r A(T_1 - T_2) \tag{2.69}$$

Substituting for Q from Eqn 2.67 gives:

$$h_r = \frac{Q}{A(T_1 - T_2)} = \frac{5.669\varepsilon\left[\left(\frac{T_1}{100}\right)^4 - \left(\frac{T_2}{100}\right)^4\right]}{(T_1 - T_2)} \tag{2.70}$$

Eqns 2.69 and 2.70 also apply to non-black body surroundings, provided that none of the radiation from the grey body is reflected back to it. A typical example of this is the radiant heat loss from a hot pipeline to its surroundings (air).

In the above equations it was assumed that the black body surroundings completely envelop the radiating body, that the two surfaces can see one another completely and that any gases between the radiating body and the surroundings do not absorb energy.

As a first approximation, it may be assumed that, for a body in a large volume of air, none of the energy it radiates is reflected back to itself, i.e. black body surroundings.

It is extremely difficult to calculate the rate of radiant heat transfer between bodies that partially absorb, partially reflect and partially transmit radiation, especially if the medium between the bodies absorbs radiation.

Example 2.3

Calculate the total heat loss by convection and radiation from an unlagged, 0.100 m diameter, horizontal steam pipe at 110°C to air at 15°C. Assume that the steam pipe is a grey body with an emissivity of 0.90, and that the natural convective heat transfer film coefficient, h, is given by: $h = 1.41(\Delta T/d)^{0.25}$.

$$\textit{Radiation } Q = 5.669\,\varepsilon A\left[\left(\frac{T_1}{100}\right)^4 - \left(\frac{T_2}{100}\right)^4\right]$$

$$= 5.669 \times 0.9 \times A\ (3.83^4 - 2.88^4)$$

$$= 747\ \text{W m}^{-2}$$

$$\textit{Convection } h = 1.41(\Delta T/d)^{0.25}$$

$$= 1.41[(110 - 15)/0.100]^{0.25}$$

$$= 7.83\ \text{W m}^{-2}\ \text{K}^{-1}$$

$$Q = hA\ T$$
$$Q = 7.83A\ (110 - 15)$$
$$= 744\ \mathrm{W\ m^{-2}}$$
$$Total = 747 + 744 = 1\ 491\ \mathrm{W\ m^{-2}}$$

2.6 HEAT TRANSFER EQUIPMENT

There are basically two types of heat transfer equipment used in the process industries – direct contact and surface. In direct-contact exchangers the two fluids are brought into physical contact. Typical examples are the condensers used to condense contaminated steam and water cooling towers, in which water is cooled by direct contact with air. In surface exchangers the two fluids are brought into thermal contact by the surface which keeps them physically separated. The many different types of surface exchanger are known generically as heat exchangers.

Shell-and-tube Heat Exchangers

The fundamental equation for the thermal design of a heat exchanger operating under steady state conditions, and assuming no heat losses, is:

$$Q = UA\Delta T_m = [\dot{m}\,c_p(t_i - t_o)]_H = [\dot{m}\,c_p(t_o - t_i)]_C \qquad (2.71)$$

where Q = rate of heat transfer, W
U = overall heat transfer coefficient, defined by Eqn 2.38 and based on a surface area A, W m^{-2} K^{-1}
A = surface area for heat transfer, m^2
ΔT_m = logarithmic mean temperature difference, defined by Eqn 2.40, K
$\dot{m}$ = mass flow rate, kg s^{-1}
c_p = specific heat capacity, J kg^{-1} K^{-1}
t = temperature of fluid, °C
i and o refer to inlet and outlet conditions
H and C refer to hot and cold fluid.

If the hot fluid supplies heat by condensation at constant temperature, the heat transferred will be $\dot{m}\lambda$, where λ is the specific latent heat.

If, in any particular system, the overall heat transfer coefficient may be regarded as being essentially constant, then the rate of heat transfer depends only on the product of the area, A, and the mean temperature difference, ΔT_m. A large area term involves relatively high initial capital expenditure, but relatively low operating costs. A large temperature-difference term involves high operating costs, but relatively low capital costs. A prime objective in heat exchanger design is to optimize the area and temperature terms.

The shell-and-tube heat exchanger is the most important type of surface exchanger and it will be used to illustrate the basic principle involved in the thermal design and operation of heat exchangers in general.

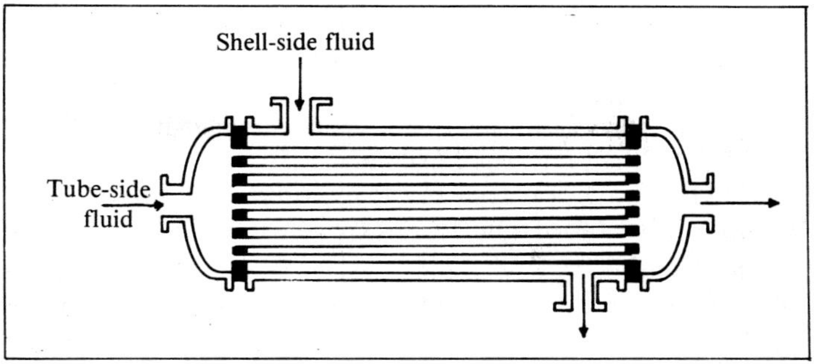

Figure 2.15 – 1–1 Shell-and-tube heat exchanger

Figure 2.15 shows the simplest form of shell-and-tube heat exchanger. It consists of a bundle of parallel tubes, the ends of which are fixed into two tube plates or sheets. The tubes may be on a square-pitch or triangular-pitch configuration. The tube bundle is surrounded by a cylindrical shell provided with two distribution heads. The heat exchanger, having one tube pass and one shell pass, is known as a single (tube) pass or 1–1 heat exchanger. A heat exchanger system having two shells and four tube passes in total is known as a 2–4 heat exchanger.

The mechanical design and construction of shell-and-tube heat exchangers are described in British Standard BS 3274:1960, 'Tubular heat exchangers for general purposes'.

Example 2.4

Calculate the number of 3.00 m long, 10×10^{-3} m internal diameter tubes required in a shell-and-tube heat exchanger to increase the temperature of 2.02 kg s^{-1} of an aqueous process liquid from 15°C to 75°C. Condensing steam at 115°C is the heat transfer medium. Assume that the mean density and specific heat capacity of the process liquid is 1 030 kg m^{-3} and 4.07×10^{3} J kg^{-1} K^{-1} respectively, and that the mean overall heat transfer coefficient, based on the inside area of the tubes, is 800 W m^{-2} K^{-1}.

$$\Delta T_1 = 115 - 15 = 100 \text{ K} \qquad \Delta T_2 = 115 - 75 = 40 \text{ K}$$

$$\Delta T_m = \frac{100 - 40}{\ln\left(\frac{100}{40}\right)} = 65.5 \text{ K}$$

$$Q = UA\Delta T_m = [\dot{m}\, c_p(t_o - t_i)]_C$$

$$= 800 \times A \times 65.5 = 2.02 \times 4.07 \times 10^3 (75 - 15)$$

$$A = 9.41 \text{ m}^2$$

Surface area per tube $= \pi d L = \pi \times 10 \times 10^{-3} \times 3.00 = 0.0942$ m^2

Number of tubes required $= 9.41/0.0942 = 100$ tubes

One disadvantage of installing a large surface in a shell-and-tube heat exchanger is that the cross-sectional area of the tubes is also comparatively large and the velocity of the fluid through the tubes is relatively low. The velocity and hence the overall heat transfer coefficient may be increased by using a multi-pass shell-and-tube exchanger. In such a unit, the tube-side fluid is diverted by baffles placed in the distribution heads so that it enters only a fraction of the tubes and thus passes back and forth through the exchanger several times.

If, in Example 2.4, a velocity of 1.00 m s^{-1} is recommended for optimum conditions of heat transfer, fouling factors and pumping costs, the number of tube passes required is calculated as follows:

Total cross-sectional area of one hundred, 10×10^{-3} diameter tubes

$$= 100\pi d^2/4 = 100 \times \pi(10 \times 10^{-3})^2/4 = 7.85 \times 10^{-3}\ \mathrm{m}^2$$

$$\text{Volumetric flow rate} = 2.02/1\,030 = 1.96 \times 10^{-3}\ \mathrm{m}^3\ \mathrm{s}^{-1}$$

$$\text{Fluid velocity} = 1.96 \times 10^{-3}/7.85 \times 10^{-3} = 0.250\ \mathrm{m\ s}^{-1}$$

$$\text{Number of passes} = 1.00/0.250 = 4\ \text{passes.}$$

The tubes would be arranged to have nominally twenty-five tubes per pass.

When the shell-side and tube-side film heat transfer coefficients are of similar magnitude, it is important to optimize the velocities and hence film coefficients of both the shell-side and tube-side fluids. As already mentioned, the velocity of the tube-side fluid may be increased by using a multi-pass heat exchanger. The shell-side velocity may be increased by installing baffles in the shell side to lengthen the path of the shell-side fluid.

The centre-to-centre distance between baffles is called the baffle pitch or spacing; it is usually between 20% and 100% of the shell diameter. The most common variety of baffle is the segmental disc type, illustrated by Figure 2.16. These are discs of sheet metal having a diameter just smaller than the shell diameter. A portion of the disc is removed, the remainder is drilled to accommodate the tubes. When 25% of the baffle is removed it is known as a 25%-cut baffle. The baffles may be arranged in the shell to give an 'up-and-down' flow or rotated through 90° to provide 'side-to-side' flow. The latter is employed when a mixture of liquid and non-condensable gas, or condensing vapour, flows through the shell side.

The shell and tubes of a shell-and-tube heat exchanger frequently have to withstand different corrosion rates and are therefore fabricated from different materials of construction. Thermal strains can occur in heat exchangers due to differential thermal expansion under both steady state, and start-up and shut-down conditions. These strains may be sufficient to buckle the tubes or pull them from the tube plates. A common method for avoiding damage due to thermal strain is to use a floating-head construc-

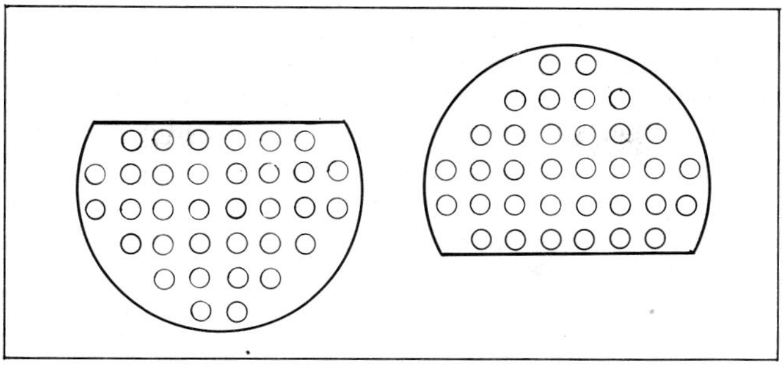

Figure 2.16 – Segmental baffles

tion. The tubes are fixed at one end into a stationary tube plate. At the other end, the tubes are attached to a floating head, which is structurally independent of the shell; the tubes are therefore free to expand and contract without damage. A two-pass, floating head, baffled shell space, shell-and-tube heat exchanger is illustrated by Figure 2.17.

In U-bend exchangers, which always have an even number of tube passes, the tubes are bent to form U-bends, and secured into one tube sheet. The tubes can expand and contract freely, thus eliminating the necessity for a floating head. This advantage must be balanced against several disadvantages. The tubes must be bent individually to fit the shell; this is costly. It is difficult to clean the inside of the tubes by mechanical means.

In Figure 2.15 the hot and cold fluids may flow in the same direction, co-currently, or in opposite directions, counter-currently. The flow direction has a bearing on the calculation of ΔT_m and the capacity of the unit. Except in the case where one fluid is isothermal, such as a condensing vapour or boiling liquid, there is a distinct thermal advantage to using counter-current flow in a single-pass heat exchanger.

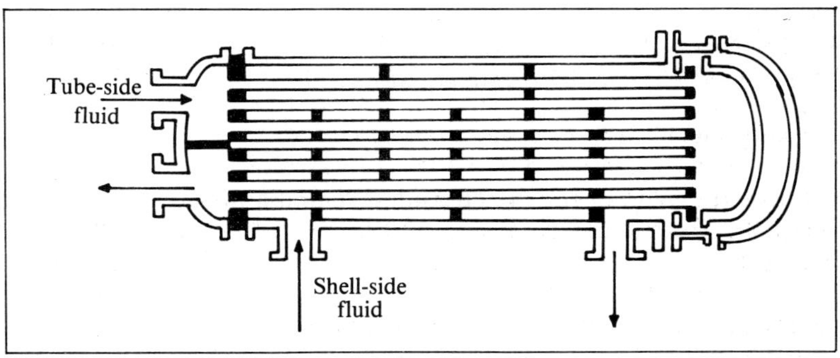

Figure 2.17 – 1–2 Floating head, baffled shell space, shell-and-tube heat exchanger

Example 2.5
Calculate the logarithmic mean temperature difference (LMTD) for both counter-current and co-current flow heat exchangers when a hot fluid is cooled from 150°C to 100°C by a cold fluid which enters at 20°C and exits at 60°C.

Counter-current		Co-current	
150°C ——→——	100°C	150°C ——→——	100°C
$\Delta T_1 = 90°C$	$\Delta T_2 = 80°C$	$\Delta T_1 = 130°C$	$\Delta T_2 = 40°C$
60°C ——←——	20°C	20°C ——→——	60°C

$$\Delta T_m = \frac{90-80}{\ln(90/80)} = 84.9°C \qquad \Delta T_m = \frac{130-40}{\ln(130/40)} = 76.4°C$$

The LMTD is greater for counter-current flow and hence, for a fixed thermal duty, a smaller area would be required for the counter-current heat exchanger.

A serious limitation of a co-current heat exchanger is that the exit temperature of the hot fluid can never be below the exit temperature of the cold fluid. To achieve 'temperature-crossing' a counter-current or multi-pass heat exchanger must be used.

Co-current exchangers are favoured for heating viscous liquids as the arrangement frequently results in a higher overall heat transfer coefficient than is the case for a counter-current exchanger.

In a multi-pass shell-and-tube heat exchanger, the shell-side and tube-side fluids do not flow in either true co-current or counter-current flow, but in cross-flow. This is especially true if shell-side baffles are employed. In multi-pass exchangers, the principal path of the shell-side fluid is at right angles to, or across, the tubes.

Bowman, Mueller and Nagle developed an analysis for complete cross-flow and derived the modified rate equation:

$$Q = UAY\,\Delta T_m \tag{2.72}$$

The appropriate value of the dimensionless temperature correction factor, Y, is obtained from standard graphs. When the temperature of the shell-side fluid is constant throughout the exchanger the factor Y has a value of unity.

When calculating the economic number of shells in a heat exchanger system, the value of Y should be greater than about 0.80. A value of Y less than 0.80 is considered to be uneconomical and usually means that a temperature-cross condition is being approached in one shell. If a temperature-cross is forced into a single shell, Y rapidly approaches zero and the exchanger transfers only a small fraction of the normally expected load.

2.7 STEAM AS A HEAT TRANSFER MEDIUM

Steam is normally generated at constant pressure in boilers. The heat content per kilogram, or specific enthalpy, of steam depends on pressure and temperature. Steam may be either saturated or superheated. Saturated steam is a vapour and, at a particular pressure, has a definite temperature, the saturation temperature. Superheated steam is a gas and, at a particular pressure, may have any temperature in excess of the saturation temperature.

When water is heated at constant pressure, heat may be added in two or three distinct and separate stages. In the first stage, the heat which is added (sensible heat) increases the water temperature to the boiling temperature or saturation temperature corresponding to the pressure of the steam.

In the second stage, the heat is added to convert the liquid phase water at saturation temperature to saturated steam. This is an isothermal process, the heat associated with the phase change of 1 kg of water is the specific latent heat of vaporization, λ. The magnitude of the latent heat depends on the relative intermolecular distances in the liquid and vapour phases. It decreases with increase in pressure until, at the critical point 22.1×10^6 N m^{-2} and 374°C, it becomes zero. In practice it is not possible to complete the second stage, to produce dry saturated steam, when the vapour is in contact with its own liquid phase.

At a particular pressure there are an infinite number of conditions of wet saturated, or simply wet, steam. Liquid phase water at the saturation temperature has a dryness fraction of zero, dry saturated steam has a dryness fraction of unity. Wet steam, which is a mixture of water vapour (steam) and liquid phase water droplets, has a dryness fraction between zero and unity. The dryness fraction of saturated steam from a boiler is typically between 0.95 and 0.97.

The total enthalpy of saturated steam at a particular pressure, and therefore definite temperature, depends on its dryness fraction or quality. It will have the full sensible heat contribution but only that latent heat contribution corresponding to the dryness fraction. The specific enthalpy of dry saturated steam, h_V, is the sum of the sensible enthalpy and latent heat.

If additional heat is to be added to the saturated steam at constant pressure, the third stage of heating involves passing the steam through a superheater. In the superheater the wet steam is completely dried, and then its temperature is increased as sensible heat is added.

The total specific enthalpy of superheated steam, h_G, depends on its pressure and temperature, or degrees of superheat. Standard steam tables show that, at a particular pressure, the enthalpy increases with increasing temperature.

The heat required to change from one condition of water or steam to another may be calculated from the steam tables.

Example 2.6
Calculate the sensible heat required to increase the temperature of one kilogram of water from 20°C to 70°C.

$$\text{Heat} = (h_L)_{70°C} - (h_L)_{20°C}$$

$$= 293 \times 10^3 - 83.9 \times 10^3$$

$$= 209 \times 10^3 \text{ J}$$

Example 2.7
Calculate the heat required to convert 1.25 kg of water at 25°C into saturated steam at 212°C with a dryness fraction of 0.95.

$$\text{Heat} = m\ [(h_{212°C} + x\lambda_{212°C}) - h_{25°C}]$$

$$= 1.25\ [(909 \times 10^3 + 0.95 \times 1890 \times 10^3) - 105 \times 10^3]$$

$$= 1.25\ (2705 - 105)10^3$$

$$= 3.25 \times 10^6 \text{ J}$$

When the pressure of saturated steam is reduced by passing it through a reducing valve or orifice, no external work is done and therefore there is no change in heat content of the steam. As the pressure of the steam has changed, its condition must also change to accommodate its unchanged heat content. If dry saturated steam has its pressure reduced adiabatically, the steam will become superheated. If wet steam is expanded adiabatically, its dryness fraction will increase and the steam may become superheated depending on its heat content and final pressure.

Superheated steam, being a gas, is a poor heat transfer medium and has a low film heat transfer coefficient. The heat transfer surface area associated with superheated steam is comparatively large and hence, for heat transfer purposes, the steam should be desuperheated. Steam is desuperheated by mixing it with sufficient condensate or treated water to produce saturated steam. In practice, for control purposes, the desuperheater system is designed so that the steam is not totally desuperheated.

The three common methods for desuperheating steam using water in the form of condensate or treated water are:

(a) Spraying the water into a vessel through which the superheated steam is passing, the excess water which is not vaporized is recycled to the vessel via a make-up tank and pump

(b) Injecting the water directly into the pipeline carrying the steam
(c) Bubbling the steam through a pre-set and maintained level of water.

The vaporization of the water requires a finite time and sufficient length of pipeline, say 10 m, to be installed between the desuperheater and the heat exchanger where the saturated steam will be utilized. Failure to allow enough time for water to vaporize may mean that the steam with suspended water droplets at a velocity of up to about 100 kilometres per hour could impinge on the heat exchanger tubes causing erosion.

If there is no control on the quantity of water mixed with the superheated steam, the steam may be not only desuperheated but saturated to an uncertain dryness fraction. To prevent this happening a temperature-sensing element is generally installed downstream of the desuperheater. This element should be set to control the quantity of water added to ensure a residual amount of superheat, say 1–5°C above the saturation temperature.

3. MASS TRANSFER

3.1 INTRODUCTION

General

Mass transfer or diffusion is concerned with the molecular and eddy (bulk) movement of mass within a system; it occurs as a result of a concentration gradient. When one or more concentration gradients exist within a mixture consisting of two or more components, each of the components will tend to move in such a direction as to equalize the respective concentrations, thereby nullifying the gradients.

Mass transfer may occur within a gas phase, a liquid phase or a solid, it may also occur between two phases. When two immiscible phases are contacted, components are transferred across the interfacial boundary according to the concentration gradients that are established. In certain processes, heat and mass transfer occur simultaneously.

A typical example of mass transfer is the evaporation of a liquid from a surface. Near the liquid surface a dynamic equilibrium is established in which the rate of vaporization of liquid molecules equals the rate of condensation of vapour molecules; there is no net evaporation (bulk movement away from the surface). When the liquid – vapour system is in contact with a stagnant gas, such as air, a concentration gradient is established. This gradient has a high concentration of vapour near the liquid surface and a low concentration of vapour in the bulk air. As a consequence of this concentration gradient, molecules of vapour will drift away slowly by molecular diffusion from the liquid/vapour region to the bulk air. The rate of condensation of vapour molecules decreases, thereby producing a net evaporation of liquid. If, however, the air moves over the surface of the liquid at a sufficiently high velocity to produce turbulence, then vapour molecules which have diffused by molecular motion into the laminar sublayer will be swept away by eddy diffusion and the rate of the net evaporation process will be increased.

Mass transfer is involved in physical processes such as air conditioning, where water vapour and air are mixed. It is involved also in chemical reactions such as combustion, where oxygen must continuously diffuse to the combustible material whether this be gas, liquid or solid. It is, however, in separation processes, where the movement of mass along a concentration gradient is exploited, that a knowledge of mass transfer is most important. Components are removed from the high end of concentration gradients, thereby inducing more of those components to diffuse in order to maintain the gradients; in this way, mixtures may be separated.

The separation by mass transfer of the components of a mixture may be accomplished by:

(a) Contact of two immiscible phases
(b) Contact of miscible phases separated by a permeable or semi-permeable membrane.

(a) The three states (gas or vapour, liquid and solid) permit six possible binary combinations. However, since all gases are mutually miscible and since the diffusion rates between solids are extremely low, due to the immobility of the molecules, there are, in reality, only four binary combinations of industrial importance. A selection of the unit operations associated with these combinations is given below to indicate the scope of mass transfer.

(i) Liquid–solid: adsorption (including dyeing), drying, ion-exchange, leaching (including the removal of solutes or suspensions of solids from a textile substance by washing with water).
(ii) Liquid–liquid: solvent extraction.
(iii) Gas–solid: adsorption, desorption, dehumidification.

(b) The membrane is used to prevent the mixing of the two phases; it must be selectively permeable to the components of the mixture if diffusional separation is to be achieved. Typical examples involving membrane separation are dialysis and gas diffusion.

Modes of Mass Transfer

As already mentioned, mass transfer can occur by two mechanisms: molecular and eddy diffusion. These mechanisms depend on the dynamics of the fluid phases. In a stagnant fluid, or in a fluid flowing under laminar conditions in a direction at right angles to a concentration gradient, mass transfer is effected by random motion of the molecules on a molecular level. In a turbulent fluid, molecular diffusion is supplemented by eddy diffusion. Eddy, turbulent or convective diffusion occurs on a macroscopic level, involving the bulk mixing of eddies of fluid; it is much faster than molecular diffusion. Molecular and eddy mass transfer have similarities with heat transfer by conduction and convection.

A concentration gradient is the driving force for mass transfer. When the concentration gradient does not alter with time, the process is said to be steady state. Unsteady state mass transfer processes, where the concentration gradient changes with time, are encountered in batch processes and during start-ups and shut-downs of continuous processes.

3.2 PROPERTIES, QUANTITIES, UNITS AND DIMENSIONS

A number of the properties and quantities of fluids and solids have already been discussed in Sections 1.2 and 2.2, only the additional ones pertinent to mass transfer are discussed here.

Properties – Diffusion Coefficient

The kinetic theory of gases forms the basis of the theory of diffusion in ideal gas mixtures. A simple form of the kinetic theory postulates that the molecules of a gas are hard elastic spheres that are in a continual state of motion. Equations based on the interactions between colliding molecules can be used to calculate diffusion coefficients or diffusivities for binary ideal gas mixtures.

Individual molecules in a gas move in random directions at varying velocities due to their thermal motion. Typical instantaneous velocities vary from zero to about 500 $m\ s^{-1}$. Although the average velocity may be high, a molecule undergoes many intermolecular collisions (typically 10^{12} collisions per second) owing to the large molecular population, and hence the effective net diffusion rate is low.

The net rate of diffusion increases with decrease in pressure, which reduces the molecular population and therefore the collision rate. It increases with increase in temperature, which increases the average molecular velocity. Small molecules have larger average velocities and lower collision frequencies than larger molecules.

The diffusion coefficients for liquids and solids are much lower than those for gases. In order of magnitude terms, the diffusion coefficients at 20°C and atmospheric pressure are typically:

(a) Gases: 10^{-4}–10^{-5} $m^2\ s^{-1}$
(b) Liquids: 10^{-9}–10^{-10} $m^2\ s^{-1}$
(c) Solids: 10^{-12}–10^{-14} $m^2\ s^{-1}$

Comprehensive lists of equations, numerical values and references for diffusion coefficients in gases, liquids, solutions, electrolyte solutions, porous materials and polymers are included in Perry, and Sherwood, Pigford and Wilke. The difficulties encountered in predicting or evaluating diffusivities when an active solid phase is present in the system are outlined in Section 3.5.

Quantities – Concentration, Transfer Rate, Mass Transfer Coefficients

In mass transfer, it is customary to deal in molar concentrations, c. In the SI system this means moles per cubic metre of solution, mol m^{-3}. Other concentration units, such as gram/litre and lb/ft^3 occur in the literature and this leads to difficulties, especially when dealing with the units of mass transfer coefficients. In gas phase diffusion, frequently it is convenient to express the gas concentrations in terms of partial pressures, p, the units of which are N m^{-2} in the SI system. Once again other units, such as atmospheres and millimeters of mercury, occur in the literature and this tends to lead to some confusion in the units of mass transfer coefficients.

For an ideal gas mixture, the partial pressure of component A, p_A, is related to the molar concentration of component A, c_A:

$$p_A = \frac{n_A RT}{V} = c_A RT \tag{3.1}$$

where n_A = number of moles of component A, mol
R = universal gas constant, 8.314 J mol^{-1} K^{-1}
T = absolute temperature of mixture, K
V = volume of mixture, m^3.

The rate of mass transfer, N_A, is defined as the number of moles of component A transferred per unit time, mol s^{-1}. The mass transfer flux, N_A/A, is the rate of mass transfer per unit cross-sectional area, mol m^{-2} s^{-1}.

Mass transfer coefficients are analogous to heat transfer coefficients. The individual or film coefficients and the overall coefficients are designated k and K respectively. The symbols are given subscripts to indicate the type of phase to which they are associated or refer. Thus the subscripts G and L refer to gas (or vapour) and liquid phases respectively. In some literature the coefficients in counter-diffusion are given a prime, ′, to distinguish them from the coefficients associated with the diffusion of one component along a concentration gradient through a non-diffusing component or components. Thus, the film mass transfer coefficient for water vapour diffusing from a moist surface through unsaturated air, which does not diffuse towards the moist surface, is designated k_G. If the moist surface were in contact with a gas that was soluble in the water, then if concentration gradients were established, the water vapour would diffuse from the moist surface counter-current to the soluble gas, which would diffuse towards the moist surface. The film coefficient for the water vapour could then be written as k'_G.

The rate of transfer of water vapour from a moist surface into air, in terms of the film mass transfer coefficient or conductance, is:

$$N_A = k_G A \, \Delta c \tag{3.2}$$

In resistance terms, the equation is:

$$N_A = \frac{\Delta c}{\frac{1}{k_G A}} \tag{3.3}$$

where Δc = concentration difference or driving force
$(1/k_G A)$ = resistance to mass transfer.

There is a physical analogy between the film coefficients and the overall coefficients for heat transfer across a plane solid surface of negligible resistance and the corresponding coefficients for mass transfer at an interphase boundary when the resistance is generally assumed to be zero. The overall rate of mass transfer across an interphase boundary depends on the equilibrium of the system, which itself changes with mass transfer. Therefore, the mathematical relationship between the film and overall coefficients contains an equilibrium term which is not included in the heat transfer equation. Some mathematical relationships are developed in Section 3.4.

Units and Dimensions

TABLE 3.1
Units and Dimensions in Mass Transfer

Property or quantity	Symbol	Dimensions	Units
Diffusion coefficient	D	$L^2 T^{-1}$	$m^2 s^{-1}$
Concentration (liquid phase)	c	ML^{-3}	$mol\ m^{-3}$
Concentration (gas phase)	p	$ML^{-1} T^{-2}$	$N\ m^{-2}$
Rate of mass transfer	N	MT^{-1}	$mol\ s^{-1}$
Mass transfer flux	N/A	$ML^{-2} T^{-1}$	$mol\ m^{-2} s^{-1}$
Film mass transfer coefficient	k	LT^{-1}	$m\ s^{-1}$
Overall mass transfer coefficient	K	LT^{-1}	$m\ s^{-1}$

3.3 MOLECULAR DIFFUSION

Rate Equation

The fact that molecular diffusion occurs by random molecular motions and is therefore directly analogous to heat transfer by conduction was first recognized by Fick in 1855. He put molecular diffusion on a quantitative basis by adapting the equation for conduction that had been derived by Fourier in 1822. It was later recognized that both processes are analogous to momentum transfer in laminar flow that also occurs by random molecular motion.

The basic rate equation for the steady state undirectional molecular diffusion of component A in a mixture, Fick's law of diffusion, is:

$$\frac{N_A}{A} = -D \cdot \frac{dc_A}{dy} \tag{3.4}$$

where N_A/A = mass flux or the rate of transfer of component A in the y-direction, through unit area at right angles to the direction of diffusion, mol m^{-2} s^{-1}
D = mass diffusivity, m^2 s^{-1}
c_A = molar concentration of component A, mol m^{-3}
y = distance in the direction of diffusion, m.

The Fourier equation for conduction, corresponding to Eqn 3.4, is:

$$\frac{Q}{A} = -K \cdot \frac{dt}{dy} \tag{2.10}$$

where Q/A = heat transfer flux, W m^{-2}
K = thermal conductivity of material, W m^{-1} K^{-1}
dt/dy = temperature gradient through material in y-direction, K m^{-1}.

The analogy between Eqns 2.10 and 3.4 is impaired by the fact that the units of K and D are not identical. The defect can be rectified by substituting the heat content per unit volume, $c_p\rho t$, for the temperature, t, in Eqn 2.10:

$$\frac{Q}{A} = -\frac{K}{c_p \rho} \cdot \frac{d(c_p \rho t)}{dy} = -\alpha \cdot \frac{d(c_p \rho t)}{dy} \tag{3.5}$$

where c_p = heat capacity of material, J kg^{-1} K^{-1}
ρ = density of material, kg m^{-3}
α = thermal diffusivity, $K/(c_p\ )$, m^2 s^{-1}.

The units of thermal diffusivity, α, and mass diffusivity, D, are the same, m^2 s^{-1}.

The Newton equation for momentum in laminar flow:

$$\tau = -\mu \cdot \frac{dv}{dy} \tag{1.5}$$

may be written in terms of momentum per unit volume, ρv:

$$\tau = -\frac{\mu}{\rho} \cdot \frac{d(\rho v)}{dy} = -\nu \cdot \frac{d(\rho v)}{dy} \tag{3.6}$$

where τ = momentum flux (shear stress), N m^{-2}
μ = viscosity of material, N s m^{-2}
v = fluid velocity, m s^{-1}
ρ = density of material, kg m^{-3}
ν = momentum diffusivity (kinematic viscosity), m^2 s^{-1}.

Eqns 3.4, 3.5 and 3.6 are all of the form:

$$\text{flux} = \text{diffusivity} \times \text{concentration gradient} \tag{3.7}$$

where the momentum concentration = ρv
the thermal concentration = $c_p \rho t$
the mass concentration = c_A.

The corresponding diffusivities, τ, α and D, have the same dimensions, and the concentration gradients are linear for a homogeneous medium and steady state conditions.

Steady State Molecular Diffusion in the Gas Phase

Assuming that the gas (or vapour) mixture contains only two components, A and B, and that it behaves as an ideal gas, then, from Eqn 3.1, Fick's law may be rewritten:

$$\frac{N_A}{A} = -\frac{D_{AB}}{RT} \cdot \frac{dp_A}{dy} \tag{3.8}$$

Steady State Molecular Diffusion in the Liquid Phase

The rate equations for molecular diffusion in the gas phase may be used for molecular diffusion in the liquid phase. However, since the nature of the interactions of liquid molecules is less well understood than those of gas molecules, and since the liquid mass diffusivity of a system is concentration dependent, the resulting equations are less reliable than those for gases.

3.4 CONVECTIVE DIFFUSION

Effective Film Thickness

Convective mass transfer, where molecular diffusion is supplemented by eddy diffusion, is similar to convective heat transfer.

When a fluid flows in turbulent flow past a surface or an interface, a boundary layer develops, similar to that illustrated by Figures 1.7 and 1.8.

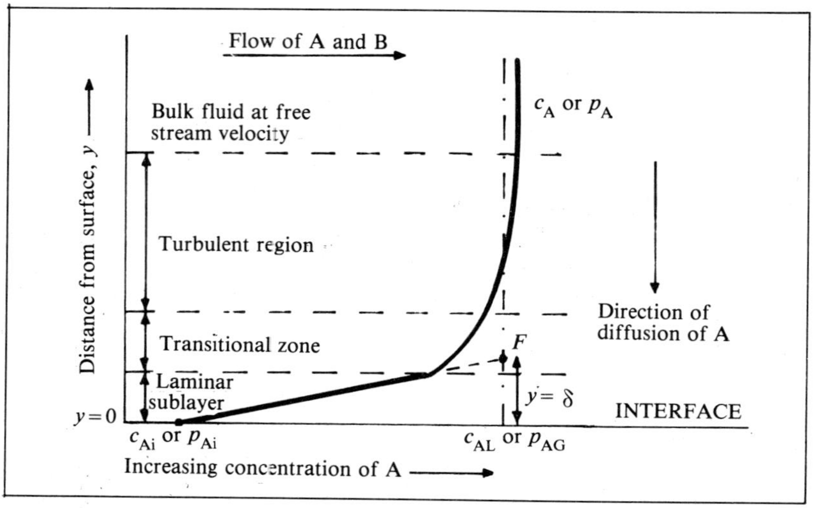

Figure 3.1 – Concentration gradient for convective mass transfer to a surface

Within this boundary layer there are three regions: the laminar sublayer, the transitional or buffer zone and the turbulent region. If mass transfer is occurring from the fluid to the surface, then a concentration gradient exists where the concentration of the diffusing component varies from the free-stream value to the interface value. The major portion of this concentration gradient lies within the laminar sublayer. As eddies from the turbulent and buffer regions do not penetrate into the laminar region, mass transfer in this region occurs exclusively by molecular diffusion. In the turbulent region, mass transfer occurs mainly by eddy diffusion. In the buffer region, both modes are equally important.

The concentration gradient for the mass transfer of a solute, A, *to* a surface or interface is illustrated by Figure 3.1. If the flowing fluid is a liquid then the concentration of solute A would be designated by c_A. At a particular distance along the interface (in the x-direction) the concentration of A at the interface would be c_{Ai} and the corresponding *average* concentration of A in the flowing liquid would be c_{AL}. The concentration driving force would be $(c_{AL} - c_{Ai})$. If the flowing fluid were a gas, the concentration of A would probably be measured in partial pressure units, p_A. The corresponding interface, average and driving force concentrations would be, respectively, p_{Ai}, p_{AG}, $(p_{AG} - p_{Ai})$.

The situation illustrated by Figure 3.1 is typical of:

(a) A soluble dye in solution being transferred to a substrate
(b) A solute being transferred to a solid surface (e.g. crystallization)
(c) A soluble gas being absorbed in a liquid or a gas or vapour being adsorbed onto a solid

(d) A solute being transferred from one solvent to a more preferred solvent.

Although Figure 3.1 illustrates convective mass transfer, it also, in general terms, represents convective momentum transfer and convective heat transfer. In momentum transfer towards the surface or interface the gradient would refer to velocity, when the interface 'concentration' value would be zero (if the interface were stationary) and the average 'concentration' value would be the average volumetric velocity measured across the entire fluid stream. In terms of heat transfer from the fluid stream to a surface, the gradient would refer to temperature. The interface value would be the surface temperature and the average value would be the 'average fluid temperature'.

In convective momentum and heat transfer to a stationary surface, the velocity at the surface is zero. In the case of convective mass transfer this clearly is not the case, since molecules actually move *to* or *from* the actual surface. If, however, the mass transfer flux is low, it can be assumed that the velocity at the surface is essentially zero. This assumption forms the basis of the mathematical relationships between momentum, heat and mass transfer.

It was mentioned in Section 2.4 that if, in convective heat transfer to a surface, the Prandtl number of the fluid is unity, then the temperature gradient and the velocity gradient are coincident. A similar situation exists in convective mass transfer to a surface; if the Schmidt number of the fluid is unity, then the concentration gradient and velocity gradient are coincident.

Figure 3.2 illustrates the concentration gradient for the convective mass transfer of a solute *from* a surface or interface. This is typical of:

(a) A soluble component of a solid being transferred to a liquid (leaching or dissolution)

(b) The vaporization of a liquid or the evaporation of a solvent into a gas stream

(c) A solute being transferred from one solvent to a more preferred solvent

(d) The desorption of a solute from a solid into a fluid stream.

Although Figure 3.2 refers to convective mass transfer from a surface or interface, it could equally well represent convective heat transfer from a surface to a fluid or convective momentum transfer from a surface. In the latter case the surface would be moving through a 'stationary' fluid, i.e. the free stream velocity would be zero.

Figure 3.3 illustrates the convective equimolecular counter-diffusion of the two components. Component B is diffusing *from* the surface or interface and component A is diffusing in the opposite direction. If the flowing fluid were a liquid, Figure 3.3 could represent:

(a) A reactive dye system

(b) An ion exchange system.

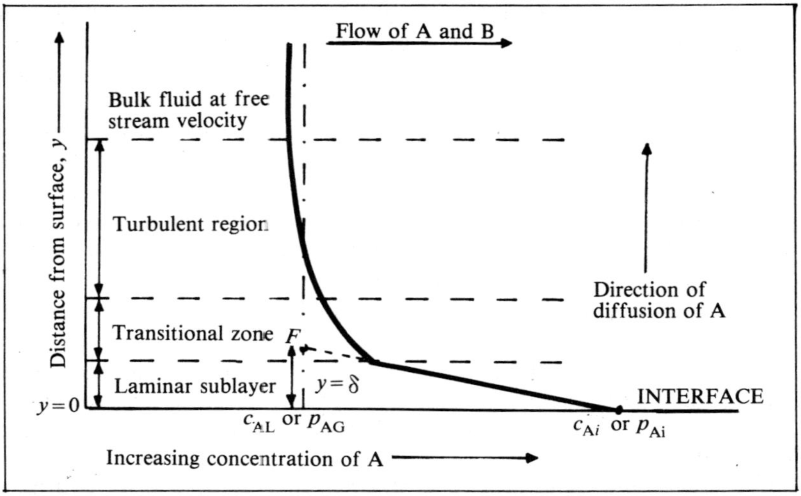

Figure 3.2 – Concentration gradient for convective mass transfer from a surface

The concentration gradient for A in Figure 3.3 could, in general terms, also represent a velocity gradient and a temperature gradient if simultaneous heat transfer were occurring.

If convective equimolecular counter-diffusion were occurring simultaneously with heat transfer to the surface or interface in a fluid with a Prandtl number and Schmidt number of unity, then the velocity gradient,

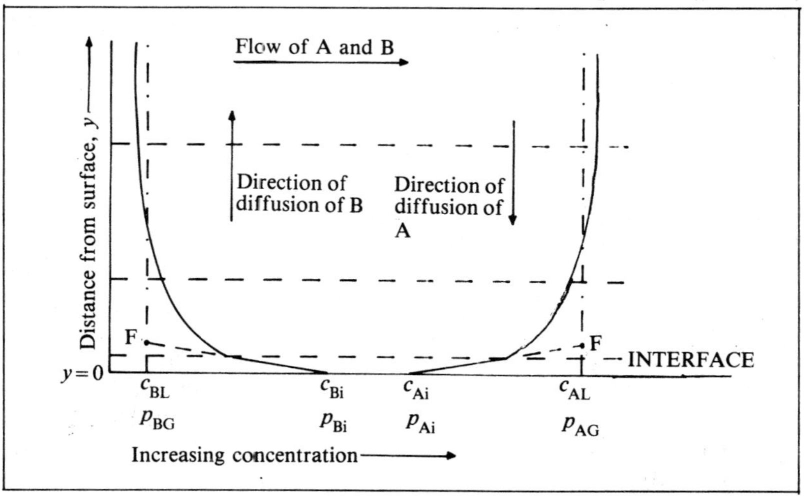

Figure 3.3 – Concentration gradients for convective equimolecular counter-diffusion

temperature gradient and the concentration gradient for A would be coincident. This situation may occur in the hot-air drying of a porous material, when the liquid phase vacates pores in the material which must be filled by air.

In Figure 3.1 the major portion of the concentration gradient lies within the laminar sublayer. Within this region the concentration gradient is linear and the rate of mass transfer by molecular diffusion can be calculated. The rate of mass transfer due to eddy diffusion within the remainder of the boundary layer cannot be calculated directly and therefore simple models have been developed to enable the rate to be estimated. One such model borrows the concept of effective film thickness from convective heat transfer.

In mass transfer a fictitious film thickness, δ, is chosen such that this fluid film offers the same resistance to molecular diffusion as would be encountered in the combined processes of molecular eddy diffusion within the entire boundary layer.

In Figure 3.1 the effective film thickness, δ, is obtained by extending the laminar concentration gradient to cut the average fluid concentration c_{AL} or p_{AG} at point F. The distance from the surface or interface to this point of intersection is the film thickness, $y=\delta$. Similar constructions are shown in Figures 3.2 and 3.3.

Evaluation of the Film Coefficients

The parameters which can affect the film mass transfer coefficients are: the bulk velocity, v, (m s^{-1}); density, ρ, (kg m^{-3}); viscosity, μ, (N s m^{-2}), mass diffusivity, D, (m^2 s^{-1}) and a characteristic linear dimension, say 'diameter', d, (m). The relationship of the film coefficient, k, (m s^{-1}) to these parameters may be written:

$$k=\mathrm{f}(v, \rho, \mu, D, d) \tag{3.9}$$

Dimensional analysis yields the dimensionless equation:

$$\mathrm{Sh}=C\,\mathrm{Re}^x\,\mathrm{Sc}^y \tag{3.10}$$

where Sh = Sherwood group, kd/D
Re = Reynolds group, $dv\rho/\mu$
Sc = Schmidt group, $\mu/\rho D$.

Eqn 3.10 is directly analogous the the dimensionless equation for convective heat transfer, Eqn 2.43.

Sherwood Group

$$\mathrm{Sh} = \frac{kd}{D} \tag{3.11}$$

Eqn 3.11 is the generalized form of the Sherwood group, which is analogous to the Nusselt group in convective heat transfer. In gas phase mass transfer the units of the film coefficient, k_G, would normally be expressed as mol m^{-2} s^{-1} (N $m^{-2})^{-1}$. These units are not consistent with a dimensionless Sherwood group and they must therefore be corrected to mol m^{-2} s^{-1} (mol $m^{-3})^{-1}$ or simply m s^{-1}. This correction may be achieved by using Eqn 3.1 to convert the pressure concentration units (N m^{-2}) into liquid concentration units (mol m^{-3}). Thus, the corrected film mass transfer coefficient, k_{GC}, is $k_G RT$.

Traditionally, the generalized Sherwood group is defined in terms of the liquid phase film mass transfer coefficient for counter-diffusion, k_L'. Hence, the Sherwood group may be defined:

$$\mathrm{Sh} = \frac{k_L' d}{D} = \frac{d}{\delta_L} \tag{3.12}$$

The Sherwood group for the various mass transfer situations considered in this section may therefore be defined:

$$\mathrm{Sh} = k_G RT \cdot \frac{p_{BM}}{P} \cdot \frac{d}{D} = k_{GC} \cdot \frac{p_{BM}}{P} \cdot \frac{d}{D} \tag{3.13}$$

$$\mathrm{Sh} = k_L \cdot \frac{c_{BM}}{c_{total}} \cdot \frac{d}{D} \tag{3.14}$$

$$\mathrm{Sh} = k_G' RT \cdot \frac{d}{D} = k_{GC}' \cdot \frac{d}{D} \tag{3.15}$$

$$\mathrm{Sh} = k_L' \cdot \frac{d}{D} \tag{3.16}$$

Reynolds Group

$$\mathrm{Re} = \frac{d v \rho}{\mu} \tag{3.17}$$

As the Reynolds number increases, the degree of turbulence increases and hence the overall rate of mass transfer increases.

Schmidt Group

$$\text{Sc} = \frac{\mu}{\rho D} \tag{3.18}$$

The Schmidt group, which contains only the properties of the fluid, is the ratio of momentum diffusivity, ν, or (μ/ρ), to mass diffusivity, D. It characterizes mass transfer in the same way as the Prandtl number characterizes convective heat transfer.

Stanton Group for Mass Transfer

A dimensionless group, analogous to the Stanton group in heat transfer, Eqn 2.51, may be defined, in general terms:

$$(\text{St})_m = \text{Sh}\ \text{Re}^{-1}\ \text{Sc}^{-1} = k/v \tag{3.19}$$

where the actual definition of the film mass transfer coefficient, k, is obtained from Eqns 3.13 to 3.16.

The Colburn *j*-factor for mass transfer, j_m, is defined, in general terms:

$$j_m = (\text{St})_m\ \text{Sc}^{2/3} = \left[\frac{k}{v}\right]\left[\frac{\mu}{\rho D}\right]^{2/3} \tag{3.20}$$

It has been shown by a number of techniques that, provided the rate of mass transfer is sufficiently low, and that the fluid velocity at the surface is zero, then there is an exact relationship between mass transfer, heat transfer and momentum due to wall shear. This relationship takes the form:

$$j_m = j_h = j_f = \frac{f}{2} = 0.023\ \text{Re}^{-0.2} \tag{3.21}$$

If, in a system, extra pressure loss is caused by form drag, i.e. the shape of the surface, as would be the case in flow around solid objects, then the heat and mass transfer is enhanced and Eqn 3.21 is no longer valid. The relationship between mass and heat transfer is, however, generally still valid and Eqn 3.21 may be rewritten:

$$j_m = j_h \geq 0.023\ \text{Re}^{-0.2} \tag{3.22}$$

An equation similar to the Dittus–Boelter equation is obtained by combining Eqn 3.19 with 3.21:

$$\text{Sh} = 0.023\ \text{Re}^{0.8}\ \text{Sc}^{0.33} \tag{3.23}$$

Experimental investigations indicate that the best general correlation for convective mass transfer in a pipe is:

$$\text{Sh} = 0.023\ \text{Re}^{0.83}\ \text{Sc}^{0.33} \tag{3.24}$$

$$\text{or } j_m = 0.023\ \text{Re}^{-0.17} \tag{3.25}$$

Interphase Mass Transfer

The transfer of a solute from one phase to another across an interphase boundary produces a concentration gradient in each phase because of each phase's resistance to mass transfer. The overall resistance to mass transfer may be expressed in terms of an overall mass transfer coefficient similar to the overall coefficient in heat transfer.

The mass transfer analogy of heat transfer across a thin metal surface of sufficiently high thermal conductivity, such that the resistance to heat transfer through the surface is negligible, was developed by Whitman in the 'two-film theory'. He showed that the overall mass transfer coefficient is the sum of the two individual film coefficients.

The two-film theory, which assumes that there is no resistance to mass transfer across the actual interface, is valid in most cases of mass transfer. Two notable exceptions are due to surface-active agents and high transfer fluxes. A surface-active agent tends to accumulate at a liquid interface, its presence, even as a monolayer, can reduce very considerably the rate of mass transfer. A high rate of mass transfer across a liquid interface can cause interfacial turbulence (not associated with turbulence in the bulk liquid), which tends to increase the rate of mass transfer, and is equivalent to a negative interfacial resistance. A similar situation may occur if the solute being transferred from one phase across the interphase boundary undergoes a fast chemical reaction in the second phase.

3.5 DIFFUSION THROUGH SOLID MATERIALS

Introduction

The diffusion of gases, liquids and solids through or within solid materials is important in the mass transfer operations: drying, dyeing and adsorption in general, either with or without chemical reaction; leaching and membrane processes, such as dialysis and gaseous effusion. The mechanism by which diffusion occurs depends on whether the solid material is porous or non-porous under the prevailing operating conditions.

A porous material may permit the flow of gases, pure liquids, solutions or dispersions through the holes which are an integral part of the solid structure. These holes may be pores, fissures, capillaries and interstitial voids, such as inter- and intra-fibre spacings. Diffusion that depends on the structure of the solid material may be termed 'structure-sensitive' diffusion and the diffusing molecules, solutes, molecular species or solids may be called diffusants.

A non-porous material does not have holes through which the diffusing species may move. The diffusing species must therefore penetrate the surface and move through the crystal or cellular structure of the solid as a solid solution. Diffusing by this mechanism may be termed 'structure-insensitive' diffusion. This form of diffusion is similar to diffusion through fluids. A diffusing species which penetrates the solid surface and lodges, or is adsorbed in the solid material, may be called a penetrant. A diffusing species which penetrates the solid surface, permeates through the solid matrix and emerges from the solid may be called a permeant.

A species which diffuses from a fluid to a solid surface, where it either adsorbs (physically or chemically), migrates or penetrates the surface only superficially, may be called an adsorbant.

The physical and chemical structure of a textile substrate frequently has an important influence on mass transfer operations, such as drying and dyeing.

Diffusion Through a Non-porous Material

During the steady state convective diffusion of a solute, A, from one phase to another through a flat slab of non-porous inert solid material (a solid solvent), the concentration gradient remains unchanged with time and Fick's law is:

$$\frac{N_A}{A} = -D_{AS} \cdot \frac{dc_A}{dy} \qquad (3.26)$$

where N_A/A = rate of diffusion of solute A per unit cross-section of solid, mol m^{-2} s^{-1}

D_{AS} = diffusivity of solute A through the solid S, m^2 s^{-1}.

If the diffusivity, D_{AS}, is constant and independent of concentration, then integration of Eqn 3.26 for a slab thickness, y, gives the mass transfer rate at some local position:

$$\frac{N_A}{A} = \frac{D_{AS}}{y}\left[(c_{AI})_i - (c_{AII})_i\right] = k_s \Delta c_i \qquad (3.27)$$

where $(c_{AI})_i$ and $(c_{AII})_i$ = concentration of solute at the interface between the solid and phase I, and the solid and phase II, mol m^{-3}

k_S = effective mass transfer coefficient for the solid material, m s^{-1}

Δc_i = difference in concentration of A at the two opposite faces of the solid, i.e. the driving force for diffusion through the solid material, mol m^{-3}.

In terms of the overall mass transfer coefficient, K, the mass transfer rate is:

$$\frac{N_A}{A} = K[(c_{AI})_L - (c_{AII})_L] = K\Delta c_L \qquad (3.28)$$

The local overall mass transfer coefficient may be expressed in terms of the film coefficients in the two liquid phases, k_{LI} and k_{LII}, and the slab coefficient, k_s:

$$\frac{1}{K} = \frac{1}{k_{LI}} + \frac{1}{k_S} + \frac{1}{k_{LII}} \qquad (3.29)$$

This equation is analogous to the equation for the overall heat transfer coefficient for a plane surface in terms of the thermal resistance in series.

In phase I the film coefficient, k_{LI}, is equal to D_{AI}/δ_I, where D_{AI} is the diffusivity of solute A through phase I and δ_I is the effective film thickness of phase I. In phase II, k_{LII} is equal to D_{AII}/δ_{II}. The effective film coefficient for the slab material, k_S, may be defined as D_{AS}/y.

If a gas or vapour diffuses through the solid, then partial pressure units may be used and the film coefficients would be expressed in consistent units.

In an actual item of equipment, the local overall mass transfer coefficient tends to vary with position due to the variation in effective film thickness caused by the concentration dependency of solute diffusivity. Natural circulation currents can also cause variations in effective film thicknesses, especially when the solid is vertical.

The discussion thus far has been based on a number of assumptions:

(a) The slab material is completely homogeneous
(b) There is no *flow* of fluid through the material
(c) The length of the diffusion path is equal to the slab thickness
(d) There are no slab material–solute interactions
(e) The diffusing solute permeates the material completely
(f) The process is steady state.

These assumptions will be examined with particular reference to moisture permeation through polymeric films and the drying and dyeing of textile substrates.

Diffusion Through a Polymeric Film

The steady state diffusion through a polymeric film, such as a non-porous membrane, may be described by the equations above (Diffusion Through a Non-porous Material). Problems do arise, however, in determining the diffusivity of the solute through the film, D_{AS}, and the effective diffusion path length.

Non-porous polymeric films permit the diffusion of gases and liquids which are soluble in the polymeric material. The diffusing substance dissolves in the film at one face to form a solid solution, it then diffuses through the film as if the film were a liquid, and is desorbed at the opposite face.

Diffusion in polymers is rather complex, and at present it is not possible to predict diffusivities from known properties of the polymer and diffusing material. Known variables which affect the diffusivity of a substance diffusing through a non-porous polymer are listed below.

(a) *Polymer:*
- (i) Chemical nature
- (ii) Molecular weight
- (iii) Molecular weight distribution
- (iv) Degree of cross-linking
- (v) Crystallinity
- (vi) Additives such as stabilizers and plasticizers
- (vii) Glass transition temperature
- (viii) Method of manufacture and thermal history.

(b) *Diffusing solute:*
- (i) Molecular weight
- (ii) Molecular shape
- (iii) Degree of aggregation.

(c) *Polymer–solute interaction:*
- (i) Solubility and solubility isotherm
- (ii) Chemical reactions
- (iii) Physical or chemical adsorption
- (iv) Swelling propensity
- (v) Degree of aggregation or clustering.

There are essentially two types of diffusion that occur in polymers: the diffusion of gases that have little or no interaction with the polymer and the diffusion of vapours or liquids that dissolve in the polymer and cause it to swell.

I Gases

The solubility of a gas in a polymer, defined as the maximum concentration of the gas in the polymer at the operating pressure, generally follows Henry's law:

$$c = Hp \tag{3.30}$$

where c = concentration of gas in the polymer, mol m^{-3}
H = Henry's law of solubilty coefficient, mol m^{-3} (N m^{-2})$^{-1}$
p = partial pressure of the gas in the gas phase mixture, N m^{-2}.

The diffusivity is independent of pressure and solute concentration in the polymer. The effect of temperature on diffusivity has been shown experimentally to be of the form:

$$D_{AS} = D_0 \, e^{-E/RT} \tag{3.31}$$

where D_0 = dimensional constant
E = energy of activation, J mol^{-1}
R = universal gas constant, J mol^{-1} K^{-1}
T = absolute temperature, K.

Thus values of D_{AS} may be estimated at any temperature if values are known at only two temperatures.

There is a direct relationship between the solubility of a gas and the percentage crystallinity of the polymer; gases are not soluble in the crystalline regions. For a given polymer–gas system, Eqn 3.30 may be rewritten:

$$c = H^*(1-\alpha)p \tag{3.32}$$

where H^* = Henry's law solubility coefficient for the amorphous polymer, mol m^{-3} (N m^{-2})$^{-1}$
$(1-\alpha)$ = fractional crystallinity of the polymer.

The value of H^* increases with the normal boiling point of the diffusing gas.

For the diffusion of a gas through a solid, it is common to express the rate of transfer of solute gas A in terms of its partial pressures, p_{A1} and p_{A2}, on the two sides of the solid of thickness L:

$$\frac{N_A}{A} = \frac{P_{AS}\,(p_{A1} - p_{A2})}{L} \tag{3.33}$$

The permeability constant or permeability of a solute gas through a solid, P_{AS}, may be regarded as the mass transfer flux per unit pressure gradient. It is defined as the quantity of gas (mol), at a specified temperature and pressure, diffusing per unit cross-sectional area of a specified thickness of the solid under a specified pressure gradient. Literature values of P_{AS} are commonly quoted: cm^3 of gas at STP (0°C and atmospheric pressure) diffusing per second per cm^2 of cross-sectional area through a 1.0 mm thickness of solid under a pressure gradient of 1 cm Hg. For systems that obey Henry's law and in which the diffusivity is independent of pressure, the permeability is equal to the product of the diffusivity and the solubility constant: $P_{AS} = D_{AS}H$.

Lack of solid-state theory makes it virtually impossible to predict accurately the diffusivities in solids and, therefore, values have to be determined experimentally. It is known, however, that the permeability is directly proportional to the crystallinity of a solid.

II Vapours and Liquids

The diffusivity of a solute through a polymer, D_{AS}, varies considerably with solute concentration if the solute swells the polymer. Indeed, both the diffusivity and solubility coefficient, H, increase almost exponentially with increasing solute concentration and hence the permeability, P_{AS}, may be written:

$$P_{AS} = (D_{AS})_0 H_0 e^{bc} \tag{3.34}$$

where $(D_{AS})_0$ and H_0 represent the values of D_{AS} and H for zero solute concentration; c is the solute concentration and b is a constant for the system.

A polymeric material may be regarded as consisting of crystalline and amorphous regions. The crystalline regions have inflexible structures and in general inhibit the penetration of foreign molecules. The amorphous regions are in constant thermal motion and this creates transient holes in the structure. The foreign solute molecules may either 'jump' through or become lodged in these holes. They may also create holes by rotating or moving polymer segments. The length of a jump is proportional to, and of the same order of magnitude as, the diameter of a solute molecule.

The crystalline regions reduce the volume of polymer available for the diffusion. The molecules that diffuse through the amorphous regions must detour around the crystalline regions by a tortuous path, thereby making their diffusion length greater than the thickness of the polymer material. Cross-linking between crystalline regions decreases the mobility of the polymer segments in the amorphous regions; this in turn decreases the diffusivity. The inclusion of plasticizers increases the mobility of polymer segments and hence increases the diffusivity of solute molecules.

When the first solute molecule enters an amorphous region, it loosens or swells the structure, thereby making it easier for subsequent solute molecules to enter the region in that neighbourhood. In this way, solute molecules tend to cluster and swell the amorphous region. Solute molecules that are good solvents for polymer have low clustering functions. Solute molecules which are poor solvents for the polymer tend to associate within the polymer and therefore have high clustering functions.

The above phenomena are usually observed below the glass transition temperature of the polymer. The glass transition temperature of a particular polymer is affected by the presence of foreign molecules and fillers.

Diffusion In and Through Porous Solids

A porous solid may be regarded as a heterogeneous material consisting of solid and pores or voids. These pores, which contain a fluid, may be dead-ends or they may form interconnecting networks through the porous solid.

Before diffusion within the pore system may commence, the diffusing species must first overcome the resistance to diffusion which exists in the 'film' adjacent to the solid surface. Diffusion within the pore system is affected by the size, shape and type of void and any interactions which occur between the diffusing species and the solid material.

In porous solids, diffusion may occur by one or more of three mechanisms:

(a) Molecular diffusion
(b) Knudsen diffusion
(c) Surface diffusion.

(a) Molecular Diffusion

When bulk flow of fluid occurs through a porous solid as a consequence of a pressure gradient, the mechanism by which diffusion occurs to or from the bulk fluid in a pore depends on the Reynolds number. If the Reynolds number based on pore diameter is less than about 2 100 the flow will be laminar and mass transfer will take place by molecular diffusion. If the flow is turbulent, convective diffusion supplements molecular diffusion.

When a species diffuses by molecular diffusion through the pore system of a solid from point 1 at the outside face to point 2 within a pore, it does so by a tortuous path, the actual length of which is unknown. The effective path length is greater than the straight line distance between points 1 and 2 by a tortuosity factor, τ. The correction factor takes account of the pore shape and size, variations in pore cross-sectional area and entry conditions of the pore. Except in solids of uniform pore size and shape, the tortuosity factor must be determined experimentally; it varies from about 1.5 to about 6.5.

(b) Knudsen Diffusion

Knudsen diffusion occurs when the diffusing molecules collide more frequently with the pore walls than they do with each other. This happens when the distance between molecular collisions, λ, (the mean free path) is greater than the pore diameter, d. It is important in very small diameter pores and at low concentrations (relatively large mean free path lengths). Knudsen diffusion can occur in pores under no-flow conditions; it does not depend on a pressure gradient through the pore.

The relative importance of Knudsen diffusion and molecular diffusion in a particular pore is characterized by the Knudsen group, $\mathrm{Kn} = \lambda / d$. Knudsen diffusion predominates for Knudsen numbers in excess of about 10, and molecular diffusion predominates for Knudsen numbers below about 0.01.

(c) Surface Diffusion

Surface diffusion occurs when the diffusing species that is adsorbed by the pore system is mobile and tends to move along the solid surface. This surface diffusion operates in parallel with molecular diffusion from the gas or liquid phase and therefore tends to increase the total mass transfer flux.

The numerical value of the pore surface diffusivity depends on the adsorbent–solid system, especially the nature of the adsorbent bond. If the adsorbent bond is strong, the adsorbed layer may be sufficiently thick to reduce molecular diffusion through the pore. In an extreme case, the adsorbed layer may completely fill or at least plug the pore, perhaps at a narrow constriction.

Unsteady State Diffusion in Solids

In the textile coloration industry, dyeing and drying are frequently batch or semibatch operations and consequently unsteady state conditions exist. This variation in concentration with time at a particular point is associated with a variation in the mass transfer flux, N_A/A.

The general equation for unidirectional unsteady state mass transfer is given by Fick's second law of diffusion:

$$\frac{\partial c_A}{\partial t} = D_{AB} \cdot \frac{\partial^2 c_A}{\partial y^2} \qquad (3.35)$$

Analogous equations may be written for diffusion in two and three dimensions.

Differential equations for unsteady state processes are relatively easy to establish, but difficult to solve. Analytical and graphical solutions are available in the literature for simple and regular geometrics (slabs, cylinders, spheres) assuming constant diffusivity.

Figure 3.4 illustrates the concentration profiles through a flat slab of thickness $2y$ at various times during the unsteady state diffusion of solute A from the slab. It is assumed that diffusion occurs only in the y-direction and therefore the solute diffuses from the flat parallel faces and not from the ends (x-direction) or sides of the slab (z-direction).

Initially, at time zero, the concentration of solute A in the slab is uniform everywhere at c_{A0} (profile 1). If the slab were subjected to a diffusion process and the diffusion was permitted to continue to equilibrium the final concentration of solute A in the slab would be uniform everywhere at c_A* (profile 2). The actual value of c_A* depends on the concentration of solute A in the fluid surrounding the slab. The concentration difference, $c_{A0} - c_A$*, is a measure of the total possible amount of solute removed under the equilibrium conditions.

If at time zero the surface concentration of the slab is suddenly changed to and maintained at a solute concentration of c_A*, then after some time, t, the concentration profile through the slab would be illustrated by the curve c_A (profile 3). If no further diffusion occurred *from* the slab, then the solute within the slab would redistribute by internal migration. The final average concentration at time t would be uniform everywhere at c_{At} (profile 4). The concentration difference, $c_{A0} - c_{At}$, is a measure of the solute removed in

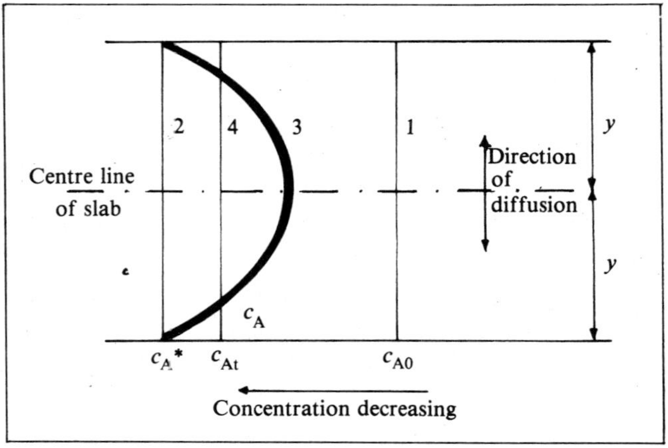

Figure 3.4 – Unsteady-state diffusion from a slab

time t, and $c_{At} - c_A^*$ is a measure of the amount of solute remaining in the slab which could be removed under the prevailing equilibrium conditions.

The fraction, E, of unaccomplished change after time t to the total possible change is obtained by integrating Eqn 3.35. For diffusion from both faces of a slab having a thickness $2y$, the fraction E is given by:

$$E = \frac{c_{At} - c_A^*}{c_{A0} - c_A^*} = \frac{8}{\pi^2}\left(e^{-\beta} + \frac{1}{9}e^{-9\beta} + \frac{1}{25}e^{-25\beta} + \ldots\right) \quad (3.36)$$

where π = numerical constant, 22/7
β = dimensionless constant, $D_{AB}t\pi^2/(4y^2)$.

Examination of the terms in brackets in Eqn 3.36 indicates that, for values of β greater than about one, E can be approximated:

$$E = \frac{8}{\pi^2}(e^{-\beta}) \quad (3.37)$$

Equations having a similar form to Eqn 3.36 are available in the literature for shapes other than flat slabs. For an infinitely long cylinder or filament of radius y, the approximated form of the equation is:

$$E = 0.692(e^{-2.345\beta}) \quad (3.38)$$

where β is defined as in Eqn 3.36.

Relationships for three-dimensional diffusion, varying diffusivity and additional diffusional resistances and chemical reactions are also available in the literature.

Eqns 3.36 to 3.38 may be used for diffusion into a solid, the fraction E is then defined:

$$E = \frac{c_A^* - c_{At}}{c_A^* - c_{A0}} \tag{3.39}$$

4 SIMULTANEOUS HEAT AND MASS TRANSFER

4.1 HUMIDIFICATION

Humidification and dehumidification are the simplest and most direct examples of simultaneous heat and mass transfer. Only two components and two phases are involved. The liquid phase is generally water and the gas phase is generally air, in which some water vapour is present. The air is assumed to be insoluble in water and hence there is no resistance to mass transfer in the liquid phase.

Properties of the Air–Water Vapour System

The humidity charts and tables prepared by different authorities vary, mainly because of their choice of reference temperature and pressure. In this section the reference temperature is 0°C and the reference pressure 101.3×10^3 N m^{-2} (atmospheric pressure).

(a) Dry-bulb Temperature

The dry-bulb temperature (DBT), t_G, is the actual or thermodynamic temperature of the gas–water vapour mixture as measured by a thermometer whose bulb is dry.

(b) Wet-bulb Temperature

The wet-bulb temperature (WBT), t_L, is the dynamic equilibrium temperature of a liquid surface when exposed to a gas–vapour mixture in such a way that the sensible heat transferred from the gas–vapour mixture to the liquid phase is equal to the latent heat associated with the vaporization of liquid into the gas–vapour mixture. It is measured by passing the gas–vapour mixture, at a velocity not less than 5 m s^{-1}, over a thermometer bulb which is covered by a cloth wick *saturated* with liquid.

(c) Humidity

Humidity is defined as the ratio of the mass of vapour to the mass of bone-dry gas in the mixture. Assuming ideal gas behaviour the humidity, w, is:

$$w = \frac{M_V}{M_G} \cdot \frac{p}{P - p} \tag{4.1}$$

where w = humidity, kg vapour per kg bone-dry gas
M_V = molecular weight of vapour
M_G = molecular weight of gas (air = 29)
p = partial pressure of vapour, N m^{-2}
P = total pressure of the mixture, N m^{-2}.

When at a given temperature, the partial pressure of the vapour in the gas–vapour mixture equals the vapour pressure of the liquid, the gas is saturated and the humidity is designated w_s, the saturated humidity:

$$w_s = \frac{M_V}{M_G} \cdot \frac{p_s}{P - p_s} \tag{4.2}$$

(d) Percentage Humidity

The percentage humidity is the ratio of the humidity to the saturated humidity, expressed as a percentage:

$$\frac{w}{w_s} = \frac{100p\,(P - p_s)}{p_s(P - p)} \tag{4.3}$$

(e) Relative Humidity

The relative humidity, RH, is the partial pressure of the vapour in the gas divided by the vapour pressure of the liquid at the same temperature and pressure, the result is expressed as a percentage:

$$\text{RH} = \frac{100p}{p_s} \tag{4.4}$$

(f) Dew Point or Saturation Temperature

The dew point, t_{DP}, is the temperature at which a given gas–vapour mixture is saturated with vapour. Cooling the gas–vapour mixture below the dew point results in the formation of droplets of liquid phase.

(g) Humid Heat or Humid Heat Capacity

The humid heat is the heat required to change the temperature of 1 kg of bone-dry gas plus its associated vapour by 1°C. The humid heat, S, is:

$$S = c_{pG} + w\,c_{pV} \tag{4.5}$$

where S = humid heat, J kg^{-1} K^{-1}
c_{pG} and c_{pV} = heat capacity at constant pressure of bone-dry gas and vapour, J kg^{-1} K^{-1}.

To a first approximation, the humid heat of an air–vapour mixture at atmospheric pressure may be taken to be:

$$S = 10^3 + w(2 \times 10^3) \text{ J kg}^{-1}\text{ K}^{-1} \tag{4.6}$$

To increase the temperature of $\dot{m}$ kg of bone dry gas plus its associated water vapour by t°C requires heat Q (watts).

$$Q = \dot{m}\, S\, t \tag{4.7}$$

(h) Total Heat or Humid Enthalpy

The humid enthalpy, H, (J kg^{-1}) of a gas–vapour mixture is the sum of the enthalpies of the bone-dry gas and its associated vapour, both being relative to a reference temperature of 0°C.

The enthalpy of the bone-dry gas at temperature t°C is:

$$H_{\mathrm{G}} = c_{p\mathrm{G}} t \tag{4.8}$$

If an air–water vapour mixture is unsaturated, then the water vapour is superheated and the enthalpy of the water vapour at t°C is:

$$H_{\mathrm{WV}} = c_{p\mathrm{L}} t_{\mathrm{DP}} + \lambda_{\mathrm{DP}} + c_{p\mathrm{WV}}(t - t_{\mathrm{DP}}) \tag{4.9}$$

where λ_{DP} = latent heat of vaporization at the dew point, J kg^{-1}
$c_{p\mathrm{L}}$ = heat capacity at constant pressure of liquid phase water, J kg^{-1} K^{-1}.

The enthalpy of 1 kg of bone-dry air plus its associated water vapour is:

$$H = c_{p\mathrm{A}} t + w\,[c_{p\mathrm{L}}\, t_{\mathrm{DP}} + \lambda_{\mathrm{DP}} + c_{p\mathrm{WV}}\,(t - t_{\mathrm{DP}})] \tag{4.10}$$

The terms within the square brackets represent the enthalpy of water vapour, having been heated as a liquid from the datum temperature of 0°C to the dew point, then vaporized at the dew point and superheated to the final temperature of t°C. This enthalpy H may be compared with the enthalpy H' for water vapour resulting from vaporization at 0°C and superheating directly to t°C. At atmospheric pressure the final enthalpy is almost independent of the vaporization path, and hence H is approximately equal to H'.

$$H \simeq H' = c_{p\mathrm{A}} t + w(c_{p\mathrm{WV}} t + \lambda_0) \tag{4.11}$$

$$= \mathrm{S}t + w\lambda_{\mathrm{o}}$$

where λ_0 = latent heat of vaporization at 0°C, 2.50×10^6 J kg^{-1}.

(i) Humid Volume

The humid volume is defined as the volume (m^3) of 1 kg of bone-dry gas plus its associated vapour. Assuming ideal gas behaviour, the humid volume, V_H, is:

$$V_H = V_G + wV_V \tag{4.12}$$

where V_G and V_V = specific volume of bone-dry gas and vapour at STP, $m^3\ kg^{-1}$

One kilomole of a gas at 0°C and 101.3×10^3 N m^{-2} absolute has a volume of 22.41 m^3. Hence the humid volume, V_H ($m^3\ kg^{-1}$) of a gas–vapour mixture at a temperature of t°C and a pressure P (N m^{-2}) having a humidity w is:

$$V_H = \left(\frac{1}{M_G} + \frac{w}{M_V}\right) 22.41 \left(\frac{t + 273}{273} \cdot \frac{101.3 \times 10^3}{P}\right) \tag{4.13}$$

The saturated humid volume is the volume of 1 kg of saturated gas.

Wet-bulb Theory

Figure 4.1 illustrates the partial pressure and temperature gradients that are established when an unsaturated gas passes over a liquid surface in an adiabatic system. As a result of the partial pressure driving force, liquid evaporates into the gas, thereby increasing its humidity. This evaporative cooling of the liquid tends to decrease its temperature to a value less than the DBT of the gas. As a consequence of this temperature driving force, sensible heat is transferred from the gas to the liquid. At equilibrium the rate of sensible heat transferred from the gas to the liquid balances the la-

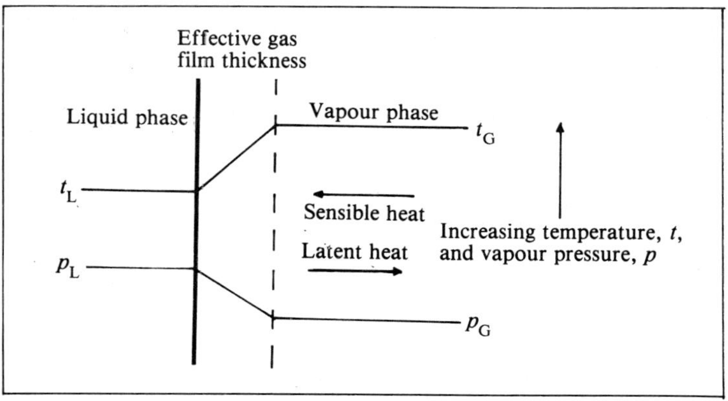

Figure 4.1 – Simultaneous heat and mass transfer at a liquid surface

tent heat required to vaporize the liquid. The equilibrium temperature of the liquid is the wet-bulb temperature. The time taken to reach equilibrium depends on the initial temperatures, the interfacial surface area and the flow rate of gas past the liquid surface. When the WBT is being determined by passing a stream of gas at a velocity in excess of about 5 m s^{-1} across the relatively small area of liquid that surrounds the thermometer bulb, equilibrium is established relatively quickly. The temperature and the humidity of the bulk gas stream are essentially unchanged. It will be appreciated, however, that, for constant monitoring, the wick surrounding the thermometer bulb must be kept saturated with liquid.

The heat transfer rate equation for the equilibrium process is:

$$\dot{m}\lambda = Q = h_G A(t_G - t_L) \tag{4.14}$$

where $\dot{m}$ = rate of vaporization of liquid, kg s^{-1}
λ = latent heat of vaporization of liquid at WBT, J kg^{-1}
Q = rate of heat transfer, W
h_G = convective heat transfer from gas–vapour mixture to liquid, W m^{-2} K^{-1}
A = area across which heat is transferred, m^2
t_G = bulk gas DBT, °C
t_L = liquid temperature and gas–vapour WBT, °C.

The simultaneous rate of mass transfer from the liquid surface to the gas–vapour mixture is:

$$N_L = k_G A(p_L - p_G) \tag{4.15}$$

where N_L = molar rate of mass transfer of liquid to gas–vapour mixture, mol s^{-1}
k_G = mass transfer coefficient, mol m^{-2} s^{-1} (N m^{-2})$^{-1}$
p_L = partial vapour pressure of liquid at WBT, N m^{-2}
p_G = partial pressure of vapour in the gas–vapour mixture, N m^{-2}.

In terms of mass flow rate rather than molar flow rate, Eqn 4.15 may be rewritten:

$$N_L M_V = M_V k_G A(p_L - p_G) = \dot{m} \tag{4.16}$$

Combining Eqns 4.14 and 4.16 gives:

$$(p_L - p_G) = \frac{h_G}{M_V\, k_G\, \lambda}(t_G - t_L) \tag{4.17}$$

If, as is often the case, the partial pressure of the vapour in the gas–vapour mixture is small compared with the total pressure, then Eqn 4.1 may be written:

$$w_L = \frac{W_V}{M_G} \cdot \frac{p_L}{P} \text{ and } w_G = \frac{M_V}{M_G} \cdot \frac{p_G}{P} \qquad (4.18)$$

Hence Eqn 4.17 can be written:

$$(w_G - w_L) = -\left(\frac{h_G}{PM_G\,k_G}\right)\frac{1}{\lambda}\,(t_G - t_L) \qquad (4.19)$$

The ratio h_G/k_G is essentially independent of the gas velocity, provided that it is in excess of about 5 m s^{-1}.

The dimensional group $h_G/(PM_G\,k_G)$ has been evaluated from experimental wet-bulb data for a number of vapours in air; selected values are shown in Table 4.1.

TABLE 4.1
Values of $h_G/(PM_G\,k_G)$ for Vapours in Air

Vapour	$\frac{h_G}{PM_G\,k_G}$	
	J kg^{-1} K^{-1}	Btu/(lb °F)
Water	1.1×10^3	0.26
Methyl alcohol	1.5×10^3	0.35
Benzene	1.7×10^3	0.41
Carbon tetrachloride	1.8×10^3	0.44
Toluene	1.8×10^3	0.44

When experimental data are not available, the dimensional group can be calculated from *j*-factor relationships to yield:

$$\frac{h_G}{PM_G\,k_G} = S\left(\frac{\text{Sc}}{\text{Pr}}\right)^{0.67} \qquad (4.20)$$

Rearranging Eqn 4.20 gives the psychrometric ratio or Lewis relationship:

$$\frac{h_G}{PM_G k_G\,S} = \text{Le} = \left(\frac{\text{Sc}}{\text{Pr}}\right)^{0.67} \qquad (4.21)$$

The Lewis group, Le, is important in direct-contact simultaneous heat and mass transfer processes. It relates the ratio of the heat to mass transfer coefficients to the heat capacity of the transfer medium (gas–vapour mixture).

If the WBT, t_L, and the DBT, t_G, of a sample of gas–vapour mixture are known, then the humidity of the mixture, w_G, can be calculated using Eqn 4.20. The values of the dimensional group, latent heat and partial pressure (and hence w_L) of the liquid can be either calculated or obtained from literature values. In the case of the air–water vapour system the value of the dimensional group is 1.1×10^3 J kg^{-1} K^{-1} and the values of the latent heat and vapour pressure of water may be obtained from published literature.

If, however, it is desired to fix the WBT and determine the temperature and humidity of the gas–water vapour mixture which will give this WBT, then only w_L, t_L and λ are known, w_G and t_G are both unknown. Therefore, although a given sample of gas–water vapour mixture (whose temperatures and humidity are known) will give a definite WBT, there are an infinite number of combinations of temperatures and humidities that will give this same WBT. Since all such pairs must satisfy Eqn 4.19, they must be on a curve on the humidity chart.

Adiabatic Saturation Temperature

When, in an adiabatic system, a gas is in contact with a relatively large surface area of liquid for a sufficient time to enable equilibrium to be established, the gas becomes saturated with liquid. At equilibrium the saturated gas and liquid phases have the same temperature, the adiabatic saturation temperature, AST. This situation is typified in adiabatic humidification processes, where gas and liquid are contacted in such a way that the exit gas is saturated with vapour at the AST. A heat balance for an adiabatic humidification process, when the gas temperature decreases from the inlet value, t_G, to the AST, t_{AS}, and the humidity increases from the inlet value, w_G, to the exit saturation value, w_{AS}, is:

$$(w_G - w_{AS}) = -\frac{S}{\lambda}(t_G - t_{AS}) \qquad (4.22)$$

In an adiabatic system, the relationship between the decrease in gas temperature, t_G, and the corresponding increase in humidity, w_G, may be represented on a humidity chart as a straight line of slope $-S/\lambda$. For a particular gas–vapour system, a range of adiabatic saturation temperatures is represented by a family of curves. These curves may be approximated to straight lines of slope $-S/\lambda$. However, because of the variation of both S and λ with temperature, the lines are neither straight nor parallel. The curves which are known as adiabatic cooling lines are essentially lines of constant enthalpy.

When the WBT and AST coincide, Eqns 4.19 and 4.22 are identical and the humid heat, S, is numerically equal to the dimensional group, $h_G/(PM_G\, k_G)$. In the air–water vapour system, the psychrometric ratio,

$h_G/(PM_G\, k_G\, S)$, equals unity when the humidity is 0.047 kg kg^{-1}. At humidities above and below this value, the WBT and AST diverge and, for accurate work, this difference should be recognized by using Eqn 4.19 for wet-bulb problems.

It is usually sufficiently accurate for engineering problems dealing with air–water vapour mixtures to use the adiabatic cooling lines for wet-bulb problems, i.e. to assume that the WBT and AST are identical.

For systems involving a liquid other than water, the difference between the WBT and the AST may be as great as 50%.

The Humidity Chart and Its Use

The properties of the air–water vapour system that are required for calculations are available in psychrometric or humidity charts and tables of many different types. These charts may be constructed to incorporate a wide range of information. A given chart is valid for one total pressure only. In charts used for engineering calculations, the adiabatic cooling lines may be drawn as straight lines by slightly distorting the ordinates.

Figure 4.2 illustrates a section of a humidity chart for the air–water vapour system. The interpretations of the various points plotted are listed below.

Point X – composition of original mixture
A – DBT of original mixture
B – humidity of original mixture
C – intersection of adiabatic cooling line and 100% RH curve
D – WBT of original mixture
E – humidity of final mixture if the original mixture is humidified adiabatically
F – dew point of original mixture
G – composition of final mixture if the original mixture is humidified at its original DBT (i.e. isothermal humidification)
H – humidity of final mixture corresponding to G above

Problem 4.1

Using the humidity chart in Appendix II and standard steam tables, determine for an air–water vapour mixture having a WBT of 20°C and a DBT of 30°C:

(a) The humidity
(b) The relative humidity
(c) The dew point
(d) The humid heat
(e) The humid volume
(f) The humidity of the mixture if saturated adiabatically
(g) The humidity and WBT of the mixture if saturated isothermally.

(*Answers*: (a) 0.011 kg kg^{-1}; (b) 40%; (c) 15°C; (d) 1.026×10^3 J kg^{-1} K^{-1}; (e) 0.884 m^3 kg^{-1}; (f) 0.015 kg kg^{-1}; (g) 0.027 kg kg^{-1})

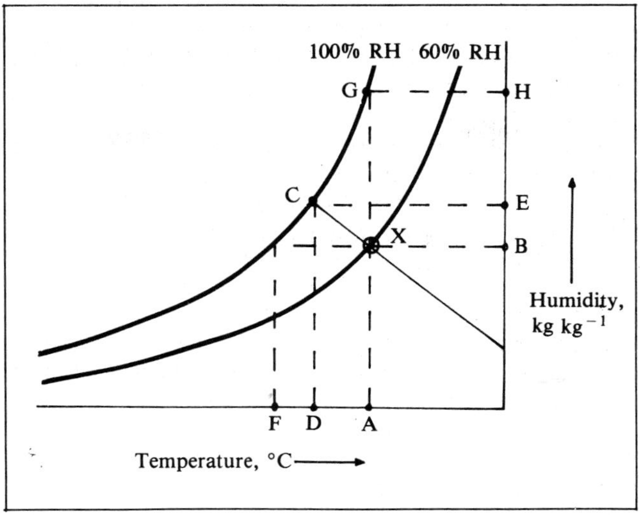

Figure 4.2 – Humidity chart

Problem 4.2

Determine the heat required to convert 5 $m^3\ s^{-1}$ of an air–water vapour mixture, having wet- and dry-bulb temperatures of 25°C and 30°C, into a mixture having the same humidity at a DBT of 70°C. Further, determine the volume of the final mixture.

(*Answers*: 234×10^3 W; 5.66 $m^3\ s^{-1}$)

Problem 4.3

Air, having wet- and dry-bulb temperatures of 10°C and 15°C, is preheated to 80°C and blown into an adiabatic dryer. The dryer gives a product containing 5% wt/wt of water at a rate of 0.25 kg s^{-1} from wet material containing 50% wt/wt water. The exhaust from the dryer is 3°C above the adiabatic saturation temperature.

Calculate:

(a) The temperature and relative humidity of the exhaust air
(b) The weight of air (kg s^{-1}) required to dry the material
(c) The heat required to preheat the air.

(*Answers*: (a) 30°C/33°C, 80%; (b) 11.2 kg; (c) 392×10^3 W)

4.2 DRYING

In order to plan drying schedules and to determine the capacity of equipment, it is necessary to know the time required to dry a material from one moisture content to another under specified conditions. The capacity of a dryer depends on the rates of both heat and mass transfer. The latter depends on the distribution of moisture on and through the solid material, and on the mechanism by which moisture migrates from the interior of the material to the vapour/liquid interface where it may be vaporized.

The energy required for drying may be in the form of conductive, convective, radiative or dielectric heating. When assessing the efficiency of a drying process it is important to consider all cost factors and not only thermal energy costs. Although the vast majority of textile dryers operate at atmospheric pressure, models are in use which operate under vacuum or above atmospheric pressure.

When a material is dried under constant conditions only the conditions of the inlet gas are constant, the moisture content and other properties of the material change. In practice, an excess of gas at constant wet- and dry-bulb temperatures is used in drying.

Normally, sensible heat is transferred from the gas stream to the moist solid. With the exception of dielectric heating (in which high-frequency waves generate heat within the moist material), heat is transferred from the gas stream to the outer surface and then into the interior of the material. Mass is transferred as a liquid or vapour within the material and as a vapour from the exposed surfaces into the gas stream. Moisture movement within the material results from a concentration gradient which is dependent on the characteristics of the solid and liquid phases. The factors governing the simultaneous heat and mass processes determine the drying rate.

The removal of moisture from a material often causes shrinkage and induces stresses; intensive drying can cause chemical change and damage to the material.

Heat and Mass Balances

For ease of calculation, the moisture content, y, is frequently expressed on a bone-dry material basis, kg of moisture per kg of bone-dry material. This is analogous to expressing the humidity of a gas–vapour mixture in terms of mass of vapour per unit mass of bone-dry gas. The percentage moisture on the bone-dry basis is $100y$. The corresponding percentage moisture content on a wet-material basis (bone-dry material plus associated moisture) is $100y/(1+y)$.

The enthalpy of a moist material, H_M, based on a kilogram of bone-dry material is:

$$H_M = c_M t_M + y c_W t_M + \Delta H_M \tag{4.23}$$

where H_M = enthalpy per kilogram of bone-dry material, J kg^{-1}
c_M = heat capacity of bone-dry material, J kg^{-1} K^{-1}
t_M = temperature of material relative to 0°C, °C
y = moisture content on bone-dry basis, kg kg^{-1}
c_W = heat capacity of liquid phase moisture, J kg^{-1} K^{-1}
ΔH_M = heat of wetting, adsorption, hydration, etc., J kg^{-1}.

The heat, Q, in watts, required to increase the temperature of a mass (kg s^{-1}) of moist material, M, from temperature 1 to temperature 2 is:

$$Q = M(H_{M2} - H_{M1}) \tag{4.24}$$

The heat capacity of some common textile materials are given in Table 4.2.

TABLE 4.2

Heat Capacity of Textile Materials

Material	Heat capacity at 25°C (J kg^{-1} K^{-1})	Material	Heat capacity at 25°C (J kg^{-1} K^{-1})
Terylene	1.26×10^3	Wool	1.38×10^3
Cotton and cellulose	1.34×10^3	Polyacrylonitrile	1.50×10^3
Linen	1.34×10^3	Nylon	$1.67\times10^3-2.30\times10^3$
Rayon	1.34×10^3	Polypropylene	1.88×10^3
Silk	1.38×10^3	Polyethylene	2.30×10^3

Figure 4.3 illustrates a counter-current dryer operating under steady state conditions.

Although the total wet material rate decreases through the dryer as moisture is removed from the material, the bone-dry material rate, M, (kg bone-dry material per second) remains constant. A similar situation applies to the gas stream, the total moist gas rate increases through the dryer as moisture is transferred from the moist material to the gas stream; however, the bone-dry gas rate, G, (kg bone-dry gas per second) remains unchanged.

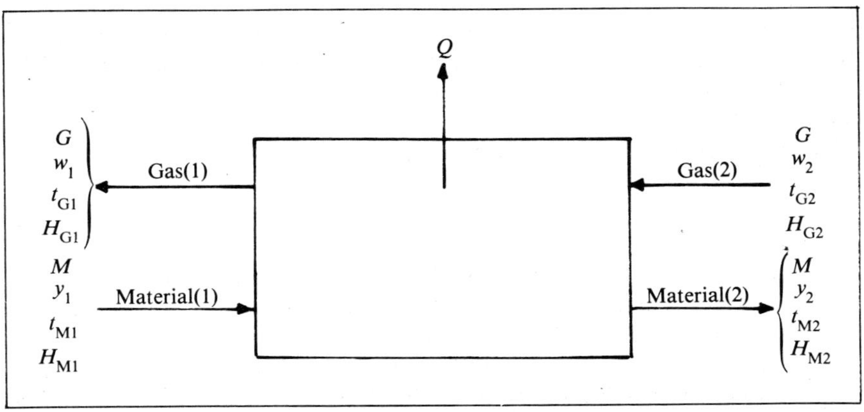

Figure 4.3 – Counter-current dryer operating under steady-state conditions

The moist material undergoes a moisture and temperature change as it passes through the dryer, the moist gas stream undergoes a corresponding humidity and temperature change.

The overall mass balance for the dryer illustrated by Figure 4.3 is:

$$My_1 + Gw_2 = My_2 + Gw_1 \tag{4.25}$$

or

$$M(y_1 - y_2) = G(w_1 - w_2) \tag{4.26}$$

For a co-current dryer the corresponding balance is:

$$My_1 + Gw_1 = My_2 + Gw_2 \tag{4.27}$$

or

$$M(y_1 - y_2) = -G(w_1 - w_2) \tag{4.28}$$

If, as shown in Figure 4.4, a proportion of the exhaust gas is recycled to improve the thermal efficiency, such that the recycle ratio is R (mass of bone-dry air recycled per total mass of bone-dry gas passing through the dryer), the new exit humidity is w_1^+, and the mass balance is:

$$My_l + (1-R)Gw_2 + RGw_1^+ = My_2 + Gw_1^+ \tag{4.29}$$

or

$$M(y_1 - y_2) = Gw_1^+ - RGw_1^+ - (1-R)Gw_2$$

$$= (1-R)\,G(w_1^+ - w_2) \tag{4.30}$$

Comparing Eqns 4.26 and 4.30 shows that:

$$(w_1 + w_2) = (1-R)\,(w_1^+ - w_2) \tag{4.31}$$

or

$$R = (w_1^+ - w_l)/(w_1^+ - w_2) \tag{4.32}$$

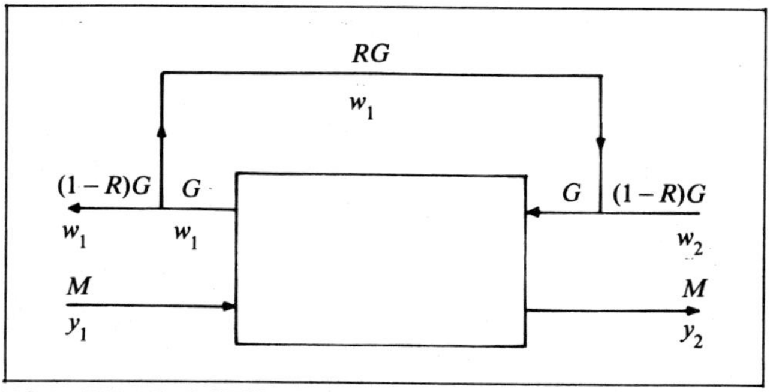

Figure 4.4 – Counter-current dryer operating under steady-state conditions with gas recycle

Hence, as the recycle ratio increases, the exhaust gas humidity, w_1^+, increases and the fresh gas is more diluted with recycled, moisture-laden gas. The gas passing through the dryer has, therefore, a lower potential for moisture uptake and a larger dryer will be required. The greater capital cost of the larger dryer is offset by the lower operating cost for heating the air, and hence there will be an optimum recycle ratio which yields a minimum total cost.

The enthalpy balance for the counter-current dryer illustrated by Figure 4.3 is:

$$MH_{M1} + GH_{G2} = MH_{M2} + GH_{G1} + Q \qquad (4.33)$$

or

$$M(H_{M1} - H_{M2}) = G(H_{G1} - H_{G2}) + Q \qquad (4.34)$$

For adiabatic operations the heat loss, Q, is zero. If heat is added within the dryer to an extent greater than the heat loss, then Q is negative.

When a dryer is direct-fired, water is one of the products of combustion and this must be taken into account in the moisture balance.

The thermal efficiency of a steam-heated dryer may be expressed as the specific steam consumption: kg of steam used per kg of water vaporized.

Under steady state drying conditions in a stenter, heat is required for three purposes:

(a) To increase the temperature of the fabric and its associated water from the inlet temperature to the exit temperature and to supply the latent heat of vaporization of the water (see Example 4.1)

(b) To heat fresh air to compensate for the exhaust air lost when partial air recirculation is used (see Example 4.2)

(c) To compensate for convective and radiative heat losses from the stenter.

The heat used in (a) above represents the essential heat requirement. Whilst the actual essential steam consumption is a function of steam pressure, fabric type, initial and final moisture, it can be shown by a heat and mass balance that, in general, the essential specific steam consumption varies between 1.3 and 1.4.

The amount of fresh air added to a stenter can have a profound effect on the steam consumption. Each 1% of air ingress can increase the specific steam consumption by about 0.25, i.e. a 10% air ingress can correspond to a value of 2.5. The air ingress for optimum efficiency corresponds to a specific steam consumption of about 0.4.

The thermal insulation of a stenter should be such that the convective and radiative heat losses are at an economic minimum. A typical value of the specific steam consumption for a well-insulated stenter is about 0.2.

Thus, although the essential heat requirement corresponds to a specific steam consumption of about 1.4, a target figure for the actual practical consumption is about 2.0.

Example 4.1

A fabric is dried from 70% moisture to 5% moisture (both on a dry basis) in a stenter at a rate of 0.15 kg s^{-1} (bone-dry basis). If the inlet and outlet fabric temperatures are 25°C and 100°C and the mean operating temperature of the stenter at atmospheric pressure is 130°C, calculate:

(a) The essential heat requirement
(b) The specific steam consumption when steam at (i) 0.7×10^6, (ii) 0.8×10^6 and (iii) 1.2×10^6 N m^{-2} absolute is used as the heating medium.

Assume that the mean heat capacity of the fabric is 1.38×10^3 J kg^{-1} K^{-1} and the water vaporizes from the fabric at 100°C.

Quantity of water entering stenter $= 0.15 \times 0.70 = 0.1050$ kg s^{-1}
Quantity of water leaving stenter $= 0.15 \times 0.05 = 0.0075$ kg s^{-1}
Quantity of water to be vaporized $= 0.0975$ kg s^{-1}

Heat required to increase the temperature of the fabric is:

$$0.15 \times 1.38 \times 10^3 \, (100 - 25) = 15.5 \times 10^3 \text{ W}$$

Heat required to increase 0.105 kg s^{-1} of water from 25°C to 100°C is, from steam tables:

$$0.105 \, (419 - 105) 10^3 = 33.0 \times 10^3 \text{ W}$$

Heat required to vaporize 0.0975 kg s^{-1} of water at 100°C is, from steam tables:

$$0.0975 \times 2257 \times 10^3 = 220 \times 10^3 \text{ W}$$

Heat required to superheat 0.0975 kg s^{-1} of water from 100°C to 130°C is, from steam tables:

$$0.0975 \, (2737 - 2676) 10^3 = 5.9 \times 10^3 \text{ W}$$

The total heat required $= 274 \times 10^3$ W
Heat required per kg of water vaporized $= 2814 \times 10^3$ W

Steam pressure (N m^{-2})	*Latent heat* (J kg^{-1})	*Specific steam consumption*
0.7×10^6	2067×10^3	1.36
0.8×10^6	2048×10^3	1.37
1.2×10^6	1986×10^3	1.42

Example 4.2

In the previous example the air recirculation rate was 45.0 kg s^{-1} and 0.75 kg s^{-1} was exhausted to maintain the optimum conditions for drying.

Calculate the heat required and the corresponding specific steam consumptions at the three steam pressures to heat 0.75 kg s^{-1} of air with a dry-bulb temperature of 25°C and a humidity of 0.010 kg kg^{-1} to 130°C at constant humidity.

The humid heat of inlet air at 25°C is, from steam tables:

$$S = 1.00 \times 10^3 + 0.010\ (1.88 \times 10^3) = 1.02 \times 10^3 \text{ J kg}^{-1} \text{ K}^{-1}$$

The humid heat of the air at 130°C and a humidity of 0.010 kg kg^{-1} is, from steam tables:

$$S = 1.01 \times 10^3 + 0.010\ (2.15 \times 10^3) = 1.03 \times 10^3 \text{ J kg}^{-1} \text{ K}^{-1}$$

The heat required to increase the temperature of the air from 25°C to 130°C is:

$$0.75\ [(1.03 \times 10^3 \times 130) - (1.02 \times 10^3 \times 25)] = 81 \times 10^3 \text{ W}$$

Heat required per kg of water vaporized $= 834 \times 10^3$ W.

Steam pressure (N m^{-2})	*Latent heat* (J kg^{-1})	*Specific steam consumption*
0.7×10^6	2067×10^3	0.40
0.8×10^6	2048×10^3	0.41
1.2×10^6	1986×10^3	0.42

Moisture–Material Relationships

The extent to which a material can be dried depends upon the way in which the moisture is associated with the material.

It has already been mentioned that the moisture content, y, can be expressed on the dry or the wet basis. The former is generally more convenient since the moisture loss is obtained by direct subtraction of the moisture content before and after drying.

A number of the more important terms used in drying are mentioned below and illustrated in Figure 4.5.

(a) The total moisture content, y_T, is the total mass of moisture per unit mass of bone-dry material.

(b) The equilibrium moisture content, y^*, is the limiting moisture content to which a given material can be dried under given conditions of gas temperature and humidity. The moisture in the material is in equilibrium with the moisture in the gas–vapour mixture.

(c) Bound moisture refers to moisture which is chemically or physically bound to the material. It exhibits a heat of wetting and exerts a vapour pressure less than that of the pure liquid at the same temperature.

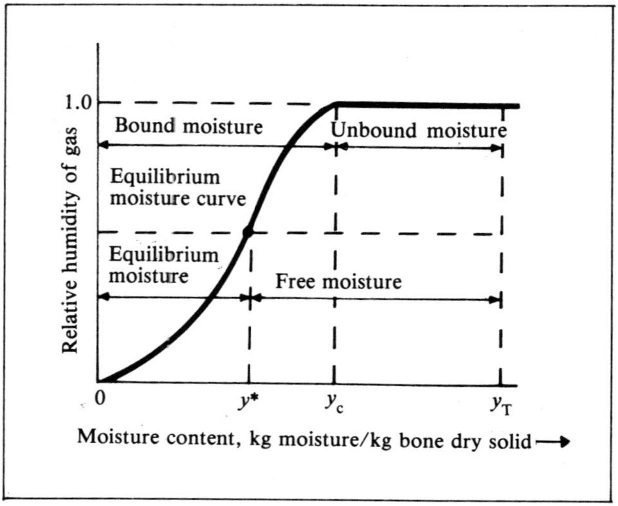

Figure 4.5 – Moisture contents

(d) Unbound moisture is the moisture in excess of the equilibrium moisture content corresponding to 100% relative humidity. It exerts the same vapour pressure as the pure liquid.

(e) The critical moisture content, y_c, corresponds to the transition between bound and unbound moisture. It marks the end of the constant-rate drying period.

(f) The free moisture content is the moisture which can be removed at a given gas temperature and humidity. It is the difference between the total moisture and the equilibrium moisture contents. It may include both bound and unbound moisture.

(g) The commercial dry moisture content is the moisture content of a finished product, expressed in terms of the mass of moisture per unit mass of product (bone-dry material plus residual moisture content).

The moisture in a moist material exerts a vapour pressure which depends on the nature of the moisture, the nature of the material and the temperature. If a moist material is exposed to a continuous supply of a gas–vapour mixture having a definite temperature and humidity (i.e. having a definite vapour pressure) the material will either lose moisture to or gain moisture from the gas–vapour mixture, until equilibrium is achieved.

Figure 4.6 illustrates the equilibrium moisture contents of a number of textile materials; it shows that, irrespective of its initial water content, a sample of wool in contact with air, having a temperature of 25°C and a relative humidity of 65%, will eventually achieve its equilibrium moisture content of 16%. Thus, if the wool is to be stored at a relative humidity of 65% and a temperature of 25°C it is wasteful, from an energy viewpoint, to

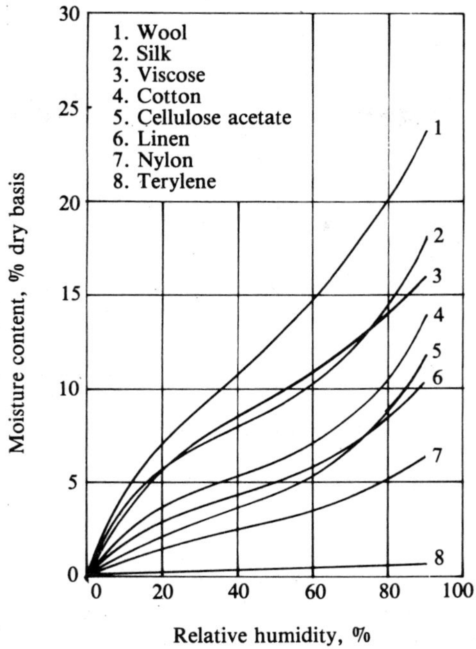

Figure 4.6 – Equilibrium moisture curves at 25 °C

dry the material below a moisture content of about 16%.

For a given relative humidity, the equilibrum moisture content of a material decreases with increase in gas temperature.

Rates of Drying

The drying rate, R, may be defined as the rate of change of moisture content from unit surface area, A, of a mass of material, M:

$$R = -\frac{M}{A} \cdot \frac{dy}{dt} \tag{4.35}$$

where R = drying rate, kg m^{-2} s^{-1}
M = mass of bone dry material M, kg
A = surface area, m^2
dy/dt = rate of change of moisture content, kg kg^{-1} s^{-1}.

Figure 4.7 illustrates the changing moisture content of a given mass of material as it is dried either in a batch process or as it passes through a continuous process. The total, critical and equilibrium moisture contents are represented by y_T, y_c and y^* respectively.

Although Figure 4.7 illustrates that the drying rate is a function of time or

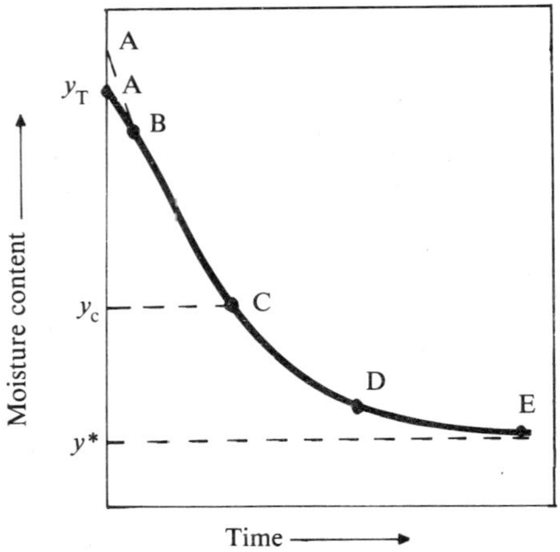

Figure 4.7 – Drying curve

moisture content, this variation can be demonstrated more clearly by graphically or analytically differentiating the curve and plotting the drying rate, R, as a function of moisture content, y, as in Figure 4.8, or time, t, as in Figure 4.9. The resulting rate curves show that the drying process can be divided into three periods:

(a) Warming-up period, AB
(b) Constant-rate period, BC
(c) Falling-rate period or periods, CD and DE.

If the moisture content is less than y_c then there will be no constant-rate period. If the final moisture content is in excess of y_c then there will be no falling-rate period demonstrated.

(a) The *warming-up period* is the time required for the moist material to attain the operating temperature. In most commercial dryers this time is generally a small proportion of the total drying time.

(b) During the *constant-rate period* the moisture vaporizes from the saturated surface of the material. The rate is independent of the moisture content and is essentially that of moisture vaporizing from a free liquid surface under the same conditions of gas velocity, temperature and humidity. The drying process involves the transfer of heat from the bulk gas to the material and the simultaneous transfer of mass from the saturated surface of the material to the bulk gas stream, across a 'stagnant film' of gas.

From wet-bulb theory it can be shown that, for steady state conditions, the equation for the simultaneous heat and mass transfer for the vaporization of water is:

$$R_c = h_G(t_G - t_i)/\lambda = 18k_G(p_i - p_G) \qquad (4.36)$$

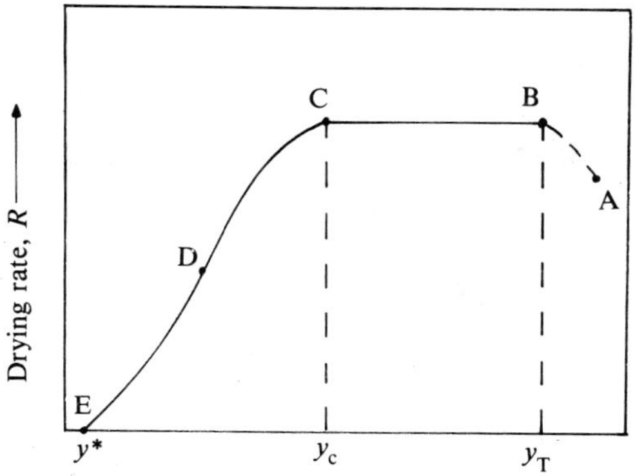

Figure 4.8 – Drying-rate curve

where R_c = drying rate during constant rate period, kg m^{-2} s^{-1}
h_G = convective heat transfer coefficient, W m^{-2} K^{-1}
$(t_G - t_i)$ = temperature difference between bulk gas and interface, K
λ = latent heat of vaporization, J kg^{-1}
k_G = gas mass transfer coefficient, mol m^{-2} s^{-1} (N $m^{-2})^{-1}$
$(p_i - p_G)$ = vapour pressure difference between interface and bulk gas, N m^{-2}.

As it is usually assumed that there is no resistance to mass transfer at the interface, p_i in Eqn 4.36 may be taken as the vapour pressure of the liquid phase at the interface temperature. From wet-bulb theory, the wetted surface temperature should be the same as the WBT corresponding to the bulk air conditions. In practice, however, heat transfer to the wetted surface occurs by radiation and conduction as well as convection, and hence the actual

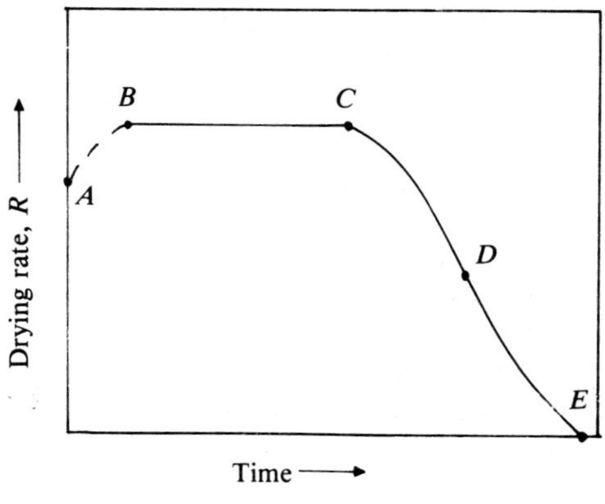

Figure 4.9 – Drying-rate curve

surface temperature is between the WBT and the boiling temperature of the liquid phase.

Either form of Eqn 4.36 may be used to determine the drying rate. However, as an error in determining the interfacial temperature, t_i, affects the driving force $(t_G - t_i)$ less than it affects the term $(p_i - p_G)$, it is more reliable to calculate the drying rate from the heat transfer equation.

The convective heat transfer coefficient, h_G, depends on the geometry of the system, the gas velocity, and the physical properties of the gas. To a first approximation, h_G may be expressed:

$$h_G = \alpha G^m d^{-n} \tag{4.37}$$

where h_G = heat transfer coefficient, W m^{-2} K^{-1}
G = mass velocity of gas, kg m^{-2} s^{-1}
d = characteristic linear dimension, m
α, m and n = empirical constants which are reported in the literature, see, for example, Perry's Chemical Engineers' Handbook.

The drying rate can be increased by increasing the gas flow rate, increasing the gas temperature or decreasing the gas humidity.

The constant-rate period ends when the rate of vaporization of liquid from the surface of the material exceeds the rate at which liquid can migrate from the interior of the material to the surface. The critical moisture content corresponding to the end of the constant-rate period cannot be predicted; its value depends on:

(i) The properties of the material being dried
(ii) The thickness of the material being dried
(iii) The factors which influence the moisture migration within the material
(iv) The velocity, temperature and humidity of the gas stream
(v) The type of dryer.

It is known that the critical moisture content increases with increase in drying rate and with increase in the thickness of the material being dried.

(c) During the *falling-rate period* or periods the drying rate decreases with decrease in moisture content. The actual shape of the drying-rate curve depends on the structure and composition of the material being dried and on the mechanism by which the moisture moves within the material. The moisture migration process, which produces a moisture gradient through the material, is the rate-determining step.

When the average moisture content of the material falls below the critical moisture content, the surface of the material is no longer completely wetted. As the average moisture content is decreased, the dry areas occupy an increasingly large proportion of the surface. The first falling-rate period corresponds to drying from an unsaturated surface. In some cases, the drying rate in the first falling-rate period is a linear function of the average moisture of the material. In this event it would be represented by a straight line from point C to point D in Figures 4.8 and 4.9.

The first falling-rate period continues until the entire liquid front recedes into the interior of the material and leaves the surface completely dry. This corresponds to point D on the drying-rate curves. The drying-rate curves may now exhibit a further discontinuity representing the start of the second falling-rate period. If the change from a partially wetted surface to a totally dry surface occurs very gradually, no change in drying rate may be detected and hence the second falling-rate period may not be apparent.

When the liquid phase recedes from the surface of the material, drying proceeds by the transfer of heat from the gas stream to the dry surface by convection and then through the material to the liquid phase by conduction. The moisture is vaporized within the material and the resulting vapour migrates by diffusion through the material to the surface. The rate of diffusion of the vapour is essentially independent of the external conditions and determines the rate of drying. When drying to low moisture contents, the second falling-rate period usually predominates in determining the overall drying time.

The falling-rate periods are characterized by increasing temperatures, both at the surface and within the solid. The factors which affect the falling-rate periods are, in essence, those which control the rate of internal moisture movement.

At point E on the drying-rate curves, the moisture content of the material has decreased to the equilibrium value, y^*, for the prevailing gas humidity, and no further drying is possible.

Dielectric Drying

A dielectric is the electrically non-conducting medium between the two electrodes in an electric capacitor. Dielectric drying refers to the removal of moisture from a wet material by placing it between two electrodes across which an alternating high-frequency voltage is applied.

The thermal energy, E, released in a dielectric heater is:

$$E \propto f.\ V^2.\ K.\ \cos\theta \qquad (4.38)$$

where E = thermal energy released, watts
f = frequency, hertz
V = voltage, volts
K = dielectric constant, dimensionless
$\cos\theta$ = power factor, dimensionless.

The dielectric constant, K, of a medium, or dielectric, is the ratio of the capacity of the capacitor using the medium between the electrodes to the capacity of the capacitor when there is a vacuum between the electrodes. By this definition the dielectric constant of a vacuum is unity. The dielectric constant for a gas is approximately unity, for insulating materials (such as textiles) it is between 2 and 8, for polar liquids it is between 30 and 100. Water has a dielectric constant of about 80.

In a circuit containing a perfect capacitor, the current leads the voltage by 90°. In an actual capacitor the current will lead the voltage by an angle, θ, which is less than 90°. The complementary angle, $90° - \theta$, is the loss angle, δ. The 'power factor' is defined as $\cos \theta$ or $\sin \delta$ and is the ratio of the actual power produced (watts) to the apparent power delivered (current × supply voltage).

The product of the dielectric constant and the power factor, $K \cos \theta$, is known as the 'loss factor' of the dielectric. The loss factor of a dielectric is affected by temperature and frequency. When placed in a high-frequency alternating electric field, materials with a high loss factor, such as water, heat rapidly, while those with a very low loss factor, such as textile materials, are essentially unaffected by the field. The actual high frequencies employed for industrial heating and drying are allocated by international agreement. Dielectric dryers use frequencies in the radio-frequency range, hence the term 'radio frequency' (r.f.) drying. The frequency most commonly used in r.f. drying is 27.12 MHz. Microwave drying is a special form of dielectric drying in which very high frequencies, typically 10^9 to 3×10^9 Hz are applied.

In dielectric drying the wet, non-conducting, body is placed between the pair of electrodes across which an alternating high-frequency voltage is applied. The polar water molecules in the dielectric constantly realign themselves in the alternating electric field. The energy associated with these molecular-dipole oscillations is dissipated in the form of heat which increases the temperature of the water to its boiling point and then vaporizes it. As the loss factor for the water is much greater than that for textile materials, the heat is generated almost exclusively and directly in the water. The textile material will become hot by heat transfer from the water phase. The heating rate decreases as the material becomes drier and hence the system can be automated. Drying depends on the dielectric properties of the textile body; it is not limited by the physical dimensions of the body; nor does it require a temperature difference between the body and the air that surrounds the body. Thus, the drying of textile packages is even and very rapid. The quality of the finished product is more predictable and controllable than is the case with hot air drying.

Steam at 100°C and atmospheric pressure has, from steam tables, a total enthalpy of $2\,676 \times 10^3$ J kg^{-1}. Water at 20°C has an enthalpy of 84×10^3 J kg^{-1}. Hence to convert water at 20°C into dry saturated steam at 100°C requires $2\,592 \times 10^3$ J kg^{-1}. To vaporize 1 kg s^{-1} of water requires 2 592 kW and therefore 1 kWh can vaporize, theoretically, 3 600/2 592 or 1.39 kg of water. Inefficiency of power conversion reduces this theoretical evaporative capacity. Overall evaporative capacity values quoted in the literature range from 0.5 to 1.0 kg water per kWh supplied to the dryer.

The running costs of a dielectric dryer depend on the design of the dryer, the material to be dried and the electrical conversion efficency. Information on dielectric dryers and drying can be obtained from the Central Electricity Research Laboratories, Leatherhead, Surrey, UK.

It would appear that the real potential of dielectric drying is for the removal of bound water from materials that have an excessively long falling-rate period. In the textile industry this means packages and other bulky shapes where there is both a thermal and diffusional barrier to drying. It is claimed that the drying time for packages is between 15 and 45 min compared with 10 to 12 h for hot air drying. This reduction in drying time can be offset against the operating costs.

Energy Costs

The unit price of energy is an important cost factor in the processing of textiles. In many situations it would not be practicable to change the form of energy being used. However, in existing establishments it is worthwhile performing an energy audit prior to instituting an investment appraisal programme or an optimization programme to limit the amount of fuel and electricity used. With regard to electricity, care should be taken to limit maximum demand and to check that the power factor correction equipment is at its economic optimum.

The absolute and relative costs of energy change with time and geographical location. The figures given in Table 4.3 refer to average energy costs for Great Britain in November 1982. The fuel prices do not include site storage costs nor the capital and operating charges involved in converting the fuel into, say, steam. It is worth remembering that the average cost of process steam is about £7 per thousand kg (November 1982). This cost should be an incentive to check the efficient use of fuel in boilers, to repair steam leaks and, where possible, to return condensate to the boiler feed water system.

TABLE 4.3

Comparison of Energy Costs (November 1982)

Energy form	Average price, pence per gross kWh
Electricity, direct	3.62
Fuel oil – 200 s (General zone)	1.66
– 950 s (General zone)	1.46
– 3 500 s (General zone)	1.36
Natural gas	1.04
Coal	0.68

Adapted from the National Industrial Fuel Efficiency Service (NIFES)

FURTHER READING

I FLUID FLOW, HEAT TRANSFER AND MASS TRANSFER COMBINED

Badger and Banchero, 'Introduction to Chemical Engineering', McGraw–Hill, (1955).

Beek and Muttzall, 'Transport Phenomena', Wiley, (1975).

Bennett and Myers, 'Momentum, Heat and Mass Transfer', McGraw–Hill, (1974).

Bird, Stewart and Lightfoot, 'Transport Phenomena', Wiley, (1960).

Clark and Davidson, 'Manual for Process Engineering Calculations', McGraw–Hill, (1962).

Coulson and Richardson, 'Chemical Engineering', Volumes 1 and 2, Pergamon Press, (1977 and 1978).

Jordan, 'Chemical Process Development', Parts 1 and 2, Wiley, (1968).

Perry, 'Chemical Engineers' Handbook', McGraw–Hill, (1973).

Peters and Timmerhaus, 'Plant Design and Economics for Chemical Engineers', McGraw–Hill, (1968).

Welty, Wilson and Wicks, 'Fundamentals of Momentum, Heat and Mass Transfer', Wiley, (1969).

II FLUID FLOW

Giles, 'Fluid Mechanics and Hydraulics', McGraw–Hill, (1962).

Hicks and Edwards, 'Pump Application Engineering', McGraw–Hill, (1971).

Holland and Chapman, 'Pumping of Liquids', Reinhold, (1966).

Holland, 'Fluid Flow for Chemical Engineers', Arnold, (1973).

Mitchell, 'Fluid and Particle Mechanics', Pergamon Press, (1970).

Schlichting, 'Boundary Layer Theory', McGraw–Hill, (1960).

Sterbacek and Tausk, 'Mixing in the Chemical Industry', Pergamon Press, (1965).

Osborne, 'Fans', Pergamon Press, (1977).

'Technical Reference Book on Valves for the Control of Fluids', British Valve Manufacturers' Association, (1972).

III HEAT TRANSFER

Chapman, 'Heat Transfer', Collier–MacMillan, (1969).

Collier, 'Convective Boiling and Condensation', McGraw–Hill, (1972).

Dehnel, 'Fundamentals of Boiler House Techniques', Hutchinson, (1973).

Ede, 'An Introduction to Heat Transfer – Principles and Calculations', Pergamon Press, (1967).

Holland, Moores, Watson and Wilkinson, 'Heat Transfer', Heinemann, (1970).

Kays and London, 'Compact Heat Exchangers', McGraw–Hill, (1964).

Kern, 'Process Heat Transfer', McGraw–Hill, (1950).

Kreith, 'Principles of Heat Transfer', Intertext, (1965).

McAdams, 'Heat Transmission', McGraw–Hill, (1954).

Simonson, 'An Introduction to Engineering Heat Transfer', McGraw–Hill, (1967).

IV **MASS TRANSFER**

Bird and Boston, (Editors), 'Theory of Coloration of Textiles', Dyers Company Publications Trust, (1975).

Crank, 'The Mathematics of Diffusion', Oxford University Press, (1975).

Crank and Park, (Editors), 'Diffusion in Polymers', Academic Press, (1968).

Keey, 'Drying – Principles and Practice', Pergamon Press, (1972).

Peters, 'Textile Chemistry, Volume III, The Physical Chemistry of Dyeing', Elsevier, (1975).

Sherwood, Pigford and Wilke, 'Mass Transfer', McGraw–Hill, (1975).

Spalding, 'Convective Mass Transfer – An Introduction', Arnold, (1963).

Treybal, 'Mass Transfer Operations', McGraw–Hill, (1955).

APPENDIX I

All the values quoted for the various properties of fluids and gases have been taken from 'Thermodynamic and Transport Properties of Fluids' arranged by Mayhew and Rogers, Oxford: OUP (1969).

APPENDIX II

Humidity Chart for Air and Water Vapour at Atmospheric pressure

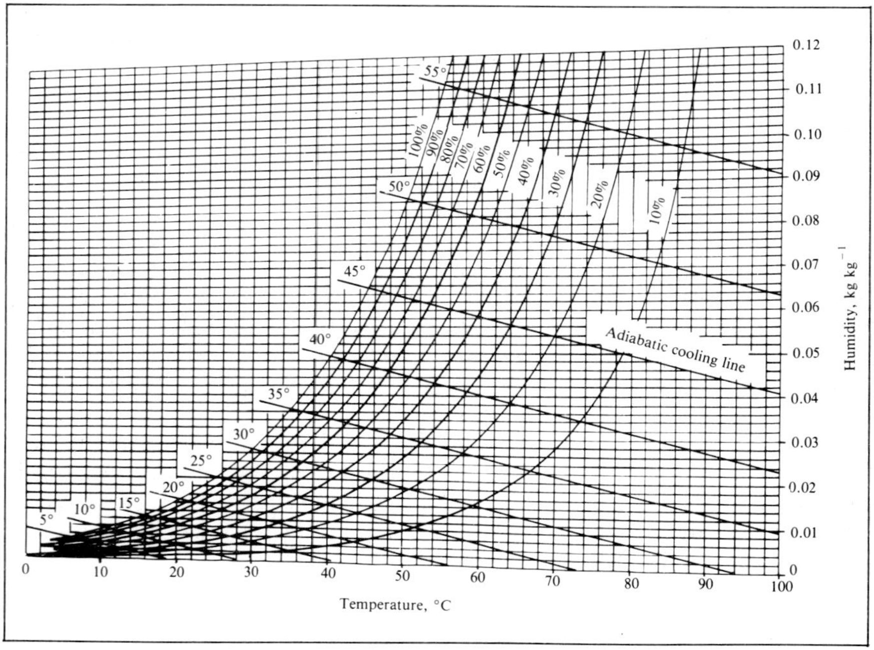

Chapter 3

Corrosion

P ELLIOTT

INTRODUCTION

Corrosion is a factor that affects every sector of industry and commerce. It is a natural phenomenon whereby a metal attempts to revert to its energy-favoured condition and, like extractive metallurgy, is a costly process, In economic terms, corrosion affects plant availability, where its implications may be much broader than many first imagine. Ideally corrosion should be considered as a compromise between a material and its environment based primarily on costs – in the broadest sense of the word. In practice, however, this ideal is not so often realized unless there exists a degree of awareness of the reasons for corrosion and a willingness to take action to prevent or minimize these effects. It cannot be overstated that corrosion failures in themselves are not important, but the costs of such failures and the ramifications associated with them are.

People have different attitudes to corrosion, which provides more complications than may be anticipated at first consideration. Some accept corrosion as an unsolvable entity, whilst others reflect that 'out-of-sight' is 'out-of-mind'. In reality much can be done to minimize such phenomena, provided that action is taken at the correct time with a suitable method. Financial savings can be considerable with total savings put close to 35% across the spectrum of British industries(*1*).

This chapter, in many ways, links all the other chapters of this book, since corrosion is related to many of the aspects or applications discussed therein. The aim is to provide the reader with a general appreciation of corrosion and to highlight techniques that might be of benefit in reducing or eliminating such attack.

1 METALLIC CORROSION

When considering corrosion, it is necessary to recognize that the term embraces a multitude of possibilities depending upon the particular material and the circumstances of the environment. Surface rusting of steels, for example, can represent one extreme whilst the complete disintegration of a sophisticated alloy by extensive internal cracking can represent another. All materials will corrode, given suitable environments, so it is useful to consider briefly what corrosion is and then highlight some of the more damaging effects associated with particular types or circumstances of environment.

When a metal is exposed to a corrosive environment it will corrode if thermodynamic factors are favourable. The kinetics of corrosion will govern to what extent the material corrodes and the type of product will obviously be significant, there being some benefit from adherent and protective products as opposed to those which are porous or non-adherent. Most common environments involve water or moisture, and much has been written with respect to these. However, briefly, if a metal (M) is corroding in water, then it is possible to envisage several reactions:

(a) Where the metal is dissolving to form ions and to generate electrons (termed the *anodic* reaction);

$$2M \rightarrow 2M^{2+} + 4e^-$$

(b) The dissolution of oxygen using electrons generated in reaction (a) (termed a *cathodic* reaction):

$$2H_2O + O_2 + 4e^- \rightarrow 4OH^-$$

(c) The combination of products from (a) and (b) to form a metal hydroxide:

$$2M^{2+} + 4OH^- \rightarrow 2M(OH)_2$$

Equations that relate to corrosive action are termed *anodic* and these need to be counter-balanced by some form of *cathodic* process, such as oxy-

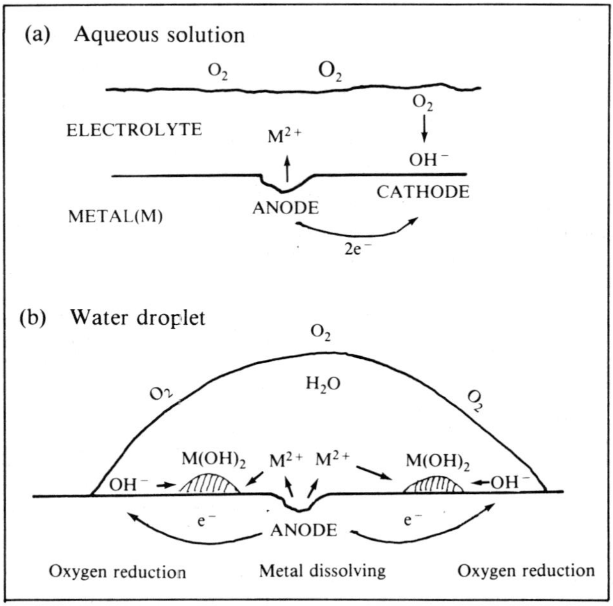

Figure 1 – Some representative corrosion processes, M = metal

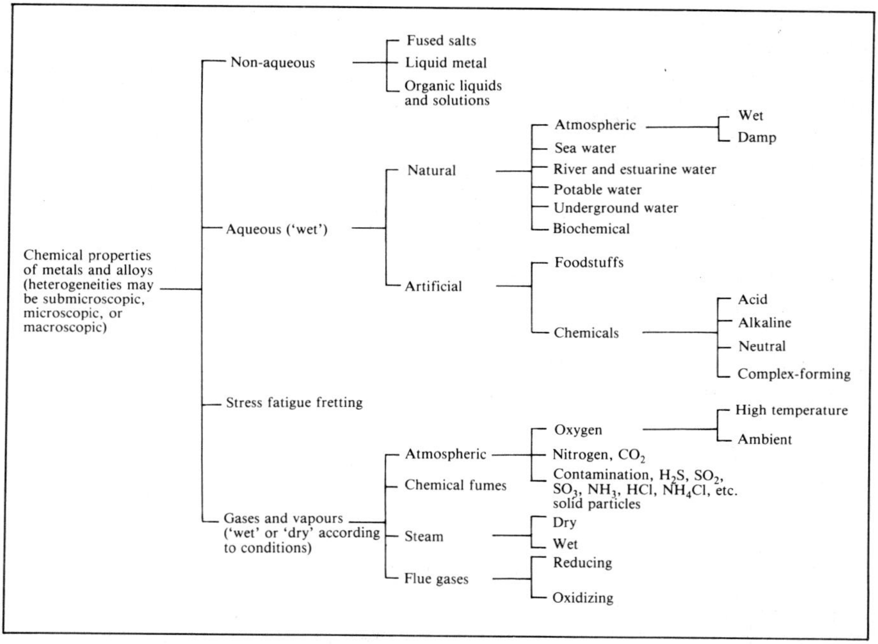

Figure 2 – Environments causing corrosion (2)

gen reduction (as in (b) above) or hydrogen gas evolution (for more acidic environments). The environment that supports these reactions is termed the electrolyte and the equations that permit electron movement are described as electrochemical. Figure 1 shows some typical simplified situations.

Corrosion rates depend upon a subtle interplay of factors controlled by the metal and the circumstances of environment. Clearly, chemical composition, metallurgical condition and method of fabrication need to be borne in mind when considering the material, but of equal importance is the chemical and physical nature of the corrosive, the temperature and the extent of fluid flow and heat flux through the metal/corrosive interface.

Corrosion classification is often based either upon the nature of the corrosive (wet or dry), the mechanism of attack (electrochemical or chemical) or upon the appearance of the corroded material (uniform or localized). There are many environments as can be judged from those summarized in Figure 2(*2*).

General Corrosion

In situations where anodic and cathodic reactions occur on adjacent sites with similar areas, the corrosive attack will be of a uniform nature, which is of considerable benefit when predicting material life expectancies. When electrons produced in the anodic reaction flow through the metal to participate in reactions at the cathodic areas, then an electrical potential exists between the anode and cathode. A complete electrical circuit is involved and the faster the metal is converted into ions the greater is the electrical current involved (i.e. the corrosion rate is greater). Corrosion 'cells', such as depicted in Figure 1, are repeated many times across a corroding metal surface, the anodic and cathodic sites switching roles continuously, hence permitting uniform or general corrosion. It is important to recognize that anodic and cathodic reactions occur at the same rate in these situations and, if *either* process stops, then so must the other – a factor of prime importance in corrosion control.

Metals will only corrode if there is a negative change in free energy (ΔG), thermodynamics being represented by the following equation relating the number of electrons associated with corrosion (z) and the cell potential (E) according to

$$\Delta G = -z E \mathcal{F}$$

where $\mathcal{F}$ is the Faraday constant. When a metal is immersed in a solution, a potential difference is set up between the atoms in the metal (M) and the metal ions in solution (M^{2+}, for the case of (a) above). This potential ($E_{M/M^{2+}}$) is called a half-cell potential at the anodic, or oxidizing site. A similar potential is present at cathodic (or reducing) sites, such as E_{O_2/OH^-} for oxygen-containing situations (like (b) above), or E_{H^+/H_2} for more acidic environments. These potentials can be easily measured using a voltmeter and an inert reference electrode and, when equated to unit activity of ions (i.e. one gram atom per litre of metal ions), they are termed standard half-cell potentials. Utilizing the hydrogen reaction ($2H^+ + 2e^- \rightleftharpoons H_2$) as a 'zero' reference for standard half-cell potential, it is possible, by comparison, to have a relative measure of other half-cell potentials. These 'redox' potentials, as they are termed (i.e. reduction–oxidation situations) are available in the literature, or from laboratory study, and the difference between them (i.e. the difference between cathodic and anodic half-cell potentials) constitutes the cell potential (E) in the above equation. If values of E are positive, ΔG is negative and corrosion can be anticipated.

The rate of corrosion is related to the specific electrolyte and the nature of the corrosion product or film present on the metal. For example, an infinite resistance in the electrolyte would mean that anodic and cathodic potentials remain the same and, using Ohm's law ($I = E/R$), there would be no current flow; such conditions arise with suitably applied protective coatings. In practice, resistance is nearer to zero and current flows between anode and cathode with a resulting change in their potentials. The change in potential is termed polarization and reference to Figure 3 demonstrates how polarization effectively reduces the difference in potential between anodic and cathodic sites. The potentials rise for the anodic curve and fall for the cathodic curve and once the metal is freely corroding in its environment it will have a potential E_{corr}, which can be measured using a high-resistance voltmeter. The corrosion current density between anodic and cathodic areas is given by i_{corr}. More detailed descriptions of these phenomena and ways of measuring corrosion rates and potentials are to be found in the literature(*2–8*). In many cases metals that thermodynamically should show rapid corrosion rates do not. Such metals are said to be passive (i.e. they are corroding at a very slow rate). The phenomenon of passivation is usually associated with oxidizing environments and with the formation of thin (~3 nm) protective oxide layers. Typical examples include chromium, nickel, titanium and alloys that contain such elements, like stainless steel.

The tendency for passivation is a property of the anode and this can be reflected in terms of potential/current density curves. Figure 4, for example, contrasts the anodic polarization curve for metal, M, without passivation (cf. Figure 3) and with passivation. The region where high polarization occurs (XY) corresponds to the passive region. Obvious differences in behaviour can be anticipated depending upon the relative position of the cathodic polarization curve on these plots. Clearly, if the cathodic curve follows that shown in Figure 4c, then the corrosion may be very low (i_L) or relatively high (i_H). This represents one of the disadvantages of passive materials, since slight changes within the environment can give rise to active corrosion, often in the form of localized attack.

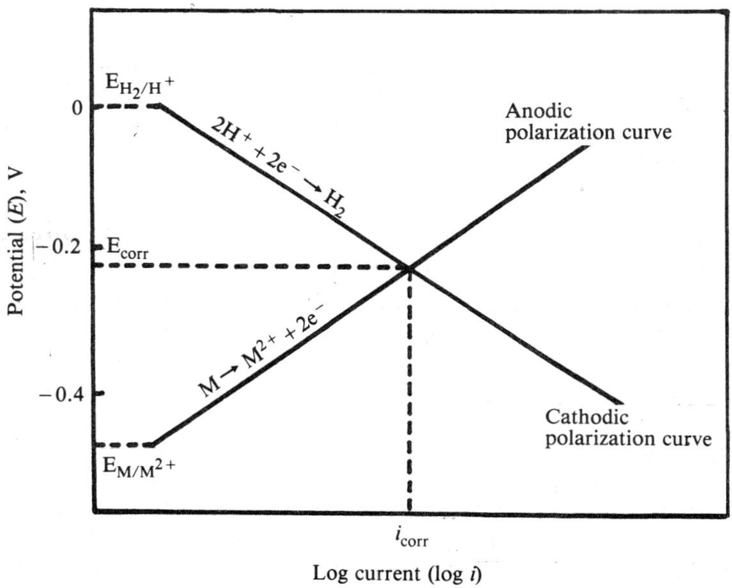

Figure 3 – Polarization for a metal in an acidic solution

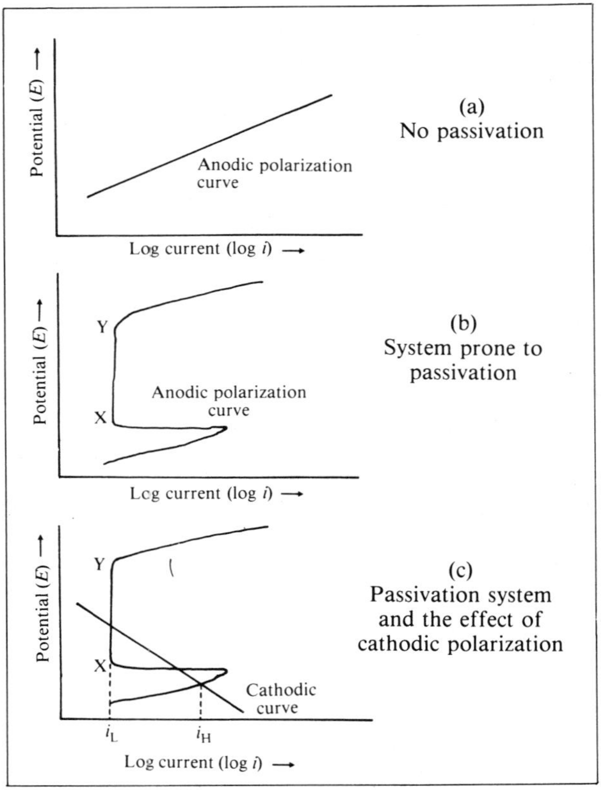

Figure 4 – Anodic polarization curves for metal and the effect of cathodic polarization and passivity

Localized Corrosion

The most prevalent cause of premature corrosion failure in a metallic structure is the occurrence of unexpected localized, rather than uniform, corrosion. Such attack leads to a disproportionate penetration of the metal and in these circumstances the direct translation of weight-loss corrosion rates into penetration rates will be unreliable (a limitation in data charts or corrosion tables – see below).

Localized corrosion represents the selective removal of metal by attack confined to small zones exposed to a corrosive environment. Such damage is often associated with particular environmental circumstances and is exacerbated by such effects as flow or stress. For localized corrosion, the anodes and cathodes will be separated, the anode area will be much less than the cathodic area, and the anodic potential will be less than the cathodic potential.

The types of localized corrosion to be described later in this section are either initiated or developed within the metal (pitting, intercrystalline corrosion, selective dissolution, stress corrosion cracks, corrosion fatigue, etc.) or associated with some aspect of the environment, design or fabrication (crevice corrosion, deposit corrosion, filiform corrosion, erosion corrosion, cavitation, impingement, etc.) Other problems arise due to impressed or 'stray' currents (dissimilar metal contact, effect of external electrical circuitry, e.g. overhead power lines). Many localized corrosion possibilities could be eliminated with adequate inspection and quality control during fabrication. The extent of damage may be macroscopic or microscopic and it should be noted that, although the latter may relate to a minute loss of metal, considerable damage can occur before the problem becomes visible.

Galvanic or Bimetallic Corrosion

The preferential dissolution of a metal, when two dissimilar metals are in electrical contact in an electrolyte solution, is referred to as galvanic corrosion. The important factor is the difference in electrode potential between the two metals, because this provides the 'driving force' for corrosion. The effect of galvanic corrosion is not necessarily greater if a larger potential difference exists between the metals because the open-circuit potential difference does not accurately reflect the rate of corrosion. Attack is usually confined to the more 'active' metal of the couple and is governed greatly by the relative areas of the two metals, a small anodic area with a large cathodic area prompts severe damage (e.g. a small steel bolt in a copper plate should be avoided). Figure 5a demonstrates features of galvanic corrosion, and Figure 5b shows a situation from service where an aluminium

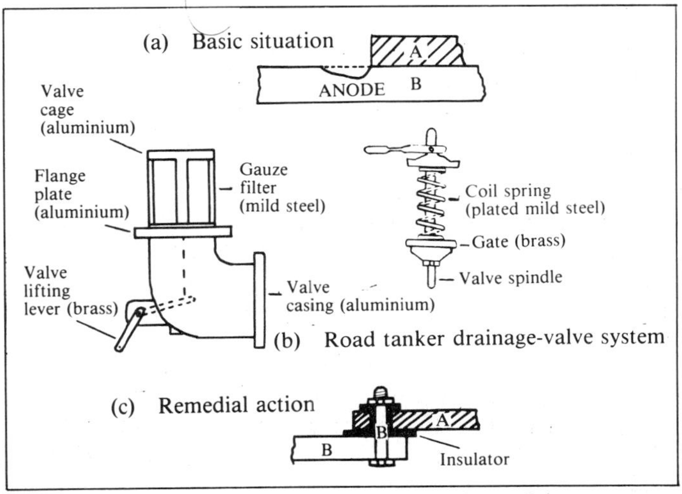

Figure 5 – Effects of galvanic corrosion on joining dissimilar metals in contact with an electrolyte

drainage valve casing was in direct contact with brass and ferrous components. Failure had occurred within months because of the very acute effect of aluminium (more anodic) in contact with copper (more noble) – a feature clearly apparent from so-called galvanic series, which are tables of potential differences between different metal couples measured in given environments. Table 1 shows values of mean electrode potentials for metals and alloys immersed in dilute chloride solution at ambient temperature(*9*).

TABLE 1

Galvanic Series of Dilute Chloride Solution at Ambient Temperature(*9*)

	Metal/alloys	Electrode potential (V)*
	Platinum	+0.26
	Gold	−0.10
more active (anodic) ↑	Stainless steels (passive)	−0.10
	Monel (Ni–Cu)	−0.11
	Titanium	−0.13
	Copper	−0.29
	Brasses	−0.33
	Stainless steels (active)	−0.53
	Lead	−0.55
more noble (cathodic) ↓	Carbon steel	−0.70
	Aluminium and alloys	−0.88
	Zinc	−1.03
	Magnesium	−1.73

*Mean electrode potentials on the saturated calomel scale

Galvanic series are directly related to the particular environment and, as such, the tables so produced are restrictive in that the ranking order of the metals and alloys will vary if changes occur in the environment. Of particular significance are flow rate, temperature, chloride concentration and aeration (i.e. oxygen concentration). Galvanic corrosion occurs when (i) the two metals are in electronic contact, and (ii) when there is a conducting electrolyte path between the two metals. Accelerated attack is significant when the combined resistance of these two parts of the circuit decreases and corrosion may be eliminated or avoided by completely 'breaking' either of these parts of the circuit involved. For example, no galvanic corrosion occurs in a dry system.

Bimetallic corrosion can be reduced or eliminated in several ways, for example, by the use of compatible metals, of one metal as a coating for an other, or of paints or polymeric coatings on one or both metals. It is usually more sensible to coat the cathode rather than the anode, since discontinuities in the coating would not lead to such serious localized attack. Hence, in galvanized material, the zinc will corrode preferentially and, if a small hole is present, the exposed steel will not be corroded, provided good electrical contact is maintained from the base steel through the electrolyte to the zinc. Insulation provides an alternative way to break the circuit between the two metals, as is shown in Figure 5c. Further information is available in the literature(*2,3*) including a table that provides a guide to such effects(*9*). Of particular importance is the use of galvanic corrosion as a means of corrosion control.

Microgalvanic corrosion can also arise when a more noble metal is deposited from the electrolyte onto the other metal. Figure 6 shows an example of this so-called 'deposit' corrosion in which an aluminium plate stirrer has become pitted and perforated. The electrolyte was 'pure' water,

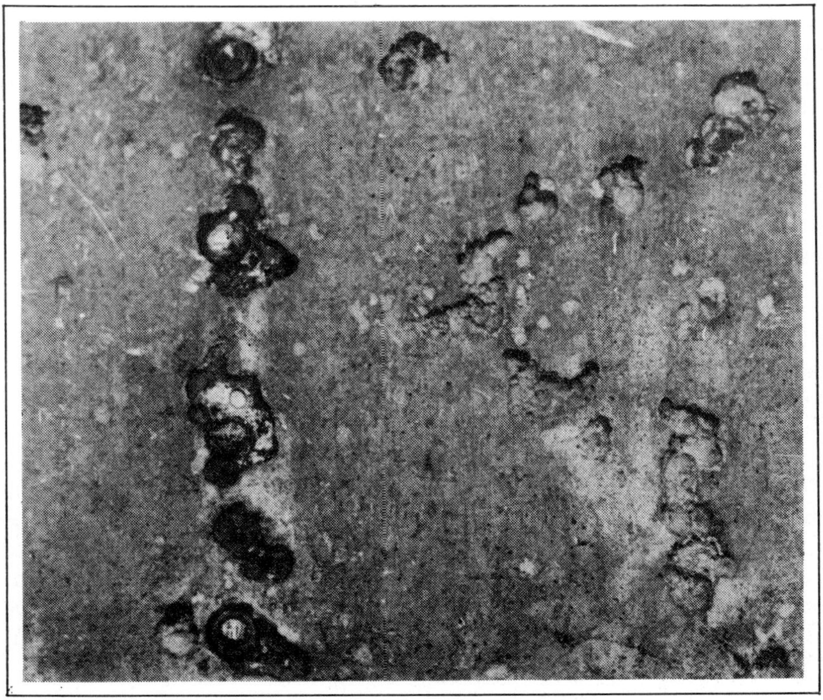

Figure 6 – Pitting corrosion in an aluminium plate stirrer as a result of deposited copper (i.e. galvanic corrosion); the damage was noticed after about 6 to 9 months (note also the effects of crevice corrosion in the rivet regions)

which would normally have no corrosive attack on aluminium, but the system was heated with copper coils, whereby copper ions entered solution and subsequently plated out onto the aluminium surface to form intense local galvanic cells. Other forms of localized corrosion relate to local changes in composition and these will be dealt with below (e.g. selective dissolution, weld decay and intergranular corrosion).

Pitting
Pits are holes that form in an otherwise relatively unattacked surface (e.g. Figure 6), although the shape, size and distribution of these holes can vary considerably as a result of the material and/or the particular environment. Pits arise in a variety of ways, but are often associated with local cell action arising from differences in the metal composition and from disturbances in the surface film (e.g. galvanic cells or active–passive situations that were mentioned previously). Changes within the environment can also result in pitting damage as will become apparent in this chapter. Pitting arises on metals that readily form adherent protective surface films and the overall process may occur over a long period or can be unexpected and severe, especially if undercutting has occurred to any significant degree. Pits may also be associated with intergranular, cavitation, crevice and erosion corrosion (see below).

Clean surfaces, free from defects or local differences in metal composition, can help reduce the onset of pitting, and conditions that permit the build-up of undesirable elements in the environment (e.g. stagnant corrosives) should be avoided. Alternative materials, coatings or the use of electrochemical protection provide ways of reducing such damage.

Crevice Corrosion
This particularly serious form of localized corrosion can occur within cracks or crevices that form between mating surfaces of metal (e.g. bolts, rivets, expansion joints) or between metal/non-metal surfaces (e.g. seals, gaskets, poor coatings). The extent of damage is usually not appreciated because the most severe attack occurs within the crevice (Figure 7).

The environmental conditions within a crevice, with time, become different from the conditions on the free surfaces and local cell action arises such that the *interior* surfaces become anodic and suffer from extensive corrosion. Metals that form adherent protective surface films are particularly prone to such damage and much care is needed with design since crevice corrosion is a common weakness in many components and units, in particular those materials that form passive or natural oxide films.

A common situation arises when the supply of oxygen within a crevice is interrupted and less is available compared with the outer surfaces of an assembly. Such a situation, where corrosion results from a lateral oxygen concentration gradient, is termed 'differential aeration', there being differences in concentration of dissolved oxygen within the electrolyte. In these

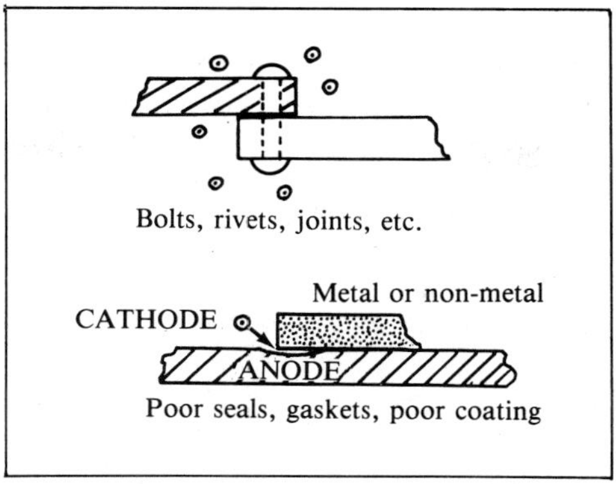

Figure 7 – Crevice corrosion of metal assuming film or solution of electrolyte present (⊙ = possible sites)

cases the cathodic reaction is reduction of dissolved oxygen, which increases the pH value (more alkaline). Areas offering good access to oxygen act as cathodes and acute corrosion occurs if a three-phase boundary (metal–electrolyte–atmosphere) results from the geometry of the system (Figure 8). A shorter oxygen-diffusion path exists across the meniscus zone resulting in more pronounced corrosion just below the 'waterline' of the electrolyte solution. Differential aeration effects are also responsible for deposit and poultice corrosion.

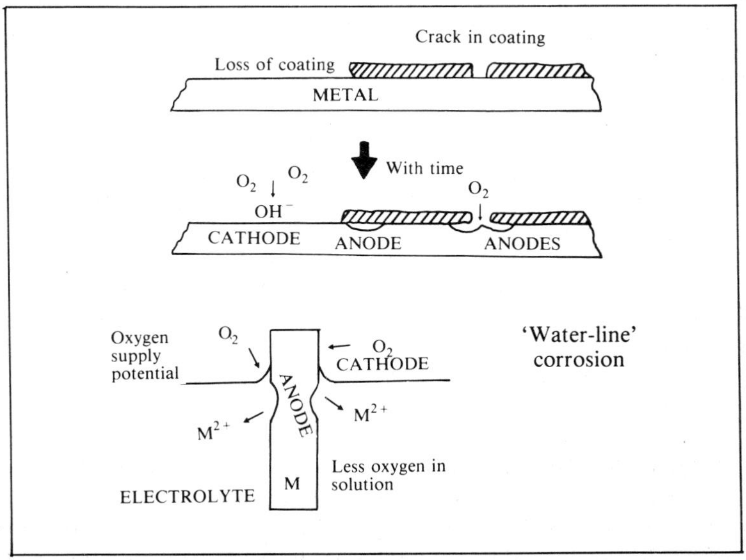

Figure 8 – 'Differential aeration' effects

Deposits
Dirt, sand or other debris that restricts the availability of oxygen to the metal provide ideal sites for local pitting corrosion. A lack of awareness at design and construction or during operation can create conditions ripe for such corrosion, and it is necessary to avoid situations that permit the build-up of stagnant residues (Figure 9). Clean surfaces, wherever possible, help to minimize such effects and suitable measures with process fluids (e.g. filters) can also be beneficial.

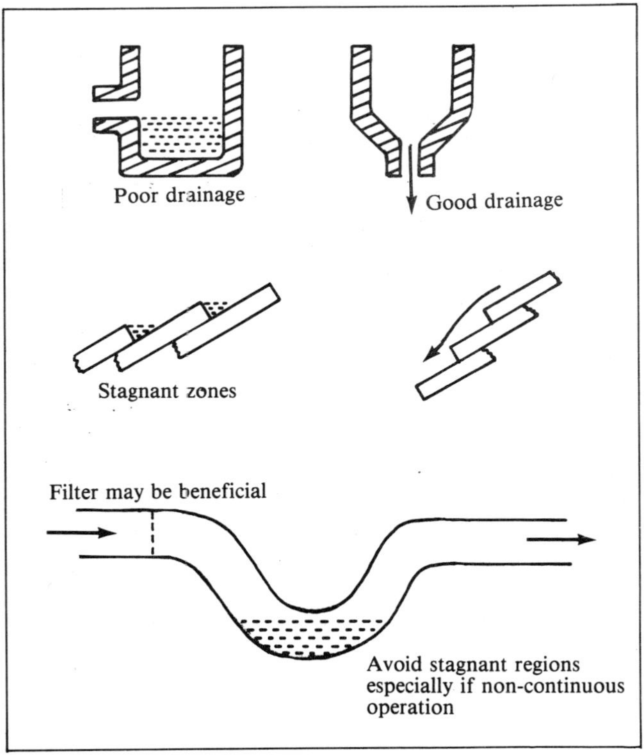

Figure 9 – Situations that create stagnant residues, which can concentrate harmful species

Poultice Corrosion
Poultice corrosion results in pitting of a metal after contact with some form of absorptive material (e.g. lagging), which has been wetted periodically. During drying-out periods any harmful corrosives can become concentrated and attack is most pronounced when dry/wet areas induce differential aeration effects. The remedy in such cases is to avoid metallic contact with the absorptive materials or to utilize some form of coating.

Filiform Corrosion

Filiform corrosion damage is associated with differential aeration beneath coatings, such as lacquers, enamel, paint or rubber. 'Hair-line' products develop under capillary forces, there being an active 'front-line' and a cathodic periphery where oxygen is in greater supply. The metal in itself is not destroyed, but coating adhesion and general surface appearance are affected. Moisture is of particular importance for such damage and if the relative humidity is kept below 65% then damage is avoided.

Selective Dissolution

Selective corrosion (i.e. the preferential removal of one constituent from an alloy) gives little visible evidence of metal loss, pitting, cracks or dimensional change; there is, however, a significant reduction in mechanical properties. Two common examples of this phenomenon are the loss of iron from grey cast iron (graphitization) and the loss of zinc from brasses containing more than 15% Zn (dezincification). The mechanisms are still subject to discussion, but the damage can be avoided by using more-resistant alloys.

Intergranular Corrosion

A highly selective form of corrosion which occurs at or near grain boundaries in a material is termed intergranular corrosion, the result ultimately being a marked change in mechanical properties with complete dislodgement of grains in severe cases. Such corrosion arises from a difference in potential that develops between grain boundaries, there being a local difference in chemical composition between the two zones.

A particularly aggressive type of grain-boundary attack, usually, but not necessarily, intergranular in nature, is common to aluminium alloys that have been heavily worked. Grain boundaries become elongated in the direction of working and corrosive action occurs along such regions producing laminated corrosion products. The phenomenon is termed exfoliation and is particularly severe in aluminium alloys containing copper, magnesium, or zinc and magnesium. During attack the corrosion products force the material apart and cause the alloy to break up into flat sheets. Figure 10 shows premature failure of an aluminium alloy used in an electrical instrument by exfoliation, as a result of poor seals that have permitted moist, polluted air to enter the instrument body. Heat treatment or alternative materials can combat the problem, although the situation in Figure 10 could have been resolved by using drying agents or by improving the seals.

In some cases intergranular corrosion is a result of impurity elements that segregate at the grain boundaries, for example the presence of iron in aluminium alloys can result in grain-boundary intergranular attack in alkaline environments. In other cases, second-phase constituents, such as carbide, nitride or intermetallic precipitates, are responsible for this type of corrosion. The extent of damage can be clearly visible to the naked eye on occasions, as typified by the austenitic stainless steel plate removed from

Figure 10 – Exfoliation of aluminium alloy plate from within an electric instrument removed from site after about 1½ years

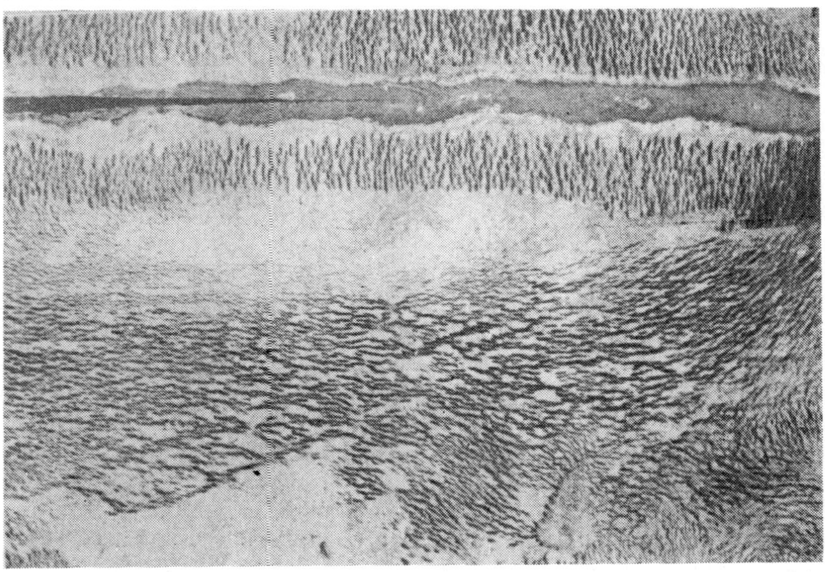

Figure 11 – Intergranular corrosion in austenitic stainless steel plate used in a process plant

service shown in Figure 11. Such damage often arises because of poor heat treatment and fabrication, which obviously affect the shape and size of grains as well as the size and distribution of second-phase constituents. If alternative heat treatments are not possible, it may be necessary to adopt alternative materials to minimize or overcome such corrosion. Although only a small volume of metal is corroded, there is a marked effect on tensile properties and percentage of elongation.

Sensitization

Sensitization is intergranular corrosion associated with the formation of chromium carbide ($Cr_{23}C_6$) in austenitic stainless steels containing more than 0.03% carbon. Sensitization occurs in the temperature range 425–900°C, where carbon migrates to grain boundaries, precipitating chromium carbide, which depletes the adjacent steel of chromium. In some cases the corrosion resistance locally can be similar to that of carbon steel and the ability of the base material to passivate is lost. If such material is exposed to an electrolyte, then corrosion will proceed very rapidly in the chromium-depleted areas (Figure 12). The degree of damage is dependent upon time, a few minutes at the critical temperature range of 750°C being equivalent to several hours at lower or higher temperatures. In welds, the phenomenon is called 'weld decay'. It can be decreased by (a) using 'stabilized' steels that contain small quantities of stronger carbide formers, such as titanium and niobium (stainless steels types 321 and 347, BS 970), or (b) using steels with carbon contents below 0.02% (types 304L and 316L), or (c) post-sensitizing heat treatments, designed to allow the chromium content in the depleted zone to be replenished by diffusion and, during subsequent cooling, to minimize the time at which the material is in the critical temperature range.

Sensitization can also occur in ferritic stainless steels (type 430 for example) at temperatures above 930°C.

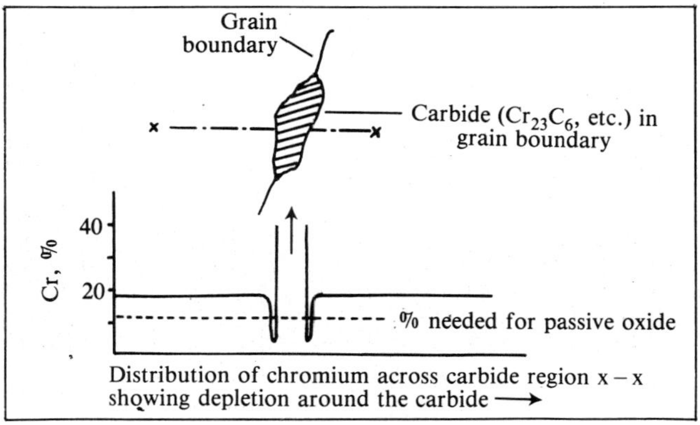

Figure 12 – Corrosion associated with 'sensitization' of austenitic stainless steel

Stress Corrosion

Stress corrosion causes failure of alloys by cracking when the effects of an applied or residual tensile stress and a specific environment are combined. Stress corrosion cracking (s.c.c.) can follow intergranular or transgranular paths, there often being a tendency for branch cracking (Figure 13). The phenomenon is complex in nature, being related to mechanical, metallurgical and electrochemical aspects; it is more generally found in alloys, but high-purity metals are not necessarily immune. Table 2 shows some common examples where stress corrosion can be anticipated. The effects of stress corrosion are normally combatted by stress-relieving procedures or removal of the corrosive environment. Alternatively, materials of better resistance may be a solution, for example the use of duplex ferritic–austenitic stainless steels, which give better general corrosion resistance than ferritic steels and greater resistance to stress corrosion cracking than normal austenitic stainless steels. As with many localized phenomena, stress corrosion is not fully understood, although it can perhaps be regarded as the most extreme form of localized corrosion. Further information is given in the literature(*2–8*).

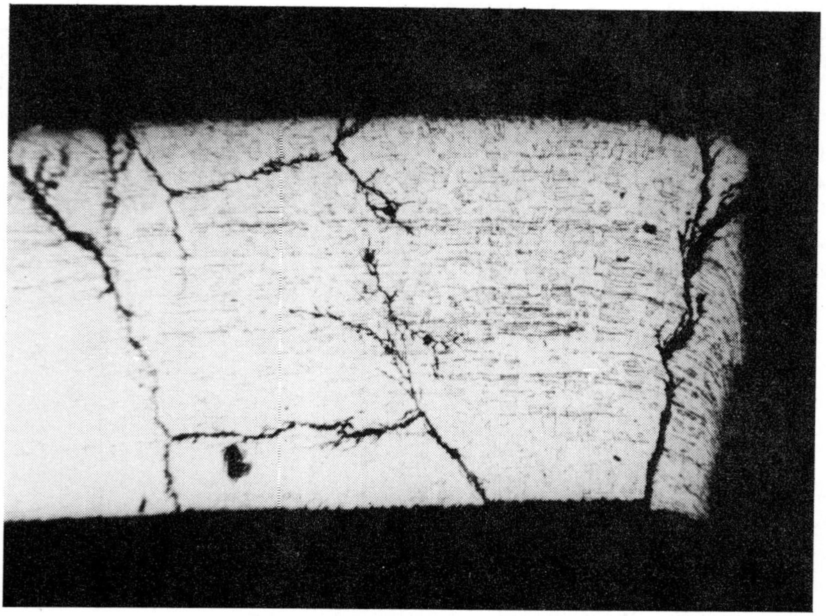

Figure 13 – Stress corrosion cracking in an austenitic stainless steel plate due to the combined presence of tensile stress, chloride ion and temperature above 70°C

TABLE 2
Typical Systems Prone to SCC(*5,9*)

Alloy	Susceptible environment	Comment
Aluminium alloys		Al–Cu particularly for
Al–Cu, Al–Mg	Moisture + Cl^-	certain tempers and high
Al–Zn–Mg–Cu	Organic solvents	strength alloys, Al–Mg > 3% Mg
'Alclad' alloys		'Alclad' – at edges or if damaged
Copper alloys		
Cu–Zn, Cu–Zn–Sn	NH_4^+ + moisture (solution/vapour)	
Cu–Al	Moist SO_2	
Cu–Si (α)	Some solutions, amine	
Mild steel	Hot conc. caustic soda, conc. nitrates H_2S, anhydrous liq. NH_3, some carbonates/phosphates	
Low-alloy steels (quench/tempered)	Certain solutions e.g brines +	Martensitic steels esp. susceptible
Ferritic stainless steels, martensitic steels	H_2S, dil NaCl, dil H_2SO_4	
Austenitic stainless steels	Cl^- + moisture + O_2 > 65°C Hot conc. caustic soda	Cracks have been noted down to 10 ppm Cl^- and residual sress levels of as low as 15 MPa.
Duplex ferritic–austenitic steels, high Ni/Cr + Cu + Mo	Service experience limited	Better than austenitics
Nickel alloys		
Ni–Cr–Fe	HF vapour, fused caustic soda, H_2O > 300°C	
Ni–Cu Ni–Fe–Cr–Mo–Cu	HF vapour, fused caustic soda	
Titanium	Methanol + halide ions	
Commercial purity	(Some types in sea water and aqueous Cl^- soln) N_2O_4	
High-strength alloys	N_2O_4 and fused salt	
Ti–Pd	Limited data (probably as above)	
Zirconium	Cl^- ions	
Commercial-purity alloys	$FeCl_3$ soln., I_2 at 350°C	

Corrosion Fatigue
Corrosion fatigue is a form of stress-assisted corrosion with attack dependent upon fluctuating stresses causing surface cracks that penetrate the material, leading to ultimate failure. As with stress corrosion, there is a need for a reactive environment, which includes air for certain materials,

especially if it is polluted with ammonia or sulphur dioxide (copper alloys) and water (with chloride ion, which affects almost all metals). Corrosion fatigue can be considered as the lowering of the fatigue resistance of a metal in a corrosive environment, as can be noted from a so-called *S/N* curve, depicted in Figure 14, where *S* is the mean applied stress and *N* is the number of cycles to failure.

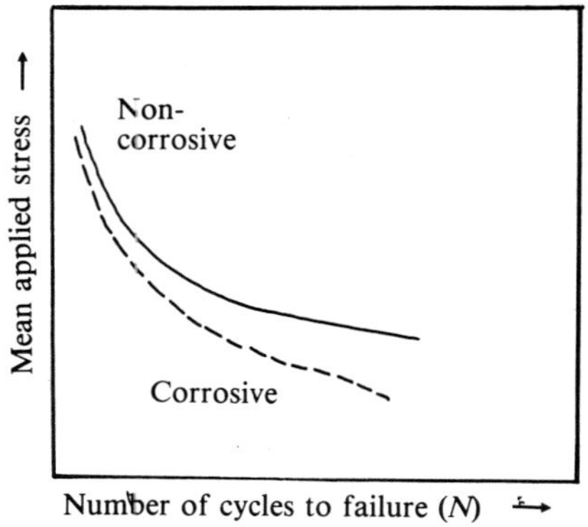

Figure 14 – Corrosion fatigue as reflected in stress vs number of cycles; curves for a metal in corrosive and non-corrosive environments under identical test conditions

In practice, there are many examples of corrosion fatigue, including failures of steels in boiler waters, steam condensates or in certain natural waters. Life expectancy can be increased by reducing the applied stress, by stress-relieving heat treatments, using inhibitors or by applying coatings (e.g. zinc on steel, aluminium 'cladding' of aluminium alloys or painting). Fatigue is also prolonged by putting the surface under a compressive stress, e.g. by shot peening, rolling, nitriding, etc.).

Fretting Corrosion

Fretting corrosion occurs on areas of contact between metals subjected to repeated relative displacement under load. Damage often appears in the form of pits, although there is usually an accumulation of oxidized debris. Attack is more evident in gaseous environments and is especially damaging because the corrosion-product debris can detach and act as an abrasive. The phenomenon is common to vibrating machinery, where components such as bolted or riveted joints, bearings and keyed gears, etc. are all possible areas of weakness.

Fretting corrosion can be reduced or eliminated by careful design (e.g. increasing the load to reduce slip in the metal, increasing the relative motion to reduce attack), through material selection (stainless and chromium steels and copper and its alloys are claimed to have a good corrosion-fatigue resistance), by increasing the hardness of one, or both, of the fretting materials, by lubrication (high-tenacity oils or phosphate conversion coatings plus lubricants) and by the use of gaskets.

Erosion Corrosion

When corrosion processes are associated with a moving environment there can be an increased rate of attack as a consequence of erosion corrosion. In simple terms, mechanical wear can remove protective or passive films and hence expose bare metal directly to the corrosive species. Such problems arise when high-velocity fluids (turbulent flow) impinge on metallic surfaces giving rise to deep, smooth-based pits, the products being removed by the flowing media. Entrained air bubbles or suspended debris in fast-moving slurries or fluids are prime causes of impingement attack, a form of localized damage often with directional features. Sharp contours in pipework (elbows, T-sections, bends, etc.) are frequent targets for attack. Clearly, designs that reduce flow and turbulence, eliminate particulate matter and avoid sharp geometrical profiles are to be favoured. Harder materials can also be of use in reducing such damage.

Cavitation is another form of erosion corrosion experienced in rapidly moving liquids. Damage is associated with the formation and collapse of vapour bubbles on the metal surface, the collapsing bubbles introducing local high pressures to the metal (i.e. local deformation effects by work hardening, etc.). Protective films or products are destroyed and corrosion increases in severity.

Figure 15 contrasts two service failures, where premature failure has been caused by erosion corrosion, (a) by impingement and (b) by cavitation.

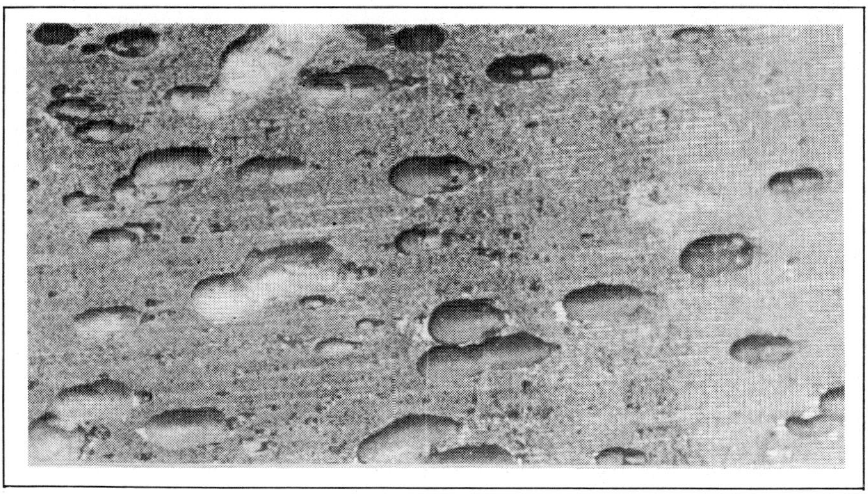

(a)

(b)

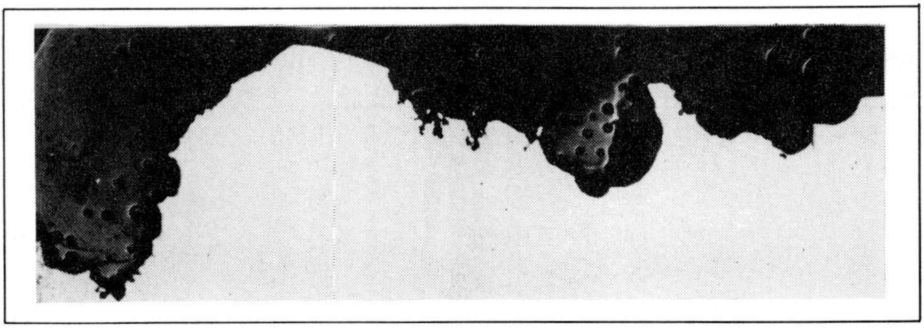

(c)

Figure 15 – Damage by erosion corrosion

(a) Impingement corrosion of an aluminium alloy due to direct impact of steam spray

(b) Cavitation damage in a rapidly rotating pronged impeller used with an aqueous dispersion of dyes and pigments, the nitrided cast steel has reached this condition in about 9 weeks

(c) Section through (b) showing the effect of the cavitation attack

2 NON-METALLIC MATERIALS

Non-metallic materials as linings or as materials in their own right have found use in many corrosive environments despite the obvious limitations in mechanical and thermal properties, such as poor creep deformation, low modulus of elasticity, high notch sensitivity, etc.

Polymeric materials are not subject to electromechanical corrosion as are metallic materials, however they do degrade, as will be noted below. Plastics show excellent resistance in many environments, in particular inorganic chemicals, but they are not so resistant to organic chemicals. Temperature is a critical factor for polymers, there being relatively low maximum safe design temperatures associated with softening points often below 100°C, as can be judged from some typical materials included in Table 3.

TABLE 3

Safe Working Temperatures (Guide)

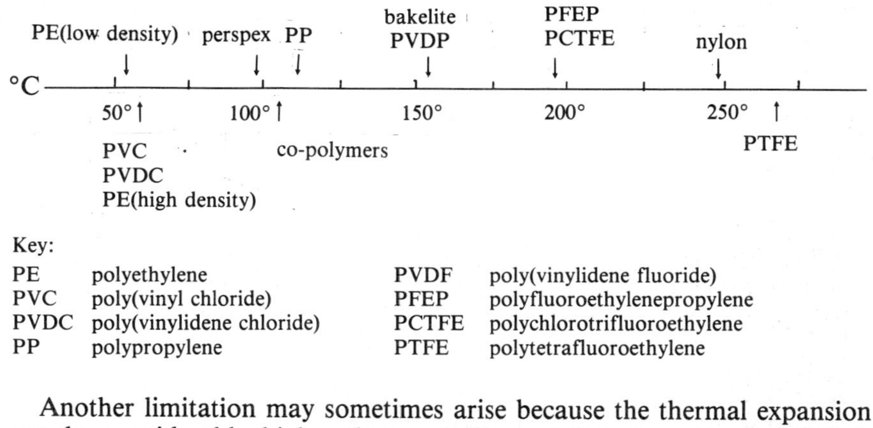

Key:

PE	polyethylene	PVDF	poly(vinylidene fluoride)
PVC	poly(vinyl chloride)	PFEP	polyfluoroethylenepropylene
PVDC	poly(vinylidene chloride)	PCTFE	polychlorotrifluoroethylene
PP	polypropylene	PTFE	polytetrafluoroethylene

Another limitation may sometimes arise because the thermal expansion can be considerably higher than metallic materials (e.g. Cu: 1.7×10^{-5}, polythene: 18×10^{-5}, etc.). Polymers tend to be brittle at low temperatures, some being unsuitable for temperatures ~ 0°C; polypropylene and PVC, for example, should not be installed in frosty weather.

Other classes of non-metallic materials include rubber, ceramics, carbon and wood. In general, rubber, like plastics, is more resistant to chloride ions and hydrochloric acid, but is less resistant to strong sulphuric acid and oxidizing acids, such as nitric; it has less resistance to solvents. Ceramics and carbon/graphite have good corrosion properties, but tend to be brittle or fragile. The use of wood is often limited in strong, oxidizing acids and dilute alkalies. Wood also suffers from biological attack(*3*). The remainder of this section is specifically concerned with the degradation of plastics, and further information can be obtained elsewhere(*2,3,11–15*).

Degradation of Plastics

Plastics tend to oxidize at much lower temperatures than do metals, the so-called degradation often being associated with the breakdown of the long-chain molecules that comprise these materials, unless certain additives have been introduced to offset such phenomena. Unlike metals, the most common failure mode is dissolution or liquid absorption.

Swelling

Certain polymers swell in particular environments, for example, certain polyolefins exposed to chlorinated solvents or benzene. Absorption of moisture when immersed in liquid or from humid environments is not unusual, and the extent of this effect is used to define permeability.

Ultraviolet Degradation/Weathering

Light, heat, rain, oxygen and ozone can contribute to the degradation of many polymeric materials. In less-extreme situations fading or discolouring may occur, but in some cases constituents added for particular benefits may also deteriorate with subsequent loss of plastic quality. Ultraviolet radiation is particularly harmful because it initiates breakdown of the polymer chains that are the basis of plastic materials (chain scission). Carbon black is a common additive used to help overcome this problem.

Environmental Stress Cracking

Many plastics fail at relatively low stress levels when exposed to certain active environments, failure being of a 'brittle' nature. The stress source is often a residual, locked-in stress remaining from the moulding operation during manufacture. Surface-active agents, such as detergents, are common causes of cracking, and ethylene-type materials are particularly prone to such damage. Good designs that minimize stress-concentration effects are recommended, as are alternative materials.

Crazing

Surface cracks, which rarely penetrate structures, are less catastrophic than environmental cracking, the phenomenon being termed crazing. Such damage may result from moulding or applied stresses and it is common with polystyrene, polycarbonate, etc. exposed to solvents such as kerosene.

3 IMPLICATIONS OF CORROSION

The implications of corrosion are much broader than many at first imagine. A small hole in some component can result in an unsightly panel with relatively cheap repair or replacement costs. A similar hole in a process pipeline carrying corrosive fluids can be much more serious; safety, health and pollution aspects may be involved, in addition to a possible complete plant or equipment shut-down with significant economic overtones.

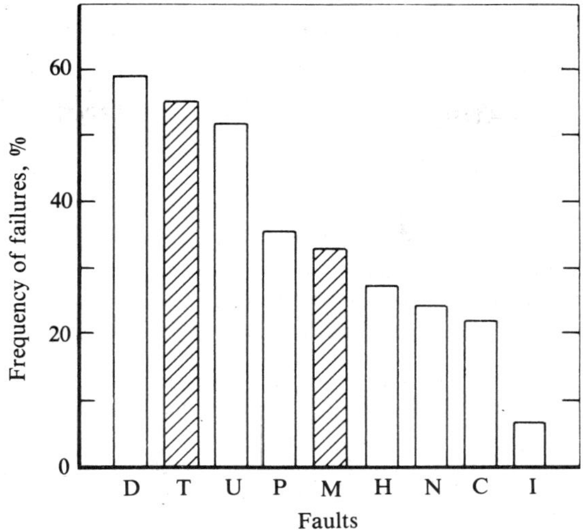

D : Plant design faults
T : Incorrect application of protective treatment
U : Unforeseen operating conditions
P : Poor process control
M : Material faults
H : Human errors
N : Unaware of a corrosion risk
C : Contamination of product
I : Instruments at fault
(Shaded sections relate to the findings of the Hoar Committee)

*Figure 16 – Reasons for corrosion failures or events in the oil and chemical industry(*17*)*

Corrosion is claimed to be the biggest single cause of plant or equipment breakdown in the chemical and allied industries(*16*). Such phenomena arise for a variety of reasons, high among which is poor design, incorrect application of protective treatment and a lack of awareness of operating conditions and control (Figure 16)(*17*). The oil and chemical industries are a corrosion-conscious sector, so there is a need to improve the degree of communication in other types of industry, with a view to drawing attention to the types of corrosion described earlier and some of the principal ways of combatting attack. It is, however, necessary to appreciate that other factors also affect plant availability including friction, wear, fracture, degradation and fouling. Such factors form part of the wider subject of loss prevention, which in turn has economic and safety overtones.

Economics

UK corrision losses in 1971 were put at £1,365m of which £310m could have been saved through better awareness, communication and implementation of existing knowledge(*1*). Inflation has probably put this figure nearer to £4,000m and information from the USA shows corrosion losses there to be $70 billion ±30% (in 1975)(*18*). The surveys undertaken to produce these costs estimates, though extensive, do not necessarily include other factors such as wear and fracture, nor losses associated with ceramics, polymers, timber, rubber etc.; hence the figures are conservative.

All corrosion can in practice be avoided – but at a cost. In many circumstances, it is necessary to have 'controlled' corrosion – provided that it remains under control. 'Economics' dictate, therefore, the ultimate choice in material, preventive technique or maintenance/replacement scheduling(*19*). Other implications, which include health and safety, may make more reliable and more expensive materials a possibility. Table 4 portrays some of the broader aspects of corrosion as it affects plant or process – the economic and safety aspects are clear.

TABLE 4

Broader Implications of Corrosion

	Implications
Economic aspects	Production or product losses Shutdown Material replacement Equipment replacement Overdesign – corrosion allowances (loss in efficiency – heat transfer, etc.) Preventive maintenance – cathodic protection/paint
Safety and health aspects	Product contamination – (health) Leakage – equipment and/or personnel Fire Toxic wastes Explosion Collapse Appearance

Safety

This implies reliability, which in turn demands efficient and economic operation of plant or equipment. For safe operation it may often be necessary to introduce some extra cost factor by ensuring better standards of reliability. This frequently entails more sophisticated materials, improved design specifications and more regular inspections.

Safety in terms of hazards, such as leaks, fires, explosion or collapse of plant, is affected primarily by a weakening of the materials. In situations under stress or flow there is a need for extra concern or some form of monitoring, since failure can be sudden and often catastrophic, as, for example, the classic bursting of a boiler tube due to overheating and the

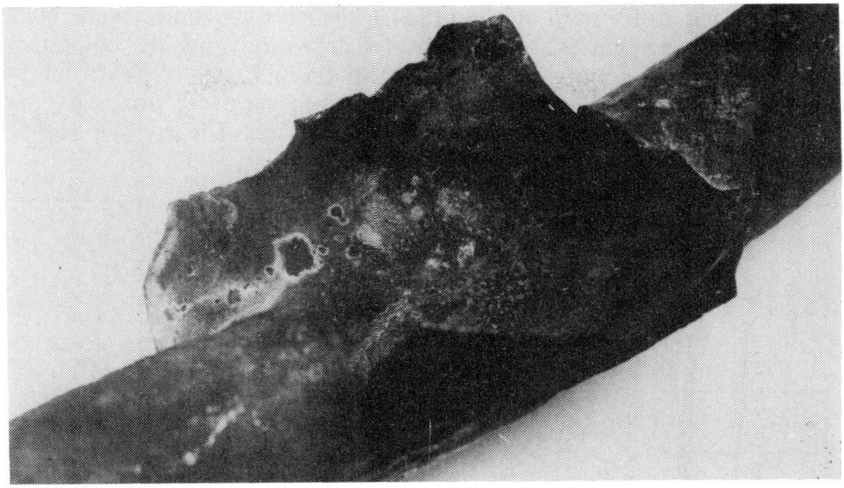

Figure 17 – Burst boiler tube due to overheating as a result of a loss of water supply in the system

associated sharp reduction in creep strength (Figure 17). Localized corrosion is a very common reason for premature failure and the serious implications on safety need to be recognized.

Reliability

This is a function of age, there being early periods wherein excursions or 'teething' troubles are dominant and a later period when plant is reaching obsolescence. It is necessary to take account of age and service hours in particular when evaluating a material or considering a failure. 'Off-load' conditions need to be fully appraised because these may be the primary cause of failure; for example, stand-by pumps may suffer vibrational effects from running neighbours that render them useless when required.

Design

Design has already been mentioned as a prime reason for corrosion failure (Figure 16). Plant design is a complex entity, there being many different aspects, ranging from technical and economic to legal and safety as can be noted from the 'wheel' outline included in Figure 18(*20*). Whilst 'corrosion' as a word does not appear in the various categories, it is clear that certain technical factors are directly related to corrosion and need to be considered in addition to the economic and safety factors already described above. Customer satisfaction falls within the area of market quality, there being no need for premature corrosion effects if due care is given to the materials and their compatibility with the environment. Rust stains, for example, around non-galvanized bolts need never arise if the correct choice were made earlier. Other categories that relate to corrosion include equipment (materials of construction, specifications), building (types of construction, specifications), building (types of construction, hazards, drainage) and development (materials and process research), etc. Clearly the subject becomes one of management policy.

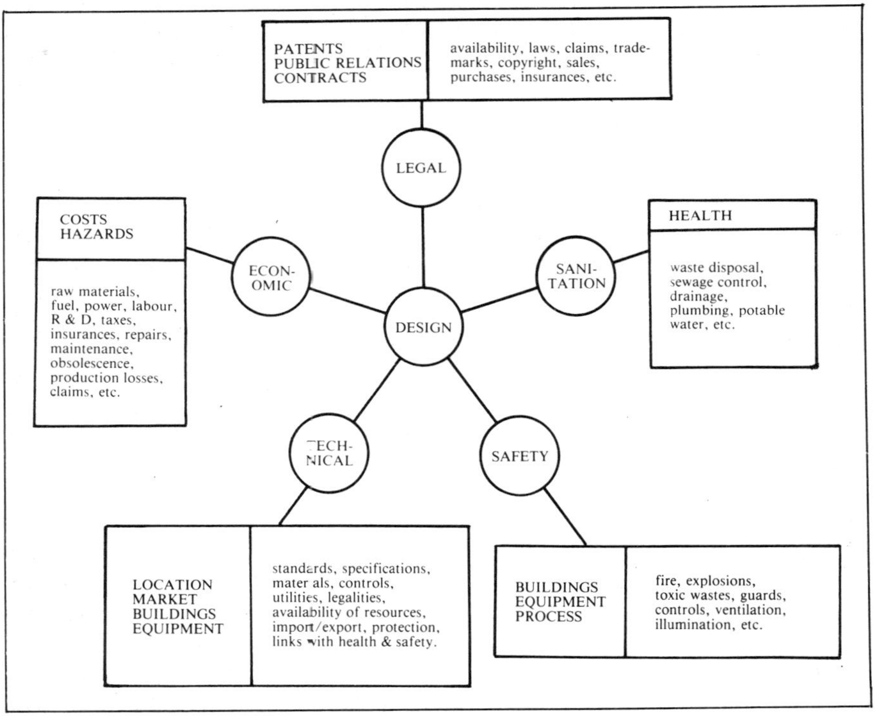

Figure 18 – Plant design factors(20)

Legal Aspects

The first legal duty placed on an employer is the common law duty of care, which is perhaps the widest duty covering all employers irrespective of whether any further statutory duties are laid upon them. This covers employees, neighbours and visitors (even trespassing children), and the employer must provide safe equipment, instructions for its safe use and, where appropriate, adequate supervision. There are obviously areas within these categories that will be affected by corrosion damage and the need for efficient inspection is warranted.

The second duty placed by the law upon an employer is to comply with any relevant statutes. The main statute prior to 1974 was the Factories Act (1961) with its various associated regulations. The main statute since then is the Health and Safety at Work Act (1974), which now also incorporates the common law duty of an employer to provide safe equipment. If an employer is found guilty of breech of his common law duty of care, he can face damage claims of negligence, although minimum insurance cover is a statutory obligation since the Employers Liability (Compulsory Insurance) Act (1969). By contrast, an employer found guilty of breech of a statutory requirement has committed a crime and is liable to a fine and/or imprisonment, against which he cannot insure.

Insurance Aspects

It is not possible to cover corrosion losses *per se* by insurance. However, mechanical damage, such as cracking or bursting of tubes, can be covered and generally is, for example, where pressure vessels or hoisting gear are involved. Insurance companies accumulate knowledge statistically so that cover can be arranged, and there is often a link to inspection, which is really a separate entity in itself. For example, certification of equipment will follow mandatory inspection, and it may be necessary to have such documentation before insurance cover is available. In cases where *uniform* corrosion is well established, it may also possible to have insurance cover, but again this will relate to the mechanical integrity of the material, which will also be subject to regular inspection.

Maintenance

As a general rule, manufacturers recommend operational procedures for unit or machine maintenance. It is therefore of particular importance to establish correct maintenance strategies for each machine or unit in order to keep components in a servicable condition for as long as possible. Short intervals between inspections are often warranted where inspection is easy (e.g. where parts are accessible) and hence time and costs involved are relatively low. Designers should, by providing designs which readily lend themselves to maintenance, make short-interval inspections possible. In cases where inspection requires extensive disassembly work, there may be justification for alternative techniques (e.g. borescope inspection).

There is evidence that corrosion hazards occur between the design office and the field execution. Poor fabrication, lack of qualified supervision and inefficient inspection are typical reasons for premature failures, and some of these are discussed elsewhere in this chapter.

4 THE EFFECT OF FABRICATION METHODS ON PLANT LIFE

The preferred manufacturing technique can have a significant effect upon plant or component life, the choice being related to cost, function and material property. Fabrication includes all forming and finishing processes associated with materials ranging from castings, wrought products, sintered-powder components, etc., but also including machining and joining operations. Plant availability, as indicated earlier, is related to the reliable operation of units or equipment, there being several significant factors (Section 3). Defective or poor fabrication will be one such factor that can result in premature failure. Likewise, naturally undesirable situations (not necessarily of a chemical or corrosive nature), which arise for particular materials, can be significant. For example, mild steel is difficult to cold work in thick section, it embrittles at temperatures below about 10°C and loses creep properties at temperatures around 450°C.

Some materials are traditionally used in their cast condition (certain brasses, cast irons, gunmetals), whilst others can readily be cold- or hot-worked to particular advantage (thin-sheet mild steel, copper and certain of its alloys, aluminium bronzes, aluminium, stainless steels, etc.). Welding technology has improved over the years so that materials which once presented certain difficulties can now be fabricated (e.g. cast irons, aluminium, etc.), provided that due care is given to weld preparation, procedure and filler-metal selection; British Standards are available(*21*). Particular problems that can arise during fabrication are summarized below; corrosion implications are considered where appropriate.

Castings

Historically, many components were used in cast form and there is some cost advantage in using castings for many current components. Defects include the presence of voids or holes, due to shrinkage effects or gas–liquid metal–mould reactions, and these could act as sites for pitting and/or crevice corrosion, or as stress raisers (notches) for crack initiation, given the necessary geometric and/or environmental circumstances. Local galvanic corrosion can also result in non-homogeneous castings due to segregation effects. Corrosive effects will obviously be significant when defects are encountered on the metal surface, but mechanical properties may also be much affected. Selective dissolution has already been described; for example, the loss of iron from grey cast iron (graphitization) or of aluminium from certain cast aluminium bronzes (dealuminization). There is some indirect advantage to be gained from cast sections because they are usually thicker than wrought sections, so can tolerate more general corrosion.

Wrought Materials

Material properties can be improved by cold- or hot-forming operations, although corrosion problems may arise in certain environments, especially with respect to passive or surface films that may be formed on the metal. Cold working leads to a flattening or elongation of grains within a metal in the direction of fabrication. Heavy cold working results in a ‘fibrous’ structure that can lead to particular problems, especially if the exposed material

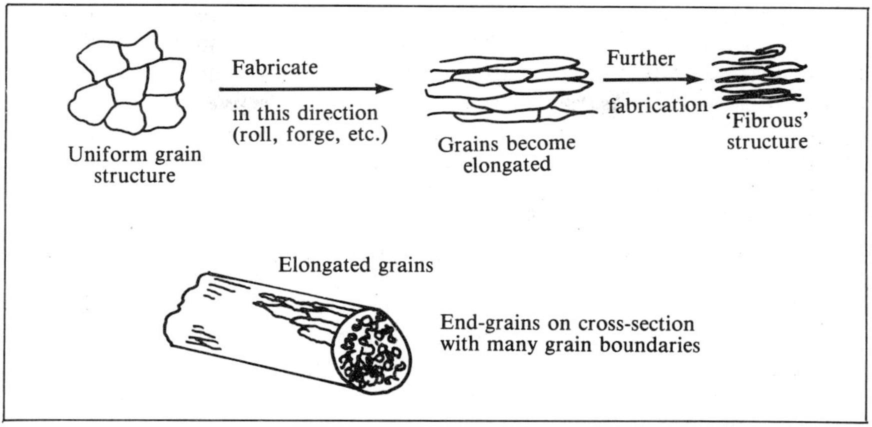

Figure 19 – Structural changes in fabrication by rolling, extrusion, etc.

crosses the section of material (Figure 19). Exfoliation or layer corrosion in certain aluminium-based alloys has been referred to earlier (Figure 10) and 'end-grain' corrosion arises on ends or cross-sections of material that are exposed to suitably aggressive environments. In such cases, grain boundaries are abundant and localized corrosion occurs, as for example in Figure 20, which shows end-grain pitting in commercial-purity titanium exposed to hot organic acids containing certain halide species.

Figure 20 – 'End-grain' corrosion in a commercial-purity titanium thermowell unit removed from a plant handling hot organic acids

Cold working can provide spring-like properties in materials such as austenitic stainless steels (iron with about 18–19% Cr, 10% Ni), but residual or tensile stresses from fabrication and/or service installations need to be controlled if stress corrosion is to be avoided in those environments specific to such attack. Excessive cold work should be discouraged in aluminium alloys containing more than 4% magnesium, to avoid the development of continuous precipitates in the grain boundaries which render the material prone to stress corrosion cracking. Severe cold work can increase corrosion of steels and aluminium in acids(*2*) and pitting of various materials, including austenitic stainless steels in ferric chloride(*3*). Initiation of such localized attack is possibly related to heterogeneities that create flaws in normally passive films. Impurities and/or non-metallic inclusions become elongated as a result of working, which may create directional properties making the material prone to local galvanic or intergranular corrosion due to differences in electrochemical potential.

Hot forming of steel can occasionally change the metallurgical structure; as a result carbides present in the steel 'ball up' to form sphereoids, and the matrix becomes less resistant to corrosion. This problem, referred to as 'ring-worm' corrosion, is discussed elsewhere(*3*). Scale formation on hot-worked materials can also be significant giving, for example, problems on painting mill-scale on steels. Codes of practice are recommended in such situations(*22*). Surface finish by working and/or machining can also be related to the onset of pitting corrosion, damage being more on rough, machined surfaces compared with smooth, polished surfaces.

Jointed Assemblies

Fastening methods can be permanent or demountable and involve mechanical fixation (rivets, screws, bolts, pins, dowels), adhesives, or solders, brazes and welds.

The components of a joint may not always be metallic; however, the corrosion behaviour is of utmost importance, the defects of the joints contributing primarily to their limitations in use.

Mechanical joints, such as those sketched in Figure 7, can give rise to several corrosion possibilities, notably crevice corrosion (as depicted) and galvanic corrosion, given the incorrect material combinations and a suitably aggressive environment. Such deterioration can be minimized by insulating the components or by applying a continuous coating (e.g. paint). However, as indicated earlier, unfavourable anode/cathode areas should be avoided if possible because small anodes, such as bolts with large cathodic areas (i.e. plate), aggravate the localized corrosion damage. Some representative defects are shown in Figure 21. Other limitations can arise because of differences in the expansion coefficient of the components, oxidation and scaling and loss of creep or strength properties at high temperatures.

Ill-fitting joints (see Figure 21) can also contribute to other corrosion possibilities, such as fretting or fatigue, dependent upon loading and other

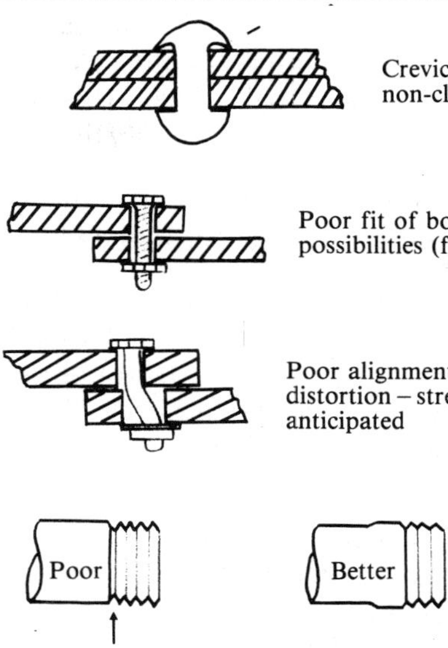

Figure 21 – Joint problems

circumstances. Furthermore, a joint will have tensile stresses which may promote stress corrosion cracking given the necessary material and surroundings. Fatigue failures may be avoided by alternative designs, for example by undercutting to eliminate sharp profiles on threaded bolts.

Solders

Solders provide more permanent types of joint, but care is needed to avoid corrosion possibilities. Solders are usually tin-base alloys and soldering is normally carried out in the range 300–450°C on materials that do not readily oxidize (e.g. tin, copper, gold), it being virtually impossible to solder materials such as titanium or tantalum. Most common engineering materials are intermediate between these two groups, so some degree of difficulty will be experienced when soldering(*23*). Fluxes are normally used to assist wetting, to remove oxides and to maintain clean surfaces during the jointing operation. Acidic inorganic chemicals are often used as fluxes (e.g. zinc and ammonium chlorides), and thorough washing is necessary to minimize subsequent corrosion possibilities. Non-corrosive fluxes are available based on natural rosin and these are especially useful where after-washing is not practical (e.g. electrical components). Poor wetting, internal cavities, porosity and flux residues may all contribute to corrosion possibilities.

Brazes

Brazes provide higher-strength joints compared to solders, being performed at temperatures from 450°C to the melting points of the materials to be joined. Brazing represents a useful technique for joining different materials such as brass to steel, copper to stainless steel, etc. Like soldering, the process relies upon efficient capillary penetration by the braze into gaps between components and the need for clean surfaces is apparent. In many instances fluxes are needed to dissolve oxide films present on both the parent metal and on the molten braze, thus promoting wetting. Similar considerations apply for certain furnace brazing operations where argon, cracked ammonia or hydrogen may be employed to eliminate oxide formation. The use of hydrogen may give rise to embrittlement effects with some materials.

Inter-alloying may occur in some cases with possible deleterious effects upon wetting and flow of braze filler. Galvanic corrosion possibilities also need careful consideration when joining dissimilar metals. Very serious problems can occur with liquid-braze filler attack if the components are under stress; intercrystalline penetration is the problem that can occur. Nickel-base alloys can be particularly prone to such damage if sufficient care is not taken. Brazing can also give rise to grain growth, which can have a significant effect on mechanical properties. Brazing is generally unsuitable for equipment exposed to ammonia because of the aggressive reactions possible with copper- and nickel-base alloys(*24*).

Welded Structures

Welding is commonly used to fabricate equipment, there being two types of process based on solid-phase welding, where pressure is applied across a joint without additional metal, and fusion welding, where the metals to be joined are locally melted, without pressure, and extra filler metal added as necessary. In each case, the effect is to produce a homegeneous material across the jointed fabrication. The primary reason for welding is to produce a shape which cannot be more satisfactorily accomplished by other methods. Examples include on-site welding of large components; assembly of vessels from combinations of tubes, forgings, plate and castings; and tank and pipeline systems.

There are many different processes described in detail elsewhere(*25,26*), but each is more suited to a particular requirement of shape, material composition and thickness, etc., and/or ease of operating or handling during the weld application. Welds may often be used to overcome design problems, such as the crevice situation (Figure 7) or end-grain attack (Figure 20). However, in welded assemblies there will be various microstructures ranging from cast and wrought to heat-treated forms, and these can present differences in physical, mechanical and chemical behaviour. Corrosion phenomena could include galvanic, crevice and intergranular corrosion, stress corrosion cracking and fatigue. Figure 22 shows some specific possibilities for corrosion attack(*7*). It is useful to keep welded structures cathodic, if this is possible, in order to avoid large cathode:anode ratios.

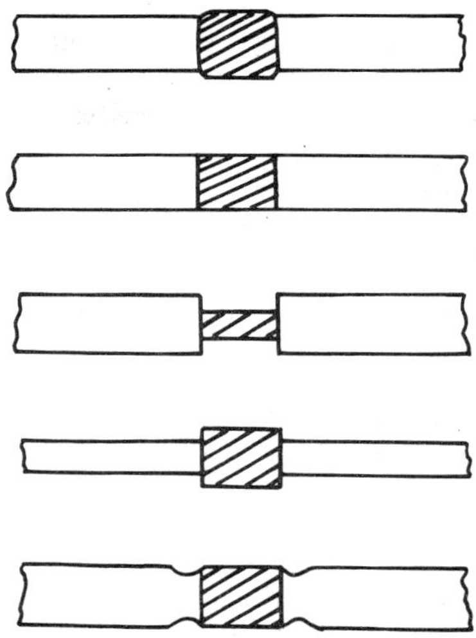

Figure 22 – Examples of weld corrosion

Welded castings will be prone to those effects associated with castings (described earlier), including microstructural hetereogeneities, secondary phases, microporosity and possible defects due to welding gases such as hydrogen, argon, nitrogen, etc. Heat-affected zones adjacent to welds can show different effects, for example increased grain size (ferritic stainless steels, Fe12–18%Cr) where the weld technique should have minimum heat input to minimize such effects, hard zones of material due to rapid cooling (martensitic structures in high carbon and alloy steels), local stress effects giving rise to cracks (hot tearing, microfissures, etc.), metallurgical changes due to temperature and material (e.g. 'weld-decay' in austenitic stainless steels, Fe18%Cr8–11% Ni, etc.).

Other effects of fabrication on plant/component life arise because of operator or machine malfunction. Included in such defects are lack of fusion between parent metal and filler, lack of penetration, excessive welding, misalignment of joint, undercutting, arc strikes, poor root profiles (ragged root), etc. (see Figure 23). Weld designs need to be carefully considered, including preparation, assembly and accessibility during assembly. Dissimilar metals can be joined by fusion welding, provided due care be taken to avoid dilution effects; discussion will be found elsewhere(*26*).

Recommended welding procedures, codes of practice and standards are available(*21*). Welder qualification tests are essential if reliability in fabrication is to be achieved.

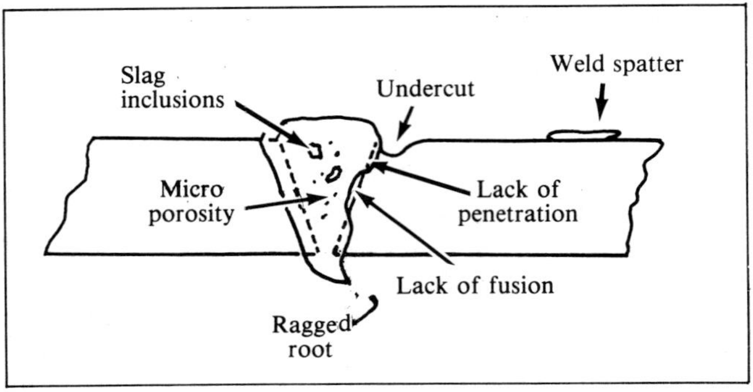

Figure 23 – Some defects associated with poor welding

5 SOURCES OF CORROSION INFORMATION

Information about materials is available from a variety of sources, including material suppliers, published technical papers, textbooks and standards such as those produced from BSI, ASTM, DIN, etc. (*27*). Some data are well defined, for example physical, mechanical and electrical properties, but materials with specified corrosion resistance are not so readily available. Hence, there will always be a degree of uncertainty when considering corrosion behaviour of materials. The situation is further complicated by the fact that each source of information has its limitations so that at best it is only possible to provide a guide to corrosion rates rather than precise information. Table 5 outlines some of the more obvious limitations on sources of information.

Corrosion Data

Experience is often the key to providing the basic corrosion information for a particular situation in plant. All sources of data are useful, but only insofar as they assist in establishing reliability of materials in the process or plant environment. Corrosion rates of 1–2 mm/year may, for example, be acceptable in cases where readily replaceable pipelines made of thick, cheap material are involved. In other cases rates of 0.1 mm/year can be too severe, for example where close tolerances are needed in machinery such as pumps or compressors, or valves in pipelines, etc. Further complications arise from fabrication and construction (see Section 4) and such factors need to be taken into account when considering corrosion data; these points will be discussed further in Section 6.

A large volume of information has been published on the corrosion resistance of materials, but this is often under very specific conditions. It is rarely an easy task to relate results from technical papers to conditions in a plant, for a variety of reasons as contrasted in Table 6.

TABLE 5
Sources of Information and Limitations

Source	Typical limitations
Textbooks	Too general for designers, good basis if time to read and understand, more case histories needed, require 'rule-of-thumb' presentations (e.g. booklets or guides) (listed in references of this chapter)
Technical papers	Wide dispersion of information which may be overlooked, data often of little use in solving practical problems, data may be too fundamental and rarely precisely related to operational plant, information often in journals not consulted by plant engineers or designers, non-standardization of units
Abstracts	Collected information solves some of the problems referred to in published papers, still abundant information, good basis for company to build up specific data relevant to plant, may need computer link-in
Material suppliers	Major source of technical information, sometimes excessive, may overplay advantages and underplay limitations, data often related to plant conditions, useful guides(*28–33*)
Data sheets/pamphlets	Often too vague, at best simple guides but need to be examined and used with care
Corrosion charts	Good sources of quick reference, only guides that may not take account of other factors (flow, temperature, etc.), need to be kept up-to-date(*34*)
Corrosion manuals	Ideal if kept up-to-date and relating to specific plant, need a company policy for implementation, technical manuals provide comprehensive information but still cover a broad field(*9,35–39*)
Standards and codes of practice	Generally do not refer directly to corrosion data, often relate to protection from corrosion, may be too comprehensive for user or may cover several grades of treatment or process, choice is difficult for user especially if non-expert
Advisory services	Can provide independent advice but at a cost which is often recovered in ultimate savings, need to provide sufficient background information, useful if contacted at design stage to introduce suitable controls or anti-corrosive measures

TABLE 6

Differences in Fundamental and Industrial Data

Fundamental studies	Industrial studies
Pure metal Experimental alloy(s) (controlled composition)	Complex alloys Commercial specifications (range of composition)
Simple shape	Complex shapes (large and welded, etc.)
Isothermal Temperature controlled Environment controlled (simple composition, constant) flow, etc)	Cyclic (on/off load) Temperature fluctuations Environment variable (complex composition, variable flow, etc.) shut downs, etc.
Often simple processes Monitored (electrochemical, kinetics from weight changes, etc.)	Overlap of processes Monitoring not widely used (often 'post-operative' examination)
Retained product films/scales Thorough sophisticated analysis (electron-optical, X-ray, etc.)	Loss of products (site→lab) Little 'in-depth' analysis material losses, etc.)
Reports/publications issued	Restricted publicity

In most cases technical literature and data sheets only provide general answers to quite specific questions. This is often because of the presence of minor constituents in a process environment, in particular chloride ion or ammonia, which are also influenced by other contributing factors such as temperature, residence time, pH, etc. Many problems can arise during specific periods of operation (stand-by, shut-down, etc., see Table 6). Thus, passive films may be temporarily removed and may take a long time to reform, or localized corrosion may occur that may necessitate additional precautions at certain times (e.g. inert-gas flushing during shut-downs, etc.).

The degree of uncertainty and variability can usefully be demonstrated by means of an example quoted elsewhere regarding corrosion rates for various steels in different acids(*40*). Given that corrosion is of primary concern, the use of a more corrosion-resistant steel instead of mild steel is not always an advantage because the modes and rates of corrosion are not constant for the particular material. The data(*41*) given in Table 7 show that the stainless steels are preferred for hot concentrated nitric acid, whilst mild steel is better for the concentrated sulphuric acid. In hydrochloric acid only the 12%Cr12%Ni steel shows any significant improvement over mild steel.

TABLE 7

Corrosion rates in acids(*41*)

Steel type	Rate of attack (mm/year) in		
	70% HNO_3 at 60°C	20% HCl at 20°C	80% H_2SO_4 at 20°C
Mild steel	Very high	38	0.4
Fe 13%Cr	0.15	120	4.5
Fe 12%Cr 12%Ni	0.05	5	0.5
Fe 17%Cr	0.1	35	0.7
Fe 18%Cr 8%Ni	Nil	25	1.5

Such differences are largely related to the fact that stainless steels are useful, provided that a protective chromium oxide film can form on their surfaces. Hence, in the absence of adequate oxygen, as with the hydrochloric acid, the chromium content is of little benefit. It is important to recognize the importance of environmental conditions when extrapolating corrosion data for specific processes, even the data in Table 7 does not take account of other factors, such as flow, stagnation, temperature variation, etc. Corrosion information may also be hindered by lack of knowledge by plant personnel in some cases, or it may be vague when effort is made to maintain some degree of confidentiality. Such 'inadvertent' omissions can have very expensive consequences.

Data Sheets

Data sheets at best should provide a rapid guideline to the corrosion behaviour of materials. Some are too general to be useful, whereas others give very instructive information, often with references to more detailed background literature(*16*). To become beneficial, data sheets need to be upgraded into detailed manuals, preferably established at the design stage of the process. Hence, as long as information is kept up to date, a useful data source can be established. Such a scheme needs careful management and there will be a need to accummulate data from plant monitoring or failure analysis, depending upon the particular circumstances of operation or shutdown. A systems-engineering approach is an ideal way to achieve reliable information, which can be of use early in the planning or design stages of a project (see Figure 24). Some useful data sources are given in the literature(*9,35–39*).

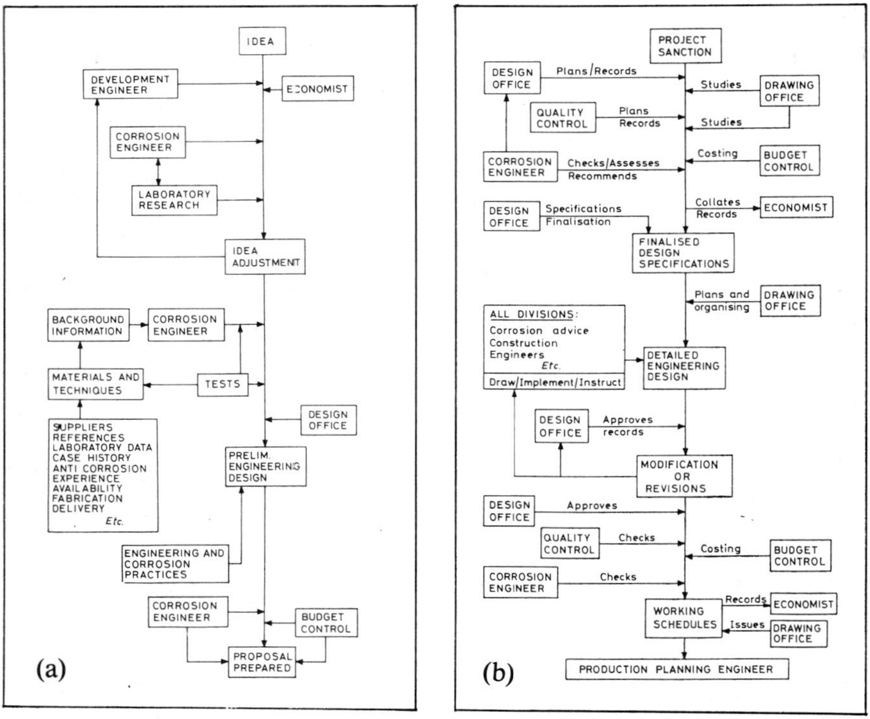

Figure 24 – Systems-engineering approach to engineering design:
(a) Preparation of proposals
(b) Actions during design stage

Codes of Practice

Various codes are available that help in choosing anti-corrosive treatments or techniques. As with most specifications, a choice of action is given as well as a precise description of the particular action, once selected. Hence, there will often be some difficulty in making a decision regarding a particular code of practice or standard(*27,43*). There are many standards that give reference to corrosion, but only one national standard is wholly devoted to design for articles to be coated (BS 4479); a standard covering bimetallic corrosion is currently being prepared(*43*).

Several standards or codes are available for protection from corrosion and these include BS CP 3003 (lining of vessels and equipment for chemical processes – Parts 1–10), BS 5493 (protection of steel structures), CP 1021 (cathodic protection).

Other standards or codes are listed elsewhere(*27,42,43*) and further information can be sought direct from the Standards Institutions or found in the literature.

6 CORROSION CONTROL

Corrosion cannot always be eliminated completely, indeed it is often economically advisable to allow materials to corrode rather than attempt to eliminate the phenomenon. In cases where corrosion is inevitable, there is therefore a need for some form of economic control. Ideally, schemes for corrosion control should be *preventive*, i.e. taken into account at the design stage as part of the capital or investment costs. Regrettably many situations needing control are recognized only after a period of operation, and curative procedures become necessary. In such cases the corrosive action or costs may be regarded as maintenance costs and hence may not be so apparent to management. Further discussion about preventive and curative schemes can be found elsewhere(*3,7,42,45,46*).

Corrosion control can be achieved by using one or more of several approaches based upon changes in design, environment or material, or by using electrochemical protection or coatings. Much has been written on these aspects of controlling corrosion(*2,8,40,42*), and only an outline approach is adopted here.

Engineering Design

The importance of correct design has been mentioned earlier, but over-design may not always be merited, since it is necessary to be aware of the particular circumstances of the process or unit involved. A systems-engineering approach is recommended where corrosion and protection are given due consideration(*17,42*). A typical scheme, given in Figure 24, shows how background information, literature, material properties, etc. are incorporated as basic design concepts. There is a need to ensure that design specifications are understood and acted upon at construction and during operation.

Designers need to be aware of localized corrosion and fabrication effects, since many problems can be eliminated by relatively simple changes in original design. When considering a design it will obviously be necessary to relate prospective materials with the process environment, paying particular attention to impurities, type of corrosion anticipated, life expectancy of units, stress and velocity effects, start-up and shut-down procedures, spillage and pollution problems, safety hazards and the cost of failure of critical components. Other factors would include ease of access for maintenance or repair.

In determining corrosion allowances it is necessary to take account of dimensional changes that occur in operation as a result of general wastage or, more significantly, due to localized effects of flow, wear, vibration, overheating, etc. The scheme outlined in Figure 25, based on published information(*45*), shows typical steps to allow for changes during operation.

More detailed information about corrosion control in engineering design can be found elsewhere(*40,42,47*).

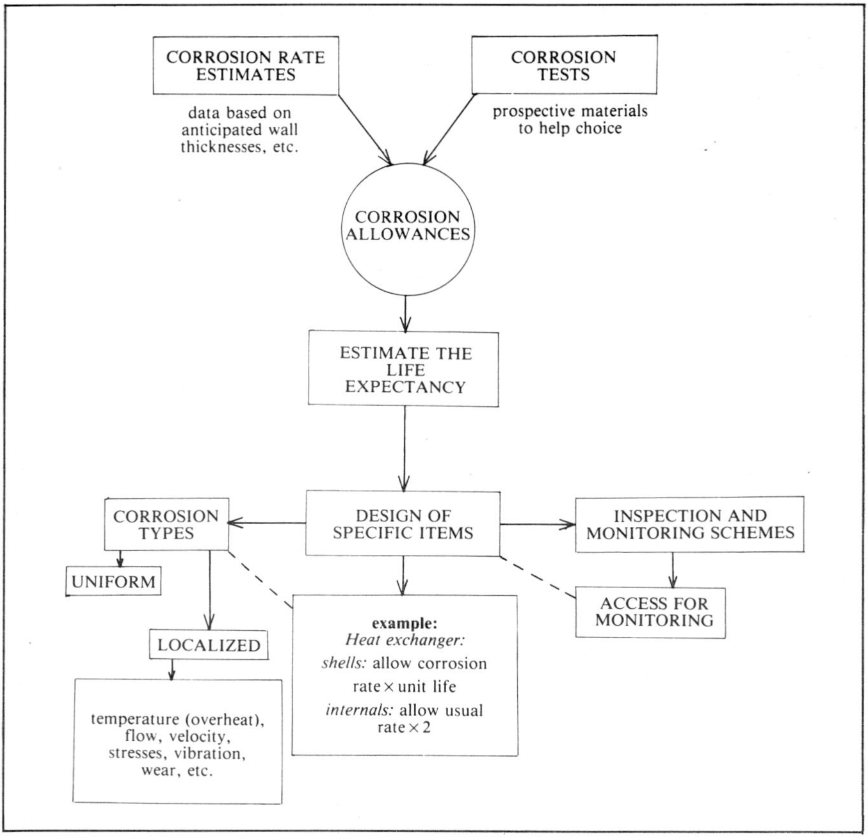

Figure 25 – Corrosion allowances and implications on design

Environmental Control

As has been mentioned already, it is necessary to ensure that the environment be maintained within certain limits, especially during periods of shutdown, start-up or testing, where corrosion data may not be easily defined. In many cases, it is the unexpected presence of minor impurities or constituents or periods of malfunctioning that create the problems. Some of the possible variables affecting the environment are outlined in Figure 26. With advances in monitoring techniques it should now be possible to introduce better control and, consequently, much lower maintenance costs. Such monitoring procedures are to be recommended(*7,48*).

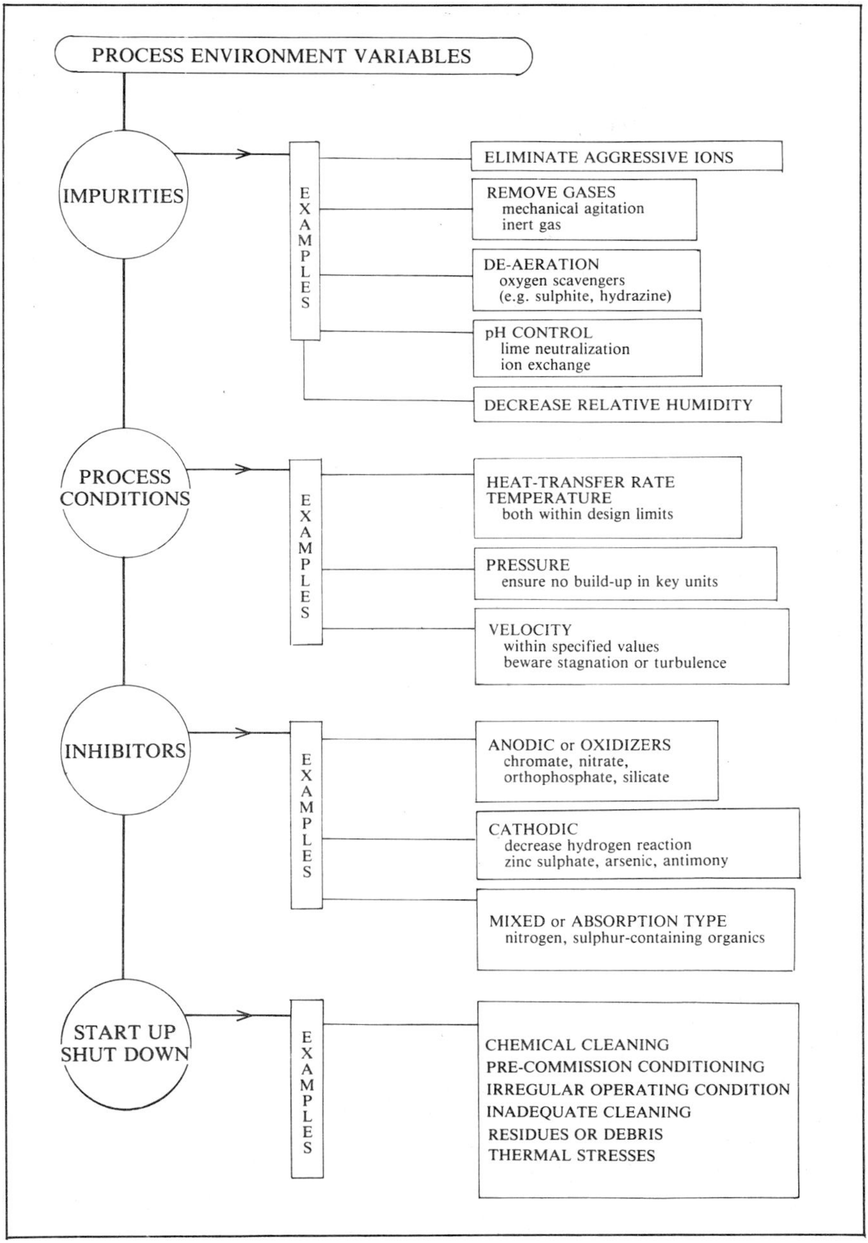

Figure 26 – Typical variables that affect a process environment

Material Selection

Alternative materials, or sometimes heat treatment, can provide a way of controlling corrosion, although direct and consequential costs need to be fully appraised when making a selection. In selecting a corrosion-resistant material it is necessary to be aware of various mechanical, physical and engineering properties, the ease of fabrication, delivery time and availability in the shape or section required, and, of major importance, the cost. More-resistant alloys are available, but these tend to be expensive and their cost needs to be justified. The use of plastics is still developing in the chemical process industries, although the limitations of certain plastics (Section 2), and the problem of inspection for weld structural integrity, have restricted development in some areas.

The data sheets and sources of information referred to in Section 5 can provide guidelines for corrosion resistance, but care is needed to allow for local effects where wastage or cracking can be significant. In alloy selection there are several 'acceptable' combinations of metal and environment, which often reflect maximum corrosion resistance for lowest expenditure(*3*); Table 8 lists such combinations.

TABLE 8

Metal/Environments for Corrosion Control(*3*)

Material	Environment
Stainless steel	Nitric acid
Nickel and alloys	Caustic soda
Monel (Ni–Cu)	Hydrofluoric acid
Hastelloys (Ni–Mo, Ni–Cr–Mo)	Hot hydrochloric acid
Lead	Dilute sulphuric acid
Carbon steel	Conc. sulphuric acid
Tin	Distilled water
Titanium	Hot, strong oxidizing solutions
Tantalum	Exceptional resistance in many environments

In addition to the 'natural' combinations given above, there are also some 'rule-of-thumb' guidelines for material resistance to corrosion. These are summarized in Table 9 which is based on published information(*3*).

TABLE 9

Environments for Material Protection(*3*)

Environment	Material
Reducing or non-oxidizing (e.g. air-free acids and aq. sol.)	Nickel and alloys Copper and alloys
Oxidizing conditions	Chromium-containing alloys
Extremely powerful oxidizing conditions	Titanium and alloys

Electrochemical Protection

Electrical techniques for corrosion control involve changing the potential of a material. For instance, an increase of potential into the passive region (see Figure 4b) provides anodic protection, the rate of corrosion being much reduced because of passive-film formation (e.g. anodizing of aluminium). A lowering of potential reduces the energy difference available for metals forming ions (see Section 2) and results in cathodic protection. Zinc coatings on steel, for example, confer cathodic protection because the zinc acts as a sacrificial anode and the protection arises because the electrode potential of the steel is reduced to a value where the steel becomes immune (i.e. it acts as a cathode). Zinc does not of course need to be in physical contact with the steel, provided there is electrical contact between the two metals and the resistance of the circuit is not too high. Similar protection can be achieved using inert anodes and a source of direct current (i.e. battery, generator, etc.). Table 10 contrasts some applications and principles of electrochemical corrosion control(*46*).

TABLE 10

Electrochemical Protection

ANODIC PROTECTION	
Primarily for chemical process plant	Used to maintain metal passivity
Used for road tankers	Potentiostatic control of metal potential with respect to a reference electrode
Used on shell internals	Operating conditions determined by electrochemical measurement
CATHODIC PROTECTION	
Used on pipelines, etc. Marine structures, pilings, water boxes	Galvanic couplings to Zn, Al,Mg (sacrificial anodes)
(Not suitable for tube side of heat exchangers)	Use impressed power from external d.c. source with an inert anode

Codes of practice(*49*) and further information is available in the literature(*2,8*)

Coatings

Coatings represent a most important aspect of corrosion control, with options ranging from metallic and inorganic to organic types. Many different techniques are available and, as with material selection, it is important to be aware of the limitations or disadvantages during planning or design. It is often stated that a coating is only as good as the initial preparation of the surface and much care should be given to this aspect. Grit- or shot-blasting is generally accepted as most effective for large structures where accessibility is also a vital factor in design. Acid pickling is another technique that results in the removal of reactive species prior to coating. Successful coatings, therefore, depend upon surface preparation (for which a British Standard is available(*50*)), and a regular 'defect-free' covering(*42*).

Electrodeposited coatings reveal aspects of the geometry of a system, being thicker at sharp edges and corners and thinner in recesses where current densities are higher or lower respectively. Paint films will tend to drain-off and be thicker due to gravity effects in certain locations. Surface tension and shrinkage during drying can also present problems for paints on sharp edges and corners. Smooth contours or suitable elastic fillers (mastic, plastisols, etc.) can be of benefit on occasion. Further design aspects are described elsewhere(*40,42,47,50*).

Metallic Coatings

These can be noble or sacrificial from the corrosion point of view, there being bimetallic coupling effects on occasion (see Section 2). Higher-temperature deposition techniques, such as hot dipping, may often give rise to beneficial metallurgical bonding, but there will also be limitations in all techniques, as noted elsewhere(*40,42*). Noble metallic coatings for steel include chromium, nickel, lead, gold, titanium, platinum, etc., the last three systems being employed when unique corrosion properties are needed. Sacrificial metal coatings for steel include zinc (the most widely used system), cadmium, aluminium (in externally aggressive situations like sea water) and tin. Chromium electroplating is most often applied to steel over a base coat of nickel, since thick coats ($>3\mu m$) tend to crack and thin coats ($<0.3\mu m$) tend to be porous. The duplex coatings comprise a thin layer on the metal and a thicker, cracked outer layer to a total thickness of about$5\mu m$(*42,51,52*).

Inorganic Coatings

These include chemical conversion coatings, where conditions are designed to develop surface films that might normally be achieved by alloying or where the natural products are allowed to form more rapidly. Chromate treatments for metals such as aluminium, zinc, cadmium, magnesium and copper are possible, but do not provide protection against atmospheric attack on steel. Phosphate treatments remove surface rust and provide reasonably protective coatings, which are frequently used prior to painting.

Phosphate treatments are also available for zinc alloys and cadmium-plated articles. Anodizing has been referred to previously, the technique (for aluminium) being to grow protective aluminium oxide to a considerable thickness in a controlled way by electrolytic methods. Thicker films of alumina develop in sulphuric acid whilst thinner films, up to 6μm, form in chromic acid; film thicknesses in the range 15–25μm are preferred for atmospheric conditions(*42*). Other metals, such as magnesium, titanium and tantalum can also be anodized.

Organic Coatings

These include elastomers and plastics, but most important of all are paints, which represent the most common way to prevent or control corrosion. Paints are mixtures of insoluble particles of pigment (e.g. metal oxides TiO_2, Pb_3O_4, Fe_2O_3, or other compounds such as $ZnCrO_4$, $PbCO_3$, etc.) suspended in an organic vehicle (e.g. natural linseed or tung oils). Exposure to air causes these oils to oxidize and polymerize into solids, the process being hastened by catalysts such as lead, manganese or cobalt soaps. Synthetic resins are often used where conditions are more aggressive, e.g. continuous contact with water, acids, alkalies or elevated temperatures.

Paints show permeability to oxygen and water, and therefore do not act solely as physical barriers between a metal and its environment. Paints have low diffusivities of moisture and oxygen and the diffusional paths are lengthened and convoluted by the presence of the solid pigments and fillers incorporated therein. Characteristics of binders and pigments, etc. are given elsewhere(*2,42*).

The majority of paint failures, blisters and peeling, etc. arise because of inadequate surface preparation and improper application, rather than paint composition or formulation. Etch primers, containing zinc chromate or calcium plumbate, are often employed to prevent corrosion after initial surface derusting of steel, aluminium and zinc(*42*), and factory-applied precoats are very desirable provided that transit and storage are properly executed.

An economic answer for many applications is a combination of coating and cathodic protection, as long as the coating is also resistant to the alkalinity that develops at the metal interface. The cost advantage is demonstrated in the following example (quoted elsewhere) of a buried pipeline(*40*):

Bare steel pipeline – current consumption $\sim$ 5–500 mA/m^2 (dependent on soil)

Organic coat and cathodic protection – current requirement now $\simeq$ 0.01 mA/m^2.

7 CORROSION IN TEXTILE PLANT

It is easy to conceive of situations in textile plant where corrosion processes can be significant. The aggressive liquors and gases associated with bleaching processes, represent one such area. By contrast, it is difficult, if not impossible, to provide a solution to all such corrosion processes unless relevant information is to hand. No overall data book exists at this time for the textile industry and perhaps there is some justification in compiling such a volume in the future. Meanwhile, the effort should be towards a proper consideration of each situation – as or, preferably, before it occurs – by better appreciation of design and control of the various processes or units involved in the industry. This book covers many different aspects of engineering technology, but all have relevance to corrosion and protection, as can be readily appreciated from the following examples.

Power generation/transmission – wear in bearings, loss of inhibited lubricant, stray currents, etc.

Boiler plant – boiler tube failures (water side/fire side) heat-transfer effects, high-temperature scaling, overheating, etc.

Heating and ventilation – humidity, temperature effects on environmental corrosion, etc.

Civil engineering – constructional materials, drainage, cleaning, painting, fabrication and joining, etc.

Fluid flow – erosion corrosion, cavitation, impingement, etc.

Water supply/effluent control – scaling effects, flow, chloride/sulphides, etc.

Instruments/control – meters for pH, temperature, flow, corrosion monitoring, etc.

Health/safety/economics – collapse, explosion, fire, etc.

In principle, all corrosion processes can be eliminated or restricted by utilizing one or more of the approaches outlined in this chapter, bearing in mind that an economic solution is always desirable, but not always achieved. This chapter has attempted to highlight the ways by which corrosion occurs and has given some attention to areas that are not always taken into account, such as fabrication and/or design.

REFERENCES

1. Report of the Committee on Corrosion (London: HMSO, 1971).
2. Shrier, 'Corrosion', Vol.1&2 (London: Newnes–Butterworth, 2nd Edn, 1976);'Electrochem Princ. of Corrosion' (London: DoI, 1982).
3. Fontana and Greene, 'Corrosion Engineering' (New York: McGraw–Hill, 2nd Edn, 1978).
4. Evans, 'The Corrosion and Oxidation of Metals' (London: Arnold, 1960); Supplements in 1968 and 1976.
5. Scully, 'The Fundamentals of Corrosion' (Oxford: Pergamon Press, 2nd Edn, 1975).

6. Wranglen, 'An Introduction to Corrosion and Protection' (Stockholm: Inst. Metallskydd, 1971).
7. Henthorn, 'Corrosion – Causes and Control' (Reading, Pa.: Carpenter Tech. Corpn); Originally published as Chem. Engng Refresher Series, Chem. Eng., 1971–2.
8. Uhlig, 'Corrosion and Corrosion Control' (New York: J. Wiley & Sons, 1963).
9. Smithells (Ed), 'Metals Reference Book' (London: Butterworths, 1970).
10. Waterhouse, 'Fretting Corrosion' (Oxford: Pergamon Press, 1972).
11. NACE, 'Process Industries Corrosion' (1975).
12. Engineering Design Guide, No. 27, 'Introduction to Materials Science' (Oxford: OUP, 1978).
13. Engineering Design Guide, No. 17 & 19, 'Engineering Properties of Plastics' and 'Selection and Use of Thermoplastics' (Oxford: OUP, 1977).
14. Evans, 'Selecting Engineering Metals for Chemical and Process Plant' London: Business Books Ltd, 1974).
15. Dept. of Industry, 'Guides to Practice in Corrosion Control': No. 2 'Corrosion of Metals by Wood' (London: DOI, 1975).
16. Evans, 'Corrosion in Chemical and Process Industries', The Chem. Eng., No. 331 (1978) 263.
17. Elliott, 'Corrosion Survey Report', Inst. of Chem. Engrs (Supplement 1973).
18. NBS Special Publications 511–1 & 511–2, 'Economic Effects of Metallic Corrosion in the United States' (Washington DC: NBS, 1978).
19. Dept. of Industry, 'Controlling Corrosion: No. 3 – Economics' (London: DoI, 1975).
20. Vilbrandt and Dryden, in 'Chemical Engineering Plant Design' (New York: McGraw–Hill, 1959) Chapter 2.
21. BSI, Sectional List – SL7, 'Welding' (London: BSI, 1970).
22. BSI, 'Cleaning and Preparation of Metal Surfaces', CP3012 (London: BSI, 1972).
23. Engineering Design Guide, No. 7, 'Soldering' (Oxford: OUP, 1975).
24. Engineering Design Guide, No. 10, 'Brazing' (Oxford: OUP, 1975).
25. Engineering Design Guide, No. 6, 'Welding Processes' (Oxford: OUP, 1975).
26. Houldcroft, 'Welding Processes' (Cambridge: CUP, 1972).
27. Dept of Industry, Guides to Practice in Corrosion Control, No. 1, 'Sources of Corrosion Information' (London: DOI, 1978).
28. 'Wiggin Nickel Alloys in Chemical and Process Plant' (3681); 'Design of Chemical Plant in Wiggin Nickel Alloys' (7011), (Hereford: Henry Wiggin & Co.); Books from INCO Europe Ltd, Millbank, London, SW1P 4QF.

29. 'Corrosion Resistance of Hastelloy Alloys', F30/131F, Cabot Corpn (1978).
30. 'Guide to Corrosion Resistance of Alloys containing Mo', Climax Molybdenum Co., New York.
31. 'Corrosion Resistance of Titanium', IMI Ti Ltd, Birmingham, England.
32. 'Langley Corrosion Resistance Guide', Langley Alloys Ltd, Slough SL3 6EA.
33. 'Stainless Plate', BSC Stainless, Shepcote Lane, Sheffield S9 1TR.
34. IPC Corrosion Chart, (London: Engng. Chem and Marine Press Ltd, 1971).
35. Nelson, 'Corrosion Data Survey' (1967); Hammer, ibid., (Houston, Texas, USA: NACE, 1974).
36. Dechema – Werkstoff-Tabelle, Dechema Deutsche Gesellscaft für Chem. Frankfürt a.M., W. Germany.
37. Rabald 'Corrosion Guide' (Amsterdam: Elsevier, 2nd Edn, 1968).
38. Uhlig, 'Corrosion Guide', (New York: J. Wiley and Sons Inc., 1948).
39. Schweitzer, 'Corrosion Resistance Tables' (New York: Dekker, 1976).
40. Dept of Industry, 'Corrosion Control in Engng Design' (London: HMSO, 1978); 'Corrosion Control Checklist for Design Offices' (London: DoI, 1981).
41. Keating, 'Chromium–Nickel Austenitic Steels' (London: Butterworth, 1956).
42. Engineering Design Guide, No. 21 (Oxford: OUP, 1977).
43. Dept of Industry, 'Controlling Corrosion' No. 4, 'Specifications and Standards' (London: HMSO).
44. BSI, Sectional List, 'Codes of Practice' (London: BSI, 1975).
45. Anti-Corrosion Methods and Materials (1972) 5.
46. Dawson, 'Methods of Corrosion Control' The Chem. Eng., No. 331 (1978).
47. Pludek, 'Design and Corrosion Control' (London: Macmillan 1977).
48. Callow, Richardson and Dawson, Brit. Corr. J., **11**(1976) 123.
49. BSI, 'Cathodic Protection' CP 1021 (London: BSI, 1973).
50. BSI, 'Protective Coating of Iron and Steel', BS 5493 (1977) 'Methods of Protection against Corrosion' DD24: 1973 (London: BSI).
51. Gabe, 'Principles of Metal Surface Treatment and Protection' (Oxford: Pergamon Press, 1972).
52. 'Canning Handbook on Electroplating' (Birmingham: W. Canning and Co., 21st Edn, 1970).

Chapter 4

Fabric Transport and Fabric Control

J HEATON

INTRODUCTION

It has been said that if fabric can be handled there need be no difficulty in the other aspects of preparation, dyeing and finishing. Like all generalizations, this contains a grain of truth – the subject is indeed of great importance. Yet it is the one aspect of textile processing that does not lend itself to organized study; the writer has been able to find little or nothing in the literature which would assist the practitioner. The subject is tremendously complicated, due to the idiosyncracies of a multitude of fabric constructions and processing machinery.

The tremendous development of warp- and weft-knit fabrics in the last few years, together with the availability and application of modern electronics, has had a notable effect on the sophistication of fabric handling. Examples are in the uncurling of difficult warp-knit fabrics, the control of selvedges and the development of automatic weft straighteners.

Having paid respects to the awkwardness of the subject, the following notes constitute an attempt to consider some practical aspects.

1 ROPE PROCESSING

Though fabric is produced and used in open width, the easiest method of processing is in the form of a rope. Handling is easy, as is storage, and a rope of cloth can be transported virtually any distance and in any direction by passing it through circular guides.

Preparation

The original sequence for cellulosic fabrics was based on the rope method. This allowed easy storage in kier, pit or pile for long reaction times. A typical tight- or slack-rope washer could give the rope five or six dip–squeeze sequences in a compact space, and this is particularly well suited to washing out chemicals and colloidal impurities. A high running speed allowed good productivity at low capital cost.

Some fabrics, such as the heavier 'ducks' or tight constructions such as two-fold poplins and many synthetic-fibre/cellulosic blends, are inherently unsuited to such a process because of the formation of permanent creases, but suitable fabrics can be put through continuous versions of the old methods, in which reaction delays are provided by J-boxes adapted from the Gantt piler, while washing and impregnation can be carried out in refined, tension-reduced units such as the Tensitrol machine (Figure 1).

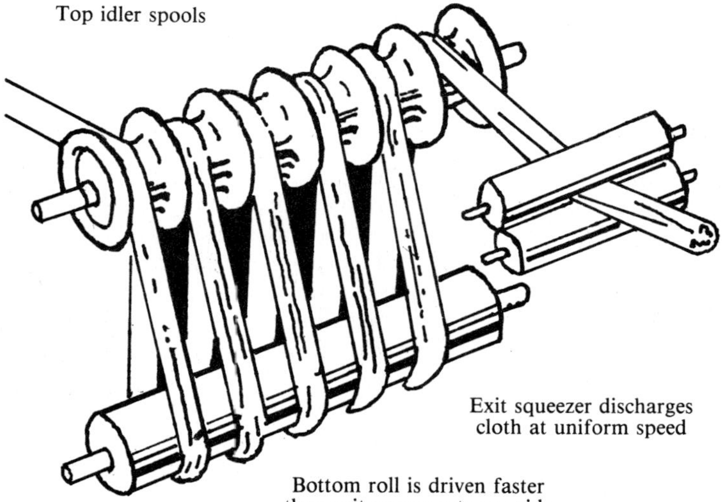

Figure 1 – Tensitrol principle

CONDITION 1 – *When cloth shrinks*
- A Traction on bottom roll increases
- B Overdrive decreases
- C Result: cloth speed upward increases

CONDITION 2 – *When cloth stretches*
- A Traction on bottom roll decreases
- B Overdrive increases
- C Result: cloth speed upward decreases

In either condition – and at all times – the cloth travels the full distance through the machine virtually 'as limp as a rag'.

Dyeing

The rope method does not fit well into the time and reaction sequence of most continuous dyeing processes, but for batch dyeing the winch and its modern successor, the jet machine, are pre-eminent, particularly for high-temperature work and for tension-sensitive fabrics, such as tubular knit. In its usual form, a winch drops the fabric into the back of the dyebath and picks it up from the front, over an auxiliary or 'jockey' roller. A typical piece of fabric, perhaps 100 m in length, is loaded by throwing an end over the reel, bringing the end to the front, and securely sewing the two ends together to form an endless rope. A large machine may well hold twelve or more such ropes. The unloading procedure is the reverse of loading, so it can be imagined that loading and unloading are slow and laborious. The time has been greatly reduced by a device which allows the cloth to be processed as an endless rope, yet maintaining the essential characteristic of a winch process(*1*).

Dyers have experimented for many years with spiral loading of normal winch vessels, the two opposing fabric ends (head and tail) being sewn together and caused to traverse the width of the vessel by appropriate 'pot eyes' or guide rollers. Some fabrics could be handled successfully in such a manner, but the more common result was that one or more 'drafts' or loops would end up as a pile of excess fabric, whilst adjacent loops tightened up on the winch reel itself.

The Autoloda, for continuous spiral loading and subsequent processing, is a remarkably ingenious device and deceptively simple. The following diagrams and instructions have been abstracted from an Operations Manual, by courtesy of Leemetals of Macclesfield.

'This basic method of loading is illustrated in four diagrams of the Operating Instructions.

Detailed instructions are as follows:

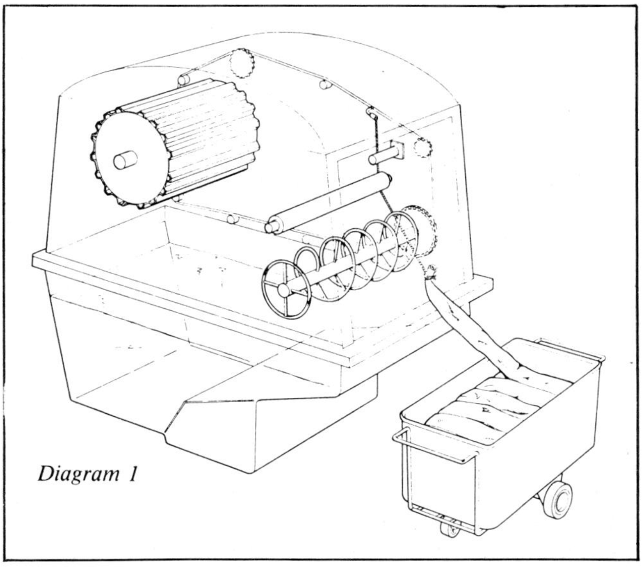

Diagram 1

Diagram 1

This diagram shows the correct path of towing cord and fabric when starting to load. The fabric is prepared in one single length, by stitching shorter lengths together if necessary, and should be presented to the machine in such a way that it can pull easily from its container. Although the equipment will tolerate a certain amount of "slip" on the winch, this should be avoided as much as possible by wrapping the winch with a suitable cloth cover if necessary.

Before starting to load the machine, the counter on the Autoloda control must be set to the required rope length. The figures on the counter indicate metres per rope, and the setting is calculated by dividing the total length of the fabric by the number of ropes which the Autoloda spiral will make. For this purpose, do *not* count the first and the last spaces of the spiral. The counter indexing fingers may be moved up or down, according to the direction required.

One end of the fabric is attached to the towing eye on the conveyor chain by means of a towing cord or tape about two metres long. The correct length of this cord is important. Underneath the bottom length of the chain, about 1½ metres from the front edge of the machine, is the tow cord "rest bar". The purpose of this bar is to lift the leading end of the fabric out of the dye-liquor (about 20 cm), to prevent it being dragged forward by the moving fabric next to it. When the conveyor chain has completed a revolution and stopped, the joint of the towing cord and the fabric should be hanging a few centimetres under the "rest bar". There is a provision for lateral movement of this bar – and if necessary it should be adjusted to ensure that the fabric falls clear of it when the conveyor chain starts to run.

It is important to avoid straining the conveyor chain, and it is possible for heavy or very absorbent fabrics to impose too much load on the chain. This condition should be avoided by having a "lead cloth". This should be a length of lightweight fabric about six metres long, stitched to the fabric at one end and fastened to the towing cord at the other end. In this case it is the end of the "lead cloth" which is held in position by the "rest bar". Lightweight and mediumweight fabrics do not require a "lead cloth".

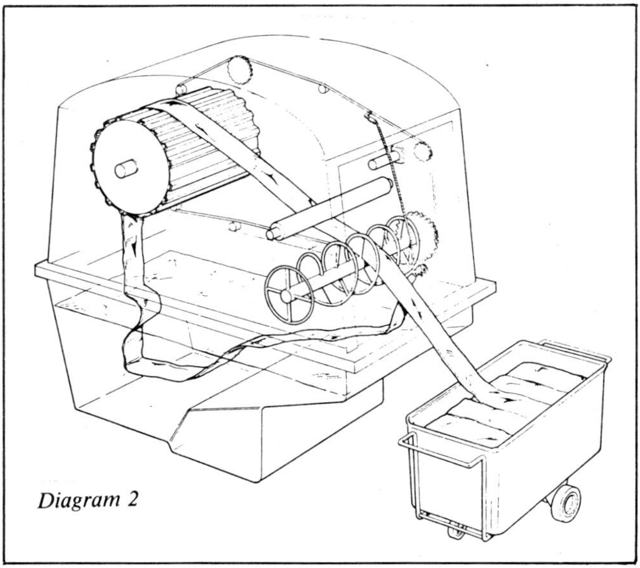

Diagram 2

Diagram 2

This illustrates why it is advisable to load the fabric *between* the spiral and the front roller. The action of the spiral maintains the fabric load in its correct location during loading.

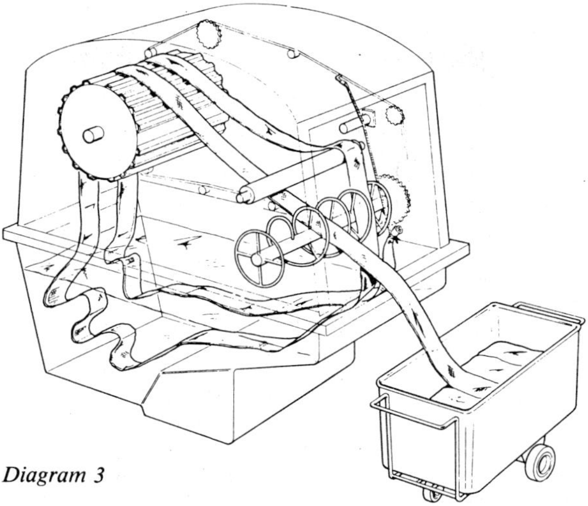

Diagram 3

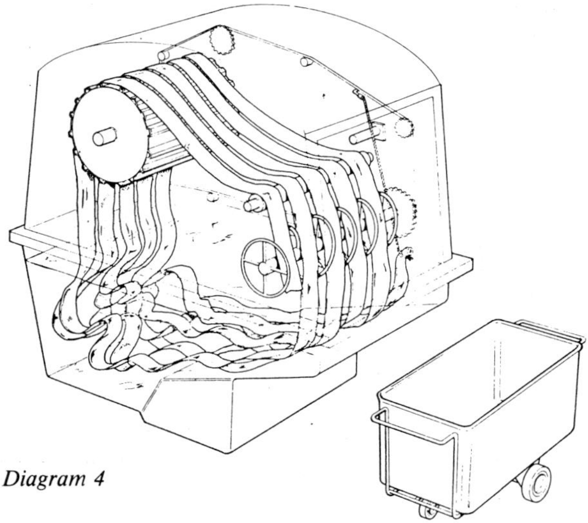

Diagram 4

Diagrams 3 and 4

These illustrate the loading action, which is fully automatic. It is only necessary to ensure that the fabric is leaving the container without strain.

Running

If the original calculation of rope lengths was correct, the fabric will fill the spiral except for the first and last spaces. If in doubt it is advisable to calculate for a few spare spaces, these will occur at the opposite end of the conveyor chain. The counter can be altered whilst the machine is running and the spare spaces filled up. Remember that the principle of the counter is to set the rope length at each revolution on the conveyor chain, this length is then fed to each succeeding rope in turn, so it will take some time for the result of any alteration to take effect at the other end, it is always advisable to have one spare space at the end of the spiral. Remember also, that some fabrics alter their length during dyeing, and if this occurs the counter should be altered to suit.'

Seeing an Autoloda in action is convincing, but many questions may be asked, such as:

How does it really work?
If the spiral loads fabric step-by-step leftwards, why does that action apparently cease when loading itself is complete?
Why doesn't fabric end up as a pile at one end?

Before attempting to explain the principles involved, it should be pointed out that with a conveyor chain on the right-hand side of the vessel, the spiral must have a left-hand thread. If the chain is on the left then the spiral would be a right-hand thread.

The diagrams show that there are two main features – the conveyor chain with its 'towing eye' and the 'spiral'. It may be deduced from the diagrams and operating instructions that the chain and spiral are geared together in such a way that the towing eye makes one complete revolution for each turn of the spiral. The chain/spiral drive is linked to the winch reel drive by a variable-speed gear, which has a counter device for measuring the length of fabric required in each draft or loop.

The method of loading is generally understood at once, but it is less clear why rope maintains its symmetrical arrangement during continuous operation.

Consider, then, three vessel arrangements.

1. A normal vessel, spirally loaded manually, with head and tail ends free. It is immediately obvious that, on continued operation, the leading end would drop into the vessel bottom and the rest of the rope would follow.
2. A normal vessel with the addition of chain and towing eye only. The fabric would be loaded manually in a spiral manner and the leading end (head) attached to the towing eye. However, on continued operation, through a normal 'pot eye' or 'peg rail', the loops would gradually move to the right, leaving an increasing number of peg-rail spaces vacant. This would cram the loops to to the right until continual operation proved impossible. This happens because any *peg-rail* space, guiding fabric to the reel, draws its supply of fabric from the vessel bottom to the immediate left of that peg-rail space. (This statement is conveniently verified by following the position of the loose tail end.)

3. An Autoloda vessel. This now has the driven spiral and it is the operation of the latter that overcomes the rope progression to the right, by simultaneously traversing the rope to the left.

Thus, during continuous operation, the path of any seam or marker point on a loop of fabric remains substantially at the same distance from one end. The symmetry of loading is therefore maintained during continuous operation.

Finally, it is usual for the drive to the chain and spiral to be set automatically to give intermittent rotation. A knock-off lever stops the chain and spiral when the towing eye has just moved beyond the spiral. This leaves the leading end of fabric (head end) immersed in the dye-liquor. When the reel has delivered its preset length of fabric per loop or draft, a timer restarts the chain and spiral to give a single revolution. This revolution is performed rapidly so that the head end of fabric is quickly lifted from the dyebath, over the spiral and reel and back into the dyebath.

The original patent provides for alternative continuous rotation of the chain and spiral, but it will be clear that the rotation must then be relatively slow, so that the head end of fabric would spend too much time out of the dye-liquor.

In any dye vessel, the creased rope is not likely to alter its configuration significantly during its slow passage from back to front. In the hot, wet conditions, these creases sometimes become permanent, particularly in the case of circular-knitted synthetic-polymer-fibre fabrics. This problem can frequently be overcome by persuading the tube to inflate into a 'balloon'(*2*). Given the correct machine shape, in which the mass of fabric is held to the back of the vessel, the rope is pulled forward over the closed-coil liquor heater. The liquor entrained in the rope boils, and the tube of fabric is inflated. When it collapses again, the creases are redistributed and thus do not have sufficient time to become permanent.

Opening

For drying and further processing, the rope of cloth must be opened out to full width. This process, 'scutching', was at one time carried out by hand, and later by an automatic scutcher. The method depends on the fact that, before entering the rope process, the cloth is flat. Thus, if it has acquired several twists in one direction at one position, there must be somewhere the same number of twists in the opposite direction. The cloth is drawn out on a very long free run and, on its return, is vigorously agitated by a rapidly rotating beater (Figure 2). The twists resolve themselves, and the cloth is expanded to full width by a series of driven scroll rollers.

A more modern, beaterless type is available that is suitable for delicate fabrics. Twist in the rope is detected by feelers, and the rope is mechanically rotated as it passes through a guide (Figure 3). Alternatively, a waggon containing the cloth may be rotated.

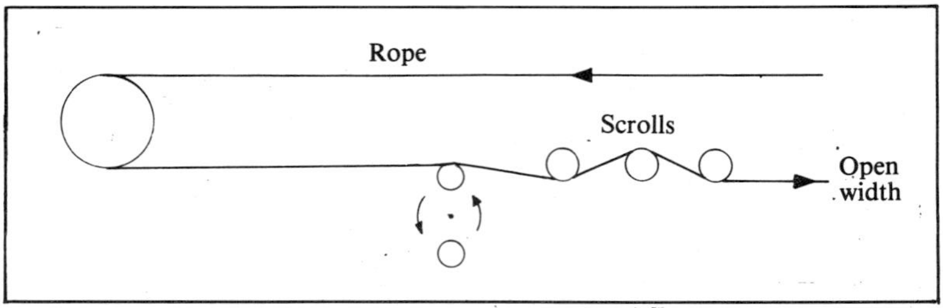

Figure 2

Figure 3 – Automatic detwister (roof installation)

2 OPEN-WIDTH PROCESSING

The construction of some fabrics demands open-width processing. The heavier types may be impossible to put into rope form. Tightly woven fabrics, made from two-fold yarns, such as the traditional shirting poplins, are prone to form permanent 'ghost' creases, due to minor physical changes in the region of a rope crease. Others again, with Ventiles as an excellent example, become extremely stiff when wet, due to swelling of the yarns which results in a tightly packed structure. The drying and heat setting of woven fabrics are examples of processes which are inherently open width. However, open-width methods have been increasingly applied in recent times, to facilitate continuous processing in preparation and dyeing.

Tension Control

With a few exceptions, where the fabric is required to have little or no residual warp crimp, the intention is to process with the minimum lengthwise tension that will allow smooth running. It is not possible to forecast how great this will be. There is a very indirect relationship between dimensions in the warp and weft directions, so that warp tension tends to extend the length and cause width contraction. Nevertheless, in cases where fine warp-direction creases have caused difficulty (suggesting width contraction) the cure has been found to be in increasing the lengthwise tension. Clearly, experimentation is necessary.

Compensator rollers perform the dual function of synchronizing machinery units coupled in line, and applying a desired load to the cloth. The older type, using dead weights (Figures 4 and 5) which could be added or removed, had the great advantage of simplicity and visibility – one could see that they were working correctly.

Pneumatic loading, which offers easy and fine adjustment, is now very popular (Figure 6).

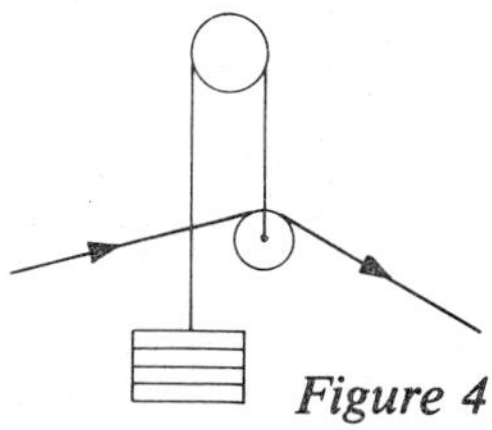

Figure 4

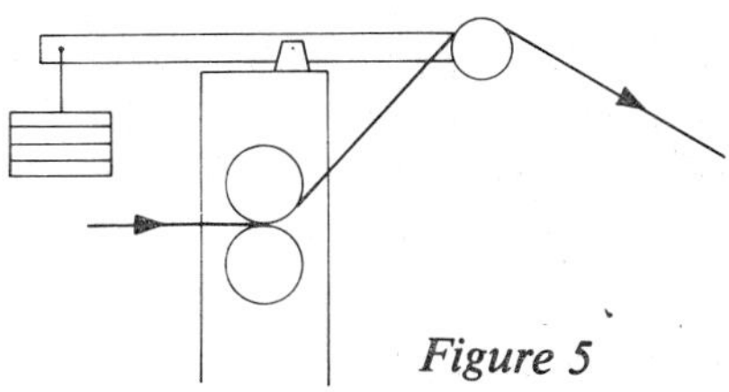

Figure 5

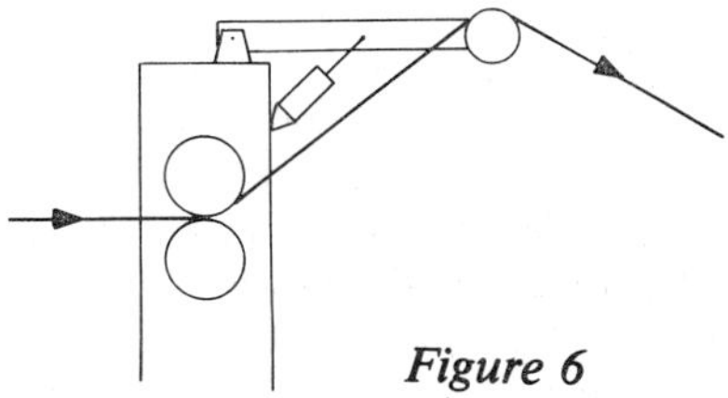

Figure 6

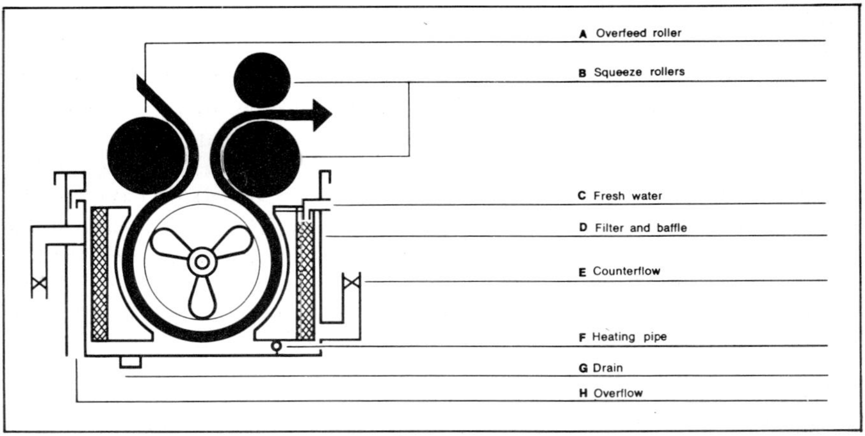

Figure 7 – Sieve-drum principle, the rotation of the propellors draws the washing liquor through the fabric and the sieve drum, expelling the liquor back into the compartment via the ends of the drum

In practice, once set up, compensator loadings are rarely changed, and the pneumatic-cylinder type can cause trouble that is difficult to trace, due to the development of a stick–slip effect. Probably the best method of loading is by a pneumatic diaphragm. In spite of the inertia effect of the added weight, compensator rollers should be sufficiently robust to prevent deflection and consequent creasing. The arms must be stiff in the lateral direction, to avoid a shake developing during running.

If the compensator roller rises or falls, then the speed of its associated unit must be decreased or increased. This is achieved by connecting the compensator to a speed-adjusting device. There may, for example, be a direct linkage to a variable-speed gearbox, or a rheostat may be moved to affect the speed of a d.c. motor or the slippage of a magnetic clutch.

Cylinder drying machines have presented a difficult problem in obtaining minimum tension, because the cloth tends to alter its dimensions as it dries. The cylinders are usually driven at the same speed, and will not permit relaxation. However, an ingenious solution is available, each cylinder being driven by a spiral spring 'belt' which allows a degree of independence(*3*).

Open-width washing machines are frequently offered with tension-reducing drives to the upper guide rollers, sometimes with light auxiliary pressing rollers, or 'jockey' rollers to reduce slippage. This can be a good system, particularly at modest running speeds, but is less successful when run at speed. Maintenance must be meticulous. Jockey rollers are sometimes claimed to have the additional advantage of forming a light nip, so holding back contamination in the washing range. This is true only while the rollers are in good condition, but the small circumference means that wear is rapid. Unless frequent regrinding of the many rollers is carried out, the effect rapidly becomes unlevel.

A curious additional hazard is that swarf and other particles of metal have been found to embed themselves preferentially in jockey rollers, with consequent fabric damage.

Sieve drums are offered as a means of low-tension support in washing machines (see Figure 7). The liquor may be sprayed vigorously on the cloth, or positively sucked through by a propellor at each end of the drum. This latter method tends also to keep the fabric at full width and to reduce selvedge curl. However, drum washers usually operate at a high liquor ratio, with disadvantages in filling and warm-up time, and with relatively high steam and water costs.

Width Control

Unless the cloth selvedges are held, as on a stenter, positive width control cannot be achieved. Much can be done to prevent width loss by employing bowed, rubber-covered roller expanders, sometimes several together, and usually immediately before nips or drying cylinders. The various types of expander have been fully described elsewhere(*4*). As in the case of rollers, the recent trend has been towards greater-diameter low-friction bearings and wear-resistant coverings, to obtain longer life.

Rollers

Rollers are uncomplicated devices that do not receive the attention that they deserve. For this reason, there was a period when manufacturers installed small rollers, so that they could offer a machine at an attractive price. This is a false economy. The required diameter depends upon the width of the machine, but, in the common 2 m range, the preferred diameter is about 12 cm, with substantial wall thickness and large journals. Such rollers can withstand shock loadings from the inevitable lap-rounds, and can be straightened if bent. Strong rollers resist deflection under load, which is a cause of creasing little acknowledged.

The spacing of rollers in a multi-roller machine is a matter still to be resolved. Clearly, the less the free travel of the cloth, the less is the chance of creases appearing. A preferred maximum spacing is 1 m and this is a common arrangement in washing machines, where there may be turbulence. In smooth-running machines, such as steamers, a greater spacing is frequently allowed. Intermediate static bars ('steady rods' or 'breaker bars') are sometimes fitted between widely spaced rollers. This method should be used with caution, as there is a tendency for the accumulation of lint and other debris, which may mark the cloth.

3 SYNCHRONIZATION OF MACHINERY

The fine trimming of the speeds of machines run in tandem is achieved by compensators, as described in Section 2. However, it sometimes happens that considerable buffer capacity between machines is required, as for example when changing batches of finished cloth from a stenter.

Two methods are in common use. If the introduction of superficial creasing can be tolerated, as at a stage earlier than finishing, a small J-box can be used. The cloth is allowed to pile into the J-box while the full batch is removed and replaced by an empty unit; the new batch is then run overspeed until the contents of the J-box are taken up.

If even light creasing is not permissible, as in the case of a finishing stenter, then a roller-type accumulator can be used. The cloth is laced between two banks of rollers and, when delivery from the machine is stopped, the banks are allowed to move further apart to provide a lengthening cloth path. Again, when the new batch is in position, it is run over-speed until the roller banks come together (see Section 8).

4 CONTROL DEVICES

Various auxiliary devices are in use to assist in the positioning of fabric as it passes through machinery. Most of these have been developed in relatively recent times in response to the need to save labour, to cope with increased running speeds and to handle the elastic knitted structures that have been developed.

Entry Frames

Cloth entering a machine, typically from a waggon or batch, must do so smoothly, under a constant, controllable tension. The methods by which this is achieved vary a great deal from the simple to the elaborate, and are subject to local preference. Examples are

(a) A simple arrangement of square-section wooden or stainless-steel-clad bars as shown in Figure 8
(b) Twin bars, which can be rotated by a worm and gear, to increase or decrease the tension on the cloth, as shown in Figure 9
(c) A roller with an adjustable brake (see Figure 10)
(d) An arrangement of supporting bars, as illustrated in Figure 11.

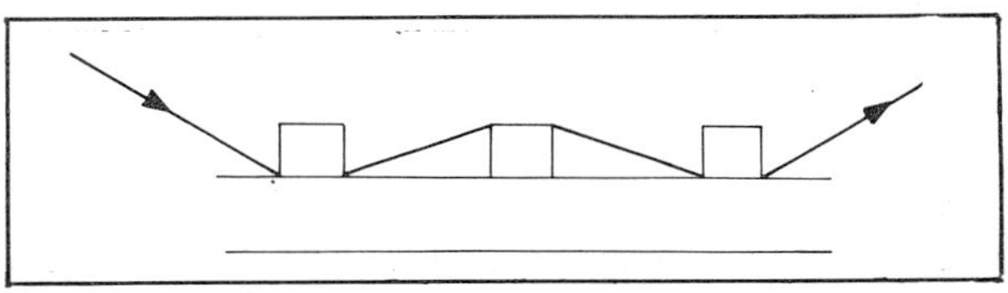

Figure 8

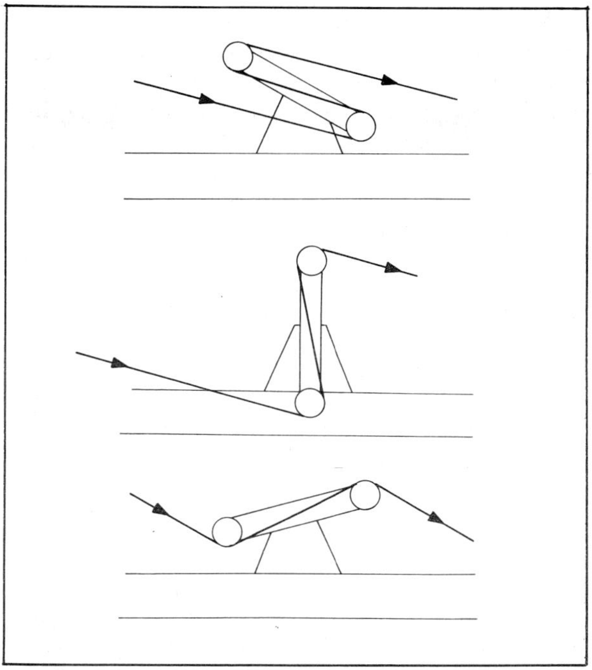

Figure 9

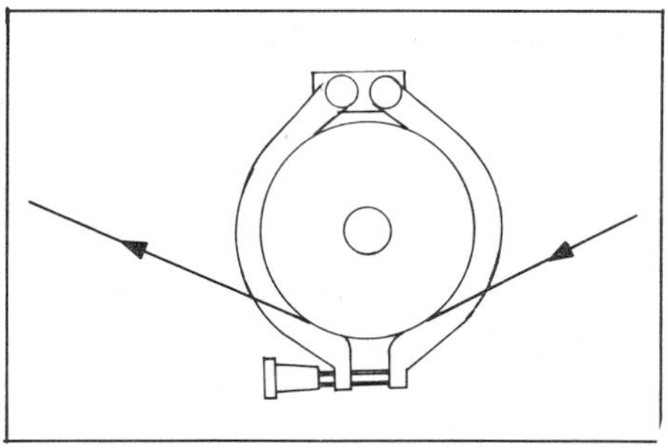

Figure 10

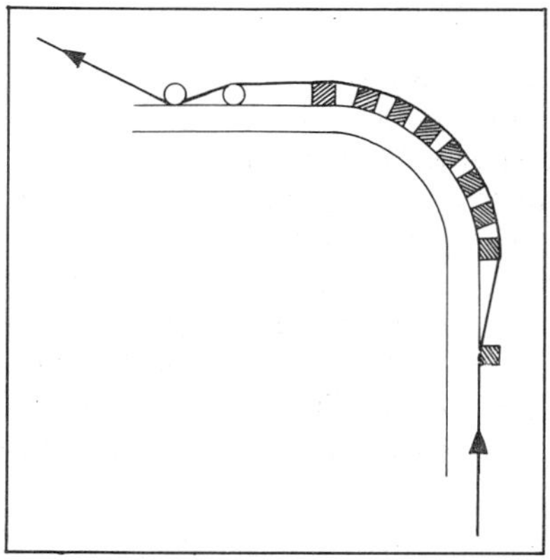

Figure 11

Guiders

The relatively coarse positioning of a running web of cloth is achieved by running each edge through a pair of light nip rollers. When the detecting element, either a photoelectric device or a mechanical feeler, shows that the edge has moved inwards, the nip operates in such a way as to pull the cloth outwards. The nip may either be in constant engagement and caused to incline when the edge moves (Figure 12a) or the inclination may be permanent, and the normally open rollers caused to close (Figure 12b). The rollers may be rotated by the movement of cloth, or positively driven (Figure 12c) to reduce cloth tension. Guiders can be obtained, at a price, which are highly resistant to chemical attack and which can be used, for example, inside a J-box (see Figure 13).

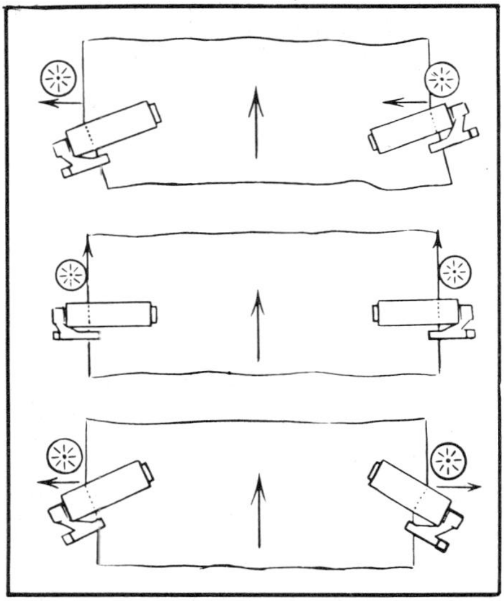

Figure 12a – Swivelling guiders

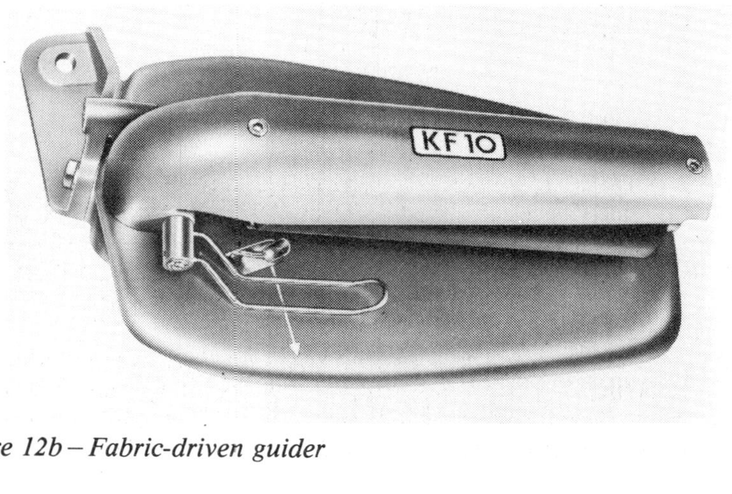

Figure 12b – Fabric-driven guider

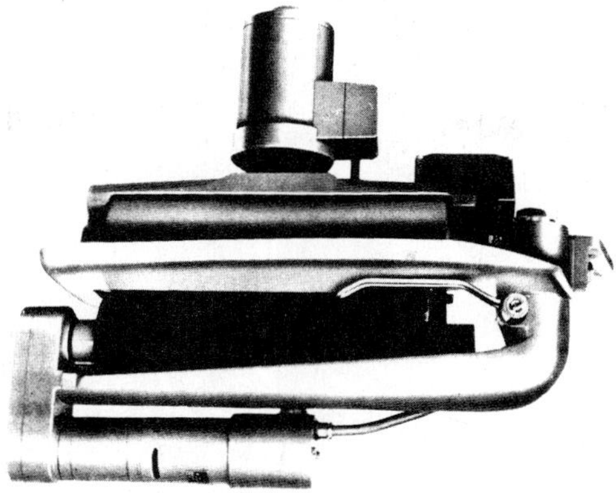

Figure 12c – Motor-driven guider

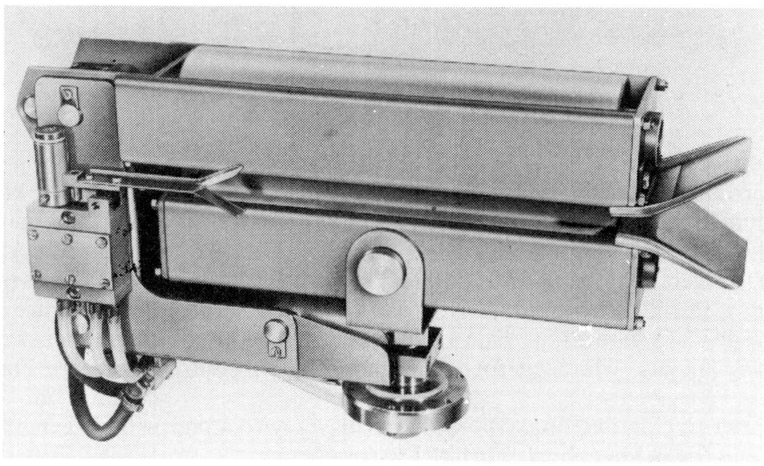

Figure 13 – Guider resistant to chemical attack

An interesting type of guider has been developed to deal with certain knitted fabrics, which exhibit severe edge curling. The edge passes over a large, rapidly spinning disc, on which it is pressed if deviation occurs (Figure 14). In this way the fabric is guided, while at the same time opening up to 18–20 cm of curl. A similar effect is obtained by locating a pair of rapidly rotating scroll rollers immediately before the guiders.

Guiders are always used in pairs, so that they perform the additional function of pulling the cloth to width, as, for example, when following a J-box. To allow the cloth to move more easily, it is usual to allow a long free run, frequently a vertical drop of about 3 m into the guiders.

Figure 14 – Spinning-disc uncurler

Selvedge Uncurlers

The problems posed by selvedges vary from the simple turn-over, which particularly affects some twill fabrics, to the extreme of certain knitted structures, which will form a rope if permitted. An arrangement of spring-loaded fingers may be used to correct the simpler faults, while more severe curling is unrolled by motor-driven pairs of spiral scroll rollers (Figure 15).

For delicate fabrics, a pair of plates may be used, having outward-pointing air jets, which perform the uncurling without contact (see Figure 16).

Plates carrying diagonal ribs and intermeshing to an adjustable extent are available for more robust fabrics (Figure 17).

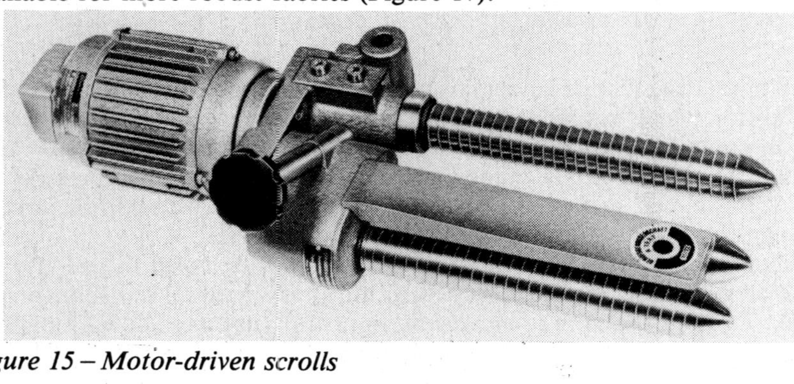

Figure 15 – Motor-driven scrolls

Figure 16 – Pneumatic selvedge uncurling unit and an edge sensor, for guiding woven and knitted fabrics into stenters without creasing or distortion ▷

Figure 17 – Uncurling, using diagonal-ribbed plates

▽

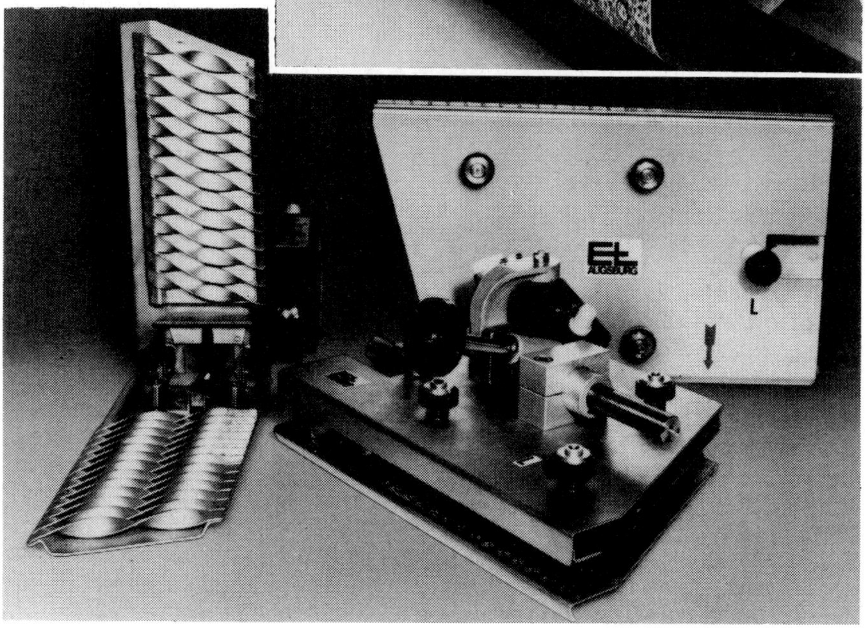

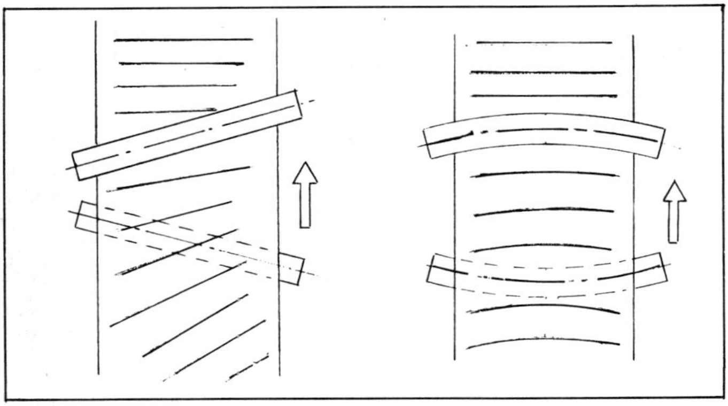

Figure 18

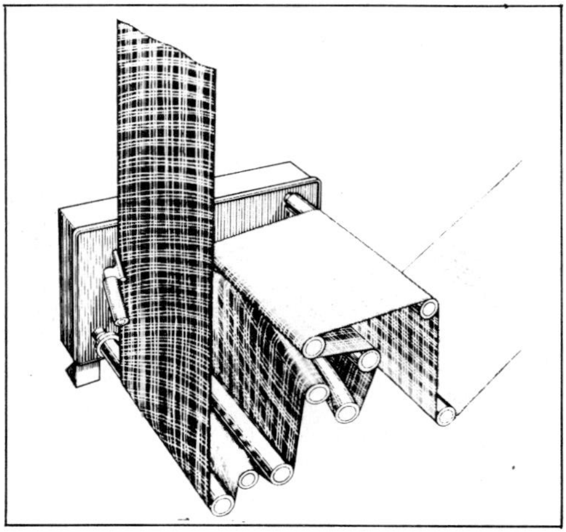

Figure 19 – Weft straightener

Weft Straighteners

Distorted weft has long been corrected manually by applying a roller eccentrically to the cloth. Thus, a skewed weft requires a skewed roller, and a bowed weft requires a hollow, or barrel-type, roller. In all cases, the principle employed is that those parts of the weft that are advanced are made to follow a longer path than the rest. This is illustrated in Figure 18.

Nowadays the detection and correction must be automatic, in order to accommodate high running speeds and the demand for accuracy. Detection is optical or mechanical, both methods being usable over a surprising range of fabric constructions and colours. Correction is carried out automatically by an array of skewed and bowed rollers mounted in the apparatus (see Figure 19).

Typically, the process is applied at the stenter, and is usually most effective when the cloth is damp from the pad mangle. Occasionally, for greater accuracy, a second correction is applied to the dry cloth at the exit.

5 NIPS

These simple mechanisms, fully described in Chapter 1, can give rise to running difficulties not easily traced. It is not always realized that a nip can cause a cloth to extend or shrink, depending on the path it takes. Ideally, the cloth running through a nip should be at right angles to a line joining the bowl centres. If this is not so, the deformation of the bowl coverings can act in the same way as a roller shrinkage machine. Particularly in the shrinking configuration, this can give rise to crease formation as the cloth relaxes and in the case of a double nip can actually cause transverse creases to be pressed in by the second nip. These points are best illustrated by referring to Figure 20.

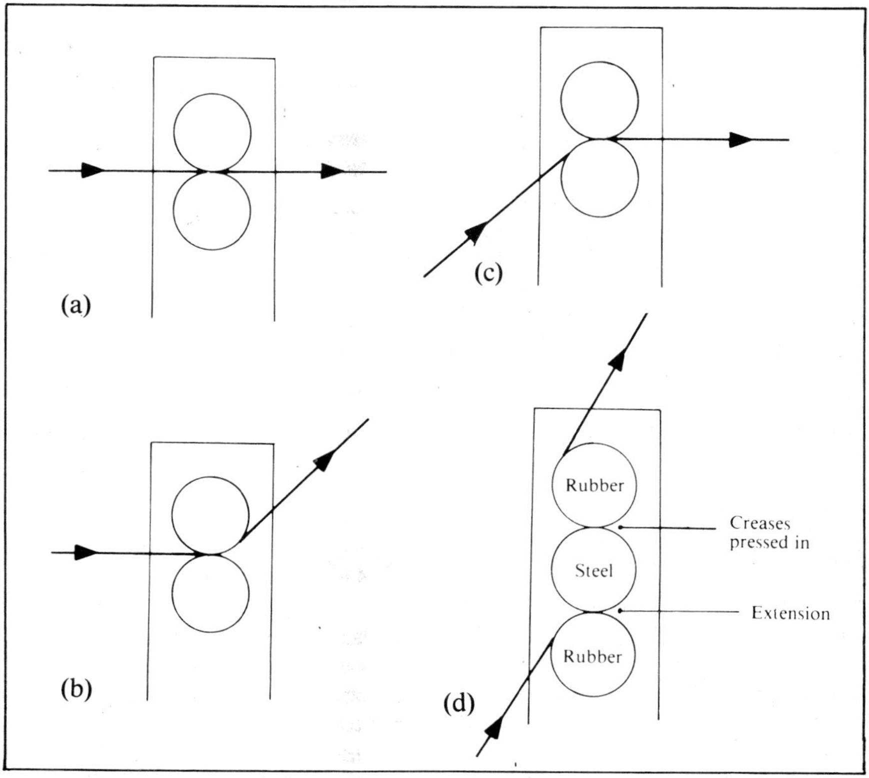

Figure 20

Back lapping, i.e. when cloth follows the bowl round instead of going where intended, is always a potential problem. Unfortunately, there is no simple method of avoidance, other than the provision and positioning of carrier rollers, together with the adjustment of tensions. If it should occur persistently, then the situation can often be improved by fitting a positive drive to the carrier roller after the nip, so geared that the surface speed is slightly faster than that of the cloth. Doctor ('stripper') blades have been used on nips, but are not popular because they need careful maintenance and tend to collect lint or dig in on the roller surface.

6 DELAY UNITS

Short-time devices have been discussed in Section 3. However, the wet processing of textiles is essentially a matter of chemical reactions, some of which take a considerable time, for example desizing, scouring, bleaching and dye fixation. The interactions of time, temperature and concentration apply, if loosely; frequently the processer would choose to extend the time, if the chemical usage were thereby reduced.

The earliest attempts to provide a substantial delay period while maintaining continuous processing were the Gantt piler, and its successor, the J-box. These are perfectly satisfactory in handling many fabrics, but a reasonable delay requires a large box with considerable fabric compression at the bottom; consequently, they are unsuitable for pressure-sensitive fabrics, such as some cotton poplins, or polyester blends.

Some interesting devices have been designed in the attempt to provide a delay period when processing in open width, and some of these are considered in detail below. The so-called 'shock' processes are a consequence of the difficulty in providing an extended delay, compensating as they do for a reaction time of only 1–2 min by very high concentrations of chemicals. In many cases, the result is inferior and the chemical cost is extremely high. Good reviews on the development of delay equipment have been published(*6*). However, it is worth emphasizing that the success of a particular type of machine is dependent upon the type of cloth being processed. The ideal does not yet exist.

Pad–Batch (Pad–Roll)

In this basically simple process, illustrated in Figure 21, the cloth is padded through the liquor and taken up on a roller, often in a steam-filled chamber, while the reaction occurs.

A typical batch contains 2 000 to 4 000 m of fabric, and will take perhaps 30 min to complete, and the same time to unroll afterwards. Thus, the inside end will always receive a longer delay than the outside and, in principle at least, the result can never be uniform. Problems can also arise from temperature gradients between the inside and outside of the batch, and between the sides (selvedges) and the centre.

Twin Batch

In the pad–roll system, as shown in Figure 21, the batch chamber is usually a separate 'wheel-away' chamber with the steam preheater section fixed adjacent to the pad. In the twin-batch system, the batch rollers are enclosed in a fixed chamber, usually in line with aftertreatment wash boxes; the steam preheating section is also part of the main chamber.

The batch rollers are mounted upon a carriage or 'turret' so that the batches may be rotated through 180° at intervals, the axis of rotation being mid-way between the two batches.

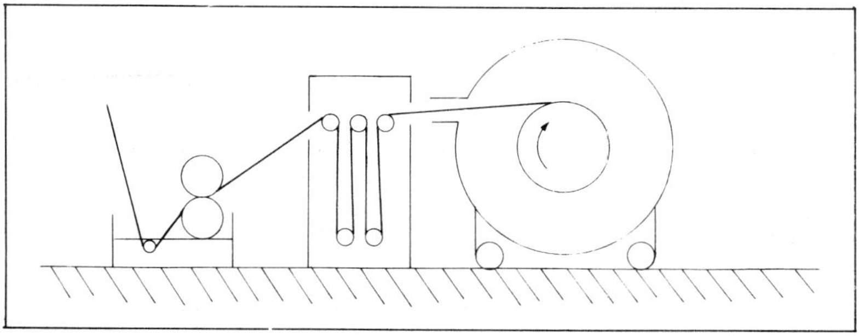

Figure 21

Figure 22 and notes below explain the operating sequences by which continuous operation is achieved. They are extracted from the Autobleach Manual by courtesy of Smith Textile Machinery Ltd.

1 Startup, with both rollers empty

Wind half the total yardage required for a full batch on to the top roller (Figure 22a). Stop the machine and rotate the batch arms to bring the empty roller to the top position (Figure 22b). Hand tuck a loop of fabric from batch A to be trapped on batch B (Figure 22c).

Restart the machine at slow speed and commence double thickness batching on the top roller. Stop the machine again when 5 or 6 yards of fabric remain on the bottom roller. Unwind this remaining yardage and 'thread' the loose end under the trip bar, over the guide roller, through the exit water seal and to the swing-arm compensator and draw nip by which fabric is pulled from the chamber (Figure 22d).

2 Automatic continuous run

Set control to 'Auto' and restart the machine; immediately, the fabric line from the top batch roller to the exit water seal (again Figure 22d) will tighten. The tension on the trip bar operates the batch rotating motor, revolving the batch arms 180° (Figure 22e). A loop of fabric from batch B is blown by a steam jet to be trapped on batch A (Figure 22f). Batch A then fills up in double thickness, taking the upper layer from fabrics entering the chamber and the lower layer from batch B, whilst the other layer from batch B is pulled to the exit water seal.

When batch B becomes empty, the sequence d, e, f begins again automatically, the machine slowing to a crawl at each change point.

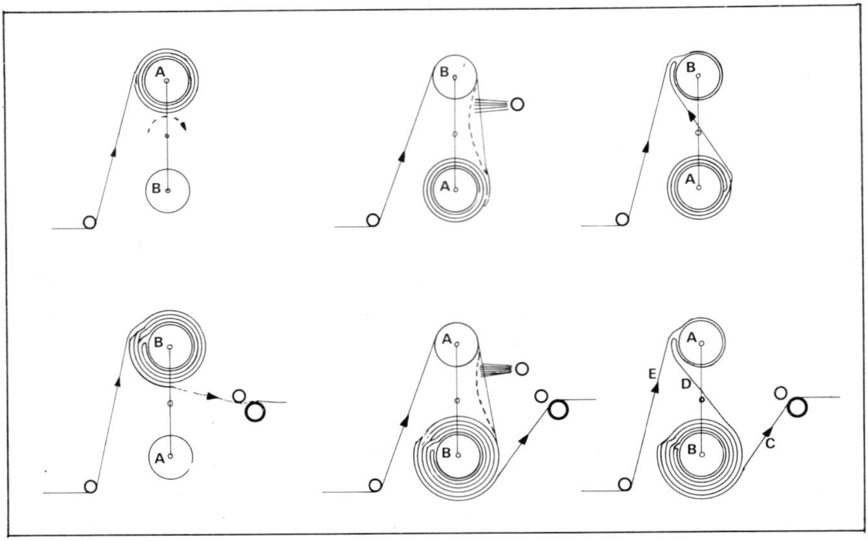

Figure 22

Practical difficulties which may arise with this type of machine are:

(a) Coping with small lots or orders which include marked changes in fabric width

(b) Fabric creases which tend to take place at the time of a batch changeover or whenever fabric tension is unduly tight or slack.

Roller Steamers

The cloth is laced between two horizontal sets of rollers, typically 1.5 m apart (Figure 23). If the rollers are spaced at 25 cm centres, one can readily deduce that this is an extremely bulky piece of equipment which, within a reasonable size and running speed, is capable of giving a delay reckoned in seconds rather than minutes. This is suitable for dye fixation, but not (with exceptions where cloth construction makes alternatives impossible) for preparation processes.

Even if double lacing is used (with consequent difficulties in threading up), the fundamental drawback is not overcome.

It is of paramount importance that there should be no creasing during dye fixation, as there is no access to the cloth during the process. The emphasis is on prevention and the comments in Section 2 are of particular importance, as is the accuracy of construction and erection of the machine.

During the passage of the cloth through a multi-roller machine, tension will inevitably build up. This is frequently reduced by driving some or all of the top rollers by means of a geared shaft, by belts or by hydraulic motors that can allow a degree of slippage.

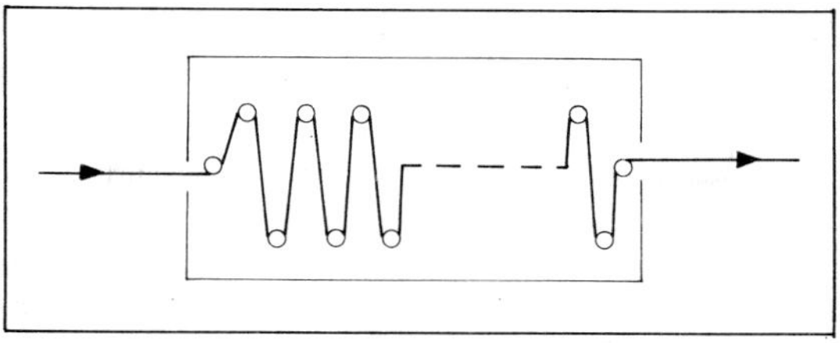

Figure 23

Loop Steamers

The cloth is caused to hang freely in loops between rods or rollers. The process is ideal for fixing prints, which require a long fixation period and one-sided contact. However, a huge steamer is required even at modest running speeds. The subject is fully discussed in 'Textile Printing'(*7*).

J- and U-Boxes

As mentioned previously, these are satisfactory for the preparation of fabrics which are not pressure sensitive. Various shapes have been proposed in attempts to obtain the smoothest passage of the cloth. The wet-bottom type is an attempt to reduce friction and to exclude air. A roller bottom may be used to reduce friction in the bend. The U-box is intended to provide a full dwell time, while reducing the static pressure on the cloth at the bottom of the pile. Given the appropriate cloth, all these configurations can work well, but there are always some cloths that will be permanently creased. Figure 24 illustrates these various configurations.

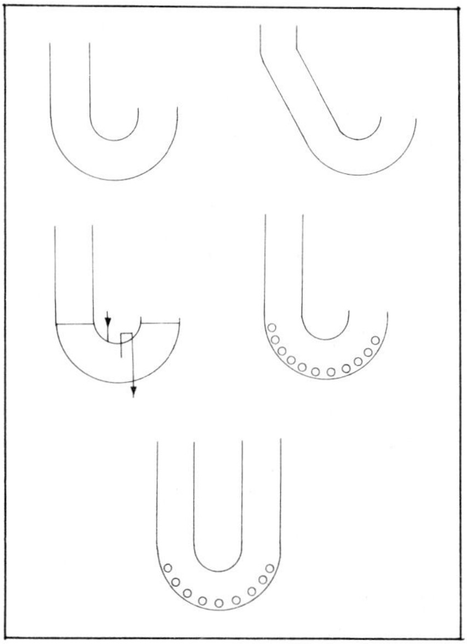

Figure 24

Relaxed Conveyor Units

While clearly unsuitable for dye fixation, these units do offer the delay periods needed for preparation. The earlier types were based on a slow-moving plain or woven-mesh belt, upon which the fabric was allowed to fall, forming a bed typically 10–20 cm deep (Figure 25). These have the disadvantage of a lack of movement between fabric and support, so that lying marks can occur. Woven belts always have the hazards caused by an unseen broken wire.

Recently the moving bed has been formed by closely set driven rollers (Figure 26). When the folds of cloth move from one roller to the next, move ment is induced within the cloth bed, promoting even treatment of the fabric.

Machines have been offered in which the space above the bed has been filled with tight-strand rollers, on the theory that preheating the cloth in the flat state tends to prevent the formation of lengthwise creases (Figure 27). This is a matter for continuing discussion and it is as well to be cautious when an additional complication is proposed.

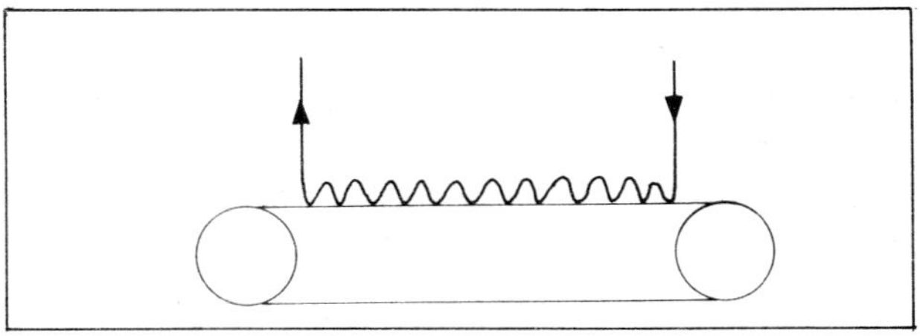

Figure 25

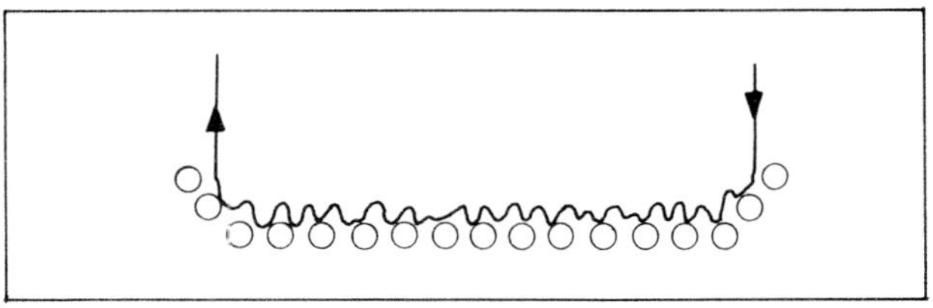

Figure 26

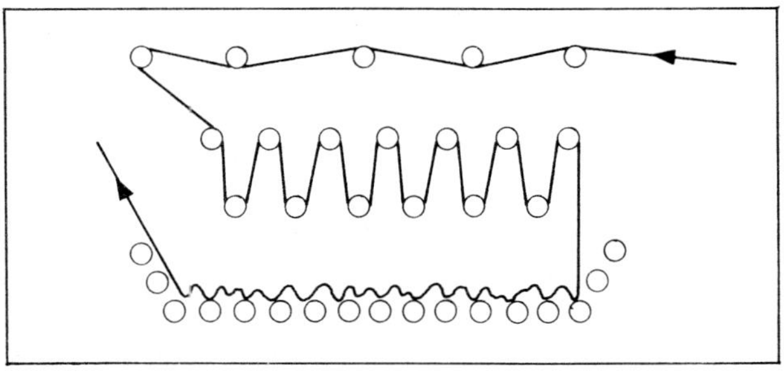

Figure 27

Problems have been encountered with slippage on the entry pull rollers. This has generally been caused by chemical factors, such as the use of certain surfactants,or over-activation of a peroxide bleach liquor, due to excessive alkalinity of the cloth, or a formulation that has been made too alkaline in the effort to obtain maximum bleaching in the time available. Various methods of plaiting the cloth on to the bed have not been proved to be of advantage – most fabrics behave satisfactorily when allowed to fall freely for 1 m or more and 'ripple' onto the bed.

The delay time available is not a function of the overall machine speed, but of the depth and length of the cloth bed. This is controlled by the speed of the bed rollers, and is a matter to be determined practically. Some fabrics form compact folds and allow a greater depth of bed; others do not. It has not proved possible to forecast the behaviour of particular fabrics.

Regaining control of the fabric as it is lifted from the bed has not proved to be very difficult if it was laid down correctly on entry. Pull rolls may be used, or the fabric may be pulled out by the subsequent nip. Any tendency to crease may be counteracted by a coarse-pitch scroll roller driven against the fabric, and a pair of guiders used as expanders.

7 OVERFEED DEVICES

When the dimensions of a cloth are being fixed, typically during stentering, it is frequently necessary to overfeed to reduce length, which has earlier been increased by running tensions. There have been various attempts to achieve overfeeding on clip stenters, but all have failed in practice, and a pin machine is always used. In a typical application, the cloth is gripped between a soft roller and a rubber belt and fed onto the pins at a rate faster than the pin chain is running (Figure 28).

It is important to realize that the indicated percentage overfeed is only an indication of the relative running speeds of the overfeed device and the chain. The overfeed achieved, which is often considerably less than that indicated, can be ascertained only by direct measurement of the cloth.

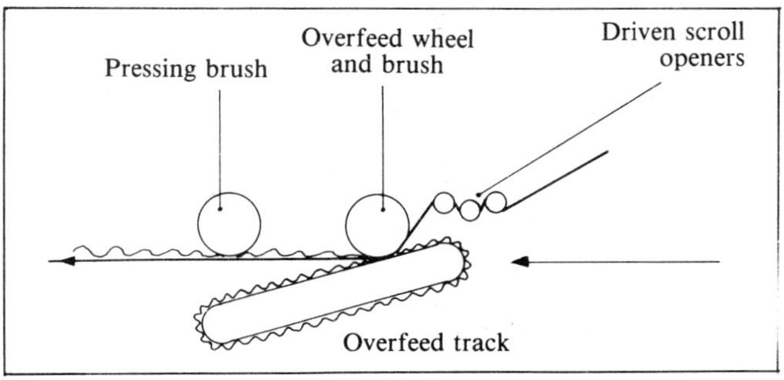

Figure 28

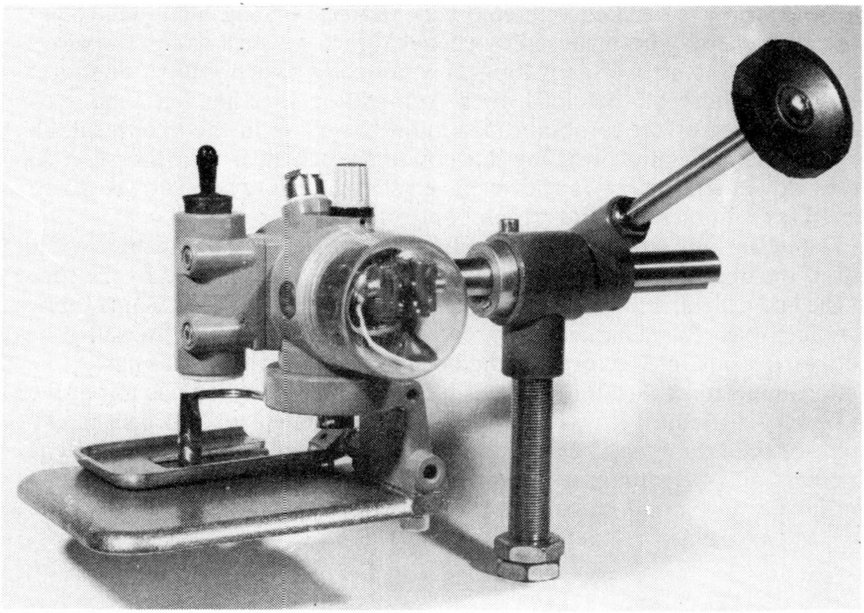

Figure 29 – Mechanical sensor

The cloth must always be pinned at the desired distance from the selvedge. Generally, this is achieved by detecting the position of the selvedge accurately by a feeler device, which may be mechanical for fabrics of firm construction (Figure 29), photoelectric (Figure 30), or occasionally, pneumatic (Figure 31).

If the selvedge deviates, the detector produces a signal causing the stenter rail to move back to the selvedge. This is often achieved by a rack and pinion arrangement, as shown in Figure 32, or by a hydraulic-pressure cylinder. The latter method has certain advantages in speed of response, and in the control of over correction or 'hunting'.

An interesting variant is the Ferranti/Dalglish guider. In this method, the selvedge is carried by a pinned disc, which is pivoted to follow it. The selvedge is carried by the pinned disc to the pins on the main stenter chain. The stenter entry rails, which possess considerable inertia, need move only for gross deviations of the selvedge.

Instruments are available, based on the degree of absorption of a beam of β-particles on passing through a fabric, indicating the weight per unit area. This is of particular value for knitted or coated fabrics, and the device can be made to control the overfeed or coating to produce fabric of a given weight automatically.

Figure 30 – Photoelectric sensor ▷

Figure 31 – Pneumatic sensor ▽

Figure 32 – Rack and pinion drive to stenter rail

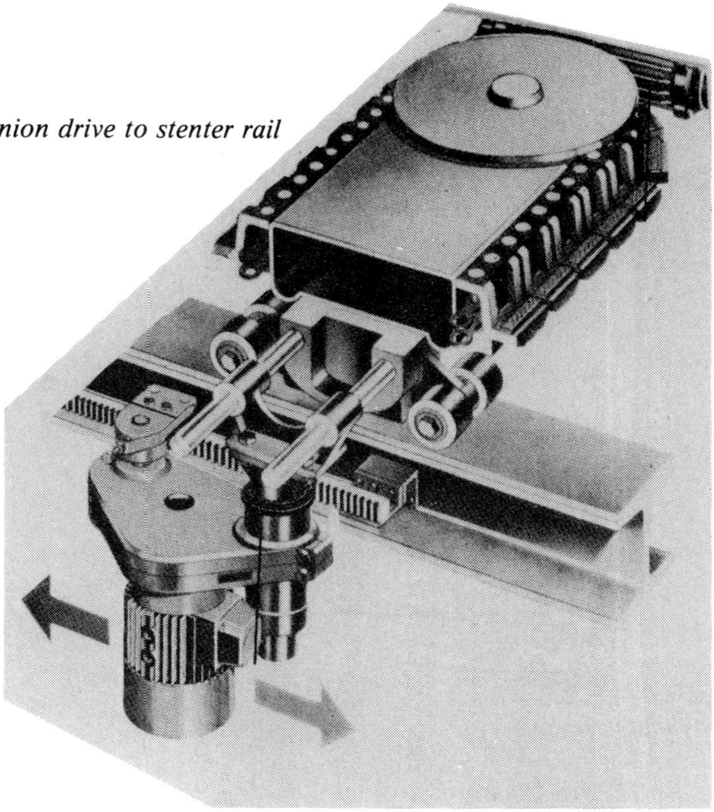

8 BATCHING AND PLAITING

The handling of fabric in big-batch form has become very popular. This convenient package allows fabric to be handled, transported, stored and unrolled in units of 1 000 to 5 000 m or more, depending on its thickness. However, the batch must be built up at substantially constant tension. As the diameter increases, the speed of rotation must progressively fall to maintain a constant surface speed. This has usually been achieved by the use of a pressing roller, laying the fabric on the batch and driving it tangentially. In this way only minimal trimming is needed between the surface speed of the batch and the speed of the associated machine (Figure 33).

Batch changing can be carried out without altering the speed of the machine, if an accumulator is provided, either in the form of a J-box or a roller type (Figure 34).

As an alternative to a pressing roller, the batch may be centre-driven. Early arrangements included d.c. motors, slipping clutches and magnetic

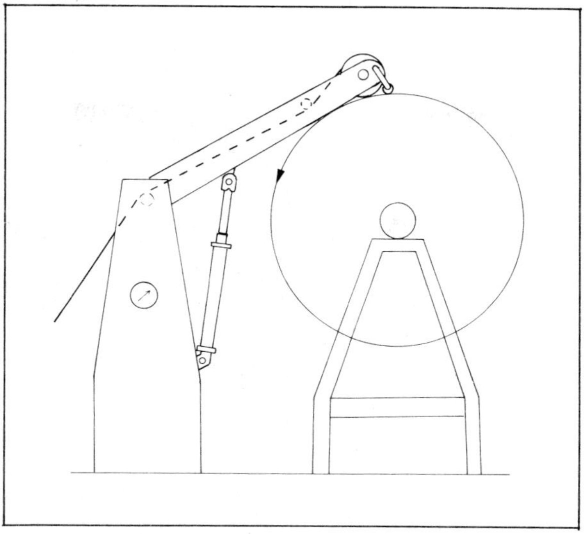

Figure 33 – Tangential drive

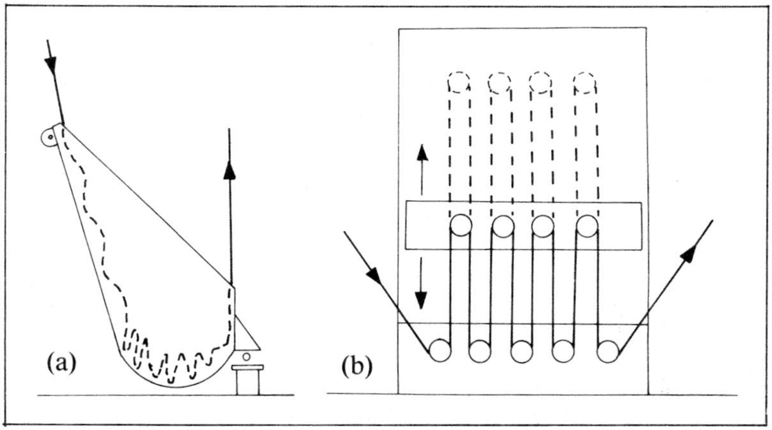

Figure 34 – (a) J-box changer (b) Roller-type batch changer

clutches. These are satisfactory for small batches, but cannot accept the combination of high torque and constant slippage required when driving a big batch. This problem has beeen solved by the use of a hydraulic motor drive, which combines high power with a controllable cut-off (Figure 35).

Even the most tension-sensitive fabrics, such as weft-knits, and woven cloths after compressive shrinking, can be batched and unbatched if a compensator, or a free loop controlled by a photoelectric cell, is included (Figure 36).

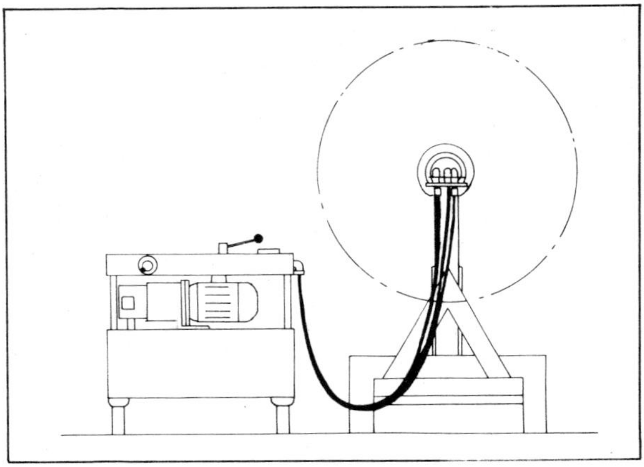

Figure 35 – Hydraulic centre-drive batching

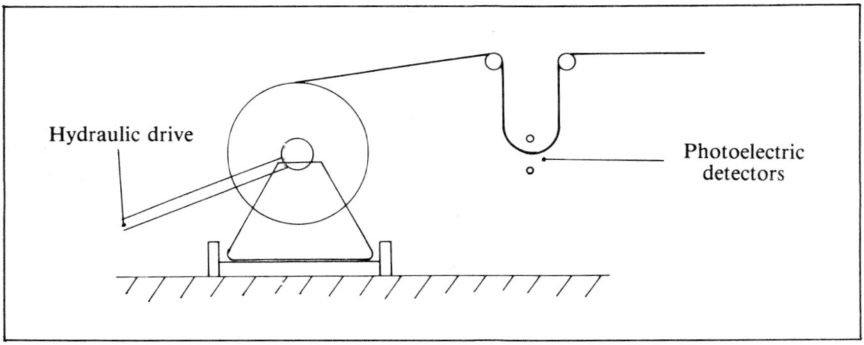

Figure 36

Fabric having particularly thick selvedges, such as the Sulzer type, can cause a problem. This can be alleviated if the cloth is caused to move slowly from side to side, the degree of traverse being equivalent to the selvedge width. A suitable method is to pass the fabric over a roller, one side of which is moved backwards and forwards by a slow-moving crank.

Another device, less likely to cause fabric distortion, is the oscillating roller (Figure 37). This is used for crease-free edge staggering of a fabric with thick selvedges during batching. The fabric is oscillated periodically thereby staggering the selvedges so that fabric distortion is avoided. The oscillating drive consists of a special rubber-covered roller, supported in spherical plummer-block bearings to accept both the rotating and axially reciprocating movement. The retractable reciprocating drive is directly coupled to the roller shaft and takes its own drive from the fabric-driven roller. The oscillating traverse is adjustable from 20–40 mm.

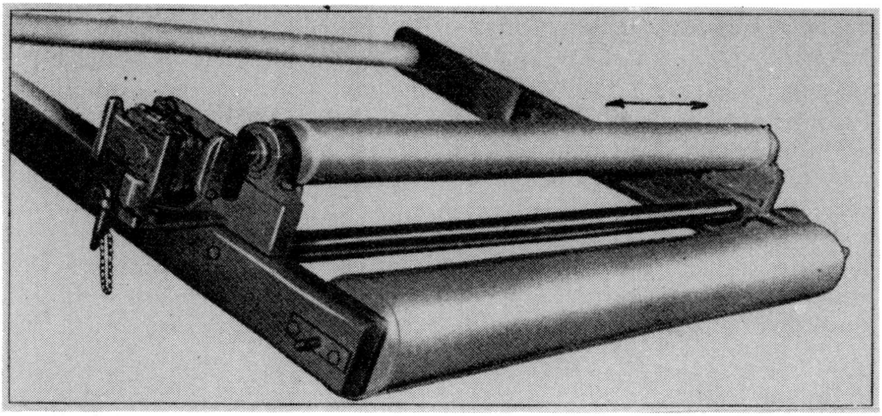

Figure 37 – Oscillating-roller batcher

Plaiting

The plaiter is a deceptively simple device which can cause much trouble. The simple box type is perfectly adequate for many dry, heavy fabrics, particularly if running tapes are threaded through slots at the bottom of the box (Figure 38a). For wet cloth, a pair of rollers, driven at a higher speed than the cloth, is generally used (Figure 38b). The length of stroke and the drop to the waggon are both important, but must be adjusted to suit local conditions.

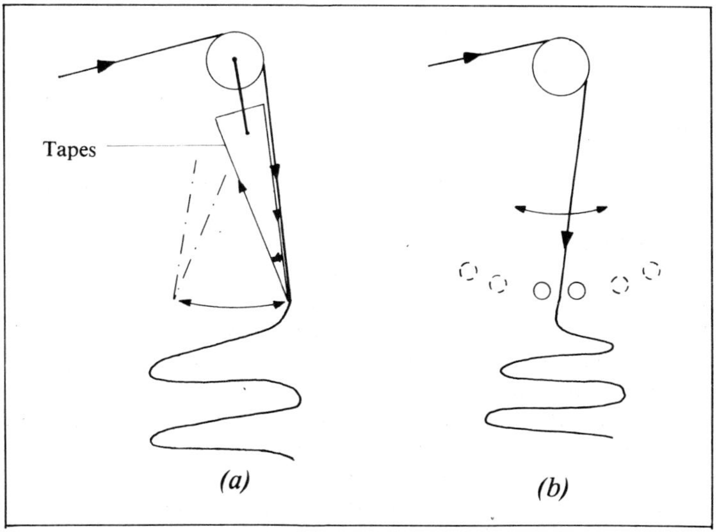

Figure 38 – Plaiting (a) for dry fabrics and (b) for wet fabrics

9 MISCELLANEOUS

Cloth frequently appears in which selvedges are woven with a higher or lower tension than the rest. Sometimes the fault is obvious, but sometimes it becomes apparent only on wetting.

Unfortunately, there is no real solution to this problem, other than applying sufficient tension during processing to stretch the tight part to the dimensions of the rest.

The tendency of knitted fabrics to curl is frequently overcome by applying strips of 'gum' from applicators, just inside the pins on stentering. The gum is fixed in the stenter, and the curled strip on the pins is cut off and discarded.

Deflecting Fabric

Fabric is generally intended to pass between machines in a straight line. Nevertheless, it is useful to remember that deliberate transverse deflections are possible, for example in accommodating the accidental misalignment of two machines. Considerable movement may be required to pass cloth between rooms and, in the extreme case, cloth may be turned back-to-face to allow the simultaneous inspection of both sides.

The change of direction is accomplished by passing the fabric round a pair of skewed rollers or bars, the exact position of which is best discovered by trial.

Seam and Metal Detectors

Mechanical finger detectors for rough, thick seams in a fabric have been available and are useful for relatively slow fabric speeds, as in schreinering, calendering and shearing. With high processing speeds and fine butt-edge sewings, more accurate detection is necessary, with a more rapid response in the ancillary control circuit. Pneumatic and optical detectors are available, but probably the most effective sensor is that based in the use of a seam-sewing thread plied with a fine wire. Passage of such a seam can then be detected electronically. The signal is amplified and used to operate the control system, which may, for example, provide automatic lift-off for calender bowls.

Metal foreign bodies present in the fabric may cause expensive damage, especially during such processes as calendering and shearing. A detector is available that detects the presence of minute particles of both ferrous and non-ferrous metals. The fabric passes over a full-width sewing head and any metal particle disturbs a magnetic alternating field. The impulse is amplified and may operate a relay to stop the machine. Indicator lights and marking devices assist in locating the metal particle.

A related device is that used to indicate the position of a seam in a dye vessel. A small PTFE-coated magnet, enclosed in a bag, is stitched to a seam. The passage of such a seam can then be indicated by means of an electronic sensor, which may be within the dyeing machine or mounted externally. This type of device is especially useful for jet dyeing machines to indicate:

1. The rope speed
2. Position of the seam for sampling
3. Position of the seam for unloading.

Fabric Slitters

These are extensively used at the delivery end of pin stenters, especially for knitted goods, where the fabric may require to be trimmed at the selvedge and/or slit into narrower fabric widths. Figure 39 illustrates the cutting principles and Figure 40 shows a typical cutting head.

To obtain a perfectly straight edge on knitted fabric the preferred methods is to lift the fabric from the pin chain before cutting. A suction device disposes of the cut edge. Figure 41 shows an installation and how it operates.

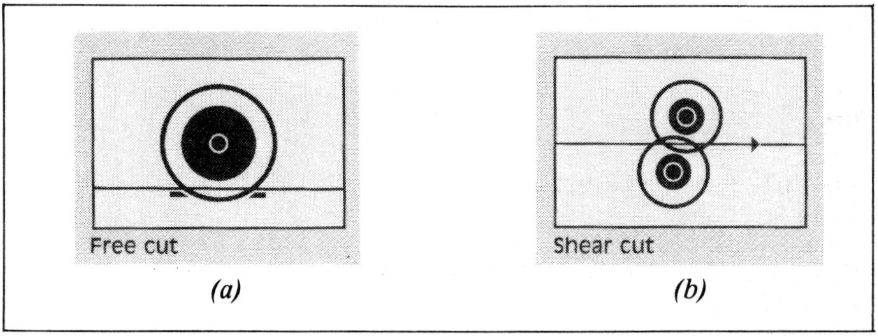

Figure 39 – (a) Circular-blade cut. Using this method a single circular blade slightly penetrates a slot in a cutting base; this simple cutting technique is only suitable for relatively thin material

(b) On the circular-knife scissor-cut principle, two circular knives operate together in such a way that the two cutting edges overlap slightly and are thereby lightly pressed against each other, this enables thin as well as thick and relatively rigid material to be cut

Figure 40 – Typical cutting head

Figure 41 – Device for lifting fabric from the pin chain prior to cutting, inset shows principle of operation

REFERENCES

1. USP 3 308 639 and BP 1 076 680.
2. ICI Fibres Manual, L11/1.
3. Deutsche Babcock 'Tattersall' type.
4. Marsh, 'An Introduction to Textile Bleaching' (London: Chapman & Hall, 1951) 417; Int. Textile Bull., (1967) 1–15.
5. Gebert, Shirley International Seminar, April 1976.
6. Anon., Int. Textile Bull., **1** (1977) 29–36.
7. 'Textile Printing', Ed. Miles (Bradford: Dyers Company Publication Trust, 1981).

Chapter 5

Instrumentation and Process Control

M G MYLROI

INTRODUCTION

Automation is no longer a 'new' subject – in fact it has not been new for some thousands of years. Looking back only one generation it is sobering to realize that if we were to rely upon manual telephone exchanges and prewar car-production techniques, the whole of our population would be employed just to provide these two facilities.

Automation is concerned with devices and systems to replace or aid human mental, as distinct from physical, effort. An ordinary machine can take the heavy work off human shoulders, but it cannot take decisions – it is this power of decision now given to a machine or device by automation that distinguishes the technology. In general, man's physical strength was augmented by mechanical devices and then power-driven machines, mainly steam power, and this was the first Industrial Revolution. As speed of manufacture and quality of product increased, man's ability to control a process operation became the limitation – man cannot observe all the details quickly enough, cannot react quickly enough, and his brain cannot make decisions quickly enough. Added to this, due to nervous fatigue, he makes mistakes, he needs an aid to his senses and brain: automation, possibly the second Industrial Revolution.

Whether we like it or not, automation is with us and its applications are inevitably increasing. The large chemical plants and oil refineries have been virtually automatic for many years, with raw materials being fed into one end of the plant and products taken off at the other, completely untouched by hand, with operating conditions automatically controlled without manual adjustment. But it is wrong to think of automation as being applicable only to large plants and used only by large organizations; in some form or other it must be applied in all industries and by all firms, large and small.

In a highly industrialized world, to survive we must increase productivity markedly and we can do this by mechanization, increasing the rate of technological change and vastly increasing the use of automation. The sophisticated control and, in particular, computer control systems now used in the leading chemical plants, oil refineries, steelworks and papermills are fine examples and clearly demonstrate the very high level of technical competence that can be reached. But these firms in total account for less than one-third of our gross national product. Over two-thirds of our wealth, if that is the correct word, is the responsibility of a host of smaller processing firms and general engineering industries, and it is these smaller firms that appear to be slow in taking advantage of automation as a means of increasing productivity. Here it is that we find fuel wastage (because quite simple automatic process controls are not applied), low yields of saleable products (because process variables are inadequately controlled) or human effort wasted because mechanization is inadequate; or we find people, capital and materials wasted because production, stocks and machine capacity are not measured and managed in a way that modern control and information processing now make possible.

There are many reasons put forward to explain why automation is slow to proceed, such as social problems and capital shortages, but an important reason is probably that automation, more than any other technology, requires a close combination of effort by people of many different skills and disciplines if it is to be used successfully and this is not easy to achieve. However, although this feature does present a major difficulty, it does at the same time make the job of the real control engineer an unusually interesting one and it is worth noting that in those industries, such as steel and chemicals, where control engineers as such are employed and have been introduced alongside production managers, process technologists and plant designers, the most significant developments in the application of automation have taken place. This suggests that the use of automation has to be stimulated from within an industry and cannot simply be applied by edict from without. The basic problem is to make those familiar with the traditional production and process methods of an industry conscious of the technology of automation and the use that can and must be made of it. Equally, the control engineer must be within the firm, as he must learn something of the business and processes of the user industry before he can apply his technology fully; this presents a particularly acute problem for the small- to medium-sized firm.

Let us now consider plant control, and I say plant control not automatic control because there is no essential difference between manual control and automatic control, the purpose and the difficulties are the same. The human operator can in fact, if really knowledgeable and efficient, use his experience and judgement so as to be more versatile and flexible in his approach to a problem than a simple automatic controller and so achieve equally good or sometimes better control than the automatic system over short periods. And this is the key, over short periods. If the process requires the operator to give constant attention then he tires and his performance will fall off. If the process is very docile then the operator gets bored or complacent and surely it is wrong to employ a human being and leave him to get bored. As like as not, if for hours on end he has no corrective action to apply to the process, he will assume that the process will continue steadily, so that when action is required of him he is not ready, but sitting quietly in some corner or other considering his football pools, or drinking his tea!

The automatic controller, even the simplest system, ensures at least adequate control of the process and maintains this consistently hour after hour, day after day.

In many modern plants this simple control is not adequate, nor is manual control. The complexity of the process and the speed and precision of the control required makes it virtually impossible for manual control to suffice, and sophisticated control systems, often using computers, are now necessary.

The following advantages might and should result from automatic control.

(a) Smoother running, increased production and improved quality
Automatic control is consistent over long periods, why is this an advantage? Obviously, if a plant is running at optimum conditions, then any variation from these must be detrimental, with lower throughput and more 'off-spec' materials.

Erratic operation allowing changes in plant conditions can cause quicker coking up of preheater tubes from temperature peaking, and temperature peaking together with variations in reaction-gas impurities could accelerate catalyst ageing.

Smoother running with less variations from set conditions can allow plant to be operated nearer to safety limits, which could give higher yields, etc., without reducing the safety margins.

(b) Savings in plant costs
If a process is designed to operate at a certain pressure and fluctuations arise because of lack of control, the plant vessels, pipework and equipment must all be designed to withstand the maximum peak pressure that might possibly occur, adding cost to the building of the plant. Similarly, if flow rates and levels in vessels are held steady by good control, then surge tanks and buffer vessels can be made smaller, giving a definite and direct saving in initial plant costs.

Again, unstable conditions lead to peaks in process temperatures, pressures, acidity, etc., and these all take their toll of plant equipment giving increased wear and tear, requiring extra shut downs and hence loss of product, replacement charges and maintenance costs.

(c) Savings in raw materials
Operations away from the defined optimum conditions must mean less efficient use of raw materials and possibly excessive use initially of heating and later cooling. Particularly with boiler-type plants, where fuel and air are used for steam raising or initial process heating, there is frequent energy wastage. These plants must run with excess oxygen present otherwise there could be a build up of combustibles leading to an explosion, but, with automatic control, this oxygen excess can be as low as 3%, whereas on

manual control excess oxygen is usually at 10%, a real wastage of fuel, with possibly 30% extra fuel being used.

(d) Increased safety
With continuous control there is less likelihood of excess peaks of temperature, pressure, composition, etc., and hence less likelihood of hazardous conditions arising; also the continuous measurements required for control give continuous monitoring of conditions and can be connected to alarms.

(e) Improved working conditions
People are not required to work continually outdoors on the plant itself, manually operating valves, reading process conditions, sampling, etc., but can work in the control room away from the noise, dirt and climatic conditions.

(f) Saving in manpower
This is seldom a major consideration in the decision to install automatic control. If the installation of automatic control decreases the number of men on a plant, the economic considerations are easy to assess. They must, however, be interpreted in the widest sense and surely it is sensible to release people for work of a more demanding nature. Frequently, additional skilled men will be required for designing the control systems and then their installation and maintenance. Usually the installation of automatic equipment proves economic by giving a better-running plant with a higher, better-quality output and lower fuel and raw material costs, not simply a saving in labour costs.

(g) Improved plant management and supervision
Invariably, an automatically controlled plant provides more and better information on plant conditions and this leads to management having greater knowledge, leading to a real improvement in plant design and operation, even if difficult to define and quantify. The process operator usually gets the plant information presented to him on a small panel rather than the scattered information given on manually controlled plants, so allowing him to assimilate plant conditions more quickly and better, again leading to an overall improvement in plant performance.

I think we can assume that more automatic control will be used on both batch and continuous processes, and that with automatic control many products at present produced only in batch processes could be manufactured on a continuous basis with marked improvements in productivity and energy saving.

1 PRINCIPLES OF AUTOMATIC CONTROL

Control is now part of our everyday lives; the word itself is used not only by engineers, but by economists, politicians, managers and a host of other people. Although these people come from different fields of work, have different backgrounds and are handling different commodities, they all exercise control to produce a desired objective. Before designing a control system one must therefore always decide what the objective is, and consider if it is attainable, otherwise intermediate realistic objectives must be sought and any constraints limiting the range of possible control actions must be taken into account.

Any control system may be divided into three component parts.

1. The sensor, which detects a change in a process variable and which produces an output signal proportional to that change.
2. The controller, which receives the signal and, after comparison with a desired set value, emits an output correction signal that is proportional to the deviation from the set value.
3. An actuator and servo device – most commonly a power-operated valve – which receives the correction signal and mechanically adjusts the process variable.

This chapter deals primarily with controllers, with appropriate references to related sensors.

In a process, whether simple or complex, there are always a number of conditions upon which the efficiency of the process depends. Some of these may be controllable, such as feed rate, heat input, operating pressure, etc., and it might be expected that if each of these were to be automatically controlled then the whole plant would run steadily and efficiently. In practice, however, this does not happen as the control of each variable may not be perfect nor will it necessarily be possible to control all of the other variables; ambient conditions, fuel and raw material compositions, etc., will vary and changes will take place within the plant itself as catalyst activity changes or dirt deposits build up in a heat exchanger.

In a simple control system, therefore, current practice is to control at a constant value all but one of the conditions that can be controlled and that materially effect the process. The remaining condition is then varied to maintain the 'most important condition' of the plant at a desired value. This most important condition could be the composition of the final product, although in practice product quality is not always easy to measure quickly and on-line, hence some related property, such as temperature, is often varied.

As a simple example let us consider a reaction where two raw materials, A and B, are heated in a reactor to produce a product P (Figure 1).

It might be thought that, if the feed rates of A and B could be held constant, and a fixed quantity of fuel supplied to the reactor, then the required product P would be produced. However, the composition of either or both

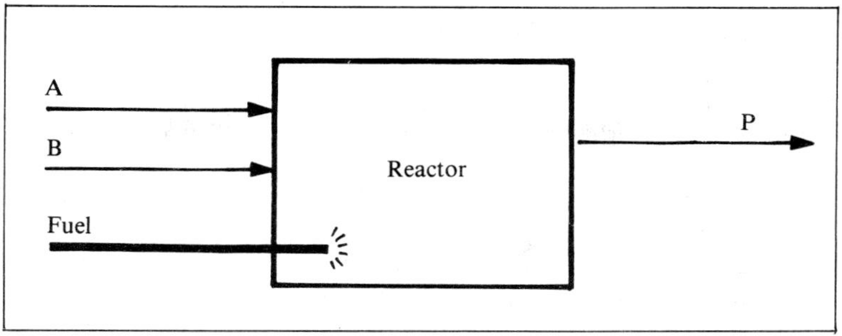

Figure 1 – Simple reaction between two raw materials

of A or B could vary as could the quality of the fuel supplied, hence the product quality would vary. The procedure would not therefore be to keep all three inputs constant, but only two, allowing the remaining one to be varied to give the required product. After consideration of the likely variations in composition and conditions of all three inputs, A, B and fuel, it might be decided to hold the feed A and the fuel supply constant and to allow B to be corrected, or varied, to give the required output, P.

If, however, we could measure accurately all the input conditions, i.e. feed rate and composition of A, B and fuel, and if we knew or measured the ambient conditions and fully understood how the reaction took place, then we could control or vary all of the input conditions to give exactly the output required.

We are beginning to talk now of open- and closed-loop control, or feed-forward and feedback control – what are these systems and what advantages or disadvantages do they have?

Let us consider from now onwards the very simple process of providing a supply of hot water.

Cold water flows into a stirred tank heated by a steam coil and hot water is drawn off. Let us assume that the water level in the tank is maintained at a steady level by a simple float-operated control on the cold-water supply.

How do we go about controlling this process to produce hot water at a desired temperature, T_d, from an inlet supply of cold water, T_i?

One way is to measure the output water temperature, T_o, and, if this temperature is lower than the desired temperature, T_d, then feed back a signal to increase the steam flow, or, if T_o is greater than T_d, then decrease the steam flow, as shown in Figure 2.

Here the output temperature is detected and measured and the measured value compared with the desired value, T_d. An output signal is then produced to cause the steam supply control valve either to open a little or to close, so providing more or less steam than previously. This change in the steam supply to the tank will eventually either raise or lower the water temperature and this change in the water temperature will again be detected at the output

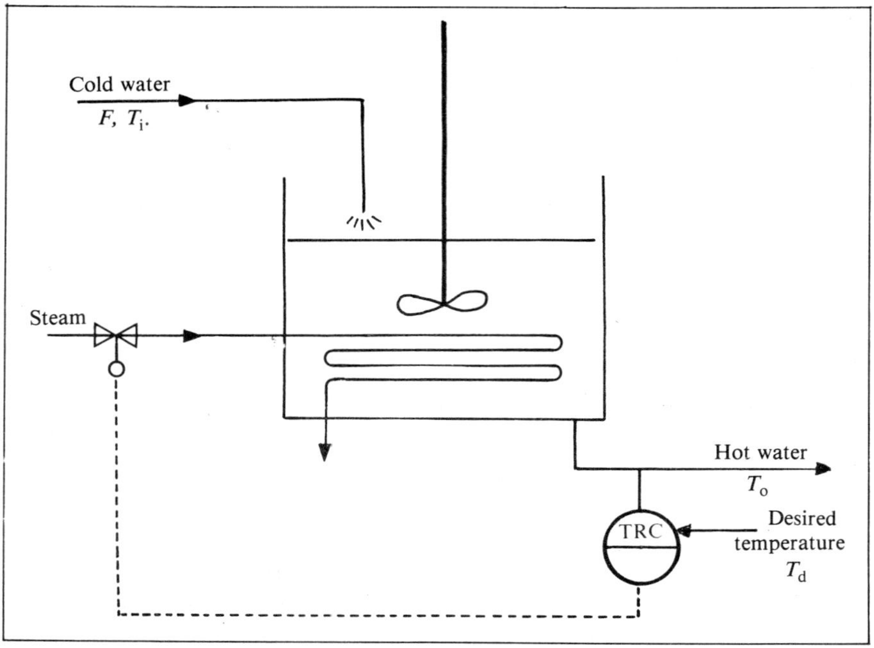

Figure 2 – Temperature-control system

and further corrective action, if necessary, will take place. We have here what we term feedback control or closed-loop control. Information is detected at the output and fed back to modify the input, causing a change in the plant condition and this changes the output, which in turn is detected as before and further information fed back to the input, hence feedback control. It is also called closed-loop control because the detector gives information to the controller, which then feeds back a signal to the input. The input is varied and so alters the plant throughput; this affects the output which is then detected again and so the cycle continues, i.e. a closed loop as shown in Figure 2.

How else could one control such a water-heating process?

Well, you could measure the inlet flow of cold water and calculate the quantity of heat required to raise this water from the inlet temperature, say T_i, to the required output temperature, T_d, and position the control valve in the steam line accordingly, as shown in Figure 3.

For success the strategy depends on the accuracy with which one can calculate the steam flow-valve position from a knowledge of the cold-water feed flow. It assumes also that outside influences are negligible and no checks are made at the outlet; this strategy is called open-loop control.

However, changes do occur in outside circumstances and these will affect

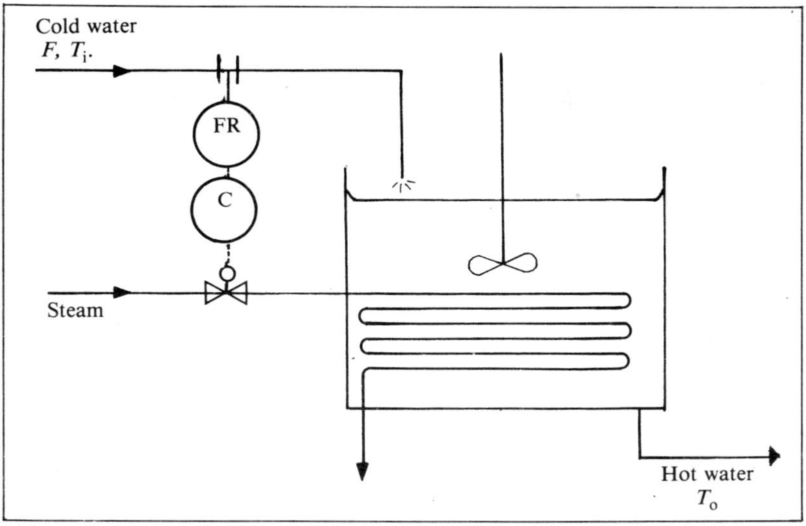

Figure 3 – Open-loop control

the system; for instance, the temperature of the inlet cold water will vary, the pressure in the steam main will vary, heat losses from the tank will vary, and so on. If we can measure these outside influences and then calculate their effect, we could possibly adjust the steam valve to compensate for them and so produce a strategy for control which takes external influences into account; this is called feed-forward control.

If, as we have said, the input measurements are correct and our knowledge of the process and valve characteristics are correct then the output temperature should be T_d, but no measurement is taken of the output and no correction supplied if $T_o \neq T_d$. The feed-forward and open-loop control systems rely upon accurate information and a correct 'model' being available of the process and control system. Is such a system likely to be used? Yes, but not in simple process control systems. In economic control systems, however, and not only these, it is essential to apply this sort of control since the effects of any changes made take a long time, possibly years, to feed through the system and hence feedback control would be impossible and some sort of control based upon forecasting or modelling, i.e. feed forward, is essential.

Let us look further at our simple system and consider the advantages and drawbacks of our basic systems, open- and closed-loop control.

In the closed-loop or feedback system, the output temperature has to vary before any correction takes place, since corrective action is initiated by there being an error, i.e. a difference between the actual output, T_o, and the desired output, T_d. Suppose that the system is running steadily with the steam valve say half open and the output temperature as required at T_d. Now let us consider what happens when more hot water than usual is drawn from the tank. The tank level would begin to fall, but the ball float valve would open and so more cold water enters the tank; this mixes with the hot water in the tank and slowly cools it and the outlet temperaure begins to fall. This temperature fall is detected by the temperature-sensing element and the controller acts upon the change and a signal is sent back to the steam-control valve, opening it a little to provide additional heating. However, this additional heating is only being supplied after the outlet temperature has already begun to fall; it is correcting an error.

How important is this? It depends very much upon the speed with which the error can be detected and the correction applied. In the case considered the temperature fall will not be detected immediately, as we shall see later in the section on controller actions, the temperature-detecting element is not immersed directly into the hot water, but in one or even two protective metal sheaths. It takes some minutes, therefore, for a temperature fall to be noted at the detecting element, which then sets the control system into action. The steam valve now opens a little and more steam flows through the heating coil, but it again takes time for the heat to be transferred through the coil to the water and this heated water to be mixed throughout the tank. During this time, additional cold water is entering the tank and cooling the water, so the water temperature is liable to fall away until the measuring and corrective action really get under way and restore the situation. This lag in the system is a major problem with feedback control, and delays in sensing systems and time lags through a plant must always be taken into account when designing a control system.

How would open-loop or the feed-forward system react in a similar situation? The answer is: much better for two reasons. Firstly, the instant the cold-water flowmeter shows an increase, the steam flow will also increase (see Figure 3), so that the additional steam heating does not have to wait for the outlet water temperature to fall before the correction is applied; it is applied immediately. Secondly, unlike temperature measurement, flow measurement is virtually instantaneous so that as soon as the cold-water flow varies it is detected and extra steam heating is supplied. Feed-forward and open-loop control, therefore, respond much more quickly than feedback control; they anticipate conditions and do not have to wait for an error to arise before starting to apply a correction. Remember of course, that the amount of this correction depends upon the validity of the model of the system embodied in the controller, there is nothing to say that the exact correction has been applied. A combination of the systems would seem the ideal answer and this is what happens in the more complex control systems, especially computer control systems. However, to operate feed-forward control successfully does mean being able to describe the plant and the control system fairly accurately in mathematical terms, i.e. being able to build a model of the system, and this is no easy task. The model can be built up theoretically from knowledge of the process and the detecting and controlling elements of the control system, or by experimental techniques performed on the plant itself. However, we are concerned with the simpler, single-loop feedback systems, which for years have carried out the majority of the automatic control used by the process industries.

2 TYPES OF CONTROLLER ACTION

By consideration of the water-heating process shown in Figure 2, we will discuss the various actions required from a controller in this single-loop, feedback control system.

On/Off Control

Here we assume that the controller can only switch the steam-control valve either to the fully open or to the fully closed position.

Consider what happens when we start with a tank filled with cold water and we require the exit water to be held at 80°C. The sensor detects a water temperature below 80°C and therefore switches the control valve fully open, the temperature of the water in the tank, and the water flowing out from it, will rise. When the water temperature at the sensor reaches 80°C the controller will close the steam valve but, because of the heat capacity of the steam coil, the temperature of the water will continue to rise for a while and then begin to fall. When the temperature falls back to 80°C the controller will switch the steam valve to fully open, and so on. The temperature will fluctuate as shown in Figure 4, not a very satisfactory state of affairs.

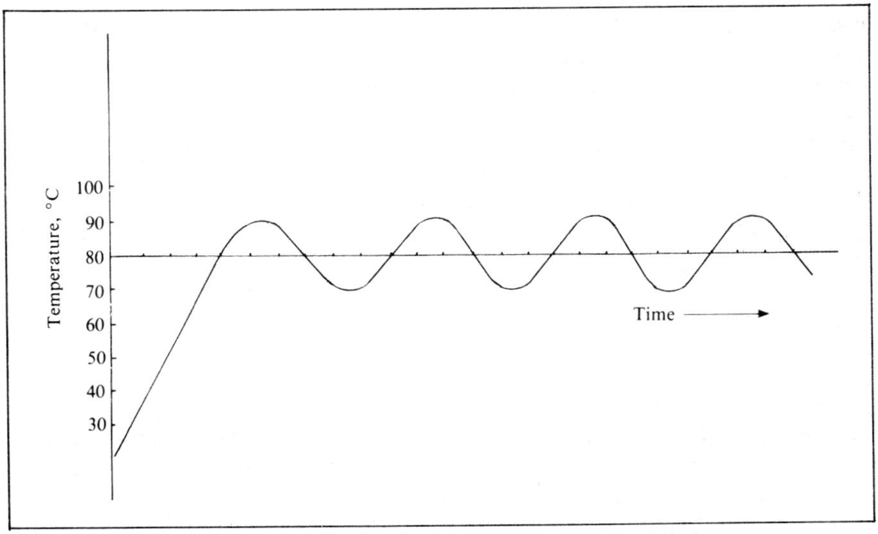

Figure 4 – On/off control

Must it always be like this? Obviously not, as anyone with central heating at home knows. Most central heating systems in the home are controlled by a simple thermostat, i.e. an on/off controller, and the home temperature does not oscillate widely. But, we must also remember, the home does not heat up quickly. If the house has been left with the central heating switched off for a week, say at Christmas time, then, on returning home and restarting the heating system, it takes a long time before the house is really warm. Then the controller takes over and the temperature will remain reasonably steady at, say, 20°C. Why is this? There is little capacity in the heating system and a large heat capacity in the house itself so that the temperature rises only slowly, but equally when the thermostat switches the heating off, the temperature only falls slowly giving a reasonably stable house temperature, as shown below in Figure 5.

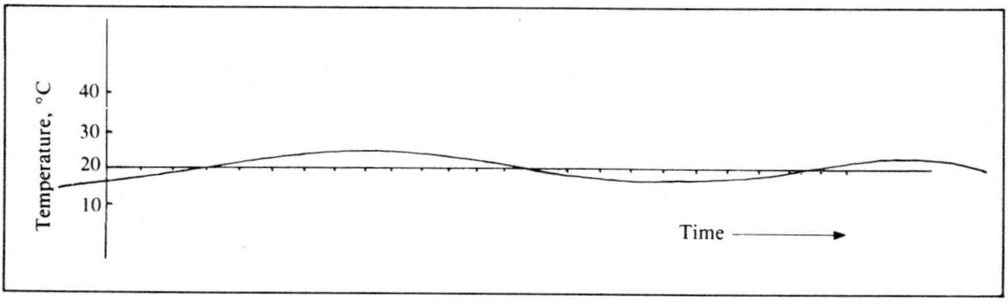

Figure 5 – On/off control

Is this type of control action of use in industry? Within limits it is. Wherever we have a system where the capacity is large and the variables that change and correct the system are small, then on/off control will provide a simple control system which, because of its simplicity, is usually cheap and reliable.

Examples are to be found in the temperature control of yarn in the textile industry, temperature control of large storage tanks, where a small heating coil is used to keep the stored liquid at a temperature above freezing, say 3°C, and many level controls in industry, particularly in intermediate storage vessels, where the exact position of the level is not critical.

Usually, however, a slightly better control is required and what is called proportional control is used.

Proportional Control

Here, in the example we are considering, the position, or the opening, of the steam-control valve would be dictated by the temperature difference between the actual temperature, T_o, and the desired temperature, T_d, i.e. the correcting action of the controller would be proportional to the error, or expressed mathematically $V_1 = -K_1 T$, where V_1 is the controller output, T the error ($T_o - T_d$), and K_1 a constant.

Let us assume that, as before, we require the temperature of the tank to be maintained at 80°C and that there is a steady flow of cold water into the tank and hot water out from it. If the temperature at the outlet, T_0, happens to rise a little, to say 81°C, then the valve will close a little to correct this error. Similarly, if the temperature falls, then the valve will open. What exactly do we mean by saying that the valve will open a little or close a little? The designer of the control system has to decide how much control action is required. He can build into the system the condition that if the temperature falls from 80 to 70°C then the steam valve will be fully open, and if the temperature rises from 80 to 90°C then the valve will be fully closed, i.e. the designer of the system arranges for the valve to move from fully open to fully closed as the temperature changes from 70 to 90°C.

If we now consider what happens when the system is started up from cold, the valve will remain fully opened until the temperature reaches 70°C, and the temperature will rise as it did in the on/off control situation. At 70°C the valve will begin to close, the temperature therefore rises less quickly, but at 80°C the valve will still be half open and only as the temperature rises above 80°C will the valve close further; hence the temperature will, in the first instance, rise higher than with on/off control as shown in Figure 6. The temperature after reaching its maximum will fall at a slower rate than with on/off control, since steam is still being supplied to the coil and, after two or three oscillations, the temperature will settle out at the required 80°C with the valve half open.

This assumes that the conditions are as calculated, i.e. that the heat input from the half open valve causes the temperature of the inflow cold water to rise to exactly 80°C and compensates for any heat losses from the tank.

What will happen if less hot water is required so that the inlet flow rate drops, say to half the previous flow rate. The temperature which has been

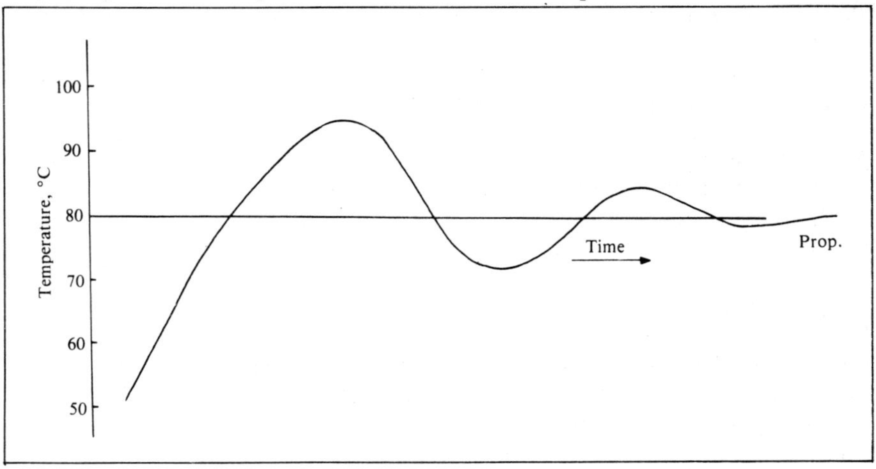

Figure 6 – Proportional control

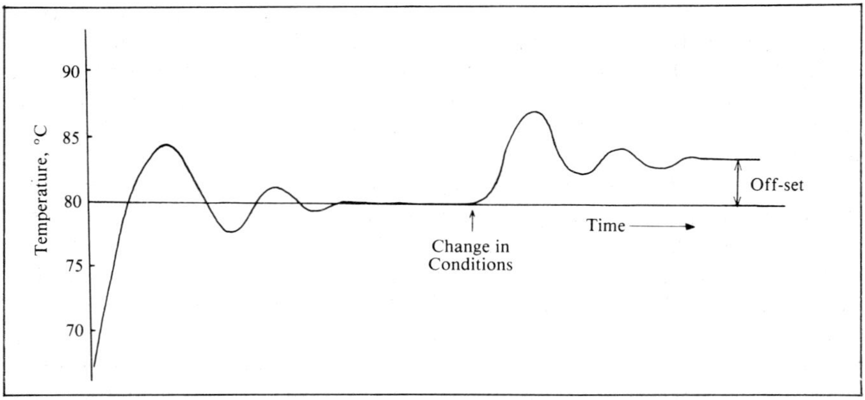

Figure 7 – Proportional control showing off-set

steady at 80°C will obviously begin to rise when less cold water enters the tank, the control valve will close slightly, the water will begin to cool and the valve re-open a little and, after a few oscillations as before, the temperature will once again settle out – but not at 80°C – at some higher temperature, see Figure 7.

Why is this?

Since we have only half the previous rate of cold water entering the tank we require only about half the heating and so we require the control valve to be at about a quarter open, not half open as before. We are *asking the controller to apply a correction*, and with proportional control the correction is proportional to the error and hence, if we must have a correction, we must settle for an error, i.e. we will be controlling at a point other than the desired point.

Let us reconsider this position. Table 1 shows design conditions in which the valve is fully closed at 90°C, fully open at 70°C and fixed at various intermediate positions depending upon the temperature.

TABLE 1
Valve Position vs Temperature

Temperature (°C)	Valve position
> 90	Fully closed
90	Fully closed
85	One-quarter open
80	Half open
75	Three-quarters open
70	Fully open
< 70	Fully open

Because the inlet water flow has changed from F to $F/2$, we wish to adjust the steam valve from half open to roughly one-quarter open. From the table we see that for the valve to be almost one-quarter open the temperature must be around 85°C, i.e. we can have steady control, but at roughly 85°C not 80°C. We have what is called an off-set and this is the problem with proportional control.

For a sustained change in conditions a correction is required, but any correction that is applied is proportional to a deviation, therefore to provide a steady correction we must have a steady deviation, i.e. an off-set.

We could, of course, have designed the proportional controller with different constants. We could have said that the valve will be half open at 80°C as before, but fully open when the temperature falls to 78°C only and fully closed if the temperature rises to 82°C. The control action will be now much more like on/off control as shown in Figure 8, but when the rate of cold water flow drops from F to $F/2$ then, after settling out, the temperature will be steady at about 81°C, i.e. we have reduced the off-set, we are nearer to the desired control condition, but the control is nearer to becoming unstable.

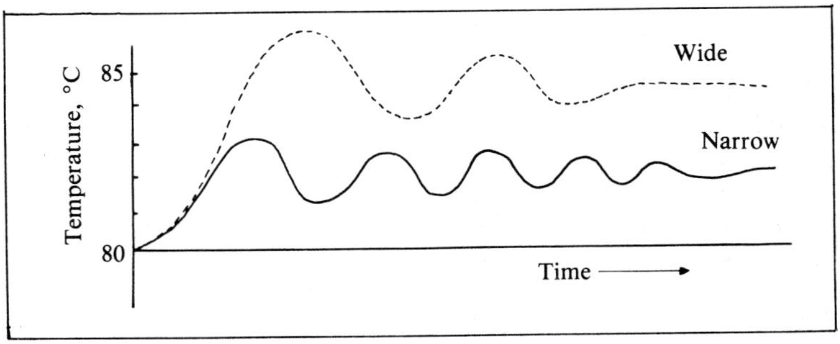

Figure 8 – Variation in proportional control action

Proportional control then applies a correction to the inlet condition that is proportional to the error between the actual value and the desired value, and provides steady control. Proportional action alone, however, only gives control at the desired temperature if the actual conditions are exactly the same as the design conditions; if there is any change from the design conditions then steady control is still given, but not at the desired value, a resultant error remains, i.e. an off-set.

Where is proportional control used?

Again, being a relatively simple system it is reasonably cheap and reliable, but it is only used on its own in industry under conditions where off-set is not important, i.e. it is used for many level controls where, by and large, it is not very important whereabouts in a tank a level is, so long as the level does remain within the tank, hence control is required, but off-set is not important. It is also occasionally used for temperature and pressure controls where the exact temperature or pressure does not matter. For instance, in the provision of an inert-gas supply to a plant it is necessary to have sufficient pressure to distribute the gas around the system, but nobody will be very worried if this pressure varies a little.

Proportional and Integral Control

To reduce to zero the off-set arising from the use of proportional control alone, a second control action is added to the controller. This action is an integral one, i.e. one which integrates the error (in this case T) with respect to time (t) and applies a correction proportional to this integral:

$$V_2 = -K_2 \int_0^t T \, dt$$

where K_2 is a constant.

The full output for the proportional plus integral controller is therefore:

$$V = V_1 + V_2 = -K_1 T - K_2 \int_0^t T \, dt$$

and the actions of a proportional controller as shown earlier and the two-term proportional plus integral controller are shown in Figure 9.

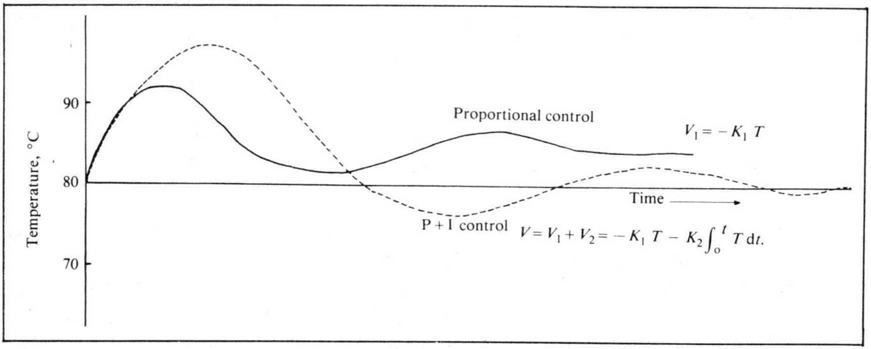

Figure 9 – Proportional plus integral control

It will be seen that a similar response occurs; there are a few oscillations, caused by the proportional term, but superimposed upon this is a steady fall back to the desired control point, caused by integrating the error and applying a further corrective action proportional to this error integral.

This proportional plus integral controller, frequently called a two-term controller or simply a P + I controller, is by far the most common one in use, especially in the process industries. It provides the necessary proportional action, which gives a correction depending upon the deviation from desired value, and, in addition, the integral action follows up this initial correction by removing any residual error (off-set) resulting from existing conditions being different from those assumed in designing the control system. For instance, in the system being considered, the water flow rate may change, or the process operator may decide he would rather have water at a temperature other than that initially considered, i.e. he may change the set point from 80°C, or of course the quality of the steam, or the efficiency of the heat exchanger, or the heat losses from the plant could have changed.

The two-term controller is used in the process industries for most pressure, level and flow control systems, but where there is a long lag in the system, as often occurs in the measurement of, say, temperature or composition, then a further complication arises that has to be corrected by adding a third action to the controller, the derivative action.

Proportional plus Integral plus Derivative Control

When measuring temperature by, say, a thermocouple device, the thermocouple is rarely immersed directly into the process fluid, but is usually supplied fitted into a protective metal sheath. Even this sheath is not usually immersed directly into the process fluid, since a faulty sheath or installation could cause a leak and hence a plant shut-down. The thermocouple sheath is therefore mounted in a protective pocket as illustrated in Figure 10, showing temperature measurement in a pipe bend. Consider now what would happen to our heating system if the steam supply pressure were to rise. Initially, additional heating would occur since, until a temperature change is

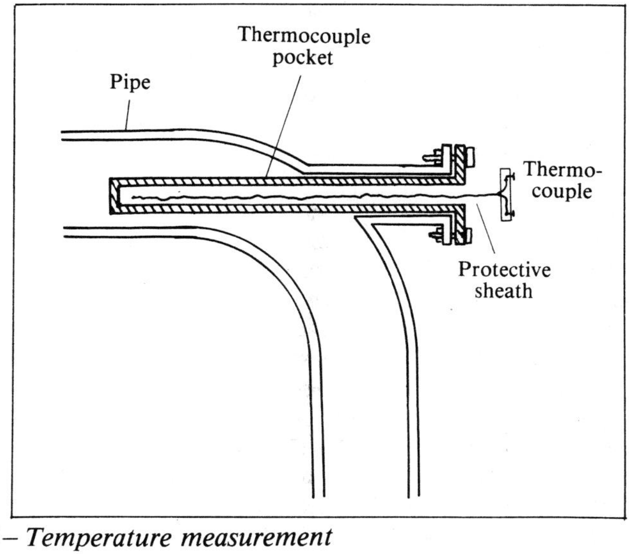

Figure 10 – Temperature measurement

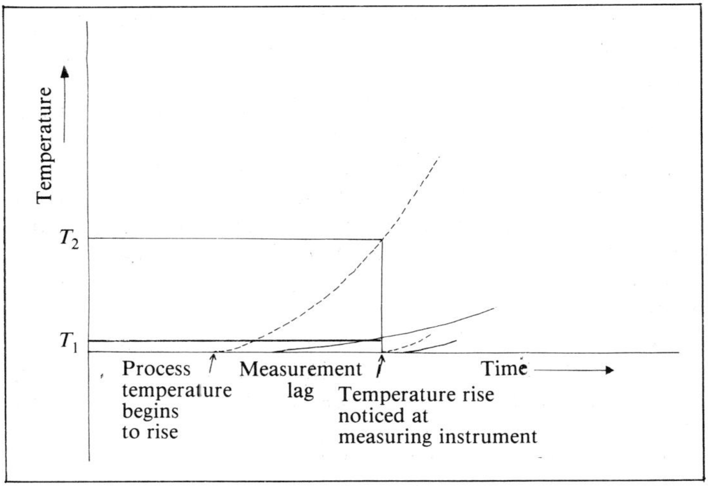

Figure 11 – Measurement lag

measured, the controller output and hence the position of the steam valve would not change. Extra steam flows through the heating coil and heats it; this then heats the surrounding water and, as mixing takes place, the water heats the thermocouple pocket, heat flows through this and heats the air surrounding the thermocouple sheath, the sheath temperature eventually rises, supplying heat to the air within the sheath and finally the thermocouple itself. This notes the temperature rise and *only then* does the controller begin to take action, i.e. after a time lapse during which excess heat has been supplied. If the temperature is rising quickly it is desirable, once noticed, to apply a large correction. When the temperature rise is first noticed at the measurement device, the actual water temperature will already have reached T_2 (Figure 11), i.e. a large error already exists and hence a large correction is required. If, however, the temperature is only rising slowly, by the time that the measurement device detects the start of the temperature change the temperature in the water will only have reached T_1 and hence less drastic action is necessary.

This third correcting action is therefore one which varies with the rate of change of the error, i.e. derivative action and

$$V_3 = -K_3 \frac{dT}{dt}$$

giving the three-term controller where

$$V = V_1 + V_2 + V_3 = -K_1 T - K_2 \int_0^t T\, dt - K_3 \frac{dT}{dt}$$

This three-term controller is usually used only when there is a long lag in the measuring system, i.e. when measuring temperature, pH, density, etc., or when there is a lag in the throughput of the plant.

Let us consider how a system would respond to a given stepwise change in conditions when using the different control actions mentioned above.

Firstly, we must consider the response time of the system, i.e. how quickly it responds to change. For example, in our heating system, if we were suddenly to open the steam valve by an amount that would eventually lift the water temperature by, say, 5°C, the water temperature would not immediately jump by 5°C, but would slowly rise, and the rate of rise would depend upon the size of the heating coil. With a large heating coil the temperature would rise quickly reaching an extra 3°C in, say, about 1 min, while with only a small heating coil it could take 20 min for the temperature to rise by about 3°C. The time it would take to reach roughly half the final level is called the system response time.

If we have a system with a response time of 1 min, and if we were to upset the system by suddenly changing one of its inputs, for example by reducing the cold-water feed rate, then it would respond, depending upon the type of controller being used, as shown in Figure 12.

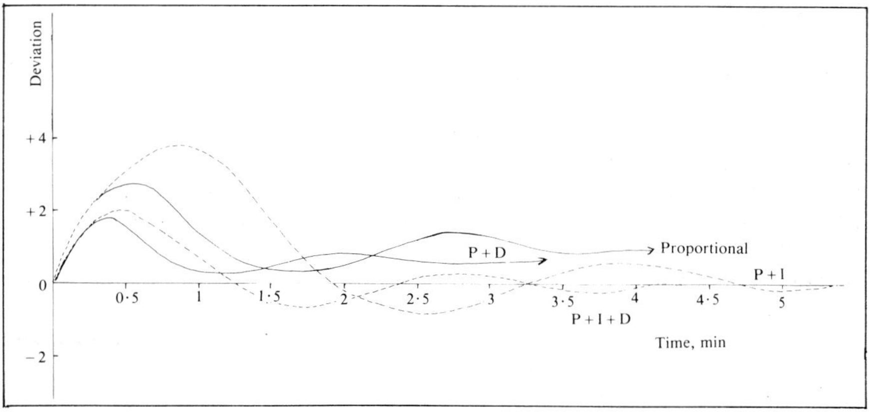

Figure 12 – Comparison of control actions

1. With proportional control alone we get a certain initial deviation, one or two oscillations, and then steady control, but off-set from the control point.
2. With P + I control we get a similarly shaped curve, but settling out with *no* off-set. However, adding the integral term gives a control action in addition to the proportional action and, so as not to let the system become unstable, we must reduce the amount of the proportional action to compensate for this added integral action. To begin with, therefore, the error is not corrected as quickly and the initial deviation is greater than for proportional control alone.

3. With P + D control, immediately the error occurs, both the proportional action and the derivative action act to reduce the error and hence the first peak is a much smaller one. After the peak the proportional term is still trying to reduce the error, but since the rate of change of error is now in the opposite direction, the derivative action is now opposing the proportional action, it is in effect trying always to prevent change; it has a stabilizing effect on a system and we can therefore use a more powerful proportional action without worrying about instability. As the curve shows, we get less initial deviation and less off-set.
4. The three-term controller is a compromise, it allows an initial deviation more than the P + D controller, but leaves no off-set.

Specification of Control Action

In describing the control actions we talk about integral and derivative action times and the bandwidth of the proportional controller. What do these terms mean?

As given above the mathematical expression for the three-term controller is

$$V = V_1 + V_2 + V_3$$

$$= -K_1 T - K_2 \int T \,\mathrm{d}t - K_3 \frac{\mathrm{d}T}{\mathrm{d}t}$$

where, as before, V = the controller output signal, K_1, K_2 and K_3 are constants, T is the error, i.e. $T_o - T_d$, and the negative sign shows that the output signal is correcting, i.e. reducing, the error T.

Now this equation can be rewritten as:

$$V = -K_1(T + \frac{K_2}{K_1} \int T \,\mathrm{d}t + \frac{K_3}{K_I} \frac{\mathrm{d}T}{\mathrm{d}t})$$

$$= -\mathrm{K}_1 \,(T + \frac{1}{s} \int T \,\mathrm{d}t + \mathrm{r} \frac{\mathrm{d}T}{\mathrm{d}t})$$

where $s = \frac{K_1}{K_2}$ (the integral action time)

and $r = \frac{K_3}{K_1}$ (the derivation action time).

What are these expressions s and r?

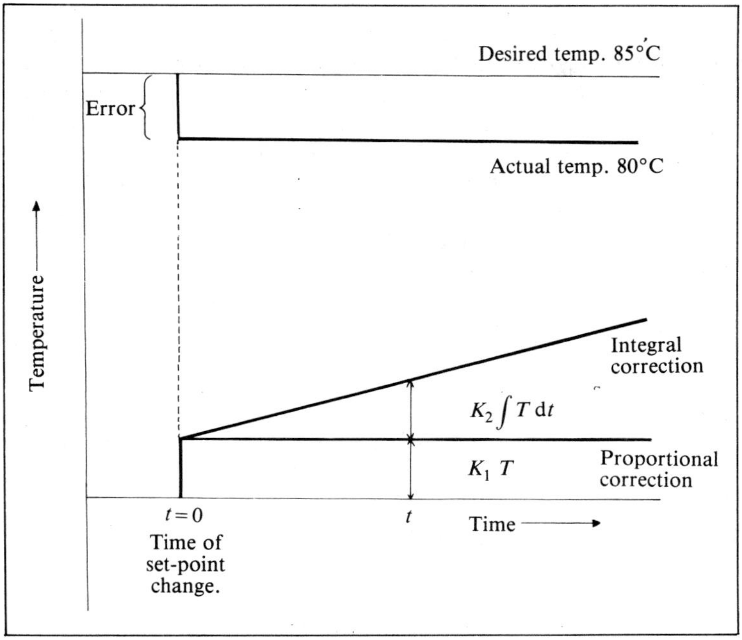

Figure 13 – Integral action time

Integral Action Time

In the heating system described earlier let us disconnect the controller output from the steam valve. If we now increase the controller set point from 80°C to 85°C then the controller will try to send a signal to the steam control valve to increase its opening. Since we have disconnected the line to the valve, however, the valve will not receive this signal and, therefore, will not move, the temperature will remain below the new set point. The output of a two-term P + I controller is shown in Figure 13.

Here we see a steady error $-T$, and the controller output from both control actions. The proportional action gives an output

$$V_1 = -K_1T$$

i.e. a steady correction as shown. In addition, since the error persists, the integral action gives an ever-increasing correction from zero, proportional to the integral of the error, i.e.

$$V_2 = -K_2\int_0^t T\,dt$$

At some time, t, we can see that the corrections from the proportional action and the integral action are the same, i.e.

$$K_1 T = K_2 \int_0^t T \, dt$$

since T is being held constant we have:

$$K_1 T = K_2 \int_0^t T \, dt = K_2 T \int_0^t dt = K_2 T t$$

$\therefore \; t = \dfrac{K_1}{K_2}$ which is what we have called the integral action time, s.

We see that the integral action time is the time it takes for the integral term to provide as much corrective action as the proportional term after a stepwise change in conditions, and on practical controllers it is adjustable from roughly 0.5 s to 50 min.

If a two-term P + I controller is then installed on a plant and the integral action time set to 1 s, this means that, when a deviation occurs, the proportional action applies a correction, the integral term begins to integrate the error and, roughly speaking, applies a further correction of the same magnitude as the proportional correction every second thereafter until the error becomes zero. Is this wanted? It depends upon the response time of the system being considered.

If we talk once again of our simple heating system, depending upon the steam supply, the heating coil, and the size of the water vessel, it may take 2–3 min to raise the water temperature by, say, 5°C. If the integral action time is 1 s then, after each second, the integral term will apply more control, each equivalent to that provided initially by the proportional controller. Obviously, long before the water temperature can begin to rise the controller will be on full output, the full heating being applied – we are back to on/off control.

Equally if the integral action term is set to the full 50 min another unsatisfactory solution results. Here the proportional term would act and correct for the deviation, but controlling at an off-set position. The integral term, which should reduce this off-set to zero, is slowly integrating the error, but it will not apply a correction equivalent to the proportional correction until approximately 50 min have elapsed, by which time conditions could well have changed once again. Therefore, in practice, all that is provided is a proportional controller. The integral action time must, therefore, be set at roughly the response time of whatever system is being considered, here, for instance, at something like 1 or 2 min.

Derivative Action Time

A similar experiment to the above would show that $r = \frac{K_3}{K_1}$, i.e. the time it would take for the proportional term of a two-term P + D controller to give the same output as the derivative term, if a steadily increasing error were present. Again, for sensible control, it can be seen that this time must be in keeping with the response time of whatever system is being controlled; usually the derivative action time is set at about one-quarter of the integral action time, or in the above case about 30 s.

Proportional Control Action

This is described and quantified by talking about a 'proportional bandwidth', which is the range of input signal required to cause the output signal to vary over the whole range. It is usually expressed as the percentage of the full scale range of the measuring instrument.

For example, if a pneumatic controller is being used to give an output of 3–15 lbf/in^2 (i.e.20 to 100 kPa)over the temperature range 0–200°C, and if the system has been designed so that the control valve is to be fully open when the temperature is 50°C and fully closed when the temperature is 150°C, then the output signal must move over the full range from 3 to 15 lbf/in^2 when the temperature rises from 50–150°C, i.e. 100°C change causes the output signal to vary over the whole output range, and since 100°C is 50% of the measurement span of 0–200°C, then the controller is said to have a proportional bandwidth of 50%.

The controller can be made more active by narrowing the proportional bandwidth. If it is reduced to 10%, this means that the control valve will move from fully open to fully closed for a temperature change of 10% of the instrument span, i.e. 10% of 200°C or ±20°C around the set point.

On most controllers the proportional bandwidth can be varied from zero to around 600%. Zero means, of course, that we have in effect on/off control, whereas 600% is a very lazy control action used in special circumstances only. Usually, for most control situations, the proportional bandwidth will be set at somewhere between 10 and 100%.

Further Control Actions

Cascade Control

As we pointed out earlier, feedback control can at times be slow to react, even using the derivative action described above. When disturbances from outside the system cause problems, cascade control can be employed.

Again this can be illustrated using the heating system described earlier and shown in Figure 14.

The controller, it must be emphasized, sets the valve position not the steam flow, the steam flow varies with the pressure in the steam main. If some of valves A, B, C or D were to open wide, then the pressure in the steam main would fall and less steam would flow through the heating coil,

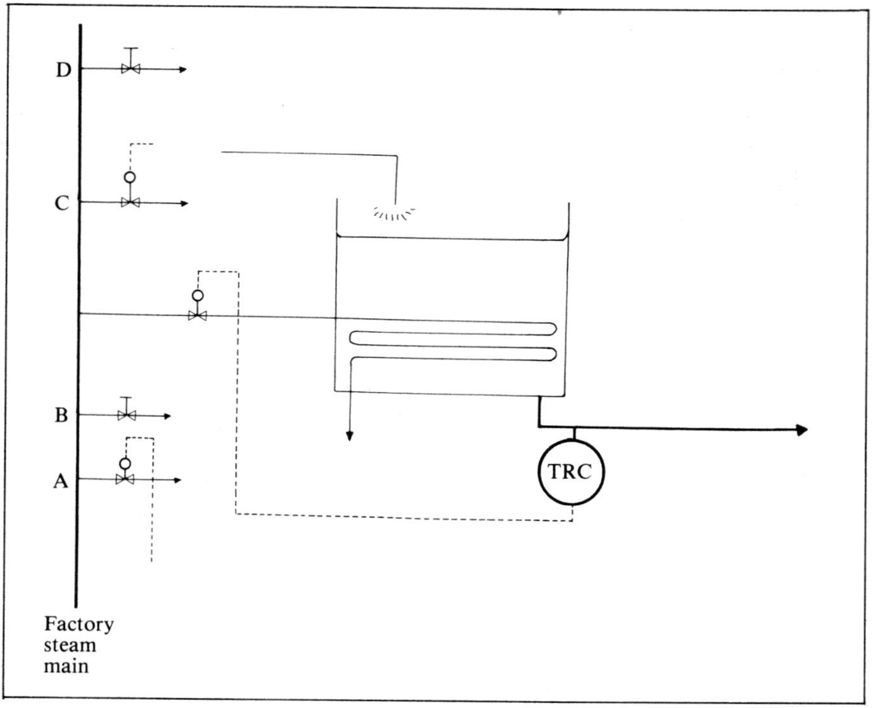

Figure 14 – Temperature control

the temperature would eventually drop and of course the controller would open the control valve to compensate. However, all this takes time. This cannot be avoided when genuine changes occur in the system itself, such as variations in cold-water flow or heat losses from the tank, but it is annoying when conditions outside our control system cause additional problems. Cascade control can help here.

The control system is altered so that the steam flow is controlled as shown in Figure 15, and the set point of the steam controller is varied by our original temperature controller.

The steam-flow controller is set to supply the calculated steam-flow rate required. If this calculation is in error, however, the temperature at the outlet will be wrong, either too high or too low. The temperature controller then gives an output signal which alters the flow controller's set point, calling for more or less steam as required, and the system is virtually as before with the temperature controller in command. What happens now if the pressure in the steam main falls? As before, the steam flow would fall, but this would be immediately detected by the steam-flow controller, the steam valve opened and the steam-flow rate restored before the heating system

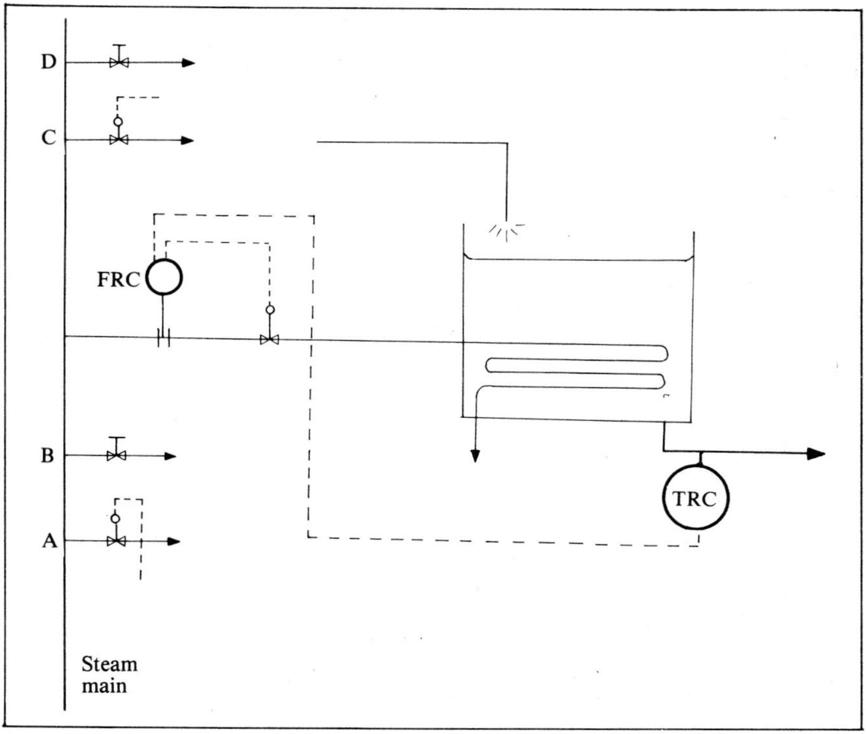

Figure 15 – Cascade control

was affected, helping, therefore, to isolate our control system from external variations.

Usually, in these systems, the slave controller, the flow controller in this case, will be a reasonably narrow bandwidth, proportional-only controller. Integral action is not required since any off-set in the flow system will result in a temperature error and the master controller, here the temperature controller, normally a three-term controller, will reset the flow-control set point.

Program Control

In batch processes we are not controlling continuously at fixed conditions of temperature, pressure, level, etc., but may wish to vary these conditions throughout the process cycle in a predetermined manner. This is easily done by incorporating a motor arrangement to move the set-point index (i.e. the desired value, or T_d in Figure 2) through a time pattern. The set-point index moves over a cam, cut to reproduce the desired sequence of operations; different control points, holding times, and controlled rates of increase and/or decrease of temperature can be provided. It is thus possible to open or close valves for fixed times to fill or empty vessels, to raise the temperature of a vessel at a preset rate, to hold a temperature steady for a given time, and so on. Throughout this cycle of events the normal control action described will maintain the condition at the value required. A two-term P + I control action is generally the best, the integral-action term eliminating the off-set which would be caused by changing the position of the set-point index; derivative action is not required as it would cause trouble at times of set-point change.

Process timers that control batch operations to preset time sequences like this are not, however, very flexible, and to adjust the time pattern means cutting and fitting a new cam. Also, in operating a time-cycle control, the maximum time required for any action under the worst conditions likely to arise must be allowed. A batch process can usually be controlled more efficiently by a sequential controller, one that moves on to the next step in a sequence immediately the previous step has been completed. For instance, in simply filling a vessel, instead of opening the filling valve long enough to ensure filling at the lowest rate, allowing excess to be returned via an overflow system, a sequential controller would open the valve until a level detector tripped and then it would immediately move to the next step in the sequence of operations and so save time. A digital computer is an ideal sequential controller and should be considered whenever programmed and sequential control is required.

Tuning a Process Controller

Unlike tuning a television set, where you can adjust the controls – brilliance, contrast, colour etc. – until a good picture is obtained, without really knowing what the controls do but without fear of causing trouble, with a process controller it is not easy to tell whether the controller is performing at its best or not. Idle turning of the control settings – proportional bandwidth, integral and derivative action times – can cause trouble.

If a plant, or process, is running steadily with no disturbances affecting it, then the controller has no action to take and it therefore does not matter what its settings are. If an unknowledgeable operator alters the control settings during such a quiet period, making the integral action time very short or very long, it will have no effect upon the process. However, if at a later time a disturbance does occur and the controller has to act, then a very short integral action time could cause the process to become unstable and oscillate, or a very long integral action time leave the plant running at an off-set position away from the desired one.

How then do we tune in our controller? Either by a thorough mathematical investigation of the process and the control system, or by experience built up over a number of years, or by one of two simple, practical experiments described below.

Closed-loop Tuning Method

This experiment is carried out with the controller fully installed and operating. The controller is set up as a proportional-only controller by

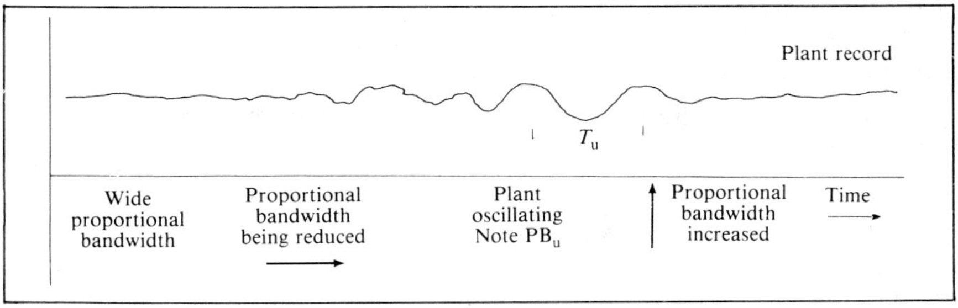

Figure 16 – Controller tuning – closed-loop method

removing the integral and derivative terms, i.e. by fully closing the integral action, maximum time, and fully opening the derivative action, minimum time. Starting with a reasonably wide proportional bandwidth, say 100%, the proportional bandwidth is successively reduced until the process just starts to oscillate continuously, a note of the proportional bandwidth setting, PB_u, and the time of oscillation, T_u, is taken, see Figure 16. The proportional bandwidth is immediately increased so as not to upset the plant further.

Good settings for the controller can now be determined, they are:

For P only controller: Proportional setting $= 2 \times PB_u$

For P + I controller: $P = 2.2\ PB_u$

$s = 85\%$ of T_u

For a P + I + D controller: $P = 1.65\ PB_u$

$s = \frac{1}{2}\ T_u$

$r = \frac{1}{8}\ T_u$

There are, however, difficulties associated with this method of determining good settings for the controller variables, because the method does necessitate disturbing the process, albeit for only a short time, and of course process managers do not like this. However, even if the manager can be persuaded to have his plant upset temporarily for better future control a problem still exists; the plant is disturbed, and there is no control over the magnitude of this disturbance. When steadily reducing the proportional bandwidth in an endeavour to start the plant oscillating, the plant response must be watched carefully, and if the disturbance becomes too large to be acceptable then the proportional bandwidth is simply increased again, the experiment sadly abandoned and the following method tried.

Open-loop Tuning Method

The controller concerned is switched from automatic to manual and the output to the control valve increased by 10% of its full value, e.g. on pneumatic control of 3–15 lbf/in^2 the output is increased by 1.2 lbf/in^2.

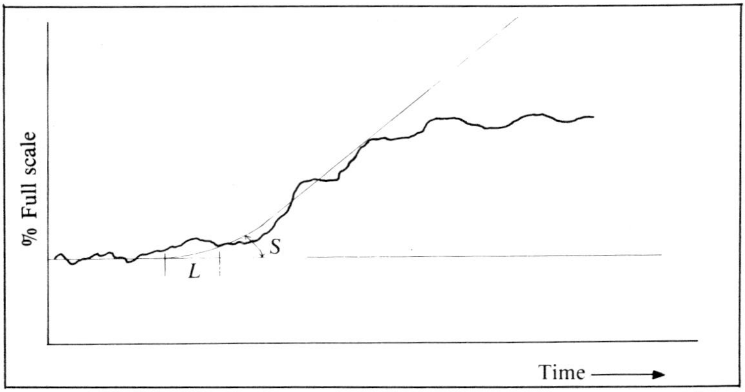

Figure 17 – Controller tuning – open-loop method

The controlled variable response is then noted and it will be similar to Figure 17, i.e. an S-shaped response.

L is a measure of the response lag, S is a measure of the rate of response, % full-scale change per minute.

Good controller settings are as follows:

For a P only controller: $P = 10SL$

For a P + I controller: $P = 11SL$

$s = 3\frac{1}{3}L$ min

For a P + I + D controller: $P = 8SL$

$S = 2L$ min

$r = 0.5L$ min

In practice, the controlled variable may not be too steady before the step change, and the rate of rise may not be too great, hence it may be difficult to determine S and L at all accurately, but at least some guide will be given.

3 CONTROLLER MECHANISMS

Pneumatic Controllers

Until very recently the vast majority of controllers, especially in chemical and petrochemical plants, were pneumatic, mainly because they could operate directly on pneumatic control valves, which are simple, reasonably cheap and, most importantly, can safely handle inflammable or potentially inflammable substances. In addition, pneumatic controllers were themselves simple, reliable and easily maintained by mechanical technicians. The action of these controllers was based upon an orifice, nozzle and

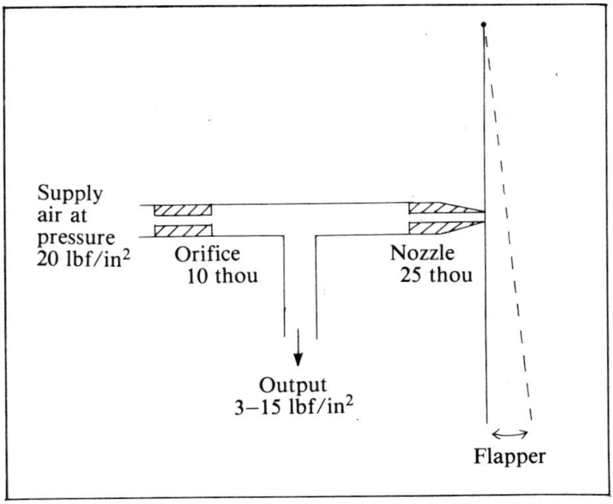

Figure 18 – Pneumatic control element

flapper combination, giving an internationally agreed control output of 3–15 lbf/in^2.

This 3–15 lbf/in^2 signal is generated, as shown in Figure 18, by moving a small, lightweight strip (1 in × ⅛ in), usually of phosphor bronze, towards or away from a small nozzle through which air is escaping. As this flapper moves to the nozzle it prevents the escape of air and the pressure behind the nozzle rises. As the flapper moves away, the air is allowed to escape and the back pressure falls.

With a supply pressure of 20 lbf/in^2 the back pressure, when the flapper just touches the nozzle, is 15 lbf/in^2 (not the full amount as some air still manages to escape); the flapper does not seal off the nozzle. When the flapper is away from the nozzle the full pressure forces the air through the orifice and then through the nozzle. Therefore:

$$\text{Back pressure} = \text{Supply pressure} \times \frac{\text{Orifice area}}{\text{Nozzle area}}$$

$$= 20 \times \frac{10^2}{25^2} \simeq 3\ \text{lbf/in}^2$$

The flapper movement, however, from closed at 15 lbf/in^2 to open at 3 lbf/in^2, is only 0.004 in or 0.1 mm, much too small a movement to be of direct practical use. The solution to this dilemma is very simple, robust and cheap. It is a small phosphor bronze bellows, illustrated in Figure 19.

How does this work? Initially the flapper is in position AB. When a change in the measured variable takes place and the flapper is moved, as above, it moves to position BC and the movement will be much greater than

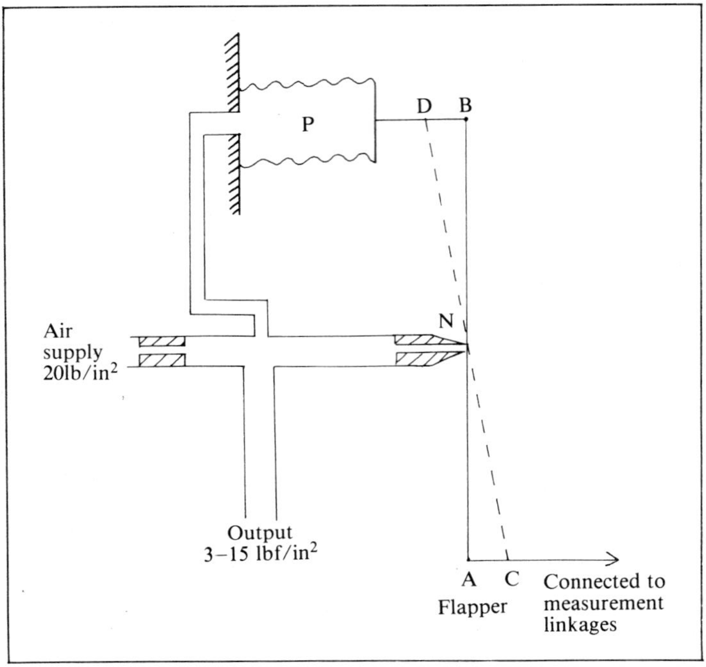

Figure 19 – Proportional controller

0.1 mm and hence the output will fall to 3 lbf/in^2. This output pressure, however, is now connected to the bellows, which will therefore collapse as the pressure inside them falls, this pulls the flapper back to position CD, i.e. the flapper is brought back close to the nozzle. This simple system is the basis of all pneumatic controllers and transmitters and enables the flapper to be positioned to millionths of an inch, much more than is required for 1% accuracy in the 3–15 lbf/in^2 output.

The system shown in Figure 19 is in fact a proportional controller. The movement of the end of the flapper can be arranged to be proportional not to whatever is being measured, e.g. temperature, pressure, etc., but, by a system of mechanical linkages, to the difference (error) between what is being measured and the desired value, i.e. to $T_o - T_d$ in our heating system. If then the flapper end moves with the error, and the output is varying with the flapper position, the output is proportional to error, a proportional controller.

The proportional bandwidth of our controller can be varied by moving the nozzle along the flapper. When the nozzle is near the bottom end of the flapper at A the resultant flapper/nozzle separation is large, almost on/off control, whereas if the nozzle is positioned near N, the flapper/nozzle separation is very small even for substantial movements of the end of the

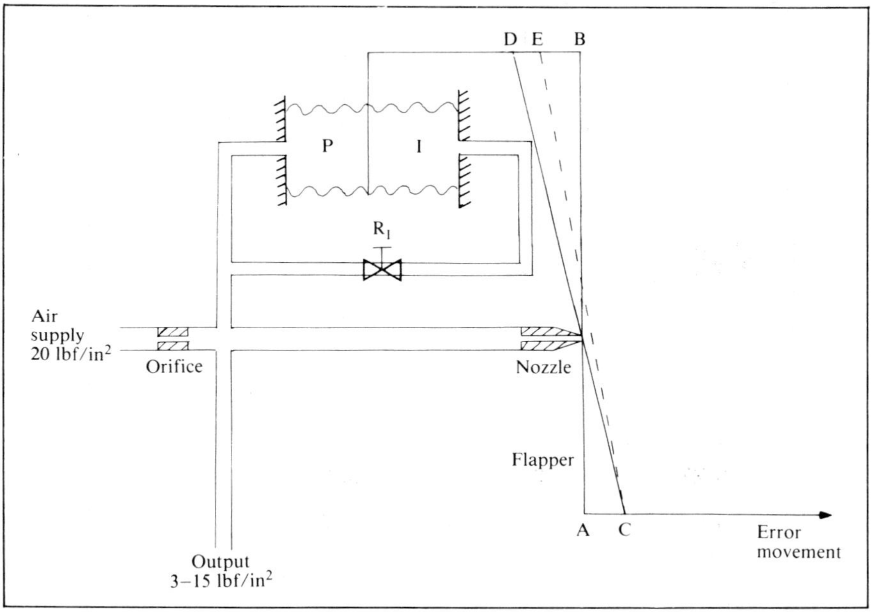

Figure 20 – Proportional + integral controller

flapper, hence there is an almost infinite proportional bandwidth.

For proportional + integral control two bellows are used, back-to-back, with a restriction valve in the feed line to the second bellows (Figure 20).

When an error arises the flapper is moved away from the nozzle, i.e. from position AB, to position BC and the output pressure and hence the pressure in bellows P will fall, the bellows contract and the flapper moves quickly to position CD, a proportional-control action operating. The pressure in bellows I will slowly leak away via the restriction R_I and these bellows will now contract and the flapper move back to position CE, with integral action taking over until the two bellows settle at the same pressure.

The rate at which bellows I move is determined by the time constant set by the bellows volume and the restriction R_I – integral action time.

Derivative action is generated by fitting a restriction R_D in the line to the original proportional-action bellows (see Figure 21).

Now when an error moves the flapper from the nozzle, i.e. from position AB to position CB, the bellows cannot collapse immediately to give proportional action only, but the output pressure will remain low until the air in bellows P can leak through R_D, and the flapper will therefore only slowly move to position CD. While it is doing so a strong correction is being supplied (derivative action) and if the error is changing a correction will be being applied trying to prevent change – the derivative-action time being set by

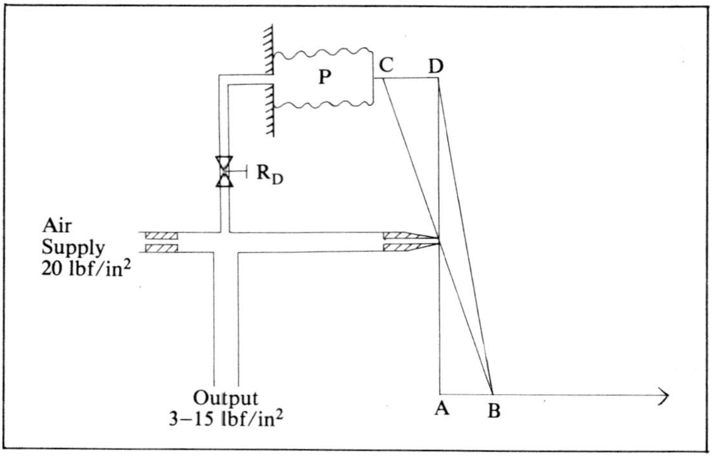

Figure 21 – Proportional – derivative controller

the time constant of bellows P capacity and restriction R_D.

The three-term pneumatic controller is as shown in Figure 22.

Electronic Controllers

In the 1950s the electric controller was introduced. Although electronic/electrical controllers had been available before, the process industry had a strong preference for the traditional pneumatic controller, which offered many important advantages when compared with the new electronic replacement. Advantages of the established system included plant safety,

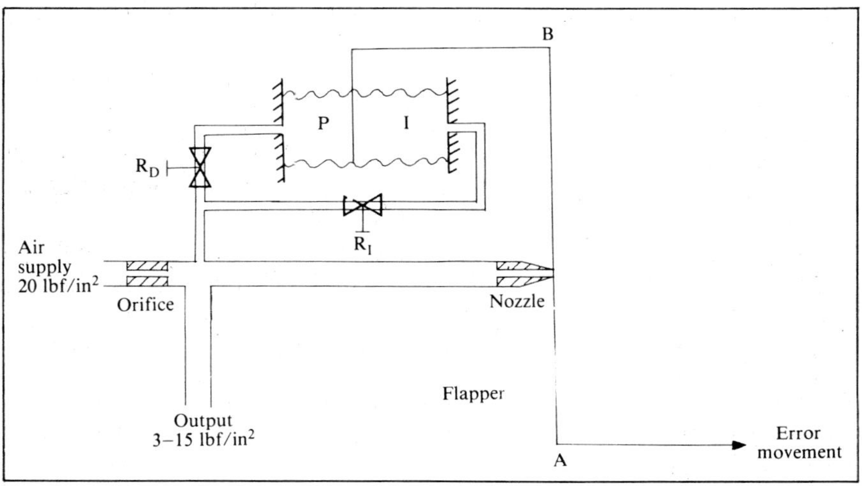

Figure 22 – Three-term pneumatic controller

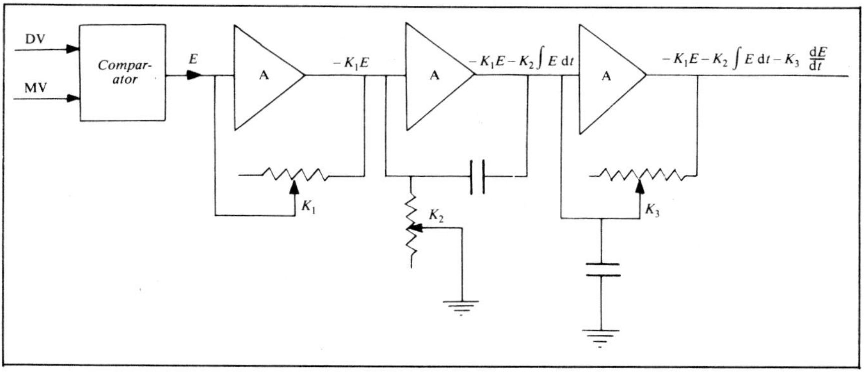

Figure 23 – Simplified circuit of electronic three-term controller

compatibility with existing transducers and a fully trained team of operators with considerable experience in pneumatic systems. However, in the 1960s, with the increasing use of the process control computer, the process industry recognized the simplicity of connecting electronic systems to the computer and consequently started to switch to the new generation of instruments.

The early electronic controllers were styled along similar lines to the pneumatic controller, the appearance of the instrument panel was similar and the only difference appeared to be that the pipes entering the back of the instrument had been replaced by wires. Inside, in its simplest form, the electronic controller was the electrical analogue of the pneumatic circuit. Capacitators replaced bellows and chambers, resistors and potentiometers replaced restrictors and needle valves. A high-quality (low-noise–low-drift) operational amplifier replaced the flapper and nozzle. A simplified circuit of a three-term controller is shown in Figure 23.

The pneumatic controller has standardized on 3–15 lbf/in^2 (20–100 kPa) for its operating signals. This is recognized and accepted throughout the world. However, in the case of the electronic system, controllers can be purchased to work on a number of different signals, including 0–10 mA, 0–10 V, 1–5 V and 4–20 mA. It can be seen that in the case of 1–5 V and the 4–20 mA system there is a similarity to the pneumatic system where the lower level is not zero, so that power failure or evidence of a broken wire can be detected. The voltage system is operated at a level sufficient to minimize the effects of extraneous electrical noise. The current system offers both the noise immunity and the advantage of its suitability to transmit signals over long distances without the usual problems of voltage drop, resulting in measurement error between the measuring head and the controller, recorder or indicator. Today the 4–20 mA signal is almost a universal standard, but the growth in microelectronics suggests that this will change and a form of digital signalling may replace it, this being either a frequency or time pulse signal of low-current and low-voltage operation.

4 INDUSTRIAL MEASUREMENT

The problems associated with industrial measurement arise from the wide range over which measurements are carried out, the harsh environments in which the equipment exists and the difficult fluids being monitored.

Temperature is frequently measured from −200°C to 1 500°C, flow in pipes from 2 mm to 2 m diameter, level from a few centimetres to 10 m and pressures from vacuum to 250 atmospheres. The fluids being monitored are frequently corrosive, dirty, solidify at low temperatures, vaporize, condense, etc.

Temperature Measurement

The major part of industrial temperature measurement is carried out using the thermocouple or the resistance thermometer, the thermocouple being most frequently used, but the resistance thermometer increasingly so, especially for temperatures below 0°C or where good accuracy is required; resistance thermometer error is roughly ±0.2% of the temperature being measured, whereas the thermocouple error is about ±0.75%.

Thermocouple Thermometers

If two dissimilar metals, A and B, are connected together and a temperature difference established between their ends, then a current will flow around the loop (Figure 24). This is the result of two effects, the Peltier effect and the Thomson effect.

Electrons are held within the crystal structures of metals and the degree of binding varies from metal to metal. If two metals are joined, as shown, then the electrons least tightly bound will flow from one metal to the other until a potential difference exists across the metal-to-metal junctions, preventing further transfer of electrons. If, however, one junction is heated with respect to the other, the electron motion will be increased and more electrons will flow across this junction. These electrons will then flow along the wire and be forced back across the colder junction as the potential builds up between the two metals, hence a current will flow – the Peltier effect.

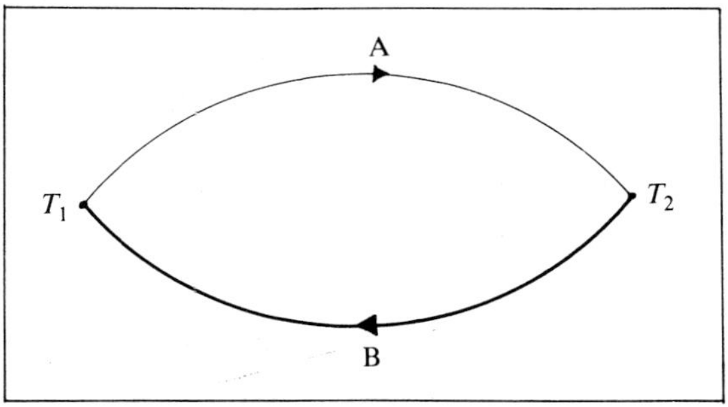

Figure 24

In addition, electrons within a single wire will flow along the wire if a temperature gradient exists, since the electrons at the hot end of the wire will be less tightly bound than those at the colder end – the Thomson effect. The result of these two effects, the Seebeck effect, can be used to measure temperature by the potential difference existing between two dissimilar wires when a temperature difference exists between their ends, i.e. $V \propto T_1 - T_2$.

The binding of the electrons in the metals depends, however, upon the strain within the wires and hence dropping, bending or twisting of a thermocouple will all have a small effect on its output and its accuracy. Annealing of the wires reduces the problem, but careful handling after annealing is required or errors of a degree or two can easily be reintroduced.

Almost any pair of dissimilar metals could be used to produce a thermocouple, but a limited number have achieved a degree of international acceptance, and these are shown in Table 2. The thermocouple output is given in BSI tables and by the manufacturers; for an Fe/Con thermocouple, Type J, the output is roughly 40 μV/°C, and thermocouple accuracies are usually quoted to $\pm 0.75\%$ of the temperature, or ± 3°C, whichever is the greater.

The important feature of the thermocouple is that it always measures the difference between one temperature and another, between the hot junction of the two thermocouple wires and the cold junction, and it is not always clear where exactly the cold junction is. In Figure 25 a typical installation is illustrated.

TABLE 2

Types of Thermocouple

Type	Composition	Normal operating range (°C)	Maximum 'spot' reading (°C)
E	Nickel–chromium/ copper–nickel or Chromel/constantan	–200 to 850	1 100
J	Iron/copper–nickel or iron/constantan	–200 to 850	1 100
K	Nickel–chromium/ nickel–aluminium or Chromel/alumel	–200 to 1 100	1 300
T	Copper/copper–nickel or copper/constantan	–250 to 400	500
B	Platinum–30% rhodium/ Platinum–6% rhodium	0 to 1 500	1 700
R	Platinum–13% rhodium/ Platinum	0 to1 400	1 650
S	Platinum–10% rhodium/ platinum	0 to 1 400	1 650

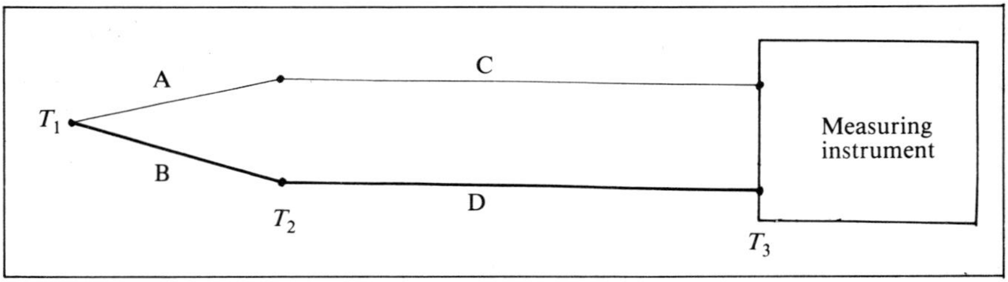

Figure 25

The working thermocouple is A/B with the hot junction at T_1, the plant temperature to be measured. One can seldom use the same wires all the way from the measuring point on the plant to the indicator or recorder on the control panel and extension leads or compensating leads will be required, C/D. These are either of nominally the same material as the thermocouple, for example, Fe/Con with a Type J thermocouple, or alloys having a similar e.m.f./temperature relationship to that of the thermocouple over a limited temperature range; these are used with Type B, R or S precious-metal thermocouples, where the high price of platinum does not allow the use of long platinum alloy extension leads.

Since the measuring instrument is measuring $T_1 - T_3$, i.e. $T_1 - T_2$ plus $T_2 - T_3$, it is essential that T_2 is as near T_3 as possible to give an accurate measure of temperature, since $T_1 - T_2$ only is being measured by the thermocouple itself, $T_2 - T_3$ is being measured by the extension or compensating leads, which have a similar, but not identical, e.m.f./temperature relationship and the variation will be more pronounced the higher the temperature difference. In practical installations, therefore, it is essential that the original thermocouple be long enough so that where it is joined to the extension leads the temperature is well below the measurement temperature; one should be able to handle the ends of the thermocouple comfortably.

Even now, however, only $T_1 - T_3$ is being measured, not, T_1 itself. Hence, even if T_3 is ambient temperature, usually the temperature in the control room, it will vary by a few degrees and hence cause errors. In early use of thermocouples the second junction was fitted into an ice bath so that $T_3 = 0°C$, hence $T_1 - T_3$ was the true temperature T_1, see Figure 26.

This, however, is inconvenient and modern instruments assume that the cold junction will be at room temperature and they include a temperature-sensitive reference junction resistor to compensate for slight variation around room temperature, and so eliminate the need for the ice bath. The thermocouple output voltage can be measured by a potentiometric circuit or by using amplifiers and a voltmeter.

Methods of construction of thermocouples are shown in Figures 27a and 27b. In Figure 27a the thermocouple wires are insulated from the protective metal sheath by ceramic beads. In Figure 27b the thermocouple wires are

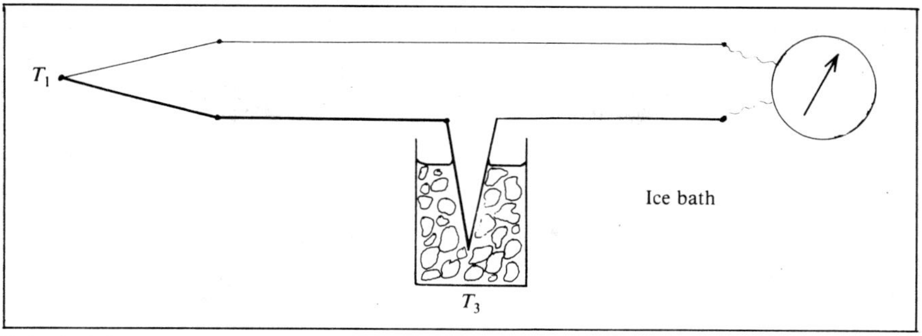

Figure 26

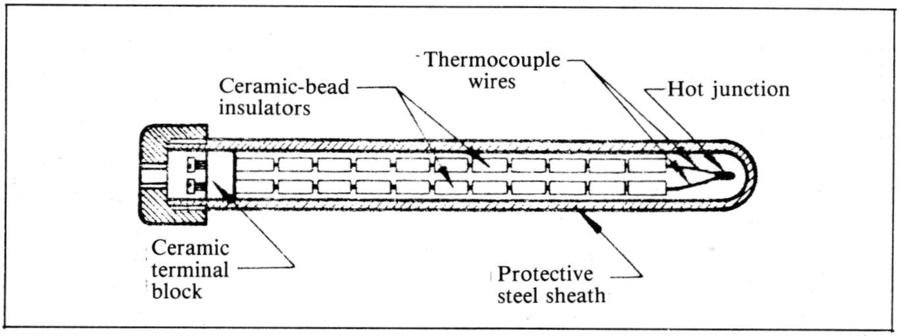

Figure 27a – Thermocouple sensing element

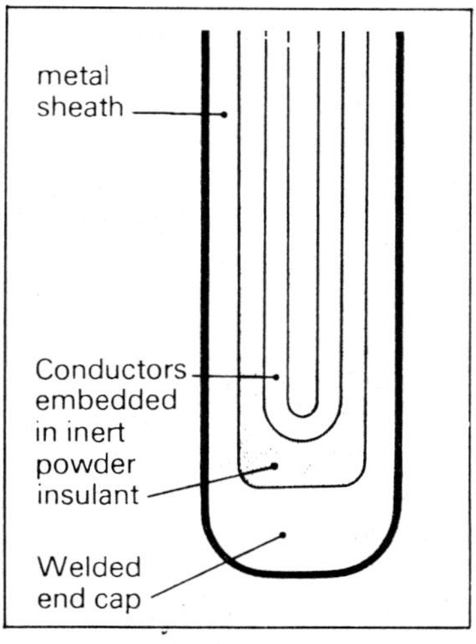

Figure 27b

embedded in an inert, powdered insulator and contained in a metal sheath hermetically sealed by a welded end cap. A more recent development of the latter type uses a single co-axial conductor instead of the two separate wires. It is claimed this type gives greater reliability under arduous conditions with faster response time.

Resistance Thermometers

Temperature-detecting elements based on the change in electrical resistivity of pure metal wires have a long history, but it is only comparatively recently that they have been used in really large numbers in industry.

Copper would appear to be an obvious metal to use as a sensing element, since it is readily available in fine wires of high purity, but, because of its susceptibility to corrosion and oxidation, it is not often used. The metal used for almost all industrial resistance thermometers is platinum, which is not subject to corrosion or oxidation in most environments; it can be obtained in fine wires of extreme purity and has a reasonably high resistivity, which allows a useful resistance to be produced without the necessity of having a bulky device.

The industrial thermometer usually has the fine platinum coil wound on a small glass or ceramic former, the whole being fitted within a metal sheath. The resistance of the coil, by near worldwide agreement, is 100 Ω at 0°C rising to 138.5 Ω at 100°C. Two grades of thermometer are available: grade I, where the tolerance is roughly ±0.2% of the temperature being measured, and grade II, roughly ±0.5%. The useable range is from −220°C to 850°C, but special care is needed above 500°C.

Measurements are made with some form of bridge circuit as shown in Figure 28a.

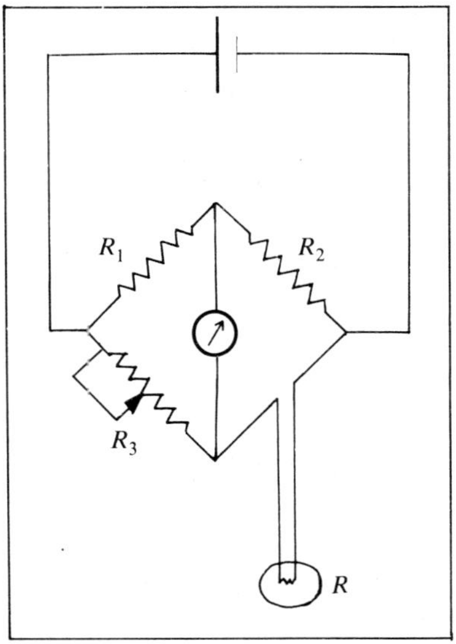

Figure 28a

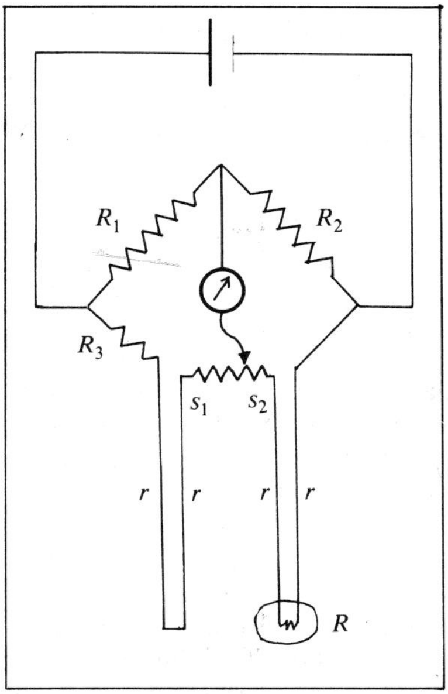

Figure 28b

Here $R = \frac{R_2}{R_1} \cdot R_3$ when the bridge is balanced.

In practice, the resistance thermometer, R, will be on site, and the measuring instrument could be some 50 m away in the control room, so that the resistance of the bridge arm shown above will not be simply R, but $R+2r$, where r is the resistance of one of the thermometer leads. Since the resistance of 3/0.029 copper wire is 0.013 Ω/m, then 50 m of lead is equivalent to 0.65 Ω or $2r = 1.3$ Ω. The resistance thermometer resistance at 0°C = 100 Ω and at 100°C it is 138.5 Ω, thus 38.5 Ω ≡ 100°C, therefore 1.3 Ω ≡ 3.4°C, i.e. an intolerable error.

For really accurate measurements, four not two leads are run to the resistance thermometer, adding cost, but also accuracy. The bridge circuit now used is shown in Figure 28b.

Here at balance $$\frac{R_1}{R_2} = \frac{R_3 + 2r + s_1}{R + 2r + s_2}$$

and if $$R_1 = R_2$$

$$R_3 + 2r + s_1 = R + 2r + s_2$$

i.e. $$R = R_3 + s_1 - s_2$$

Hence the measured resistance of the thermometer is completely independent of lead resistance r.

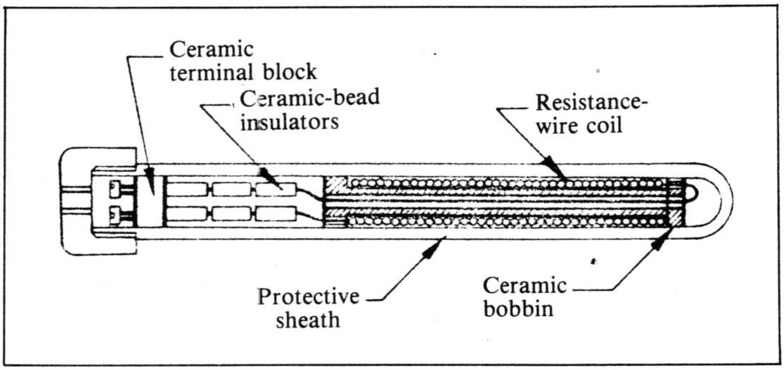

Figure 29 – Resistance thermometer sensing element

Figure 29 illustrates the construction of a resistance thermometer.

Pressure Measurement

Pressure is possibly the most commonly measured quantity in industry, especially in the process industries where pressure gauges are seen on pumps, compressors, supply lines, control valves, controllers and control-room panels. Almost all instruments today measure the difference between atmospheric pressure and the pressure of interest, in contrast, that is, to measuring the pressure as an absolute value above zero pressure. When indicating a pressure above atmospheric, the reading is indicated as *gauge pressure*, while pressures below atmospheric are defined as *vacuum*.

Bourdon Gauges

The principle pressure measuring device is the Bourdon tube pressure gauge, which in its simplest form consists of a tube of oval section, bent in a circular arc. One end of the tube is sealed and attached by a light link to a mechanism operating a pointer. The other end of the tube is fixed, and is open for the application of the pressure to be measured. The internal pressure tends to change the section of the tube from oval to circular, and this tends to straighten out the tube. The resulting movement of the free end of the tube causes the pointer to move over the scale (Figure 30).

Bourdon gauges span the pressure range 0–1 to over 500 atmospheres; in the lower ranges the Bourdon tube material is usually bronze, beryllium copper is used for intermediate pressure ranges and for high pressures the tubes are made from various alloy steels.

Care must be taken with pressure-gauge installations to ensure that the correct pressure is measured. A head of liquid in the connecting pipe between the pressure point and the gauge itself will affect the reading, and for low pressures this can be unacceptable. The gauges can be affected by excesses of temperature or by rapid pressure fluctuations, and pressure measurement installations near to injector, compressor, etc. deliveries

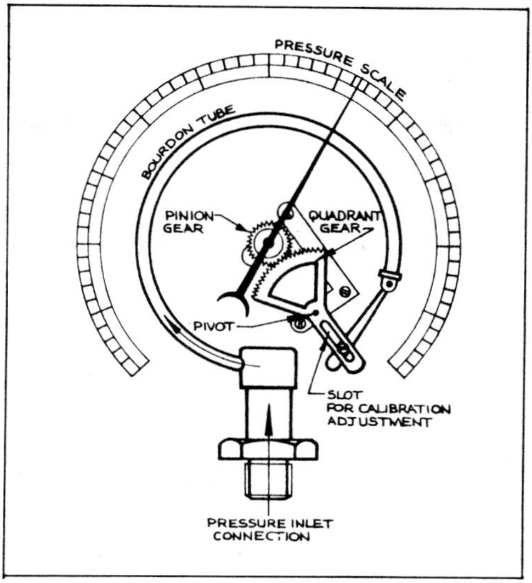

Figure 30 – Bourdon tube pressure gauge

should include a pressure snubber or needle valve restriction to dampen out the pressure fluctuations.

Other forms of pressure-measuring devices include bellows (Figure 31) for low-pressure measurements (vacuum to 5 atmospheres and in pneumatic

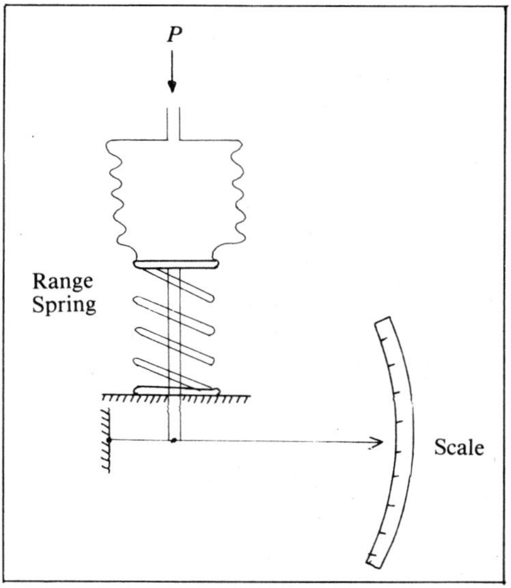

Figure 31 – Bellows gauge

controllers, range 3–15 lb/in^2) and diaphragms for pressure-difference measurement, especially in level- and flow-measuring instruments.

Level Measurement

Probably the simplest method of measuring a level is to use a dipstick, as one uses to measure the level of oil in a car's engine or sump, and as petrol tanker drivers use when delivering petrol to a garage. From the level of liquid on the dipstick, the volume of liquid in the tank can be calculated.

For precise level measurement of storage tanks in industry, the position of a float on the liquid surface is determined by suspending the float with a stainless steel wire from a winding drum. As the level changes and the float begins to move a motor rotates to raise or lower the float keeping the suspension wire taut; the revolutions of the motor are noted and hence the float position determined. This type of instrument can read accurately to ± 1 mm in tank depths of up to 20 m.

Probably the most common industrial level instrument, however, is the gas-injection type, where a gas, usually air or nitrogen, is bubbled through a dip pipe into the bottom of the tank; the pressure required to bubble the gas into the liquid is measured and hence the liquid level determined.

The dip pipe usually contains a gas reservoir at its end, as shown in Figure 32, to provide a supply of gas to counteract quick level rises and this also

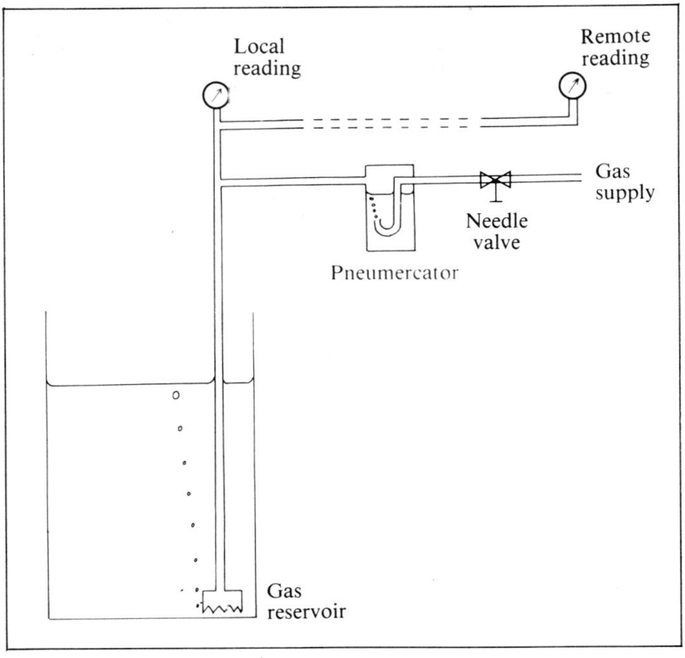

Figure 32 – Gas-injection level indicator

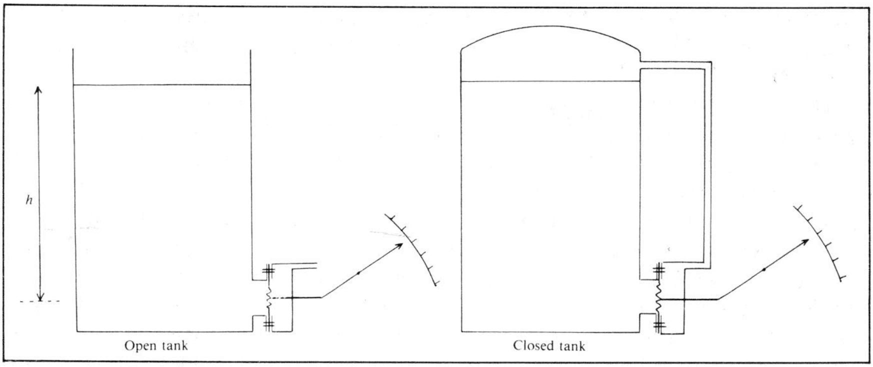

Figure 33 – Diaphragm-type level gauge

prevents the end of the dip pipe becoming blocked by dirt or dried solute. A pneumercator is usually fitted into the supply line and the restriction valve adjusted to give a slow, steady gas flow. To prevent errors arising through pressure drops along the supply line, the pressure is always measured at the tank itself and, if a remote reading is required, it is either transmitted back or a second line installed to carry the pressure reading back along a dead line.

Using the same principle as above, the pressure at the bottom of a tank may be determined using a diaphragm-type pressure gauge, see Figure 33.

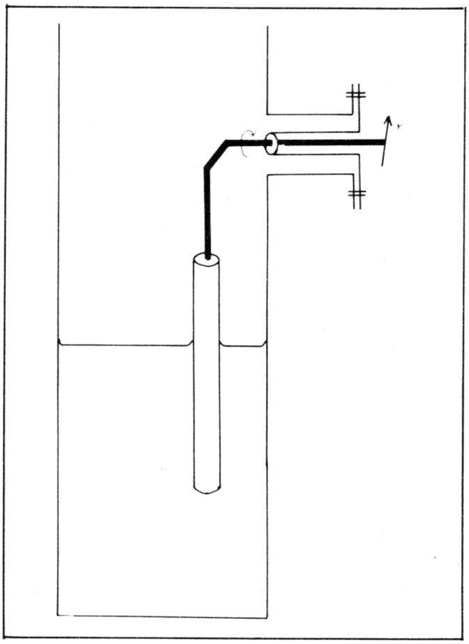

Figure 34 – Torque tube level gauge

A device frequently used for level control, where reliability rather than accuracy is the key, is a torque tube level gauge. This consists of a bouyancy float suspended inside the tank from a torque tube. As the level rises the bouyancy of the float increases and, through a right-angled linkage from the float to the torque tube, the tube is twisted and the level measured. Using this type of device, levels can be measured over the range from 1 to 20 m depth (Figure 34).

Flow Measurement

Differential Pressure

Probably the most widely used type of flowmeter is that which depends upon a differential-pressure measurement and, whatever the construction of the meter, the principle involved is the same.

The cross-sectional area of the flowing stream is reduced, causing a velocity increase and hence an increase in the kinetic energy. Since energy cannot be created or destroyed, this kinetic energy is obtained at the expense of pressure energy, so the pressure of the fluid is reduced. The pressure reduction, or pressure differential, is measured and thus the flow velocity and hence volume or mass flow can be determined.

To generate this pressure difference in a pipe a restriction is inserted, usually an orifice plate, and the pressure drop is measured (Figure 35).

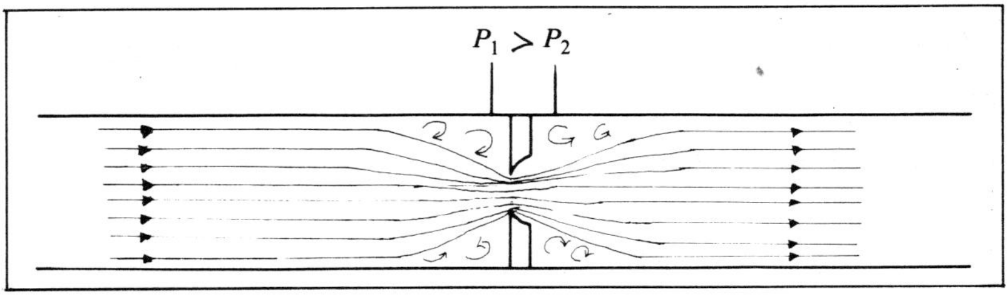

Figure 35

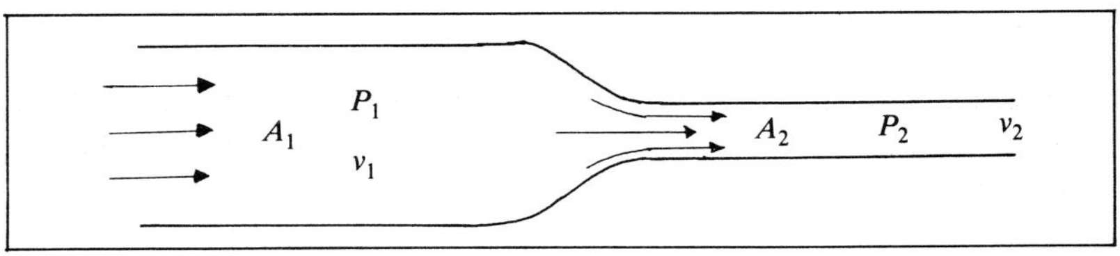

Figure 36

Before looking further at this type of meter let us derive, in a simple form, the pressure/flow relationship.

Assume that we have an incompressible fluid flowing in a smooth-walled, horizontal pipe, where the cross-section changes, as shown in Figure 36, from A_1 to A_2.

The fluid velocity will have to increase from v_1 to v_2 to accommodate this reduction in cross-sectional area, and to force the fluid along this narrower section will require p_1 to be greater than p_2.

Since energy cannot be created or destroyed, the fluid energy of both sections of the pipe must be equal.

Therefore we see that the

Kinetic Energy + Potential Energy + Pressure Energy + Internal Energy

must be equal in sections 1 and 2, i.e.

$$\tfrac{1}{2}mv_1^2 + mgh_1 + \frac{p_1 m}{\rho_1} + I_1 = \tfrac{1}{2}mv_2^2 + mgh_2 + \frac{p_2 m}{\rho_2} + I_2 \tag{1}$$

where m = mass, g
v = velocity, m/s
g = gravitational acceleration, m/s^2
h = height, m
p = pressure, N/m^2
ρ = density, g/m^3
I = internal energy, J

Kinetic, potential and internal energies are well known, pressure energy is derived as follows:

$$\text{Energy} = \text{Force} \times \text{Distance}$$

$$= F \times \frac{V}{A} = \frac{F}{A} \times V = p \times \frac{m}{\rho}$$

where V = volume, m^3
A = area, m^2

If the temperature remains the same (ideal conditions) then $I_1 = I_2$, if the pipe is horizontal, $mgh_1 = mgh_2$, and if the fluid is incompressible, nearly so for liquids, then $\rho_1 = \rho_2$.

Therefore Eqn 1 reduces to

$$\frac{v_1^2}{2} + \frac{p_1}{\rho} = \frac{v_2^2}{2} + \frac{p_2}{\rho}$$

i.e. $$v_2^2 - v_1^2 = \frac{2(p_1 - p_2)}{\rho} = 2\frac{\Delta p}{\rho} \tag{2}$$

Quantity flowing $= Q = A_1 v_1 = A_2 v_2$

$$\therefore \quad v_1 = v_2 \frac{A_2}{A_1} \tag{3}$$

Substituting Eqn 3 in Eqn 2 we get:

$$v_2^2 \left(1 - \frac{A_2^2}{A_1^2}\right) = \frac{2}{\rho} \Delta p$$

$$v_2 = \sqrt{\frac{2 \Delta p}{\rho\left(1 - \frac{A_2^2}{A_1^2}\right)}}$$

i.e. $$Q \propto v_2 \propto \sqrt{\Delta p}$$

The key point to note is that the flow rate, either velocity, mass or volume, is proportional to the square-root of the pressure difference.

Flowmeters

Orifice Plate Flowmeter

The most common and the simplest way to introduce a restriction into a pipeline is to install a plate drilled with a hole obviously smaller in diameter than the pipe diameter itself, the plate being inserted into the pipe between flanges in the pipeline. The plate is usually of stainless steel, Monel or gunmetal so that the process fluid does not corrode it, and the orifice is drilled, usually concentrically, with the leading edge square. The pressure distribution across the plate is shown in Figure 37.

The pressure drop to be measured can be taken between various points, either from well upstream to well downstream, the pressure loss being measured directly through the flanges, by corner tappings directed to either side of the orifice plate or between the *vena contracta* points, i.e. the points of maximum and minimum pressure shown in Figure 37.

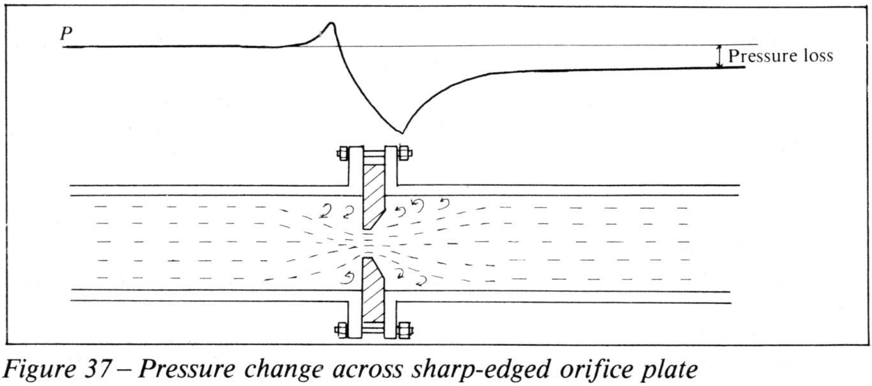

Figure 37 – Pressure change across sharp-edged orifice plate

So as not to lose too much pressure, the orifice diameter is arranged to give a pressure drop of 0–50 cm water gauge minimum to possibly 0–500 cm water gauge maximum. Smaller pressure drops than 50 cm would lead to unacceptably high pressure measurement errors, while pressure drops of greater than 500 cm would be wasteful of pressure.

The calculation of the required orifice size for a given pipe diameter and flow rate is carried out by reference to British Standards publication 1042 or the manufacturers' literature, the orifice diameter, d, being within the range $0.3D$ to $0.8D$, where D is the pipe diameter. An initial calculation is carried out, backed by many correction factors from experimentally determined graphs covering such factors as fluid compressibility, viscosity, pipe-wall roughness estimates, amount of turbulence or Reynolds Number, etc. The installation has to be carefully considered to prevent errors arising from dirt or liquid in gas streams, gas in liquid streams, the effects of swirls caused by valves, bends and elbows in the pipeline, and so on. Periodically the orifice plate must be removed for inspection, since it must remain flat, not bowed, and the orifice edge must remain sharp, not erroded by the flowing fluid. Errors of less than $\pm 1\%$ over lengthy periods are difficult to achieve, and a good installation would normally be quoted as $\pm 2\%$. To reduce the swirl effects caused by bends or valves, the orifice plate must be fitted into a long straight section of pipe, not always easy on site. There should be a minimum upstream straight section of 20 pipe diameters and a downstream straight of at least $5D$ to give reasonable measurements, and, for errors of $\pm 1\%$ or less, the straight sections for gas flows should be $130D$ and $10D$ and for liquid flows should be $50D$ and $10D$. Because of the square-root law, the rangeability is only about 3:1, often a real problem, that is, if F is the minimum flow that can be accurately measured, then the maximum flow that can be measured using this same flowmeter will be only $3F$.

Variable-area Flowmeter

Similar in principle to the orifice plate flowmeter, the rotameter consists of a tapered tube and a metal float (see Figure 38). The float can move vertically within a tapered tube so that the upward forces exerted upon it by the flow are exactly balanced by gravity. Instead of having a fixed orifice and then measuring the variable pressure differences as above, the rotameter has in fact a fixed pressure drop, i.e. the float weight, and the annulus around it varies. The position of the float gives the measure of flow through the meter. These flowmeters normally cover the range 0–0.5 m^3/s for gases and 0–0.1 m^3/s for liquids with a range of 10:1 and errors of $\pm 2\%$.

Magnetic Flowmeters

Here the velocity is determined by measuring the electromagnetic effect it produces.

Whenever a conductor cuts lines of magnetic force an e.m.f. is generated, the magnitude of which is proportional to the rate at which the lines of force

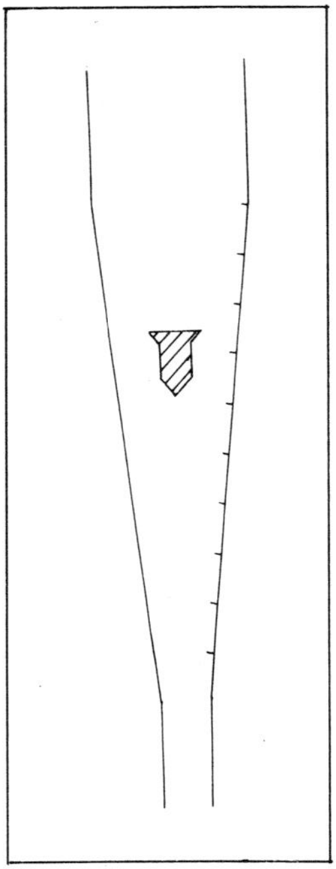

Figure 38 – Variable-area flowmeter

are cut. The e.m.f. is generated at right angles to both the velocity and the direction of the magnetic field (Figure 39).

$$\text{e.m.f.} = B.D.V \text{ , V}$$

where V = fluid velocity, m/s
B = magnetic flux density, Wb/m^2
D = pipe diameter, m

The metering tube must be non-magnetic, or lined with an insulating material, and a uniform magnetic field is induced across it. The flowing liquid, which must be conducting, is itself the moving conductor and the meter then measures the average flow velocity across the pipe. Metering tubes vary in size from 3 mm to 2 m.

Since the meter consists of a length of parallel pipe around which a powerful electromagnet is wound, the device presents no obstruction to the flow and therefore no additional pressure losses are encountered, its main disadvantage being that only conducting fluids can be metered and consequently it is not suitable for gas measurement. Errors of less than ±1% over a 10:1 flow range are attainable after calibration and the meter can work at temperatures up to 200°C.

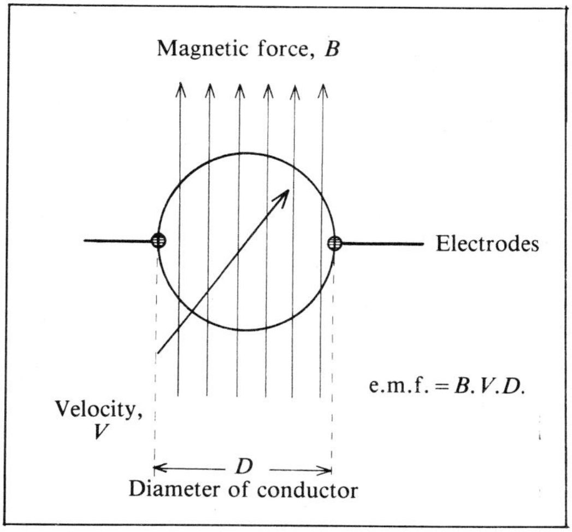

Figure 39 – Electromagnetic flowmeter

Other Flowmeters
There are a host of other types of flowmeter, including the positive displacement type, familiar at the petrol stations, where the actual volume of fluid, passed at line conditions in a given time, is measured by dividing the flow into discrete volumes and summing the volumes as they pass through the meter. Usually these meters are restricted to monitoring very clean fluids.

Turbine flowmeters incorporate a free-running bladed rotor mounted within its housing co-axially on the pipe centre line. The rotor is driven by the flowing liquid and sweeps almost the entire annular area with an angular velocity proportional to the mean fluid velocity; the speed of rotation is therefore almost proportional to volume flow rate. These meters give a fast response to changes in flow, show an excellent repeatability, have wide rangeabilities, but of course they are badly affected by slurries or liquids containing suspended matter. Flow velocities up to 10 m/s for liquids and 50 m/s for gases can be measured in pipes of diameter 5 to 50 cm.

Further meters based upon the Doppler effect, vortex shedding, thermal effects and so on are also available.

5 MICROPROCESSORS

The Microcomputer

As described earlier, much of the control in the process industries is single-loop control, where point measurements are taken and fed back to control single variables. However, processes are not as simple as this; any one output will almost certainly be dependent upon and affected by a number of inputs, and varying any one input will cause a number of outputs, not just one, to vary. We do not have independent single loops, but rather complicated multi-input, multi-output systems with variables depending upon and interacting with each other. Conditions will vary from time to time on the plant, e.g. ambient conditions, feedstock compositions, catalyst activity, heat exchange coefficients, availability of raw materials and steam supply. Also the order-book position will vary. Thus, the optimum running conditions for the plant will change and the plant control system should change to compensate for these variations.

When the computer first became available for plant control in the 1960s, people had high hopes that it would revolutionize plant control and provide fully automated plants running under optimized conditions. Since the cost of computers and their installation on site was high and the cost of writing the control software equally high, people felt that they must get maximum use from any computer installation, and possibly they were over ambitious.

During the last 10–20 years two main changes have come about. Many installations have been commissioned and common software has been written. The cost of this software, or computer programs, can therefore be shared by many users and hence the cost per installation is very much reduced. Secondly, the cost of the computer itself, the hardware cost, has fallen in absolute terms, and fallen very much in real terms as other costs have risen over these years. Computers, and especially now the microcomputers, are therefore relatively cheap to buy and install. They are also very reliable indeed; so we do not need to worry over their causing frequent plant shutdowns and the very rare fault that must still occur will, by using a number of microcomputers on one plant in place of only one computer, affect only a small part of the plant, which could go on to manual control temporarily.

What then have microcomputers to offer the control engineer, and what can microcomputers do? In fact, what is a microprocessor or microcomputer?

A microprocessor and the parts that connect to it form a microcomputer, and this computer is little different from a conventional computer except that through advances in manufacturing processes it has become smaller and much cheaper. Recent developments allow a large number of electronic devices to be built onto a single thin wafer of silicon, a silicon chip, certainly much smaller than a fingernail. This merging of several components onto one chip is known as integration, and hence integrated circuits. In the 1950s radios with six or seven transistors were available, by the mid 1960s chips were available with about 16 transistors on each, since then the number of transistors that can be put onto a single chip has grown enormously, so that by the mid 1970s chips containing 65 000 transistors could be obtained, and the cost per transistor was markedly reduced.

What does a microprocessor do?

It performs logical and arithmetical operations on binary numbers, i.e. it looks at numbers 0 and 1, and can add them, subtract them or compare them, and it can time events. It can only perform one operation at a time, but can do so rapidly obeying several hundred thousand instructions each second. It carries out its duties by following a sequence of instructions written into its memory or program store and these instructions define exactly how the microprocessor will behave. The microprocessor keeps a record, the program counter, a form of book mark, saying where it has reached in its sequence of instructions. It takes an instuction from its memory store and performs the instruction, which will be a simple one such as to take a number from store or to add a number to the total already brought from store, etc. To carry out a complex task means the computer storing many instructions and carrying out hundreds of thousands of operations each second. An electronic clock can interrupt a number of times each second (or an external signal or the operator could interrupt the computer), the microprocessor then carries out a separate list of instructions, which could be to record the time, inspect some sensor for a signal, etc., and when this sub-set of instructions is completed, the microprocessor goes back to its bookmark and continues with the list that it was executing before the interruption.

The writing up of these instructions for the computer is a specialized task, but so many jobs that the computer has to do are repetitive or similar to that required of other computers that a library of software programs has been developed that can be stored within the computer. Instructions for control algorithms, cascade control systems and such like are available in software packages and the job of installing the microprocessor is to tailor what is available to your own particular requirements. The manufacturers can provide much help, but once again it must be stressed that the successful applications come when the local control engineer knows enough programming to allow him to write the additional software for his own problem – not too difficult a task since very good manuals are now available from the manufacturers. For continuous plant control, most of the software is available and the user simply fills in a sort of questionnaire to fit the computer to his own control scheme, but for the batch plant things are still a little more complicated. Some computers require a step-by-step description of what is to happen, similar to that which must be prepared for a relay-type control system, although some later systems do accept a simpler, more descriptive set of instructions.

The microprocessor, then, is an integrated circuit made on a silicon chip and when added to other integrated circuits it becomes a computer (see Figure 40). These other circuits include a program store, which holds the instructions that organize the computer, a data store, which holds data temporarily while other data is being controlled or processed, a clock to allow the computer system to carry out time-keeping operations, such as sequencing, routine checks, etc., and interface circuits to connect the computer to the outside world or sensors, switches, keyboards, displays, printers and so on.

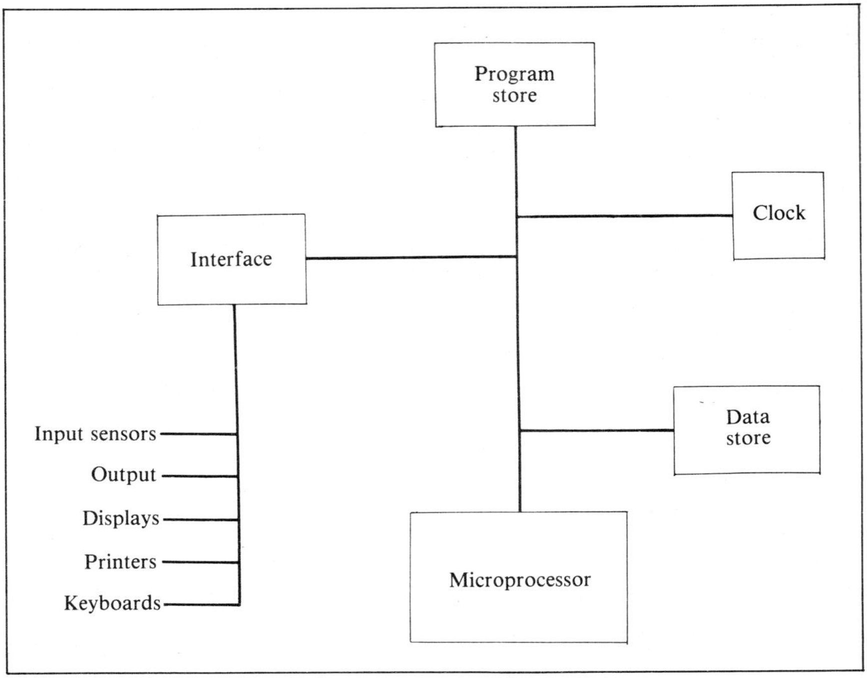

Figure 40 – The microcomputer

The data processed by a computer must be presented in groups of words, a word being a group of bits, and a bit is represented by 0 or 1. Microprocessors usually use 4-, 8- or 16-bit words.

In a 4-bit word the bits 0s and 1s can be grouped from 0000 to 1111, sixteen different combinations; ten of these combinations are required to represent the numbers from 0 to 9. With an 8-bit system there are 256 different combinations and this is necessary for measurement duties since 1 in 200 gives ½% accuracy. In addition, the standard code for representing the typewriter keyboard uses 8-bits. Using a 16-bit word gives 65 000 different combinations and is obviously very useful where high accuracies are required for measurement problems, and is certainly an advantage for process control. Variations can of course be used; for example, you can use two 4-bit words together in a 4-bit system to improve accuracies or two 8-bit words to represent a 16-bit system, but the instruction lists will be longer; if you are using a 16-bit system you could simply ignore eight bits of each word as though with an 8-bit system. The longer the word length, the more wires to carry the information and the more chip space are required, hence costs rise. Since computers handle all the bits of a word in parallel, the more bits in a word the faster the operation and this is the real difference between the 4-, 8-, and 16-bit machines.

Microprocessors in Control

Before the introduction of the microprocessor, a typical control system using a single mini-computer would be as shown in Figure 41. All the measured variables are fed to a multiplexer that switches each in turn to the analogue to digital converter (ADC), which produces a digital output equivalent to the analogue input. The other inputs to the computer are a timing pulse, from which the appropriate sampling rates are generated, and a keyboard, which allows on-line operator intervention. The numerically computed output values will then be changed via digital to analogue converters (DACs) into control signals. In the early days voltage input and output signals were quite common, but the 4–20 mA current loops are most frequently used now, since they suffer little from pick-up due to electromagnetic induction. The floating zero (i.e. 4 mA representing zero signal) means that transmission-line discontinuities cannot be confused with a zero signal.

All computer control schemes include much the same hardware and they all present similar problems to the software designer, who has to prepare the instructions or programs to run the system. It is, by and large, the software that defines the particular mode of operator interaction and that gives each system its distinctive 'personality'. The software is often written by the computer manufacturer so that the user is only called upon to configure the control scheme by specifying the control loops and the control constants for these loops. The general philosophy has been to make the computer control system look very much like the analogue discrete controllers.

Using a minicomputer for process control has one big worry, the eggs are all in one basket. To prevent a full plant shut-down being necessary if the

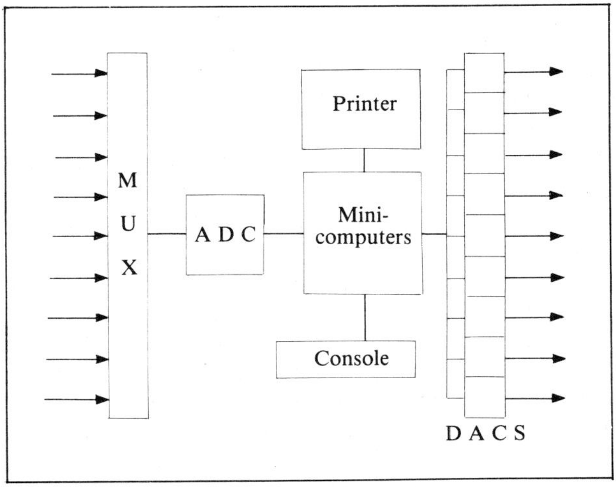

Figure 41 – Computer control system

computer fails means either using some conventional instrumentation for back-up of essential control loops or a dual processor configuration with specially written software so that the second processor can take over if the first fails. This second processor may be a stand-by for two or three-control computers, or could be used on separate tasks of data logging and analysis. However, to step in at a moment's notice and take over control automatically if the on-line computer fails means all the information instructions and so forth being up-dated constantly in both systems. Hence there are so many management and housekeeping tasks to be performed that the control algorithms are quite insignificant; programming is definitely a job for the professionals, not the control engineer.

Microprocessors have changed these considerations considerably because they are cheap enough to use in large numbers and so the processing power can be spread all round a system and things can be kept simpler by using one processor for a limited task. Also, without doubt, the reliability of these integrated-circuit systems is very high indeed.

The microprocessor will not simply be calculating, one of its major tasks will be data presentation, either textual or graphical, and by better marshalling and presentation of data the operator's task is simplified. Colour TV screens with bar graphs, flashing messages, filing of records, all help the operator and plant manager and allow control rooms to become smaller and cheaper.

One main area of control so far rather neglected is batch control. The lower cost of modern sophisticated hardware must mean that things are to change greatly in the next few years. We already have the programmable logic controller (PLC), but this is essentially a microprocessor replacing a cabinet full of relays. To program a sequence one has to describe a solution in terms of a parallel-acting set of relays or in logic units, i.e. ladder logic or Boolean logic respectively. By adding, as has been done, the three-term control algorithms giving P + I + D control for single loops, one can carry out very good batch control, but the systems are rather clumsy to set up. New software is needed so that batch processing can be specified in a much more natural way using a simple, high-level language.

As far as computer control is concerned, we are again at the beginning of a new, exciting and challenging era.

Chapter 6

Design Considerations for Laboratory and Pilot-scale Equipment

S S SMITH

INTRODUCTION

Laboratory and pilot-scale apparatus for the application of dyes and chemicals plays an important role. In this chapter the merits of individual machines are considered and, where applicable, those aspects of design relevant to the production of results comparable with full-scale machinery are assessed. Coloration and chemical or finish application is but one facet. In many instances, and in the principal problem areas, fabric handle and construction are of major importance.

Where practicable, it is not uncommon to submit for approval material samples included in a production run. This may not be acceptable, of course, when a different colour of a newly formulated fabric construction is required. In this instance the most economic and practical way is to prepare the submission on pilot-scale equipment. If this is necessary it is imperative that the end product is not only visually in accord with that produced in bulk but, equally, should possess the necessary characteristics of handle and linear specification. Such samples may be required for both design and wearing trials.

Additional requirements for such equipment are as teaching aids to simulate textile dyeing and finishing machines and also as tools in research and development by dye makers, chemical manufacturers and fibre producers.

In this chapter the following titles will be used to classify the general broad field of sample machines.

(a) *Laboratory Equipment* – units used simply for comparative, routine dye and chemical recipe formulation, fastness testing, etc.

(b) *Pilot-scale Equipment* – Those machines
   (i) intended to simulate the mechanical operation of a production unit and/or reproduce the results of such a unit
   (ii) used for production purposes, e.g. continuous processing of narrow-width fabric and tapes, small-lot dyeings, etc.

In many instances machinery of the second type can be used for the first and vice versa.

Years of development involving a degree of trial and error have shown that it is easier to scale down proportionately or miniaturize some production units more than others. In general, pilot-scale package- and hank-dyeing and beam dyeing machines are more successful than winches or jigs; specific reference is made to the subject under the appropriate headings later.

However, machines are not the only factor; the process itself may have an equally important role to play. Where a high degree of exhaustion is achieved during bulk dyeing of, say, wool, nylon, polyester or acrylic materials, good colour reproducibility may be obtained by a relatively simple pot dyeing technique using beakers, viz. in the formulation of a dye-prediction recipe for a production jet-dyed polyester fabric it is possible to obtain good correlation in a simple laboratory high-temperature machine. On the other hand, where a large sample is required or where handle and finish must also be considered there is no alternative to processing on a pilot-scale jet. It is thus clear that knowledge and experience are needed to select the right equipment for a specific purpose.

In designing a particular unit the apparently simple operation of scaling down linear size is often possible. On the other hand, textile fabrics, although readily reduced in length and width, cannot be reduced in thickness or density and still maintain the same characteristics. A wool fabric to be processed on a bench pilot winch while retaining thickness and linear weight will be subject to distortions affecting its rope-formation characteristics, angular velocity, interface frictional characteristics and consequently final handle and linear form. A wound package of yarn cannot be reduced in diameter without increasing its porosity against a given pump pressure.

Consideration of these simple examples emphasizes the problems encountered in trying to reproduce the characteristics of a production jig in a bench pilot jig. In particular, as with some other machines, the basic requirement of a scaled constancy of dyebath immersion and reaction time, liquor ratio, fabric geometry and speed is impossible to achieve. It should be noted, however, that the degree of miniaturization is also important; it should not be concluded that all 'smaller than production' machines are incapable of producing useful comparative results.

Liquor ratios are determined by technical and economic considerations, but in a given dyebath the material immersion times, time/temperature constants, fabric speeds and/or liquor flow rates are the factors which control uniformity and penetration. These, whilst carefully monitored in production, may in some circumstances of sample processing be irrelevant. In fundamental research it is nevertheless necessary to be able to evaluate the effect of time and temperature on varying concentrations of dispersions and solutions.

Experience has shown that at least two of three factors will be required to remain constant, namely the time constant of dyeing, the liquor-to-goods ratio and dye concentration. The latter two are closely interrelated and in most instances it is sufficient to control only one of them as, for example, in a sample machine where it is technically impossible or impractical for the liquor-to-goods ratio to be maintained. In this case the dye concentration relative to the substrate weight will be the crucial factor and a degree of tolerance in the liquor-to-goods ratio will have to be accepted.

Where the ion or electrolyte concentration is a significant factor in dye absorption, the linear ratios may not be acceptable and it will be necessary to use alternative apparatus. There are, therefore, a set of references which provide a guide to the construction of a pilot-scale machine. Where these cannot be maintained, ingenuity in design and application is called for. Throughout the text dimensions, various capacities, weights and volumes have been used. These are provided as a guide and should not be interpreted as limiting. The broad spectrum of machinery and the factors affecting design are considered, recalling, as concluded by Jones(*1*), that the dyeing process can be divided into three distinct steps:

1. Transport of dye through the dyebath to the fibre surface
2. Absorption of dye at the fibre surface
3. Diffusion of dye within the fibre.

1 EQUIPMENT FOR ROUTINE WORK AND FUNDAMENTAL RESEARCH

Excepting the simple hand-stirred beaker or flask, the most widely used items of laboratory equipment for dye application are those operating by

1. A vertically reciprocating agitation system commonly referred to as the Marney elevator or dunker type
2. Other self-generated liquor-circulation principles
3. The rotating-pot principle.

The majority of laboratories have occasion to use at least one as a piece of standard equipment. They are basic to their needs for routine development, comparative work and recipe formulation.

In any simple operation of dye application the following conditions are controlled or created

1. Dye and chemical concentration and/or liquor-to-goods ratios
2. Time/temperature profiles
3. Liquor or fabric agitation or movement.

Whilst the first is created by the researcher in preparing the sample for test, the latter two are generally an automated requirement of the performance of the machine.

Laboratory machines have been engineered to remove the burden of manual agitation, to provide a system of liquor or material movement and temperature control with time, thus completing these conditions.

In the dunker type, the test vessel in which samples are processed is placed in a water or oil bath, which in turn acts as the basic heat-transfer medium. The term oil bath is a generalization for media with a boiling point in excess of the desired maximum heat-transfer temperature. They are generally employed where temperatures in excess of the atmospheric boil are to be achieved. It should be remembered that at high altitudes oil baths are a prerequisite even for temperatures approaching 100°C. The sample is attached to a specially designed former whilst the reciprocating motion of the machine causes this to be agitated or liquor to be circulated within the test vessel.

The sample under test in the rotating-pot type machine is placed in its vessel, usually in such a manner that its own movement is restricted. There are several exceptions, especially in leather dyeing where movement of the material to create a pouncing effect is desirable. The test vessel is rotated in the heating medium thus creating a state in which the liquor is effectively circulated or percolated through the material sample.

The principal features are considered below in general terms, though it is not intended to compare the merits of machines of specific manufacture.

Dunker or Elevator Type

One of the main problems with this type is the difficulty of obtaining low liquor ratios. At best 20:1 can be achieved, but in general longer ratios are necessary for satisfactory operation.

A simple up and down reciprocating motion of the material sample does not always produce level dyeings and the machine manufacturer's ingenuity has been used to improve the efficiency of the agitation system by, for example,

1. Introducing an alternate spiral motion on the down and up stroke
2. Improvement in the design of the material holders to increase liquor disturbance.

Both atmospheric-pressure and high-temperature-pressurized machines have been developed. In favour of the atmospheric unit is the ease with which samples of liquor can be removed or added to the system. However, unless a cover or reflux condenser is fitted to the test vessel there is a continuous loss of liquor due to evaporation, and because the total volume of the bath is relatively small the loss can have a significant effect on the results.

In the high-temperature versions a sealed lid forms a part of some models whilst in others the test vessels are enclosed within a steam autoclave. In the autoclave type the sampling of liquor and the making of additions is not generally possible without first cooling and equalizing at atmospheric pressure. Additionally, there is a tendency for partial vaporization of the dye-liquor to occur. Nevertheless such machines have proved successful and, for most works laboratories, the extra facility of adding or sampling at high temperatures has only recently been of real importance. Where a sealed test vessel is used, methods of agitation and/or circulation fall into two categories:

1. 'Mechanical' drive to provide up and down movement of the sample holder with a seal on the shaft, or elimination of the shaft seal by an external magnet operating on the internal shaft
2. Direct liquor circulation through the sample by means of an internal 'pump', positively or magnetically driven.

Where there is a requirement to monitor the dyeing or other process whilst under pressure, this cannot be performed in a steam autoclave type. However, the sealed test vessel does lend itself to monitoring and the techniques used operate on the same principles and sampling procedures as are employed on production machines, viz. continuous visual bypass or pressure lock systems presenting a cooled extract of liquor. Further additions are similarly made from an isolated chamber either by equalizing the pressure in this chamber or by using a metering pump.

Rotating-pot Type

One of the main advantages of the rotating-pot machine is its ability to achieve relatively low liquor ratios. Within the general limits 8:1 to 40:1 and depending on the bulkiness of the material under test, this system works well for materials in all forms, although special material holders or skein carriers are necessary in some instances. Even so it is rarely possible to match liquor ratio conditions of a production jig, but the system does permit close proximity to such ratios and is suitable where high degrees of exhaustion are achieved in production.

A significant feature of a rotating-pot machine is the design of the test vessel itself and the simple way in which pressurized conditions can be created while enabling several individual tests to be carried out at one and the same time.

Polypropylene test vessels may be suitable for low-temperature work but stainless steel is more commonly used, the quality of which is important. Whilst for many applications 321 grade stainless steel may be adequate, corrosion in the presence of chloride ions will occur and to minimize this problem the higher grade 316 is usually used; but even this quality under concentrated saline conditions will suffer corrosion.

The narrow cylindrical shape of the test vessel is important and it is generally considered that for dye application purposes fabric, yarn or loose stock should be maintained in a neat, virtually stationary condition. Where squatter shapes of test vessel are used samples move more freely, but they tend to move with the liquor and, generally speaking, the uniformity of dyeing is not as good. There are exceptions, for example in leather dyeing and other instances where movement of the substrate is an essential requirement for good results; in these instances the squatter shape would be chosen.

Different, sometimes variable, rotation speeds and directions are used, it is claimed, to optimize level dyeing, but in the author's opinion, within certain speed limits, they have very little relevance in those machines using end-over-end rotation.

In addition to steam autoclave types in which sealed test vessels are rotated, many others are available and use a heat-transfer medium capable of being raised to at least 135°C. This temperature is usually regarded as the maximum required and is the minimum design temperature of both high-pressure production and laboratory machines.

Typical heating media used for achieving such temperatures are the monoethylene and polyethylene glycols, but these do have the disadvantage of breakdown and after continual use the baths emit fumes and become unpleasant to handle. A method has been developed in which the liquid heating medium has been replaced by a fluidized bed. Here a controlled volume and pressure of air is passed through a bed of sand or small glass beads known as ballotini. By controlling the air flow the bed is made to behave like a liquid and becomes the heat-transfer medium. The fluidized bed is heated and cooled in a conventional manner and serves as the heating medium in which the test vessels are rotated.

The principle advantages of such a system are self evident: cleanliness because the medium is dry, there are no fumes or odour, should a leak occur from the test vessel due to carelessness in tightening it is self apparent, and the likelihood of an operative being burned as a result of direct contact with the heating medium is small.

The conductivity of a heating medium is important and the fluidized bed, being particularly efficient, offers advantages in this respect. Further, there is evidence that the usual temperature drop betwen the heating medium and the test vessel wall is very small and therefore the temperature inside the dye tube at equilibrium is higher than when using liquid media.

The choice of liquid heating media will be influenced by the following.

1. *Miscibility with Water* The ideal should be readily miscible with water thus facilitating easy cleaning of the test vessel when the latter has been removed from the laboratory dyeing machine.
2. *Viscosity* A highly viscous medium is generally regarded as unsatisfactory because of its relative inefficiency as a conductor and because it has a natural tendency to adhere to the dye tube during rinsing. In addition it is physically more difficult to ensure temperature uniformity of the bath without imposing vigorous agitation – particularly so at lower temperatures.
3. *Conductivity* Conductivity naturally determines the medium's efficiency and is, in part, influenced by viscosity.
4. *Boiling Point* The boiling point of the medium must of necessity be greater than the maximum temperature desired. It should have good stability at the maximum operating temperature. Whilst accepting this fact, boiling point should nevertheless be as low as possible, to provide the maximum level of safety.

Should all other safety or control devices fail, the highest achievable temperature will be determined by the boiling point of the heating liquid or, alternatively, the maximum heating capacity of the heating element. For this reason the capacity of the heating element should be the minimum consistent with the maximum rate of temperature rise required for processing.

In both dunker and rotating-pot machines there is a temperature difference between the heating medium and the recipient test solution. Whilst this will depend on the forced convection of the fluid and consequently the overall heat transfer coefficient of the test vessel and heating medium, at equilibrium the difference is minimal in closed vessels but could be several degrees in an open-pot system. This difference is easily determinable by use of maximum-reading thermometers and has to be allowed for by the operative.

An important consideration in the design of closed vessels for use under pressurized conditions is the necessity to prevent excessive build-up of pressure. If a test vessel is filled to capacity with water and no air space remains on sealing, the hydraulic pressure created by expansion of water simply by raising the temperature through, say, 100°C is very considerable. Under these conditions such test vessels can explode and preventive measures are a necessary design feature.

The most satisfactory method introduced is the incorporation of a cavity in the lid to allow for the expansion. The cavity must accommodate more than the increased volume of liquor created by raising the temperature to 135°C but also allow for the increased volume developed through expansion should all the control systems fail. In this instance the determining factor will be the maximum temperature attainable.

Both dunker and rotating-pot machines are capable of giving good recipe predictions where the conditions of the dyeing trial and those in production permit a high degree of dye exhaustion. However, as previously stated, in any process demanding shorter liquor ratios the rotating-pot machines are more suitable.

A facility for the addition of dyes and chemicals is useful in certain instances and, whilst this is always easy in an open-pot atmospheric-pressure machine, the high-temperature versions are highly sophisticated(*2*). Here the progress of small sample dyeings may be monitored either visually or colorimetrically, and the liquor additions made as required. Such apparatus is necessarily expensive, which limits its widespread adoption.

It is perhaps noteworthy that a report of the Committee on the Dyeing Properties of Disperse Dyes(*3*) referred to the agreement between six different laboratories using ten machines of four different rotating-pot types – the test methods being based on the use of this type of apparatus primarily because of ready availability and general use.

2 WINCHES

The extent of miniaturization is particularly relevant in producing pilot-scale or laboratory winches. Whilst it is possible to scale down proportionately and produce a bench-standing unit, it has very little practical value as a piece-dyeing apparatus.

Rope formation, compression and drag are the most important features in trying to reproduce bulk conditions, and miniaturization to the extent that full-width fabrics cannot be handled is a determining factor. The majority of sample winches are used to process short lengths of production-width fabrics and become less of a practical tool and more of a piece of laboratory hardware when scaled to take anything significantly narrower. The exception to this requirement of rope formation and compression would be when processing open-width materials as exemplified by tufted carpeting.

Where textile materials have to be processed in rope form the width of that rope, and therefore the capacity of the sampling machine to accommodate that width, is significant. Provided that this permits one end to be accommodated widthwise, then the machine will generally produce satisfactory results, though other features such as creel diameter and speed cannot be proportionately scaled. The plaiting operation necessary for some fabrics in production is usually less critical on sample lengths and is determined largely by the length and buoyancy of the sample. Short pieces require little or no plaiting, but it is advantageous in designing a pilot winch to incorporate a variable-speed drive and variable-geometry creel, permitting adjustment in shape diameter and height above the liquor level. There is little merit in building shallow-draught winches for short sample lengths, and any plaiting on this scale is usually adequtely catered for by the variable-geometry creel. It should be noted in particular that laboratory and pilot units both in this instance and most others are not proportionate miniatures but are designed to provide a facility for dyeing samples.

The designing of a pilot winch serves as an excellent example of the sometimes impossible task of linear miniaturization whilst at the same time trying to achieve proportionate sample processing. Consider the following:

(a) If the width of the fabric is reduced significantly, the now narrower fabric will not form a rope in the same manner as a full-width fabric
(b) If the linear dimension of the unit is significantly, though proportionately, smaller than the production machine and the angular velocity of the creel similarly scaled, then the dynamic interaction of wet rope and winch creel will be different
(c) Where a proportionately scaled-down unit is disproportionately adjusted to counter any resultant processing defect of linear miniaturization, the bath volume changes disproportionately.

While it is desirable to maintain equivalent concentration and liquor ratio, the time of revolution of the rope is more of a theoretical value and in practice, both on production machines and pilot winches, control of speed is an acquired skill. Sample dyeing on pilot winches is completed in a much shorter period of time, as the action of the machine, whilst less severe mechanically, by necessity processes the shorter piece through many more revolutions. However, the final handle and finish, whilst influenced in part by interaction at the creel face and the number of passes through the bath, is often a function of some preparatory or finishing process. In practice a compromise is established. The ideal, assuming reasonable rope length, is a single-draught machine of the necessary width to accommodate that draught. The depth and breadth is calculated to maintain the liquor ratio and the winch creel diameter becomes disproportionately large. The remaining variable is the creel shape and speed, which are adjusted in the light of practical experience.

It will be seen that where a natural rope formation is a necessity of the process, directly proportional miniaturization of the machine is limited. However, it should not be forgotten that many bench-scale models do exist and, whilst these are unsuitable for processing full-width fabrics, they may be suitable for comparative laboratory work, for processing ribbons or narrow-width flat or circular fabrics.

For the sole purpose of dye-recipe prediction the sample winch has limited value, and is best replaced by other simple reciprocating or rotating-type laboratory machines.

3 PADDLE MACHINES

A provision of some pilot winches, resulting from specific design features, is the alternative use to which they can be put, for example the ability in some instances to change the shape and height of the winch creel means they can be used as overhead-paddle machines. This characteristic is a standard feature of many units and as an overhead-paddle it is suitable for dyeing hose in small quantities. However, such machines also have their limitations and probably the most suitable method for dyeing small quantities of hose is the miniaturization of what is traditionally known as a Smith drum.

Few machine makers have taken advantage of the mechanics of such a machine, but in the author's opinion this type has a number of advantages. The miniaturization of a Smith drum can only remain proportionate and be successful so long as the goods are permitted the necessary degree of movement. A production drum usually comprises three or four equal divisions, but on scaling down to the minimum a single divide is more successful. The effect is to lift and immerse the goods on each revolution, thus encouraging gravitational drainage of the dye-liquor through and over the surface of the goods. If no division is present at all, the tendency is for the goods and the liquor to move in unison and the significance of the drainage in promoting level dyeing is lost. By shaping the outer bath to follow closely the line of the drum it is possible to maintain liquor ratios and therefore such a unit with a single central division is a most practical machine.

Whilst a production overhead paddle can be proportionately reduced, the critical factor is the shape and size of the individual goods to be processed in the machine. It will be apparent that it is difficult to agitate satisfactorily and circulate on a proportionate scale a small quantity of fashioned garments. It is, of course, a question of scale, but suffice it to say that if the goods themselves are disproportionate to the size of the machine, then the miniaturized mechanics of the machine in relation to these goods will not produce a satisfactory result.

Generally speaking, knitwear goods in the form of body blanks or fashioned garments do not lend themselves to be sample-processed in pilot-scale machines where the volume of goods is small and below a minimum weight of, say, 7 kg. The same considerations apply in the Smith drum, side- and overhead-paddle, and sample winch, viz. the extent of miniaturization is limited by the relatioship between the size of the machine and the volume of goods to be processed in it. Using the side-paddle principle it is possible to process successfully somewhat smaller quantities of hose, but there are still limits, and for very small quantities the miniature Smith drum could be used.

Machinery suitable for handling two or three garments has been a constant problem, but a method that operates on the principle of the first domestic washing machines has been developed, i.e. the bath is cylindrical in shape and is fitted with a vertical centre-column paddle as the agitator. Once again, miniature equipment or pilot-scale units cannot always be scaled down proportionately from production machines.

4 JET DYEING MACHINES

For simple recipe formulation, laboratory machines of the rotating-pot type are the most widely used as they can be made to simulate the liquor ratio conditions encountered in jet dyeing machines. Where, however, it is necessary to simulate actual physical conditions of processing to study the effect of shear forces caused by fabric movement and liquor flow in, say, the jet nozzle, the pilot jet machines have to be used.

For more practical commercial uses and where sample lengths of fabric are required for submission, it has been necessary to design units capable of handling production widths. Most manufacturers offer such machines based on the operating principles of their production units, but certain minimum dimensions determine the success and degree of miniaturization. In accepting this fact the basic operating principles and features of a production unit can be maintained.

The capacity and diameter of the storage tube are first requirements and contribute to the control of fabric movement, formation and deformation, which in turn largely influence creasing. In certain respects it is possible to miniaturize jet machines proportionately, but the liquor ratios become longer in pilot units due to the disproportionate volume of pipe work. Assuming that the requirement of the unit is to process production widths, then within reasonable limits the diameter of the machine's storage tube can be from 200–400 mm. The latter dimension more nearly approaches the size used in a production machine but, from experience, it need not exceed the former. Clearly the wider the tube, the greater the storage volume. Equally, the larger the storage tube, both in diameter and length, the greater the quantity of goods that can be accommodated. With any given diameter of tube this quantity can be increased simply by extending the overall length of the storage tube.

The following summarizes the factors the design engineer must consider.

1. In pilot jets the normal liquor ratios tend to become longer. In a jet machine there is of necessity much external pipe work needed to provide circulation, and because the pilot jet nozzle must be similar to the production nozzle the pipe work to it cannot be reduced significantly in diameter.
2. Whilst in production it may be necessary to run at speeds up to 3 m/s to ensure the completion of a full cycle of a fabric length, say, every 2 min, in a pilot machine this can be achieved at a fraction of the same speed. Generally, delicate and other fabrics prone to creasing are less likely to suffer this problem on a pilot jet because the storage time, i.e. the period of restricted movement, is short.
3. Certain fabrics need vigorous and speedy traction to ensure that natural folds are physically rearranged thus eliminating the formation of set-in creases.
4. The diameter of the storage tube is determined by the bulkiness and effective diameter of the fabric to be processed in it. Production widths can be accommodated in tubes from 200–400 mm in diameter, but a narrow-width stockinette type requires something much less (90–100 mm), primarily to prevent entanglement.
5. It is a fallacy to assume that a production soft-flow machine which produces good results can necessarily only be matched by a sample machine operating on the same principles. As with many other pilot machines, practical experience is an additional prerequisite.

Dye uniformity is controlled by maintaining a relationship between liquor circulation and fabric movement. Liquor circulation is a function of pump performance but equally has a bearing (in fully flooded machines) on the speed of traction of the fabric. Delivery of liquor from the pump to the jet nozzle propels the fabric in the fully flooded types, but in some machines a bypass system operates so that the total force is not necessarily directed onto the fabric.

In those machines that are only partially flooded and where a winch creel and/or liquor-overflow system operates, the speed is controlled by the creel, whilst liquor circulation can be totally independent. The requirement that the pump be able to circulate the equivalent of the total volume of liquor, say four times per minute, is usually provided in the pilot unit, but there is no evidence to suggest that this is anything other than a safety factor rather than a fundamental requirement.

The jet nozzle itself cannot be miniaturized in proportion, for whilst its overall length may be reduced, its diameter needs to be identical to that on the production unit. In many respects the effect of the jet nozzle on fabric presents similar problems of miniaturization as does the winch creel (see Section 2).

Fabric speed and liquor circulation are optimized in production to keep fabric distortion and creasing to a minimum, at the same time ensuring level dyeing. The same is required of a pilot unit but, because of the reduction in scale, it is possible to run at slower speeds, and whilst passage through the jet nozzle occurs with greater frequency the hydrodynamic pressure created by jet impingement on the fabric is lower. Consequently a condition of circulation exists that is more frequent but less severe, and the effect on fabric linearity and appearance can be made to match that obtained in production.

5 HANK-DYEING MACHINES

In all hank-dyeing operations there is a requirement to maintain the yarn in good condition. This is achieved by keeping it in a relatively stationary position whilst the liquor is made to flow through it. In practice there is very little difficulty in reproducing production conditions on both pilot units and in basic laboratory equipment.

In dunker and rotating-pot machines, simple hank holders, on which the yarn is mounted, are used in the test vessels, whilst pilot units of proportionately scaled production units are often used. As with production machines the most important features of a pilot unit are:

1. Uniform loading to give an even pack thus eliminating channeling of the process liquor
2. Controlled or variable flow rates from the pump or impeller ensuring a minimum of turbulence but the necessary volume circulation. Many machines operate on a unidirectional flow basis but two-way flows give the operative greater flexibility.

Atmospheric pressure and high-temperature units are both feasible and, apart from a tendency to have slightly longer liquor ratios, there are no difficulties in miniaturization. The more successful pilot units are those incorporating the two-stick principle. These are usually and by necessity adjustable, catering for the varied hank lengths and, particularly in the case of acrylic dyeing, allowing for the shrinkage which takes place due to its thermoplastic properties.

Variables within the pilot machines are the same as for production units and it is a case of selecting the one best suited to the individual's needs. From a designer's point of view it is simply a question of knowing capacity requirements.

Pilot units based on production machines cater for sample weights in excess of 1 kg but samples below this weight are generally considered to be of laboratory scale. The hanks are supported on holders and processed in dunker or rotating-pot machines.

6 PACKAGE-DYEING AND PRESSURE DYEING UNITS

These machines use pump circulation as a common principle and there are both atmospheric pressure and high-temperature types for wound packages of yarn (on cheeses or cones), muffs, piece goods on beam, loose fibre and yarn in hank packed in a cage.

Provided that pack density remains constant, permeability of the pack will increase as the thickness decreases. Provision must therefore be made to cater for such a variant. Permeability of the pack should be controlled during production and, where it differs because it is smaller, the flow from the pump through the pack has to be monitored and adjusted accordingly. Excepting the need to maintain as accurately as possible the liquor to goods ratio when scaling down, the most important factor is the time required to circulate the dye solution through the pack. Control of this is a function of the pump efficiency.

A pilot machine is usually expected to perform the function of a variety of sizes of production machines and, whilst the pilot design can be calculated for a given pack size, it is not easy to cater for variants in pack size or density. The same problems arise in production, but generally the pilot package machine performs well if flow rates and the process itself are closely monitored and controlled.

Many pilot machines take only one or two single packages, though larger versions are by no means uncommon, these accommodating 5, 10, 15 and 25 kg lots. Little difficulty is encountered in scaling-down the more important features, though the point when a pilot machine becomes a production machine is very much a matter of opinion.

Clearly single- or two-package units and narrow-width beams to accommodate, say, 5–10 m are understood to be pilot scale. However, beams accommodating half a piece of full-width fabric and package machines to take 10–15 kg border on production units.

7 JIGS

Jigs cannot be miniaturized to bench size and still produce results analogous to those obtained in production. In jig dyeing the fabric picks up the liquor during its passage through an immersion trough and the passage time in the trough is at most only a few seconds. Production batch sizes can be very large, anything up to 10 000 m, and in comparison with sample lengths, which may be as small as 2 or 3 m, the time on the batch roll is consequently very much longer.

Dyeing by this method relies on maintaining uniform temperature and the influence of gravitational and centrifugal forces during and created by batching of the rolls. Movement of liquor within the rolls is an important factor and in a pilot unit it is diminutive by comparison with production machines.

Consider the apparently impossible odds of producing a scaled-down unit in an attempt to produce comparable results in the laboratory as against production.

1. If miniaturized proportionately so that fabric dwell time in the immersion tank is similar to that on a production jig, then the reaction time on the roll is still far less than on a production jig.
2. If the fabric speed is reduced, then the fabric dwell time in the immersion tank far exceeds that on a production jig.
3. If the fabric speed is increased, then both immersion time and reaction time are both far less than those on a production jig.

Clearly the numerous variables cannot be maintained as proportional constants. The ratio of time spent by cloth on the batch roll to immersion time can be as much as one hundred times greater in production when compared with a pilot-scale unit. The only possible means of bringing the two closer together would be to increase the sample lengths to something approaching production length, at the same time increasing the bath volume. Increase of the sample length, of course, increases the roll size and this is hardly practicable on a miniature unit.

The above clearly illustrates the insuperable odds in proportionate scaling, but in spite of these various problems the sample jig can be used to advantage within its own design capabilities.

Laboratory jigs are generally unsatisfactory for bulk recipe predictions, however they are widely used for preparing commercial patterns, assessing the compatibility of dyes and for comparative trials of dye behaviour. A degree of success can be achieved when processing woven nylon and when using high-exhaust dyes, but the application of vat dyes is very problematical. Under ideal conditions liquor ratios approaching 2:1 can be obtained but it is difficult to handle lengths much less than 2 m. The practical value of laboratory jigs is dependent on experience in interpretation of the results.

8 PAD MANGLES

The important features of pad mangles for the application of dyes and finishing agents are well documented(*4*); the following summarizes those factors influencing the design, application and results.

1. Loading the nip produces a flattened area of contact between the two bowls whose nip width depends on the load and the hardness and diameter of the rolls.
2. All essential nip processes occur when the textile material is within the flattened area of the nip.

3. Uniformity across the nip length is a fundamental requirement in providing an even expression or retention of liquor.
4. The pressure applied to the bowls is applied at the ends and therefore a bending moment is introduced on production machines. This bending moment, tensioning and guiding systems are not problems requiring specific attention in pilot-scale pads.

The influence of applied loads, bowl diameters and composition of the bowls is particularly relevant. Users of pilot machines are often of the mistaken opinion that given the same bowl composition with the same applied load (i.e. load per linear cm) as that of a production machine a pilot unit will produce the same effect. This is not the case, as expression or retention of liquor is largely a function of the mean effective pressure at the nip (the load per linear cm divided by the width at the nip).

The following influence the mean effective pressure (MEP).

1. Soft bowls flatten and consequently increase in radius of curvature more easily than hard bowls of the same size. In comparison, therefore, at a given load softer bowls produce a lower MEP.
2. A thin rubber covering of the bowls produces a less compressible bowl and therefore a resultant higher MEP than thicker rubber-covered bowls of the same composition. Consider the effect of applying a given load to a pad comprising two rubber-covered bowls. It will be apparent that as the load is increased distortion at the point of contact of the bowls will increase the radius of curvature, and consequently increase the nip width. It follows that there is no merit in increasing the applied load indefinitely.
3. Equally, bowl diameter affects the nip width; the larger the diameter the greater the radius of curvature, and consequently the larger the area over which the pressure is spread. At a given load the greater the bowl diameter the smaller the effective pressure and thus it requires a smaller load with smaller diameter bowls to obtain a given MEP.

Whilst reference will be made later to the effect of running speed, i.e. time in the nip, other individual features that influence liquor expression and retention are beyond the scope of this chapter and the reader is referred to published works(*6*). The main point to remember is a variable-speed drive is required to provide the necessary flexibility in a pilot mangle to match production conditions.

Difficulty in achieving complete wetting out and uniform impregnation of material by the pad liquor does influence practical application, and particularly so on the pilot scale. As far as possible, a long passage through the liquor is advantageous and the longer the time lag between liquor immersion and the nip the better. It may be necessary to give more than one immersion to obtain satisfactory impregnation, thus necessitating a second nip. In those instances where dyes have affinity for the substrate resulting in tailing problems(*7*) this has to be provided for on production runs. Fortunately, it is much less of a problem in the laboratory, but if there were a problem it would be difficult to resolve because of the much shorter lengths being processed. In producing pilot-scale equipment the bowl composition can vary; there are no hard and fast rules other than it would be usual to produce the unit of a similar composition to that of the production machine. Where the pilot is required to be multi-purpose, replaceable bowls are necessary. Stainless steel, ebonite and/or rubber-covered bowls usually within the limits 65–75° Shore are used. Such features are not predetermined by the pilot machine but are merely a requirement of the user. The ideal design demands the largest possible diameter free-running roller in the trough; although there are clearly limits to the size of this roller.

Experience has shown that whilst pilot pads may give better expression at a given applied load, in production there is a tendency to produce better penetration. In part this is related to contact area in the nip, but also to actual immersion times and the time between immersion and the nip. The effect of speed on water retention is relatively small, but it does have an effect and the slower the speed the better the squeezing action. If speed at the nip is increased then penetration of the liquor is not as good and there will be a greater pick-up of liquor. In the laboratory the immersion time may be shorter and consequently there will be less diffusion, poor uniformity and, in those cases where dyeing would take place because of the affinity, this has insufficient time to establish itself. Under these circumstances results in the laboratory may be different from those in production.

As with production machines, both vertical and horizontal models are available and the respective advantages apply. However, it should be remembered that whilst the pilot models have been designed to operate as real practical tools, a relationship and interpretation has to be established. It may be necessary, for example, to give multi-dips and nips on pilot models to equal the penetration and uniformity levels achieved in production.

Various features relevant to pilot pads have been discussed. There are many variables, but the most important and significant fact is that such pads are used for production recipe formulation and, whilst it can be shown that a production pad differs in important parameters, when the two pick-ups are made alike the resultant dyeings are identical.

9 STEAMERS, DRYERS AND THERMOSOL UNITS

Pilot steamers, dryers, baking and Thermosol units are usually used as part of a continuous or semi-continuous process. The basic requirement is to provide uniformity of controlled temperature and/or humidity-controlled steam. Dependent on the process it may be sufficient for it to be a batch operation with intermediate dwell periods, alternatively the time between impregnation on a pad and subsequent steaming may be very critical. Where this latter is the case, for example in pad steaming of vat dyes, a laboratory batch operation will not do and the steamer has to be in very close contact if not integral with the pad. The prerequisite in designing such equipment is the knowledge of what is required of it.

Steamers

Other than accurate temperature control, the fundamental requirement is to eliminate air. There were until recently no all-purpose pilot models but with 'add-on' units in series this situation has been changed. Miniaturized versions of both pressurized and atmospheric high-temperature steamers have been developed, and in the latter facilities are available to enable operatives to select controlled absolute humidity conditions at temperatures up to 250°C. This is a classic instance where the performance of the laboratory unit has exceeded that of the majority of production machines. For most applications a high absolute humidity is all that is called for, but flexibility of humidity control has proved useful for fundamental research.

In designing any pilot steamer an important feature is to prevent drops of condensed moisture falling onto the fabric. This requires the incorporation of a heated sloping roof and if necessary channelling to lead away any accumulated condensate. Atmospheric-pressure steaming in the region of 100°C causes few problems provided that an excess of steam is generated, but at higher temperatures control of the humidity is necessary. This is achieved by careful and accurate monitoring of the 'depression', i.e. the difference between the ambient and the wet-bulb temperatures. These are usually accommodated in what is called a psychrometer. A direct relationship between the depression and absolute humidity can be interpreted from psychrometric charts, thus providing a means of humidity control at selected temperatures. Some years ago the author was involved in an interpretation of the data available from psychrometric charts into an easily referable table. Table 1 shows the depression at absolute humidity at a given temperature.

TABLE 1

Oven temp. (°C)	Depression (°C) at various levels of absolute humidity									
	40%	50%	60%	65%	70%	75%	80%	85%	90%	95%
100	24	17·5	14	12	10	8	6·5	5	3·5	2
120	43	37	34	31·5	29	27·5	26	24·5	23	22
140	62·5	57	53	51	49	47·5	46	44·5	43	42
160	82	76·5	73	71	69	67	65·5	64	63	62
180	102	96·5	93	91	89	87	85·5	84	82·5	81·5
200	121	116·5	113	111	109	107	105·5	104	102·5	101·5
220	141	136	133	131	128·5	127	125·5	124	122·5	121
240	160·5	155·5	153	151	148·5	147	145·5	144	142·5	141

Pressure Steamers

Miniaturization creates few problems, as the temperature is dependent on the pressure. Pressure steamers are effectively sealed kiers in which the samples are placed or passed through and into which steam is introduced. Where the pressure steamer is designed to process fabric on a continuous basis, the most difficult area is associated with the sealing mechanism, but the principles used are the same as in production. All pressure vessels are subjected to the most rigorous design and inspection procedures, which add heavily to the total cost of manufacture.

Dryers

Pilot-scale stenters are indeed physically miniaturized production machines but, because of the complex design features, a number of aspects are simplified to keep costs down. Where full-width sample lengths have to be stentered, dried or baked the only satisfactory method is to attach the piece to a production run.

Whilst width adjustment is a requirement in production, it is difficult to justify the cost of pilot stenters simply for sample processing. Many such units are miniaturized continuous ranges in which air-flow velocities, direction and temperature control require the same detailed consideration as in the full-scale version. For laboratory purposes it is usually sufficient to have the pin chain operating at a fixed width and for the cloth to be cut to fit the machine. More commonly, small oven units have been made incorporating static pin frames, on which the samples are secured and manually adjusted to specific dimensions. In this case the pin-carrier frame is introduced into the oven where drying, setting or curing takes place. In practical terms, temperature control across the face and back of the test fabric is important, correct air circulation and a means of adjusting its direction and velocity are a basic requirement.

The following summarizes those variables which have to be considered:

1. The ability to fix or adjust fabric tensions
2. Controlled air velocities that provide independently variable flow to each face of the fabric
3. Ventilation
4. Temperature control up to 240°C.

Given these variables, laboratory ovens perform well, but ovens of different manufacture can produce completely different results primarily because of the different drying rates(*6*).

Drying Cylinders

The cylinder drying operation can be miniaturized relatively simply, though not necessarily linearly. If linear scaling is a prerequisite it will work satisfactorily but the contact time, i.e. material drying time, has to be maintained. In consequence, if the surface area of the drying cylinders is reduced then the speed of material movement will have to be reduced to match the

time of contact. Essentially, three constants have to be maintained:
1. Cylinder surface temperatures
2. Fabric/cylinder surface contact time
3. Pressure of the fabric on the cylinder.

Efficiency of drying is also determined by natural or forced ventilation in the region of the drying zone. Pilot models meeting all the parameters can be successfully constructed.

Thermosol Units

Any drying system which is capable of raising the material temperature to that required of the Thermosol treatment will perform this function, though the equipment chosen will be determined by the operative's own knowledge and experience. One interesting method not discussed above but worthy of mention is the heat-transfer capability of a fluidized bed. It is a well established and acknowledged high-efficiency system and one which has found favour in both yarn and narrow-fabric continuous processing. Miniaturization is simple and the required contact time is much less than in a hot-air oven.

A particularly important factor in continuous drying and thermofixation is a knowledge of when drying is complete and thermofixation begins. Very little apparatus is able to provide this information, though batch treatments of course permit intermediate inspection. On a production stenter careful and accurate monitoring is possible, but for laboratory and experimental purposes it frequently means either two stenter/oven chambers in series or a break in continuity, with a second passage through the oven zone, i.e. the continuous becomes a batch process.

10 DUAL- AND MULTI-PURPOSE MACHINES

Except in a few instances, multi-purpose units have not been considered, because the intrinsic features are represented under the respective headings. Multi-purpose units include:
1. Combined atmospheric steamers and ovens
2. Continuous pad–Thermosol and pad–steam ranges
3. Experimental sizing ranges, viz. pad, dry, overwax, dry splitting and beaming
4. A single unit (non-jig) providing open-width batch processing, for example, desize–scour–bleach–dye (exhaust or steam)–wash–aftertreat
5. The multi-purpose Rotonip–Rotomax combination.

The unique concept and engineering skills associated with the last-named item make it worthwhile to detail some of the operations it is capable of performing, e.g. desizing, scouring, bleaching, dyeing (atmospheric and pressure) and both atmospheric, pressure and high-temperature steaming; hot-air drying, heat setting and flame singeing; heavy mangling, calendering, embossing; sample pot, winch, package and jet dyeing, horizontal and vertical padding.

Although the normal production laboratory has limited use for such a piece of equipment, multi-purpose machines are of particular relevance and are essential for certain processes. The 'all-purpose' machine is invaluable as a teaching aid and in equipping the new research laboratory.

11 GENERAL

In constructing laboratory and pilot-scale equipment, engineering difficulties are encountered that are related to its smaller size. Weld distortion is a considerable problem with stainless steel, as is the use of thinner-gauge sheet metal. So few units are produced on a continuous basis that the use of standard 'jigs' is hard to justify.

In pilot models the reduction in size clearly poses problems of accessibility, not only for the user but also for the designer in positioning screws, fasteners and fittings, and for the manufacturer when, for example, cleaning up internal welds in a small trough.

Miniaturization demands closer tolerances for a similar fit and the manufacture of special items that on a production unit would be called from stock. Some parts of an apparatus may have to be made larger or heavier than the stresses dictate simply to accommodate standard components such as drives, chains, timing belts, sprockets and pulleys. Similarly motor-speed reducers of the required power are necessarily large in relation to the basic dimensions of many models and need early consideration in design.

Heating elements and fans are rarely available as standard components and must be specially designed and constructed. Particular attention is necessary in consideration of heat-transfer characteristics where the weight of metal in the construction of parts as a ratio to the weight of the textile sample is disproportionate to that in production machines.

The above sections have served to illustrate those aspects which influence the design and process engineer's judgement. Both W H Marshall(*8*) and H H Sumner(*9*) have considered the problems of selecting and using laboratory and pilot-scale equipment and of the significance of quality and laboratory control. Systematic human and mechanical errors are a cause for concern and should not be forgotten.

Criticism by production personnel of the laboratory's ability to produce good recipe predictions is commonplace and often related to a lack of realism by both parties.

Production departments of course require accurate information, but equally should consider whether the laboratory has the right equipment for the job and remember its own inaccuracies on the shop floor. In some instances it is not possible to provide anything more than a guide, for example the application of sulphur dyes on a jig and many operations involving continuous processing. It is also difficult to reproduce the variable effect of hydrolysis or reduction resulting in colour change; equally the laboratory must be mindful of the effect of the volatile nature of some applications. It is misleading to be critical of one parameter when another may be more important. Consider the individual influences of liquor ratio and temperature, each of which can have a significant effect on the yield of many disperse dyes applied to polyester at high temperatures.

Instrumentation as such is beyond the scope of this chapter, but, to the extent that accurate temperature control is essential, it is relevant when considering the performance of many machines and processes. The past decade has seen solid-state technology and intermittent relay systems introduced to provide a range of temperature controllers. Now chip technology has taken the lead and, for the important gamut of laboratory and pilot machines requiring time/temperature profiles, a variable two-stage rise with intermediate and top temperature hold, and single cooling cycle is generally sufficient. Present-day developments have meant that this introduction can be contained within the desirable proportions of sampling equipment.

It is not uncommon to hear of a machine in one factory producing different results to one of the same manufacture in another. Temperature or other control may be at fault, but equally the problem can be human error or a misunderstanding of the performance of the machine. Human error is not considered here, but as long as laboratory equipment is required to be multi-purpose and to be suitable for accurate control across a wide range of temperatures (say 40–130°C) then simple controllers and sensing devices will introduce their own inaccuracies.

The use of proportional band controllers to minimize overshoot of the set top temperature in laboratory machines using liquid or fluidized-bed heat-transfer media is for most purposes unnecessary. Proportional band controller is a term used to categorize those units which sense, monitor and adjust the heat-quantity input frequency to minimize fluctuation about a set point. In the case of indirect heating and control it is a commonplace and incorrect assumption that the temperature indicated by the control unit is the temperature of the process liquor. Equally it is incorrect to assume that nominal overshoot in temperature by the heat-transfer medium will cause an overshoot in the process liquor. There is both a temperature lag and a differential at equilibrium. The faster the rate of temperature rise the greater the difference when the set top temperature of the heating medium is first reached.

For the majority of laboratory dyeing trials, rates of rise not exceeding 2°C per min are the norm and at the upper limit of this range the temperature lag will be such as to cancel any significance in initial set-temperature overshoot by the heating medium. Controlled rates at 2°C or less will not cause overshoot in the process liquor itself and fluctuation of ± 2°C about the set point will cause variations in the test liquor of no more than $\pm$ 0.5°C, see Figure 1. Where very rapid rates in excess of, say, 4°C per

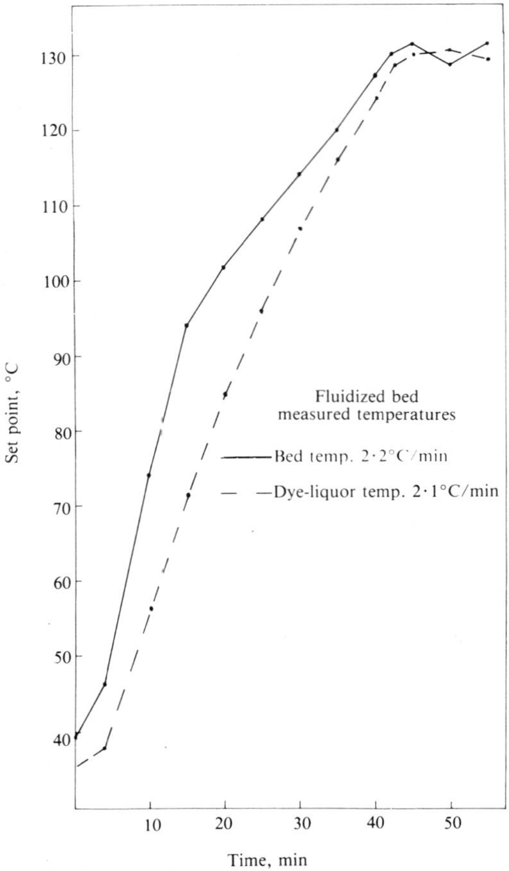

Figure 1 – Fluidized-bed heating

min are required, of necessity the heat input quantity has to be much greater and there is a case for proportional band control, not specifically to prevent initial overshoot but to minimize the degree of overshoot once equilibrium between the heating medium and the process liquor has been achieved. Even here at 4°C per min it is unlikely that the process liquor temperature will exceed ± 1°C because of the differential at equilibrium.

Where there is an initial requirement for high heat input to produce rapid rates of rise, an alternative and often more convenient form of control preventing overshoot is the inbuilt sensor, which at a predetermined temperature will switch from a high-capacity bank of heaters to another of much smaller capacity. This system whilst useful for indirect heating is widely used for direct-heating control on other machines such as miniature stenters and Thermosol ranges.

In conclusion:

1. Conditions that create an initial nominal overshoot of, say, 2–3°C in the set point will not in themselves cause overshoot of the test liquor, this is because of the temperature lag between the heat-transfer medium and the test liquor and because at equilibrium a temperature difference will always exist
2. Slow rates of rise reduce the effective heat quantity and therefore minimize or eliminate overshoot; proportional band control in this instance is not necessary
3. The need for proportional band control applies only where rapid rates of temperature rise are specified, necessitating a large heat input.

In all cases selection of the right equipment for the right job is of paramount importance:

(a) There is no case for designing a high-temperature laboratory dyeing machine if it is only required for low-temperature work
(b) There is no case for buying a bench-scale winch for routine dye and recipe formulation
(c) There is little value in having a highly sophisticated programmable temperature-control system to ensure level dyeing when the mechanics of the machine and a simple set-temperature controller will provide this facility at a fraction of the cost.

Equally, is there any merit in fitting ball races to a pilot pad whose workload and power consumption is so low that a set of bushes will perform adequately and last for years? Except on aesthetic grounds, does the operative need a highly polished stainless steel system when production is adequately served by protected mild steel? How important is design appeal and at what price?

The usefulness of any model or item of laboratory equipment demands a full understanding of its performance and a detailed knowledge of all the process mechanisms which it is intended to simulate. Some distortion of scale is necessary in most instances and it requires of the operative a degree of ingenuity and interpretation.

REFERENCES

1. Jones, Theory of Coloration of Textiles, (Bradford: Dyers Company Publications Trust, 1975) 237.
2. Marshall, Rev. Prog. Coloration, **4** (1973) 76.
3. Disperse Dyes Committee, J.S.D.C., **93** (1977) 228.
4. Moss, J. Textile Inst., **50** (1959) 654; Potter, J.S.D.C., **71** (1955) 20.
5. Speke, J.S.D.C., **70** (1954) 221.
6. Fox, Marshall and Stewart, J.S.D.C., **83** (1967) 493.
7. Marshall, J.S.D.C., **71** (1955) 13.
8. Marshall, J.S.D.C., **84** (1968) 149.
9. Sumner, J.S.D.C., **92** (1976) 84.

Chapter 7

Plant Engineering Services

W RONALD

1 CIVIL ENGINEERING

This section is not intended to be a thesis on architecture or civil engineering, but simply a résumé of problems I have encountered as a plant engineer in the textile dyeing and finishing industry. It is based on experience rather than a comprehensive knowledge of building or civil engineering, much of it gained in consultation with qualified builders, architects and civil engineers. I have found that, in the planning and construction of buildings for this industry, this is the sort of information that can be of considerable value.

The construction or reconstruction of buildings in the industry will almost certainly be determined by the financial resources available, it is often necessary to compromise between the ideal and what the budget allows.

1.1 BUILDINGS AND ROOFS

Textile bleaching, dyeing and finishing departments are not accommodated satisfactorily in conventional factory buildings such as are found in modern industrial estates. Buildings for this industry should preferably be custom-built, particularly for wet processing, where the atmosphere can be very humid and the use of highly corrosive chemicals is common.

In textile bleaching, dyeing and finishing, grey cloth is generally delivered from the weaver to the grey store or warehouse, and from there, after being sorted into orders, it goes through the various processing departments until it finally reaches the finished-goods warehouse and despatch departments. The construction of the buildings for housing the various processing units will therefore be discussed in the same order.

Grey Stores

Grey cloth is usually delivered either plaited or in rolls or bales. The packages are not very different from those of many other products, so the only matter needing special consideration is the most suitable type of racking.

Almost any orthodox type of building construction is suitable for a grey store, provided it is wind- and weather-proof. Heating and lighting standards need not be very high, and they need only comply with warehouse regulations. Floors can be of any hard-wearing material suitable for fork-lift trucks or the necessary internal transport; concrete with a composition topping or a granolithic finish, preferably sealed to prevent dusting, is ideal. Walls can be anything from corrugated sheeting on a light steel framework to double-skin brick cavity walling, or a combination of both. Any sound orthodox form of roofing is suitable. In short, the construction of a grey store is a matter of cost more than anything else.

Order Assembly Room

This might be called the grey sewing room or 'laying-out' room and is usually an annexe of the grey store. As this is a working department, heating and lighting must comply with the appropriate regulations and it is usually preferable that facilities are installed for the workers; otherwise the construction can be the same as that of the grey store.

Singeing

The singeing room often seems to be treated as the 'Cinderella' of the plant. It is often in an out-of-the-way corner, and provided it satisfies the insurance company not much thought seems to be given to it.

In recent years it has become more usual to install singeing machines in the preparation or bleaching department, in which case they must be completely housed in a fireproof chamber, usually of brick and concrete, and fitted with fire doors for access; only slots are permitted for the fabric entry and exit.

Whether the singeing machine is in a chamber in the preparation department, an annexe of the grey store, or a separate building, the housing must be made of materials that are completely fire-resistant. A concrete or granolithic floor, brick walls, and a steel or concrete flat roof is suitable.

A great deal of fumes, lint and fly is given off during singeing, so that extraction canopies and fans are needed to remove these. The result is usually a rapid build-up of lint, etc. on the roof of the singeing room. This is why a flat roof is desirable – it is usually much easier to keep clean.

The walls should also have an easily cleaned internal surface. One that can be hosed down is ideal, preferably made of enamelled brick or glazed tiles.

All electrical equipment and light fittings should be waterproof and fire-resistant.

Wet Processing

Scouring and bleaching are processes which use a great deal of steam, so the atmosphere in these departments is usually very humid. Highly corrosive chemicals, e.g. caustic soda, hydrogen peroxide, hydrochloric and sulphuric acids, are also used extensively. All can play havoc with the fabric of a building, a fact that does not seem to be recognized by architects when bleachcrofts or dyehouses are being planned.

Various protective coatings are now available, which are claimed to withstand the most hostile environments. Some of them in fact do so, provided they are applied properly and all the surfaces to be protected are accessible. A common form of fabrication for modern industrial buildings is rolled-steel framing with brick walls and asbestos-cement roof cladding. For the wet-processing departments, light steel roof members or trusses are very difficult to protect from the atmosphere, and even with heavy-section Portal frames the top flange of the roof member is not very accessible for painting; if metal purlins are used, the part of these which comes tight up against the roof cladding is impossible to paint after the roof cladding has been fitted. If the protective coating is applied before assembly, it will invariably be damaged by the steel erectors when they are bolting it together, so creating toe-holds where corrosion can start.

In a bleachcroft or dyehouse, corrosion of the roof members is not just a structural problem. In highly humid conditions, condensation will almost certainly take place on exposed steelwork and start dripping and inevitably contaminate fabric below. This is bad enough, but if the drips happen to be rusty water it is a great deal worse.

Older bleachcrofts and dyehouses generally had all-timber roofs. Trusses and purlins were of a very durable timber and clad with slates. This form of roof is ideal for a wet-process department. Possibly the only weakness in these traditional roofs was the poor insulating properties of the slates; condensation could be a real problem. In my mill, all the old slate claddings have been stripped off and replaced with two layers of asbestos-cement corrugated sheets with a 7·5 cm thick sandwich of mineral wool between them. The original timber roof members have been retained as these were made mainly from pitch pine and are in excellent condition.

Unfortunately, good quality timber is now difficult to obtain and very ex-

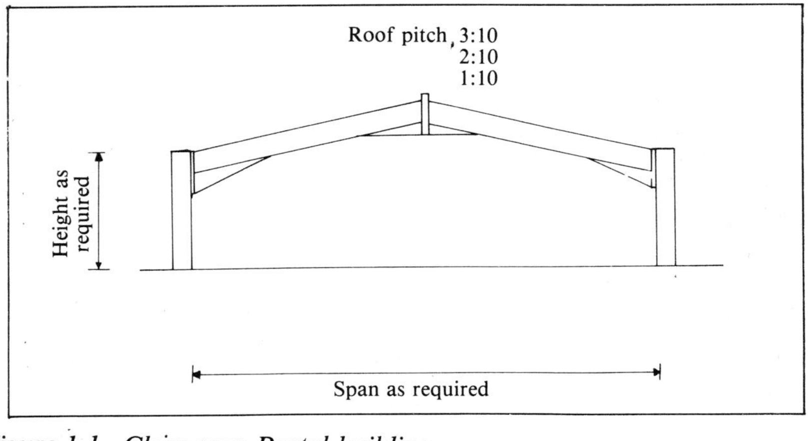

Figure 1.1 – Clear-span Portal building

pensive. The two alternatives are steel and reinforced concrete. For clear floor areas the Portal frame type of construction (Figure 1.1) is probably the best of the orthodox structures, this requires little or no light steel braces or ties and the main members are readily accessible for painting and maintenance. Cross-bracing is often used during the erection of the Portal frame but can often be removed after the vertical members have been concreted in and the roof members tied to the purlins. The purlins which carry the roof cladding are usually available in pressed-steel section, galvanized for protection in difficult atmospheric conditions. For a bleachcroft or a dyehouse it is well worth considering good quality timber purlins as these are difficult to reach for maintenance painting.

To minimize condensation and dripping, dyehouse and bleachcroft roof claddings must have very good insulating qualities; corrugated asbestos-cement sheets are the best form of cladding, and it is essential to use a double layer with a thick mat (minimum 7·5 cm) of insulating material sandwiched between the outer and inner layers (Figure 1.2). The inner layer must be coated with special asbestos-cement paint, preferably before erection. It is much easier to paint these sheets before fitting than after. Asbestos-cement is a very difficult material to treat; ordinary oil or emulsion paints will simply peel off in a very short time.

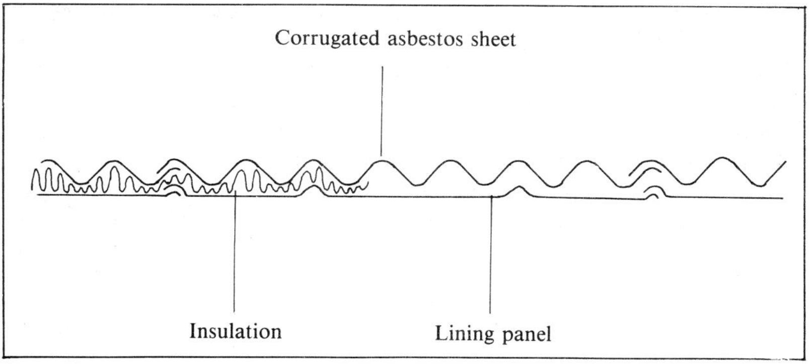

Figure 1.2

1.2 WET-PROCESS FLOORS AND DRAINS

The floors of bleachcrofts and dyehouses are subjected to spills and splashes of a variety of highly corrosive chemicals, both acid and alkaline. The problem is to find a flooring material resistant to all these chemicals as well as one that stands up to the wear and tear of cloth-transport equipment.

Concrete must be the basis of the floor; there are few alternatives. It is the topping that must stand up to the wear and tear. Many flooring materials are available, including ordinary and special granolithics, epoxy- and polyurethane-resin-based screeds, tiles or bricks.

The most durable dyehouse floors I have found were in jig and winch dyehouses. These floors were topped with 'blue' engineering brick laid on edge on top of a 15 cm reinforced concrete base. The bricks were grouted in with high-alumina cement and finally pointed with an acid-resisting cement. This proved to be an excellent floor construction but was very expensive.

Chemical-resistant tiles also make a first-class dyehouse floor, but they must be well bedded down and grouted with a suitable chemical-resistant material.

One of the most important features of bleachcrofts and dyehouses is drainage. The lay-out of drain pipes and/or floor channels for effluent must be given careful consideration.

There is now so much emphasis on energy conservation that heat recovery from bleaching and dyeing effluent is well worth looking at. The major problem in effluent-heat recovery is the segregation of the hot and cold liquors; in jig or winch dyeing a combination of hot rinses and cold rinses is not uncommon, in which case, of course, only the hot liquors need passing through a heat exchanger, the cold liquors being passed direct to the effluent drain. This is not an easy proposition, particularly with an existing plant, but might be incorporated in new plants.

Scouring, bleaching and dyeing present different problems in effluent disposal. In scouring it is not unusual for cloth to lose 5 or 6% of its grey weight, particularly with cottons and cotton/synthetic-fibre blend cloths. This loss is usually due to removal of size, spinning oil and lint, and this 'muck' has to be conveyed in the effluent liquor to the disposal point. It also has a habit of separating out and adhering to pipes and channels; if these are not cleared regularly they quickly become blocked. For this reason, effluent drains from scouring machines should be easily accessible for cleaning. Open channels with grids are widely used; if the floor is constructed with suitable falls, the grids will also take care of floor surface drainage. If the effluent is very hot, it may not be possible to use open channels because of the problem of steam rising through the grids, in which case the alternative is salt-glazed pipes. These should be very generously sized and have plenty of access points for cleaning. An inspection eye at every bend, with a manhole for access, is useful.

Dyehouse effluent is not so difficult to deal with, the major problems usually being due to the presence of chemicals and dyes. It does not contain the same amount of solids as scouring effluent. Underfloor glazed drain pipes are probably the best for conveying dyehouse effluent, but plenty of access points are needed, as such drains are notorious for becoming blocked up with rags, string and all the other debris that seems to find its way into the dyehouse drains.

It used to be common practice to lay a drain channel along the centre line of a machine (Figure 1.3).

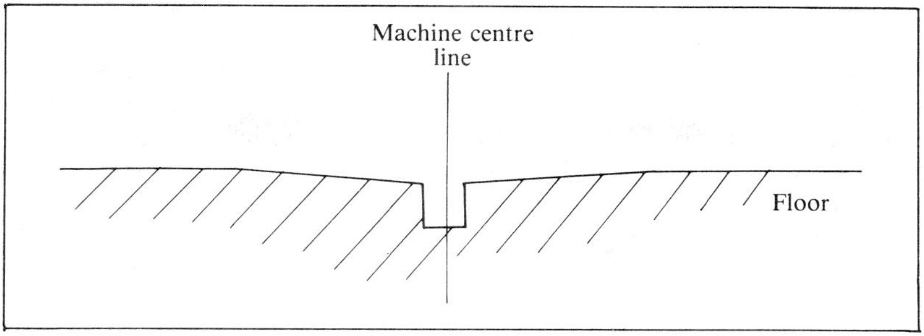

Figure 1.3 – Centre drain channel

Under a long machine these are obviously difficult to clean out and a side channel is preferable (Figure 1.4).

Ideally, the effluent liquors from a machine should be piped directly to an enclosed side drain, the dyehouse floor still being sloped for hosing down.

The chemical contents of effluent are another key factor that have to be considered in textile wet processing. Salt-glazed stoneware pipes are probably as good as anything for resisting chemical attack and are relatively cheap. The jointing of these pipes must be done with chemical-resistant materials, the conventional 'spun yarn' and mortar will soon be attacked by the effluents from textile wet processes. Glass fibre or asbestos, backed up by chemical-resisting cement, makes quite an effective jointing for these pipes. Unless salt-glazed pipes are specified, most builders or contractors will use the more common unglazed stoneware pipes with flexible joints. These are much easier to lay, but tend to pick up lint and size quite rapidly. In my opinion they are not as good as the old spigoted salt-glazed pipes.

All effluent channels or pipes should be laid with falls so that effluent flows at a good rate. If the flow is too sluggish, solid matter will tend to settle in the bottom of the channel or pipe. It is a good idea to have sumps at various points to catch the solids. These sumps are much easier to clean than are pipes.

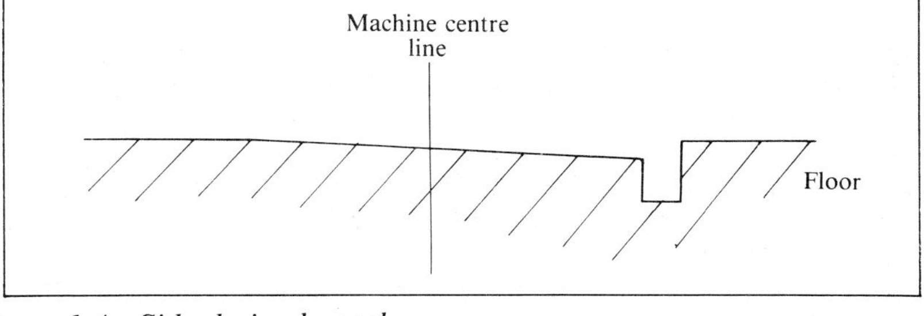

Figure 1.4 – Side drain channel

1.3 FINISHING DEPARTMENTS

In stenter rooms one usually finds that even with adequate extraction on the stenters there is always some fume emission for the entry and exit slots. Predrying equipment in front of the stenters will also create a certain amount of steam and fumes from the additives on the cloth. These fumes usually carry with them fly or lint, blown off the fabric in the stenter, or rubbed off at the padder or stenter entry rails. Condensation will occur on any relatively cold surface, including the roof structure and cladding, leaving a sticky deposit very difficult to remove. Resin finishes are particularly bad for causing this problem.

In my opinion, these deposits are the cause of the most difficult problems encountered when a stenter room requires redecorating. They cannot be painted over and can only be removed with a solvent, an operation which is very laborious and expensive. In existing stenter rooms there is not a great deal that can be done, but in new buildings the answer is to keep roof structural work to a minimum and use only hard, high-gloss enamel paint on both the structure and cladding. Again, I favour Portal-frame construction, with double-layer asbestos-cement cladding.

In other respects stenter and finishing machine rooms need only be of orthodox construction with emphasis on good natural ventilation.

Concrete with a granolithic topping is probably best for the floors of these buildings. If the floor is constructed with a reasonable thickness of reinforced concrete, it is usually possible to install stenters and finishing machinery directly without the need for any further work on foundations, etc.

Inspection and Make Up

Probably the most important considerations for inspection and make-up rooms are cleanliness and the comfort of those who work in them. The UK climate demands that these rooms require heating most of the year, and yet be cool and pleasant to work in during the summer months. For these reasons I favour artificially lit rooms with no roof glazing at all. With well-insulated roof cladding, heat is retained in the winter and kept out in the summer. Outer-wall glazing with opening windows should be used wherever possible.

The floors of these rooms should obviously be dust free, but otherwise the type of floor is largely a matter of economics. Hardwood, epoxy, and urethane resin toppings are some of the options available if the money is available, otherwise a granolithic floor will suffice. If so, this should be treated with a sealer to prevent dusting. An alternative is to use a suitable paint; this is not expensive even if the floor has to be recoated, say, every two years.

Finished-goods Warehouse

Probably the primary consideration in the construction of a warehouse is the shape and weight of the packages to be stored. For cloth made up in open width on tubes, creased and lapped on boards or plaited and bundled, some form of racking is usually required. The dimensions and capacity of this racking will determine the size of the warehouse, and as the current trend is for rack storage to go higher and higher, all these factors have a bearing on the design of the building.

To utilize the capacity of the warehouse to its maximum, a flat or nearly flat roof with the minimum of structural work is desirable. It can only be a matter of opinion which type of roof is specified. I prefer a low-pitched roof with lattice-type roof beams clad with a double layer of roof decking and about 10% of the area being roof lights installed over the racking passageways.

Walls and floors can be of virtually any approved form of construction, bearing in mind that if fork-lift trucks are to be used, the floor topping should be suitable for this traffic.

2 HEATING AND VENTILATION

The heating and ventilation of dyeing and finishing plant can present a problem, particularly in wet-processing departments and stenter rooms. In the former there is usually the problem of getting rid of steam and keeping the humidity down, and in the latter the problem of heat from drying and setting stenters. These problems are not made any easier to solve by gross climatic changes, but to be acceptable, working conditions must be of a higher standard than they used to be.

Nevertheless, short of complete air conditioning, the cost of which is prohibitive, it is almost impossible to maintain consistent atmospheric conditions in a bleachcroft, dyehouse or stenter room throughout the year. Wet-processing departments are the most difficult to deal with; in a medium-sized bleachcroft or dyehouse 5 000 kg of steam per h may be used yet space heating be required during winter. This poses the question of insulation. Some energy experts recommend that the exterior of steamers, totally enclosed wash becks and drying-cylinder ends should be insulated to reduce heat loss, but then space heaters are required to maintain a reasonable working environment. It may be better to leave these hot surfaces exposed, and use them as radiators for space heating. In summer it may be rather hot for an operator working close to a steamer or a stack of drying cylinders, but localized cooling is not too difficult to install and is much more economical than space heating.

Heating and ventilating jig and winch dyehouses are probably amongst the most difficult problems to deal with. These dyehouses used to be, and in fact some still are, recognizable by clouds of steam being ejected by fans from the gables or roofs of the building. Even with enclosed or hooded machines, the atmosphere of such a dyehouse is invariably very humid and steamy, and if the air is cold enough the moisture becomes visible in the form of fog. As conditions vary so much in dyehouses, ventilation can never be regarded as an exact science. In bygone days the principal method of dealing with the problem was to install plenty of extraction capacity by fitting large fans in the gables and roof dormers. All the air being sucked out of the dyehouse created a negative pressure in the building, and cold air rushed in from outside of the dyehouse or from adjacent departments. At openings or doorways to the outside, it is usual to find a fairly strong draught of air being pulled in, in winter this air is cold, and the whole problem of steamy fog is greater. On the other hand, if this replacement air is being drawn from adjacent rooms with a warm, dry atmosphere, conditions will usually be improved. Hot air in these other rooms can then be replaced by fresh, cool air. Good use can be made of excess hot air in, say, a stenter room to improve the atmospheric conditions in a dyehouse.

Most bleachers or dyers will have noticed that when a hot-air blower is started up in a croft or dyehouse filled with steamy fog, the fog in the vicinity seems to disappear. The hot air causes the water droplets to vaporize. Thus if hot air is introduced into a dyehouse at a high enough temperature, the atmosphere will be perfectly clear, but the humidity and temperature will make conditions intolerable.

The problem of keeping the atmosphere of a dyehouse clear of fog and condensation and at a reasonable temperature is best dealt with by extracting the moisture-laden air quite rapidly and replacing it with warm fresh air. At least 15 air changes per h are usually necessary, and ideally there should be kept a slightly positive air pressure in the dyehouse, so that air will leak out of any openings rather than cold air drawn in. The ideal method is to have fan heating units drawing in air from outside, combined with extraction fans pulling out moisture-laden air. To maintain a slight pressure, the capacity of the hot-air fans should be greater than that of the extraction fans. The hot air should be introduced at a high level; the air temperature at roof level can be fairly high without working conditions at floor level being uncomfortable. The high-level hot air also serves to keep the roof's structural work and cladding warm and prevents condensation and dripping.

It is essential that any dyehouse extraction fans and ducting should be made of non-ferrous materials. Copper was used in bygone days for fan blades and ductings, but this has now become too expensive; glass-reinforced plastic is a suitable material; PVC and acrylics usually wilt with the heat. In my experience, galvanized or coated steel is useless for this purpose; the atmosphere of a croft or dyehouse seems to be able to break down most coatings at any point of minor damage to the coating. Very durable ducting has been constructed from waterproof plywood coated with an epoxy resin reinforced with glass fibre. Some proprietary extraction ductings are made of aluminium. If these are used it is still advisable to treat them with a protective coating, as aluminium tends to oxidize quite rapidly in a dyehouse atmosphere.

Proprietary extraction and input fan systems are now available in modular form, (Figures 2.1 and 2.2) so that by selecting suitable components, hot, cold or filtered air can be introduced, and foul or humid air extracted at almost any location in the building. These units are ideal for localized air conditioning which may be all that is necessary if only a few operatives are employed, e.g. on wet-processing ranges, whereas in jig or winch dyehouses, general air conditioning will probably be required.

These modular units are usually made of aluminium and glass-reinforced plastic, both of which are quite durable in dyehouse atmospheric conditions, but in some instances it is recommended that the aluminium be given

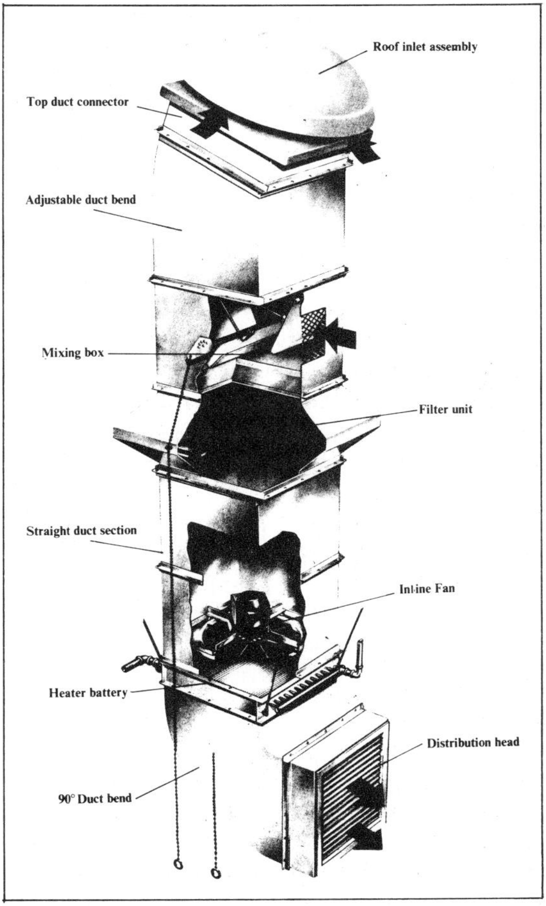

Figure 2.1 – The Modular input system

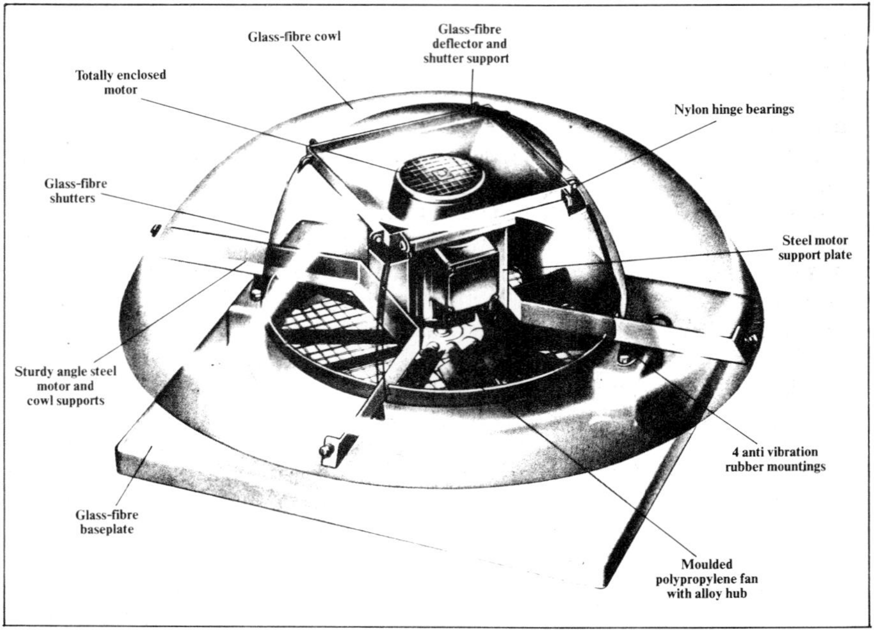

Figure 2.2 – Input or output fan

a protective coating of epoxy or polyurethane resin, to prevent oxidation.

Heating is not generally a problem in stenter rooms; it is the removal of hot air and fumes that presents the most difficulty.

The judicious use of extraction fans will usually take care of the fumes and smoke generated in stenter chambers, 'judicious' should be emphasized as the use of these extraction fans and ducts is all too often abused. If the setting of the dampers is simply left to the operator, on high-temperature machines far too much very expensive heat is extracted and blown to waste. If there is the slightest hint of hot air or fumes issuing from the cloth inlet or exit slits, the stenter operator will often open the extraction dampers much too far, with the resultant wastage of heat. This problem of excessive hot-air extraction often arises when a stenter is used for resin drying and curing, and also for heat setting. When the machine is drying and curing resin fillings the humidity will be high in the drying section and heavy and polluting fumes may be given off in the curing section. In both instances a high extraction rate from the chambers is required to reduce humidity and prevent the emission of obnoxious fumes, and a high heat loss has to be accepted. What often happens is that when the machine is changed over to heat setting a scoured dry fabric, the extraction dampers are left in the same positions, and, as little or no fumes or moisture are given off in this process, a great deal of heat is wasted.

The foregoing may appear to have more relevance to stenter thermal efficiency than to heating and ventilation, but as the primary objective in heating or ventilating a process room is the comfort of the worker, this can be achieved by more economical means than pulling large quantities of expensively heated air from the stenter chambers.

If a stenter is properly insulated and airtight, the main source of air pollution is emission of fumes or smoke from the fabric inlet and exit slits. Some, but not all, stenter manufacturers recognize this and incorporate small chambers with exhaust fans at both ends of the stenter housing. The purpose of these is to prevent the emission of fumes into the stenter room. An alternative method of dealing with these fumes is to fit an external extraction fan and suction slots completely independent of the stenter chambers.

Many stenters are fitted with some form of predryer between the filling padder and the stenter entry; a small stack of drying chambers are often employed for this purpose. These too are a localized source of atmospheric pollution, which is usually better dealt with separately. Roof-mounted extraction fans are probably the best answer, but their action must be counteracted by the use of ducting and canopies, or by boxing-in a section of the roof so that there is a high rate of extraction immediately above the predryer.

Having dealt with local sources of atmospheric pollution, the general ventilation of a stenter room does not pose too much of a problem. Ordinary ridge or roof ventilation will usually shift a lot of the excess heat by natural convection (Figure 2.3), and if the room has outside walls, opening win-

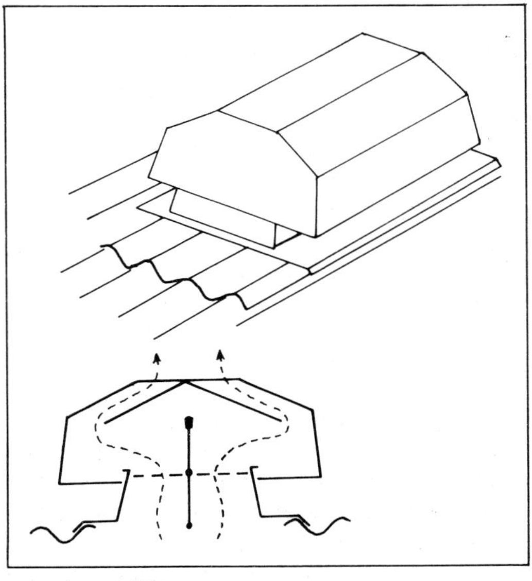

Figure 2.3 – Natural ventilator

dows may well be all that is required for the admission of fresh air. If sufficient fresh air is not available through windows in outside walls, then forced or mechanical ventilation may be necessary. In this case it is better to introduce fresh air into the building by fans and ductings, rather than resort to additional extractors. Using input fans the fresh air can be directed to where it will do the most good, whereas by using extraction the fresh air will only enter the building by openings in the walls or roof, and create uncomfortable draughts.

The heating and ventilation of working departments in which there are no steam, fume or hot process problems are essentially a matter of building design. Forced ventilation should not normally be required, except possibly on the relatively few really hot days in the UK climate. On these occasions it is often roof glazing that creates the problem rather than high temperatures. The old 'north light' type of roof was probably the best form of construction for making use of natural daylight, but few, if any, new buildings are now constructed with this type of roof, because of the cost and the excessive number of pillars required.

Modern clear-span Portal-frame construction does not lend itself to the fitting of opening roof lights. Its general ventilation is usually catered for by ridge ventilators at the apex of the roof, and can be increased by low-profile fan extraction units at localized hot spots, if required.

For working departments, as distinct from storage or warehouse buildings, and where the temperature has to be kept reasonably constant throughout the year, it is essential that the roof cladding should have good insulation properties. A popular form of cladding is double-layer corrugated asbestos-cement sheets as described and illustrated in the previous section. This form of roof cladding will effectively keep the heat in and the cold out in the winter months, and vice versa in the summer months, with the proviso that roof glazing is kept to a minimum or eliminated completely and artificial lighting is used throughout the year. Glass or clear plastic sheeting has very little insulating value and if large expanses of either are used in a roof, the problems of heating or cooling the interior of the building will be that much greater. No doubt many readers will have come up against the problem of strong sunshine beating down through a glass roof. It is a cause of many complaints and, of course, sunlight from a glass roof can be an abomination in a cloth inspection room.

All these factors have a bearing on the heating and ventilating of, e.g. grey warehouses, finished-goods warehouses, inspection and make-up rooms, in fact any room that does not require special consideration because of process plant. But how does natural daylight for part of the day via roof glazing compare in terms of cost with artificial lighting round the clock? Obviously daylight is cheaper than electricity, but this is offset to some extent by the extra cost of heating or cooling a room with a large expanse of roof glazing, not to mention the other problems direct sunlight can create. In my mill, several buildings have been reconstructed with Portal frame double-layer insulated asbestos roofs with no roof glazing whatever, and there is no doubt these are by far the easiest working departments to maintain at an equable temperature all the year round. Heating is by suspended fan heaters, using low-pressure steam; this is the only form of mechanical air movement used, extraction is by static ridge ventilators only.

It is difficult to generalize about heating and ventilating all the various process rooms in a textile dyeing and finishing plant. It is obviously much more difficult to achieve the same working conditions at, for instance, the entry end of a high-temperature stenter as at a making-up or inspection table, so that, inevitably, some jobs are less pleasant than others.

However, as previously mentioned, much can be done to improve the working conditions in specific problem areas by installing local air conditioning.

3 BOILER PLANT

Terminology

In order to discuss steam technology it is necessary to understand the basic terminology.

Absolute Pressure

This refers to the total pressure of steam, above a complete vacuum, and is measured in pascals (Pa) in the SI system, or in pounds force per square inch (lbf/in^2). It is a quantity not often used in practice, although absolute pressures are quoted in steam tables.

Gauge Pressure

This is of more practical use in steam generation, and is the pressure of steam above atmospheric pressure quoted on pressure gauges, measured usually in pounds force per square inch. To distinguish gauge pressure from absolute pressure in this chapter, I shall refer to its units as p.s.i., normal atmospheric pressure being 0 p.s.i.

Sensible Heat

This is the heat transmitted from the burning fuel to raise the temperature of the water. The amount of sensible heat that can be absorbed by the water is directly proportional to the pressure and temperature of the steam in the boiler. The temperature of steam at any given pressure can be found in steam tables.

Latent Heat

Latent heat is the heat (in joules) needed to convert 1 kg of water to 'saturated' steam at a constant temperature. The latent heat in steam can only be fully utilized by condensing the steam back into water.

Dry Saturated Steam

This is steam at the temperature of boiling water (at the particular pressure) but which does not contain any free particles of water. It is the ideal 'quality' of steam rarely, if ever, found in practice.

Superheated Steam
This is saturated steam from the boiler to which extra heat has been added in a separate apparatus known as a superheater. It is dry steam and its temperature is higher than the temperature of saturated steam at the same pressure.

Wet Steam
Should not be confused with saturated steam. Wet steam condenses as it is conveyed through the distribution pipework and is a mixture of true vapour and liquid droplets.

Textile bleaching, dyeing and finishing processes are large consumers of energy, a major proportion of which is supplied as steam. In recent years, with the advent of synthetic fibres, more high-temperature processing is required, which cannot be carried out using low- or medium-pressure steam, but in virtually all wet processing and subsequent drying, steam is the most common form of heating. As all these processes require large amounts of steam, the method of generating this steam is a very important aspect of a dyeing and finishing plant.

3.1 LANCASHIRE BOILERS

Up to the 1940s most dyeing and finishing plants used coal-fired Lancashire boilers for steam raising (Figure 3.1). This particular type of boiler was relatively cheap, easy to construct, simple to operate, did not require high-quality feed water or any sophisticated form of feed-water treatment. It had good water and steam storage capacity so that fluctuating steam demands from the bleachcroft or dyehouse could be catered for without too much variation in the steam pressure. While thermal efficiencies in the region of 75% were possible with coal-fired Lancashire boilers, coal was cheap and employers were loathe to employ skilled boiler operators, so little attention was paid to the efficiency of the boiler plant.

With coal and labour becoming more expensive and oil becoming an economic proposition, many firms either fitted mechanical stokers to their old Lancashire boilers or converted them to oil firing, rather than spend money on new and more efficient boiler plant.

3.2 ECONOMIC BOILERS

The principal successor to the Lancashire boiler was the economic boiler (Figure 3.2). This is a more sophisticated steam-raising unit, being fitted with a combustion chamber and tubes to give two or three passes to the furnace gases and therefore a far greater heat-conducting surface. For a similar

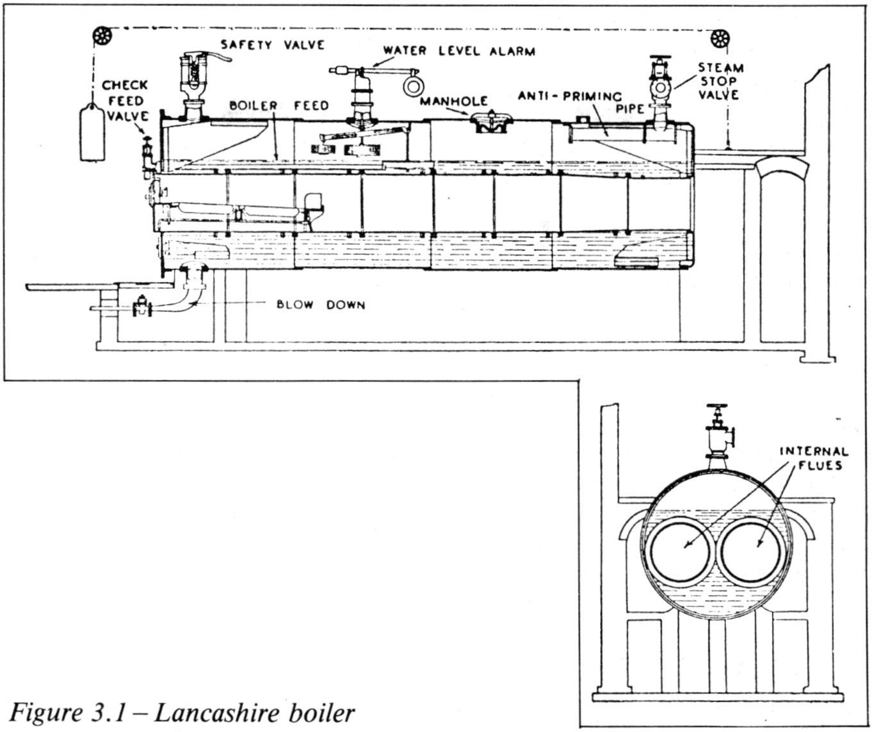

Figure 3.1 – Lancashire boiler

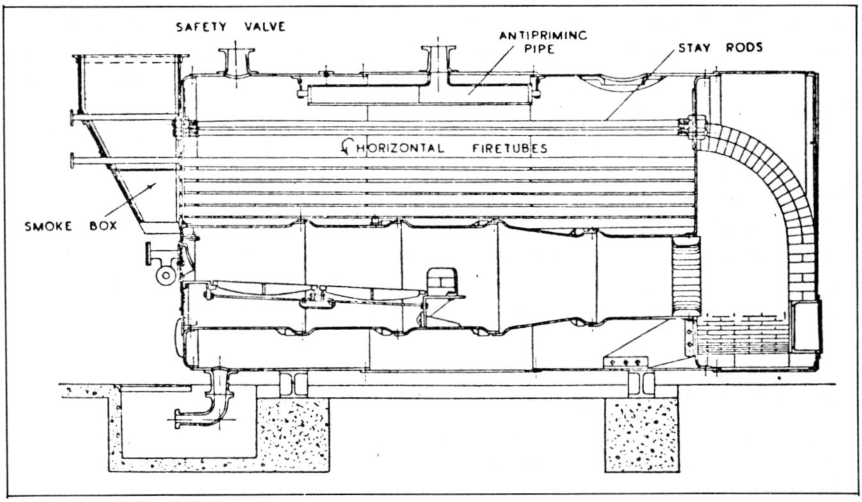

Figure 3.2 – Economic boiler

evaporation rate, the economic boiler is about half the size of the Lancashire boiler.

But if the size of the boiler is reduced and the immersed heating surface increased, the steam and water capacity is reduced. This means that the standard economic boiler is very sensitive to rapid fluctuations in steam demand, particularly with coal firing, as the fire cannot be boosted or damped down fast enough to cope with the peaks and troughs in steam demand, which are quite normal in bleaching or dyeing processes, particularly with winches and jigs. Even with automatic stokers it is not possible to vary the firing rate fast enough to cope with these fluctuations.

The use of fuel oil or gas firing has gone a long way to alleviate this problem as obviously oil or gas supply can be adjusted much faster than coal, but these are rapidly becoming premium fuels and it may well be that in years to come more and more small industrial boiler plants will revert to coal.

The evaporative capacity of Lancashire boilers usually ranged from 5 000–12 000 lb/h (0.63–1.51 kg/s) of steam at a working pressure of 150 to 200 p.s.i. Economic boilers are made with a capacity as low as 1 000 lb/h up to 25 000 lb/h, and a working pressure of 250 p.s.i. or higher.

3.3 ECONOMIZERS

As the temperature of the flue gases at the exit of a Lancashire or economic boiler can be as high as 500°C, some means of recovering at least a proportion of this heat is desirable.

The most common method of utilizing the heat in the flue gases is by fitting an economizer. This is a multi-tubular heat exchanger fitted in the gas stream between the boiler and the chimney, the hot gases passing round the outside of the tubes. It is fitted in the feed-water system between the feed pump and the boiler, the feed water passing through the tubes; the pressure in the economizer is normally a little higher than the boiler pressure.

Economizers can improve the efficiency of a boiler considerably. It is reckoned that for every 4.5°C increase in the temperature of the feed water going into the boiler there is a saving of about 1% in fuel costs. One advantage of an economizer is that, by preheating the feed water before it enters the boiler, cold-water pockets and hence stresses arising from temperature differences are eliminated. Also, if the feed water is hard, some of the scale-forming solids will be precipitated in the economizer where it is less of a problem to remove than when deposited on the boiler heating surfaces.

3.4 PACKAGED BOILERS

For small- to medium-sized steam generating plants, e.g. from as low as 1 000 lb/h up to 100 000 lb/h (0.13–12.60 kg/s), the conventional economic boiler has been largely superseded by the packaged boiler (Figure 3.3). These boilers are available with an evaporative capacity ranging from 1 000 to 25 000 lb/h and a working pressure up to 250 p.s.i.

A packaged boiler is basically a modified economic boiler and, as the term packaged implies, the boiler, together with all the necessary auxiliaries, such as the feed water pump, forced draught fan, oil heater and pump, are all mounted on a common skid so that the minimum of civil engineering is required for their installation. This is one of the principal attractions of the packaged boiler. Installation is simply a case of mounting the unit on a level hard standing, connecting it to fuel, power and water supplies and to the steam system, and the boiler is ready to operate. Provision must also be made for the disposal of the 'blow down'. The chimney, which is usually integral to the boiler, need only be as high as required by the local authorities.

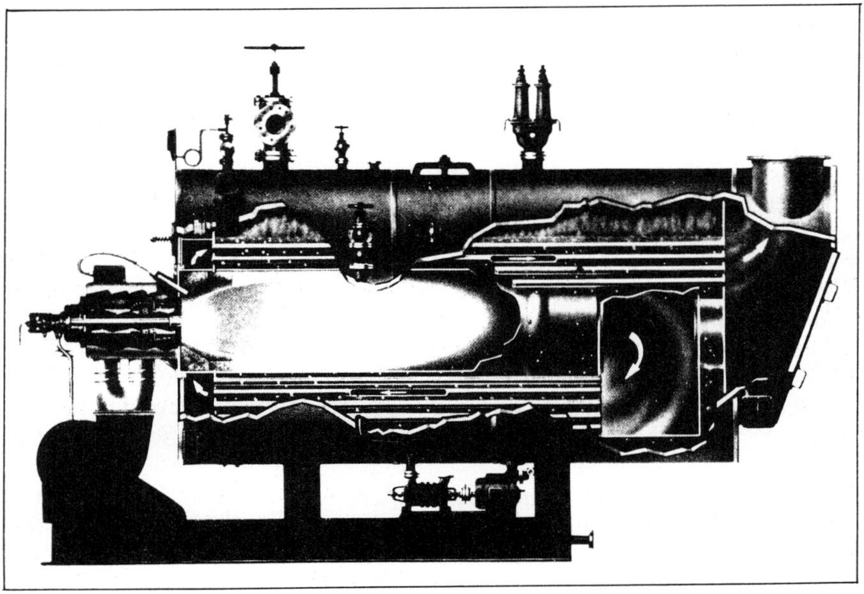

Figure 3.3 – Oil- or gas-fired packaged boiler

Packaged boilers are usually designed to be treble-pass 'wet back' boilers, i.e. the combustion chamber is completely enveloped in the boiler shell and surrounded by the boiler water, whereas in the conventional economic boiler the combustion chamber is outside the boiler shell.

The fact that there is so much heating surface in this type of boiler eliminates the need for an economizer, but it is desirable that the boiler feed water should be preheated to some extent in the feed-water tank or hot well. Preheating in the hot-water tank, which of course is only at atmospheric pressure, is limited by the height or head that the tank has over the feed-pump suction. The higher the tank, the higher the temperature of the feed water can be.

Being so compact, the packaged boiler has certain disadvantages when used in dyeing and finishing plants. The high internal heating-surface area is achieved at the expense of steam and water space, so that the steam storage capacity of the packaged boiler is very limited, and even with the fast response of oil or gas firing these boilers do not cope very well with the fluctuating steam demands of batch processing.

The quality of the feed water is also very important with the packaged boiler. It will not tolerate hard or scale-forming water to the same extent as the Lancashire boiler or conventional economic boiler. Any scale-forming matter in the feed water is rapidly precipitated in the internal heating surfaces and the efficiency of the boiler is reduced. When packaged boilers are used, e.g. in a factory heating system, most of the steam is condensed in the heaters and returned to the boiler in the form of hot condensate. Only the minimum of raw feed water is necessary to compensate for leaks, etc., and scale formation is not much of a problem. In a dyeworks where a large proportion of the steam is injected through perforated pipes into the dye vessels or wash becks, the quantity make-up water required can be 75% or higher, and it is essential that the quality of the make-up feed water be controlled.

Water treatment in not necessarily confined to packaged boiler feed-water plant, it can be beneficial with most boiler plants, to inhibit the formation of scale, to prevent corrosion and to minimize foaming or priming, but with the packaged boiler much more sophisticated boiler water treatment is required if efficiency is to be maintained.

The type of boiler water treatment used depends on the quality of the local water supply. Good soft surface water may only require a simple injected-chemical treatment, whereas hard artesian-well water will require more complex and expensive chemical treatment or possibly softening in an external plant.

As only the raw or make-up water requires to be treated, it is obviously beneficial to return as much condensate as possible from the process plant. Not only does this result in recovery of valuable heat, but the condensate is distilled water and thus completely free of scale forming matter.

Another feature of packaged boilers, which strangely enough could be considered a disadvantage, is the very simple start-up. Provided the water, power and fuel supplies are turned on, it is only necessary to close a few switches or push a few buttons to start the boiler. From then on, feed-water supply, firing rate and steam output are controlled automatically and the boiler can be left to run unattended. Further, in the event of any fault arising the boiler shuts down and a visual and/or audible alarm is given.

Problems arise when the switches have been closed or the buttons pressed and the boiler fails to start. Packaged-boiler control gear is designed to monitor all the conditions that must be met before the boiler is safe to light up. This comprises some fairly sophisticated electrical and electronic equipment, the servicing of which is often beyond the ability of the average works electrician; a qualified service engineer has then to be called in.

Any company using or contemplating the use of packaged boilers is advised to obtain as much instructional literature as possible and to carry an adequate stock of control-system spares. Many of the spares are plug-in units, which are easy to change provided the faulty one has been identified.

The normal sequence of events in the start-up of a packaged boiler is as follows.

1. The pump pumps water into the boiler until a safe working level is reached.
2. The forced-draught fan starts up and the air damper opens fully for 20 to 30 s. This is to purge the furnace and flues of any combustible pockets of gas that may be present.
3. The air damper closes and the pilot flame is ignited. The pilot-burner fuel is usually bottled gas and is ignited by high-tension electricity.
4. The pilot-flame detector signals to the control system that the pilot flame is ignited.
5. The boiler fuel valve and the air damper open and the main burner ignites.
6. The detector verifies that the main burner has ignited and the automatic control system takes over.

When heavy fuel oil is used, the oil has to be circulated through the heater before starting-up until the appropriate temperature is reached.

3.5 THERMAL STORAGE BOILERS

To overcome the problem of fluctuating steam demand, the thermal storage boiler has been developed. This is basically a jumbo-sized economic boiler with a very high water capacity.

The principle of the thermal storage boiler can be understood by reference to a technique used with the old coal-burning Lancashire boilers. Peak steam demands in a bleachcroft or dyehouse usually occur at certain times of the day, first thing in the morning and at the start of each round of dyeing. Boilermen learned by experience when to expect these peak demands and anticipated them by having their fires bright, a full head of steam and the boiler water level at the maximum. When the heavy demand came and the steam pressure started to drop, the feed pump was slowed down or even stopped. This reduced or eliminated the cooling effect of the relatively cold feed water and the sensible heat in the excess water would cause additional steam to flash off until the water dropped to a low level in the boiler, by which time, if the boilerman's judgment was correct, the peak demand would be over. This exercise was, of course, limited by the amount of excess water that could be stored in the boiler. Lancashire boilers can store a fair amount of excess water, economic boilers a little, and packaged boilers virtually none, so to cope with peak demands the packaged boiler has to depend on a rapid response to an increase in the firing rate.

This principle of storing sensible heat in the excess water in a boiler is the basis of the thermal storage boiler. This type of boiler can hold sufficient water to cope with steam demands 30% or more above its rated evaporation for up to 1 h, with minimal loss of steam pressure. Feed-water supply is automatically controlled by the steam output so that variable steam demand only reduces the feed-water supply as necessary.

3.6 STEAM ACCUMULATORS

The steam accumulator is another device for dealing with peaks and troughs in steam demand. In textile dyeing and finishing plants the working pressure of the boilers is usually from four to six times the process steam pressure, the boiler pressure range being 150 to 200 p.s.i., the process pressure range 25 to 40 p.s.i. Process steam is usually controlled by reducing valves fitted between the boilers and the plant.

There are instances where higher process steam pressure is used for high-temperature stenters, etc., but most of the steam used is at lower pressures and the steam accumulator operates by using the pressure differences between the boiler and the plant.

A steam accumulator is a large cylindrical vessel, similar in shape to a boiler. Lancashire boilers are in fact sometimes converted to accumulators, by perforating and 'blanking off' the furnace tubes, but proper accumulators tend to be considerably larger.

Starting from cold, the accumulator operates as follows. The vessel is filled to about three-quarters full with cold water. Steam is fed from the boilers and injected into the water via a series of nozzles. These nozzles are fitted at a depth well below the half-full level of the vessel and are specially designed so that the water is heated right to the bottom of the vessel.

When the water reaches boiling point, steam is produced and as the temperature of the water continues to rise the steam pressure rises until it equals the pressure of the steam being supplied from the boilers. The maximum pressure is determined by the pressure rating for the accumulator vessel. When the accumulator pressure equals the supply pressure the accumulator is fully charged.

A great deal of sensible heat is now stored in the water, and steam can now be drawn off for process work. As the pressure in the accumulator starts to drop, the sensible heat stored in the water will be converted into steam and this will continue until the pressure in the accumulator has dropped to the process steam pressure. The accumulator is then said to be discharged. The accumulator would continue to produce steam if the process steam pressure is dropped correspondingly.

In practice, accumulators are not operated over the full pressure range, but over a range which ensures that at the lowest pressure there is a reasonable head of pressure on the process steam system, and at the highest pressure the vessel is not too near the point of blowing off. Control and reducing valves are fitted to regulate charge and discharge of steam.

The accumulator is usually operated in parallel with the boiler to supply the process steam, so that when the boiler steam production is greater than the process steam demand, the excess steam charges up the accumulator, and when the process steam demand exceeds the boiler output the accumulator augments the supply by discharging. The accumulator pressure gauge is usually fitted in the boilerhouse so that the boilerman can see whether the vessel is charging or discharging, and can adjust the firing rate accordingly.

3.7 WATER-TUBE BOILERS

To achieve maximum thermal efficiency in a boiler, the steam pressure and temperature should be as high as possible. To burn low-grade solid fuel a large fire grate is necessary. To drive power-generating plant, high-pressure superheated steam is desirable.

Because of the limitations of plate thickness, the relatively small grate area, and the difficulty of fitting superheaters, shell boilers have severe limitations in meeting these criteria. Thus, when steam pressures in excess of 250 p.s.i., steam temperatures in excess of 260°C and a large grate area and combustion chamber are required, water-tube boilers are used.

The basic difference between a shell boiler and a water-tube boiler is that in a shell boiler the hot gases pass through the fire tubes and the water is inside the shell, whereas in a water-tube boiler the water passes through the tubes with the hot gases surrounding the tubes.

There are many types and designs of water-tube boilers. They are expensive and it is doubtful if one with an evaporating capacity of under 25 000 lb/h (3.16 kg/s) would be an economical proposition, although there are quite a number of boilers of this size still in operation. At the other end of the scale there are the huge boilers installed in power stations with evaporative capacities in excess of 1 000 000 lb/h at pressures of 2 000 p.s.i. and higher, and superheat temperatures in the region of 540°C.

Although there are many designs of water-tube boilers, in all of them the basic concept is more or less the same. Most modern water-tube boilers consist of a series of drums interconnected by tubes, the lower ones are the water drums and the upper are the steam drums. The interconnecting tubes between the upper and lower drums are arranged all round the furnace or combustion chamber so that the maximum surface is exposed to the hot gases. The banks of tubes nearest to the fires are the generating tubes; those away from the fires, or even outside the boiler, are the downcomer or circulating tubes. With this arrangement the heat from the fire causes the

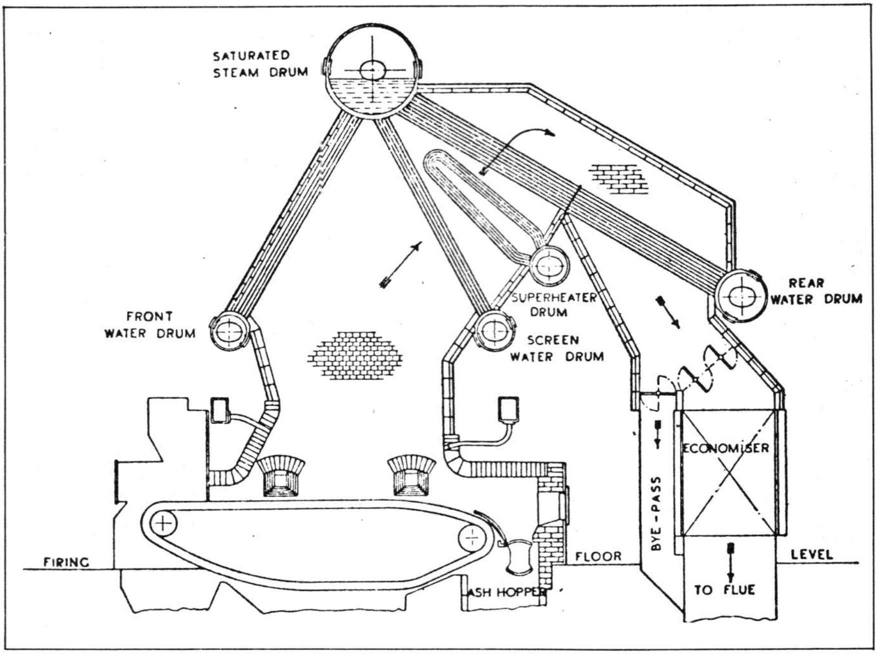

Figure 3.4 – Water-tube boiler – tubes expanded into drums

water to rise in the generating tubes and the colder water flows down the circulating tubes, see Figure 3.4.

With the volume of water in this type of boiler being quite small, the rate of circulation is very high and steam generation very rapid. Consequently, there is little or no steam storage capacity in a water-tube boiler and fluctuation in steam demand can be catered for only by firing control. With oil or gas firing and the rapid response of the water-tube boiler, fluctuating steam demands can be handled quite easily. Ships' boilers, for example, are required to cope with steam demands ranging from almost zero to maximum in a matter of minutes. To do this they have to be capable of converting the water into steam immediately it enters the boiler.

Coal-fired chain-grate water-tube boilers, as generally used in industry, are entirely different. The boilers will respond quite rapidly to a change in the firing rate, but varying the firing rate is a problem. To ignite coal and change its rate of combustion takes time. A small water-tube boiler, e.g. one which will evaporate 25 000 lb/h, will have a grate more than 20 ft (600 cm) long and a maximum grate speed of possibly 30 ft/h (15 cm/min). This speed depends on how quickly the green coal can be ignited, so it will take 40 min for the coal to travel the length of the grate. These factors limit the speed of response of the coal-fired water-tube boiler.

In the past, most industrial water-tube boilers have been installed as part of a combined heating and power plant, using back-pressure or pass-out steam engines or turbines to drive the electrical generating plant. In many cases, steam accumulators are incorporated in these plants to relieve the need for the boilers to cope with fluctuating steam demands.

One of the earliest types of water-tube boilers was the Babcock and Wilcox WIF (Figure 3.5). Many of these were built and quite a number are still in use, some to my knowledge are over 50 years old. These boilers have no water drums as such, but banks of straight tubes slung under the horizontal steam drums. These tubes are inclined with the rise from the back to front and connected to the drums via headers. The circulation is along the tubes from back to front, up the front headers into the drums, along the drums from front to back and down the rear headers to the tubes again.

The WIF boiler might well be regarded as the water-tube equivalent of the Lancashire boiler. It is fitted with large-bore tubes, usually 4 in (10 cm) diameter, and, unlike modern water-tube boilers, it will tolerate fairly hard water with the aid of relatively simple internal water treatment. With modern water-tube boilers it is essential to use fully softened water, as any

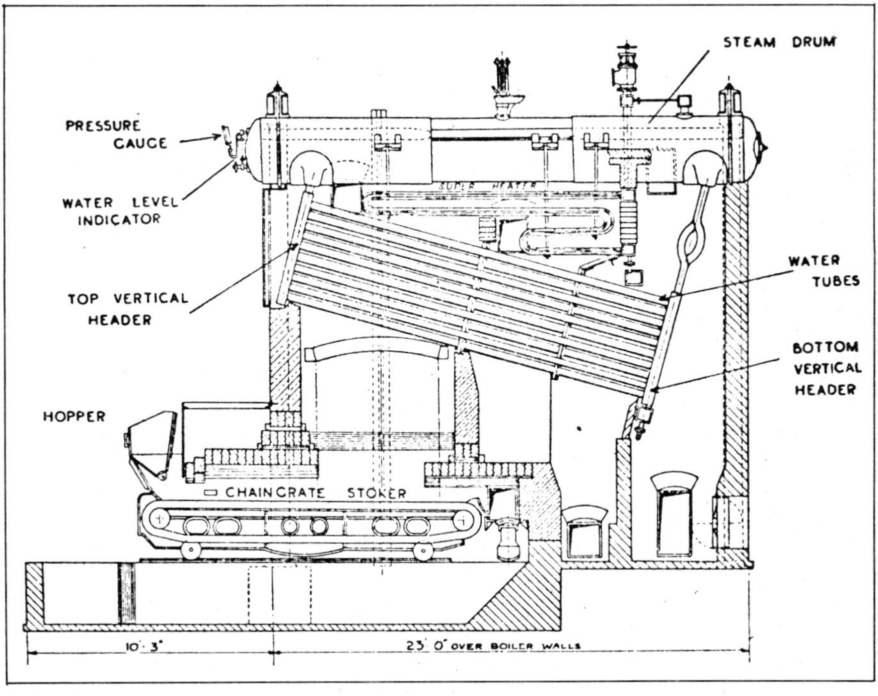

Figure 3.5 – Water-tube boiler – straight tubes with sectional headers

scale forming in the small-bore tubes is likely to cause overheating and premature tube failure.

The WIF is an excellent type of boiler for combined power and heating installations. It is a rugged piece of equipment, capable of operating at reasonable steam pressures and superheat temperatures. It can be designed to burn the cheapest coal, and with something like a 30% difference between oil fuel and coal prices, this is a big advantage.

At the time of writing, the unit costs of heavy fuel oil and coal are 8.9 and 6.2 respectively. So if we consider a company burning 300 tonnes of coal per week at, say, £30 per tonne for 46 working weeks, annual coal cost is $300 \times 30 \times 46 =$ £414,000. Then the equivalent cost for heavy fuel oil would be 414 000 $\times$ 8.9 $\div$ 62 = £594,289. The price advantage is obvious and the forecasts are that these differentials will increase.

Unfortunately the WIF is quite an expensive boiler to manufacture and appears to have become another victim of the cheap-oil era, the packaged boiler and, in the UK, the power-generating monopoly of the Electricity Boards.

3.8 FUELS

In recent years, probably most new boiler plants of less than 100 000 lb/h evaporative capacity have used packaged boilers. These boilers are much cheaper and much easier to install than conventional shell boilers.

Virtually all the early packaged boilers were designed to burn oil fuel. This was in the halcyon days when oil was cheap; in the 1950s heavy fuel oil was almost a drug on the market. Some of the smaller boilers burned light oil with a viscosity of 35 s (Redwood), and which can be stored and pumped at ambient temperatures. The larger boilers are usually designed to burn heavy fuel oil with a viscosity of 3 500 s (Redwood). This oil has to be stored at a temperature of about 40°C and preheated to above 120°C for burning. There are intermediate grades of fuel oil, but as these also require heated storage, preheating for burning, and are more expensive than heavy fuel oil they are not in general use.

Certain regulations must be complied with for the storage of fuel oil. Some are mandatory and some at the discretion of local authorities, but generally speaking it is necessary to have the storage tank or tanks installed in or over a catch pit, built of brick or concrete, with a capacity at least 50% greater than that of the tanks. The tanks should of course be sited as near the boiler plant as possible, but again, this may be at the discretion of the local authorities.

Light-oil storage tanks need only be fitted with a filling line, a drain valve, a level indicator, a vent pipe and a draw-off valve, no heating being necessary. Heavy fuel oil must be kept hot, not only in the storage tanks but also in the connecting pipework, and in the filling line, unless this is drained after each delivery. Heavy fuel oil is normally delivered and kept at pumping temperature in the storage tanks. Steam-heating coils are usually fitted in the bottom of the tank, and also combined steam and electric heaters at the outlet valves. Steam heating is usually kept on all the time steam is available, and for starting-up after weekend or holiday shut-downs electric heating can be used. Pipework should also be traced with either steam or electric heating, i.e. any tanks, pipes or valves holding or conveying heavy fuel oil should be kept warm. If this grade of oil is allowed to cool down, it becomes extremely difficult to move.

In the early 1970s North Sea gas became a very attractive boiler fuel. Its price compared very favourably with oil, but unfortunately it is now rapidly being brought into line with that of oil fuel. Not only were a large number of new gas-fired packaged-boiler plants installed during this period, but a fair proportion of the older boiler plants were converted to gas firing.

One of the major problems with what otherwise could be very attractive contracts for the supply of gas was that most of these were on what is known as an 'interruptible supply' basis. This meant that gas supplies could be shut off at very short notice for fairly long periods, so provision had to be made for burning an alternative fuel, usually oil, during the gas cut-offs. The benefits of having gas piped direct to the boiler plant are offset to some extent by the need to install oil storage tanks for the standby fuel, and these storage facilities have to be quite large as oil companies are not usually amenable to suppling oil fuel on an irregular basis. Also dual fuel-burning equipment has to be fitted to the boilers.

Burning gas in a boiler presents no particular problems except perhaps with respect to safety precautions that must be strictly enforced. These are even more stringent than those for fuel oil. Gas is an ideal fuel, it is absolutely consistent in quality, so that maximum thermal efficiencies are easily maintained once they are established, there are no soot or smoke problems and flue gases are not so corrosive as those from fuel oil.

It would appear from all the forecasts that coal will sooner or later recover at least some of the ground it has lost as a boiler fuel for private industry. The ascendency of oil and gas, it could be argued, together with the development of the packaged boiler, virtually halted any progress in the development of small to medium coal-burning units.

Coal was generally regarded by mill management as a messy fuel, it creates ash, grit, soot and dust, it takes up valuable ground space to store, it deteriorates in store, it is variable in quality and therefore creates problems in maintaining efficient combustion and the constant attention of full-time boiler house staff is generally required. But in spite of all these drawbacks, coal is being used to generate most of the electricity in this country, and the big power stations can burn it very efficiently.

There are other problems, two of which are variation in quality and variation in size. If a coal-burning plant is being considered, information on the quality and size of the coal available should be obtained.

Coal is marketed in many sizes. There are fines, slacks, smalls, peas, singles, doubles and trebles, to name only a few of the different grades, and most of them require different burning techniques. Generally speaking, the smaller the coal the higher the draught required, so that to burn fines, slacks or mixed smalls, which is a combination of both, forced and induced draught may be required, whereas the larger singles, doubles, etc. can be burned satisfactorily with the natural draught from a reasonably tall chimney. The capital cost of coal-burning boiler plant also varies widely, in general the cheaper the coal, the more expensive the boiler plant required.

When burning coal, the evaporative capacity of a boiler is directly related to the area of the grate and the thickness of the firebed. As the depth of the firebed is very limited and the grate area quite small, it is not a good idea to attempt to burn the small grades of coal in shell-type boilers, because of the limitations of grate area. In the largest shell boiler with two furnace tubes, the grate area is still limited to about 50 ft^2 (4.65 m^2) whereas in a 25 000 lb/h (3.15 kg/s) coal-fired water-tube boiler designed to burn this type of fuel the grate area is nearer 150 ft^2 (14 m^2).

Consequently, to burn fines and slack, water-tube boilers with a large grate area, forced- and induced-draught facilities and sophisticated coal- and ash-handling plant are generally installed. Such installations are very expensive and would probably not be justified where the evaporation rate is less than 100 000 lb/h. Where the rate is greater than 100 000 lb/h, the cost of water-tube boilers for burning low-grade coal, supplemented by 'back-pressure' or 'pass-out' power generation, could well be justified. Such steam and power plants are used quite extensively in paper mills, sugar refineries and many more of the large steam users, including textile dyeing and finishing plants.

In coal-burning shell-boiler plants with mechanical stokers, the larger grades of coal are preferred, fuel ignition is better, and much thicker firebeds are possible using singles and doubles.

Packaged boilers have been developed for coal firing, but, as these units are so compact and can only accommodate a single furnace, evaporative capacity is limited.

In larger power stations, the coal is dried and pulverized and blown into the furnaces where it behaves almost like a liquid or gaseous fuel. This technique has been used in smaller boiler plants, but not to any great extent. In view of the present fuel situation and the outlook for the future, the development of a packaged boiler that would burn pulverized fuel might be worthwhile. The coal could be pulverized and dried at the pit head and delivered in powder form to silos on the consumer's premises.

3.9 POWER GENERATION

It is not generally appreciated that the thermal efficiency of large steam-powered generating stations is very low, the reason being that a large proportion of the heat in the steam ends up in the sea, river, canal or wherever else the cooling water comes from.

Consider again the two forms of heat applied to generate saturated steam, the sensible heat to raise the temperature of the water, and the latent heat to convert the water to steam. Steam tables show that as the steam pressure rises the sensible heat increases and the latent heat falls, i.e. the lower the pressure the higher the latent heat content. So if a condenser is used to create a vacuum, the steam can be used to do useful work right down to a pressure of, e.g., 0.5 lbf/in^2 (3.5 kPa). This is how steam is used in large steam power stations. Steam is generated at pressures in excess of 2 000 p.s.i. and used in the turbines right down to around 0.5 lbf/in^2.

Steam at this very low absolute pressure is only at a temperature of about 27°C, so it is ineffective as a heating medium but still retains more than 1 000 Btu/lb (2 200 kJ/kg) of latent heat, which has to be removed before it will condense for return to the boiler. This latent heat is taken out by the circulating or cooling water.

The foregoing is only a very basic explanation of the steam cycle in a large turbine generating plant. In practice, the cycle is much more sophisticated, the steam is superheated in the boiler and in some cases taken out for reheating at certain points in the turbines. Nevertheless, it all ends up as low-temperature steam imparting its heat to the cooling water. The resultant thermal efficiency of the plant is only in the region of 40%.

For a private company requiring electrical power only, generation by steam is not economical, but in some industries, including textile bleaching, dyeing and finishing, where large quantities of low-pressure steam are used, the economics of power generation are quite different.

Steam pressures from 25 to 40 p.s.i. are adequate for most wet processing, and even for drying cylinders and low-temperature stenters, but boiler pressures up to 250 p.s.i. are still used, so that small volumes of high-pressure steam have to be reduced in pressure to produce large volumes of low-pressure steam.

Reducing valves are used to bring the boiler pressure down to the process pressure, and all a reducing valve does to the character of the steam is dry it or superheat it slightly. However, if a steam engine or turbine is used to carry out this pressure-reducing function, a great deal of energy is made available for driving an electrical generator. This is the principle of 'back-pressure' or 'pass-out' generators.

The following example will serve to explain back-pressure and pass-out generating plants more fully.

Steam is generated in three coal-fired water-tube boilers, operating at a pressure of 250 p.s.i., and superheated to a temperature of 320°C. The bulk of this steam is passed to turbo-alternators at this pressure and temperature, and drives the turbines utilizing a pressure drop of 210 p.s.i. The pressure of the turbine exhaust steam is thus 40 p.s.i., which is a very useful steam pressure for all wet processing, cylinder drying units and space heating.

Total potential output of the three boilers is 120 000 lb/h (15 kg/s), and each of the two turbo alternators generate 800 kW at 3 300 V. The amount of steam used in the turbines over this pressure drop is 40 lb (18 kg) per kWh, so that for 64 000 lb (29 000 kg) of steam, 1 600 kW can be generated, in exchange for a reduction in steam pressure, superheat, and some of the sensible heat. Steam at 250 p.s.i. has a total heat value of 1 203 Btu/lb (2 800 kJ/kg), steam at 40 p.s.i. a total heat value of 1 176 Btu/lb (2 735 kJ/kg) and a very useful temperature of 142°C.

Thus, while doing the same job as a reducing valve, at the same time these back-pressure generators produce 1 kW of electrical power for every 40 lb (18 kg) of steam passed through the turbine, with a heat loss of only 27 Btu/lb (65 kJ/kg) of steam, approximately 2%. The cost in fuel of producing this heat is easily calculated from the price and calorific value of the boiler fuel.

The 40 p.s.i. steam from the turbine exhaust is then used for scouring, washing and cylinder drying, also for space heating when required. In the wash becks all the heat in the steam is utilized with the exception of that lost in the effluent. In drying cylinders and space heaters, the latent heat is utilized, the steam is condensed and the hot condensate returned to the boilers. Steam required for stenters and processes needing higher temperatures is taken direct from the boilers, reduced to 150 p.s.i. and piped to the stenters, etc. There is also a steam accumulator which operates over a 40–150 p.s.i. pressure range; this levels out the peak process steam demands on the 40 p.s.i. system.

It should be noted that the foregoing boiler-evaporation and power-generation figures are potential loadings and could only be utilized if the demand for 40 p.s.i. steam was 64 000 lb/h or greater. The main factor to be considered with back-pressure power generation is the low-pressure steam demand. If the generators are exhausting more steam than the process plant requires, then the superfluous steam is blown off to the atmosphere and, instead of using the 2% of the heat for power generation, this must now be debited with something like 98% of the heat in the steam, which obviously is no longer economic.

To sum up, in a low-pressure process steam plant, back-pressure power generation has a very high overall efficency, bearing in mind that the controlling factor is the low-pressure-steam demand. Power generation is only economical within the limitations of this demand, any additional power is usually taken from public supplies. The system can operate at an overall thermal efficiency of 85% compared to 40% for a power-only condensing system.

In plants where power generation is the predominant requirement, or where higher process-steam pressures are required, the pass-out system can be used. In this system the turbine utilizes the steam from boiler pressure, right down to condensing pressure, in a similar manner to large power stations, but as the steam passes through the turbine it drops in pressure at the various stages of expansion and can be bled off or passed out at any of these stages to supply process requirements. Bled-off steam is usually passed through a reducing valve so that steam at constant pressure is supplied to the process system. The quantity being bled off is completely flexible, as the excess simply passes on through the lower-pressure stages of the turbine.

This system is a compromise between condensing- and back-pressure generation, and the overall thermal efficiency will depend on the amount of steam being passed out for process-heating purposes.

The use of combined power and heating systems in the textile dyeing and finishing industry is not so attractive as it used to be. When the fabric being processed was essentially all cotton, low-pressure steam was not only suitable for the wet processes, but in many instances it was hot enough for cylinder drying and for finishing stenters at a pressure of 40 p.s.i. and temperature of 140°C. With the advent of nylon and polyester fibres requiring temperatures in excess of 200°C for heat setting, resin curing and baking, steam can no longer be employed as a heating medium, and even jet machines normally require steam pressures in excess of 80 p.s.i. All these factors tend to reduce the demand for low-pressure steam and, consequently, the feasibility of back-pressure or pass-out power generation.

3.10 STEAM DISTRIBUTION

The proper distribution of steam is the all-important link between the boiler plant and the process plant. However efficient the steam-raising and the steam-using plant, if the steam distribution is bad, efficient operation is considerably impaired.

Pipe Sizing

It is of the utmost importance that the steam main be of the correct size for the amount of steam it has to carry. If it is too small it will lead to steam starvation at the using end; if it is too large the initial capital cost will be higher, as will be the continuing radiation losses.

There are various methods of determining the size of pipes, the most popular are:

1. By velocity
2. By pressure drop.

Having assumed a particular velocity, calculations are based on the volume of steam carried in relation to the cross-sectional area of the pipes. Experience shows that for dry saturated steam a reasonable velocity is between 80 and 120 ft/s (24–36 m/s).

Sizing on velocity takes no account of the length of travel and therefore, due to variation in frictional resistance, can give a very uneven pressure drop over a network of pipes, a situation that cannot usually be tolerated because the pressure at the process end is all important.

Therefore, accepting that a definite pressure must be maintained at the process plant it is preferable to base size on pressure drop. To do this, the effective length of the steam main should be known or estimated and account must be taken of any restrictions such as elbows, bends, valves, expansion loops, etc. For example, for a pipe of 4 in (10 cm) nominal bore, 8 ft (2.44 m) should be added for an elbow, 3 ft (92 cm) for a standard bend, 27 in (69 cm) for a parallel-slide valve and 12 ft (3.66 m) for an expansion loop. There are many formulae for the sizing of steam pipes, and graphs and charts have been produced which makes pipe sizing much simpler. Companies producing steam pipes, valves and accessories have all the necessary information and should be referred to for any assistance (see also Chapter 2).

In sizing a steam main consideration should be given to any future plans for additional steam-using plant and the siting of this in the works; it is far better to allow for this at the outset rather than to face a major re-organization when plant modifications and new installations are required.

Steam Pipes

Steam mains are usually made of hot- or cold-rolled seamless mild steel tube, which can be obtained in lengths up to 20 ft (6 m). The lengths can be joined by either screwed or welded couplings, but preferably by flanges with joints between (Figures 3.6, 3.7 and 3.8).

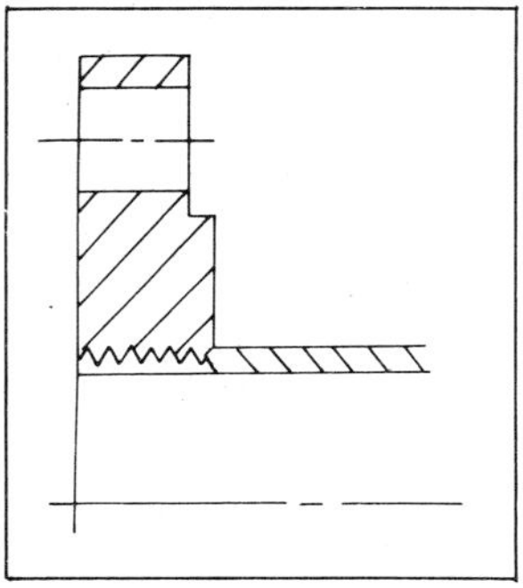

Figure 3.6 – Screwed flange

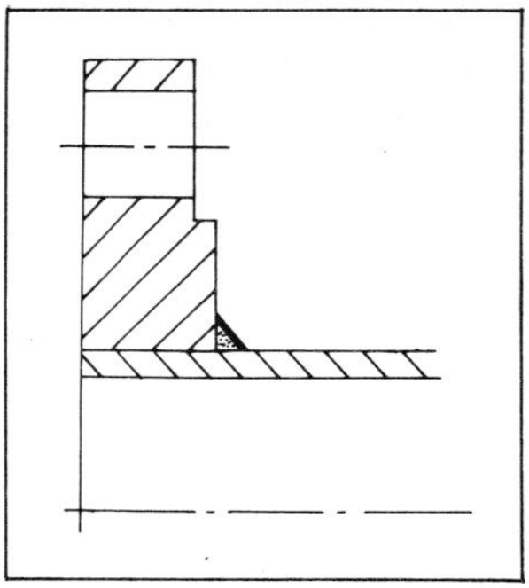

Figure 3.7 – Welded flange

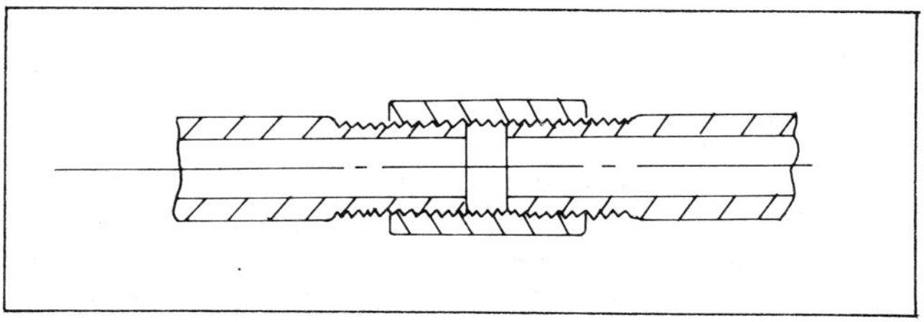

Figure 3.8 – Coupling

The thickness of the tube depends on working pressure, and to a lesser extent on working temperature. Along with standard Ts elbows and bends, tubes are made to British, US and European standard specifications. Flanges are also made to these specifications, which relate to thickness, diameter and the size and number of bolts.

Valves

Valves are a very important part of a steam-distribution system and should be chosen carefully (see Chapter 2). The steam main itself should be designed so that any section can be shut off for maintenance or when not in use.

The parallel-slide valve is possibly the most suitable because when it is fully open it offers very little restriction to the steam flow and when it is shut it is invariably tight and does not leak. A parallel-slide valve should not be used in a cracked-open (very slightly open) position and is not ideal when used as a control valve. If the requirement is to control steam flow to an item of plant then the type of valve selected must be suitable for the incorporation of automatic control, and the size of valve is determined by the steam flow required. Globe valves are generally the most suitable. The situation can arise when a particular item of plant requires steam at a pressure that cannot conveniently be provided from the main distribution system, in which case a reducing valve is fitted in a branch line from a main carrying steam at higher pressure. The reducing valve (Figure 3.9) is a valve that maintains a steady pressure on the low-pressure side, obtaining the control action from that side.

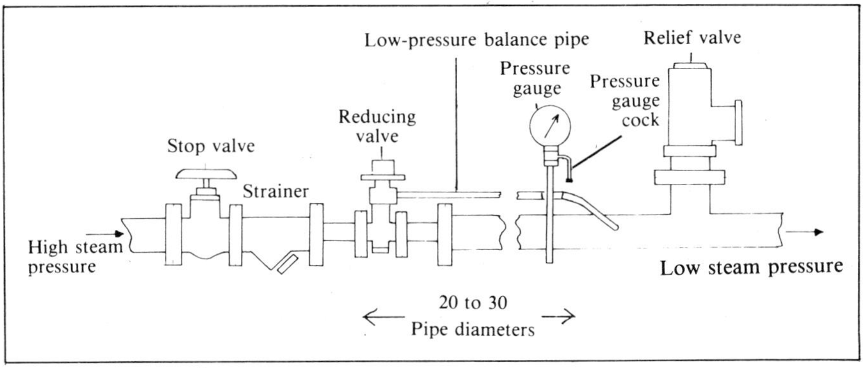

Figure 3.9 – Reducing-valve layout

Reducing valves, like any other equipment, can go wrong and stick in the open position. Thus, it is essential that plant being fed by the reducing valve be protected by an adequate safety valve.

Whenever possible steam mains and branch lines should be fitted with pressure gauges at suitable points; the provision of other instruments, such as steam meters and thermometers, depends on individual requirements.

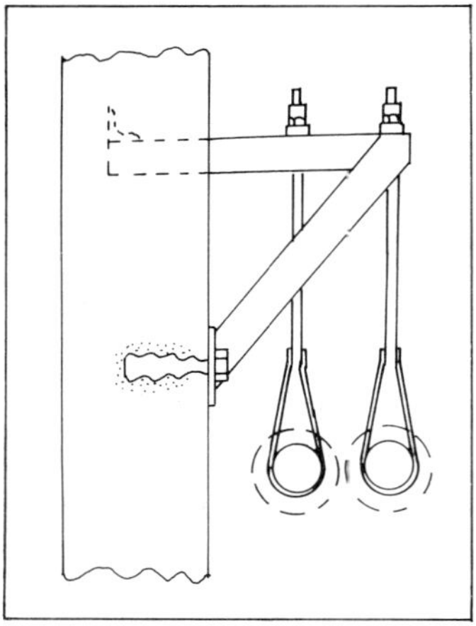

Figure 3.10 – Pipe support

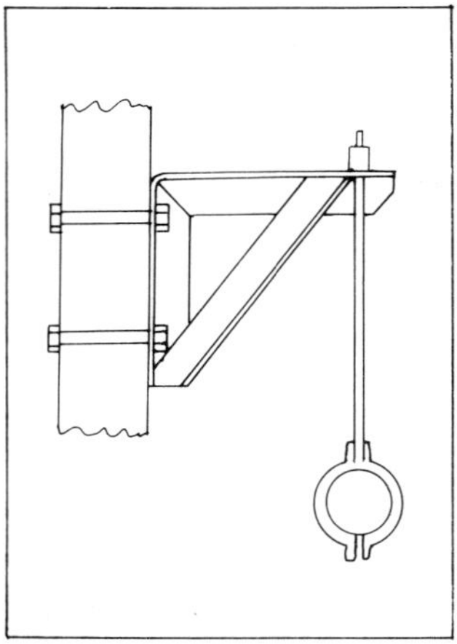

Figure 3.11 – Pipe support

When fitting a steam main, allowances must be made for expansion, means for supporting and anchoring it should be provided. Supports should, as far as possible, permit free movement of the pipe in any direction, any necessary constraint being provided by the anchors, (Figures 3.10, 3.11 and 3.12).

The spacing of supports is generally determined by the position of beams or other structural features, but the supports must not be placed too far apart, e.g. over 20 ft for a pipe of 6 in bore, and as far as possible pipe joints should be placed close to the points of support. A main for carrying steam at 150 p.s.i. will expand about 2 in for every 100 ft of length. When the pipe layout is planned the anchoring points should be decided and should be at appropriate distances along the pipe. As these are fixed points to which the pipe is anchored they should be secured to a substantial part of the building structure or, if outside, to a rigid gantry or trestle.

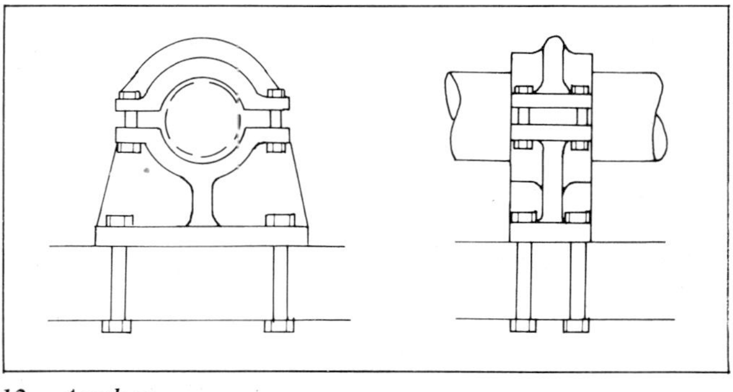

Figure 3.12 – Anchor

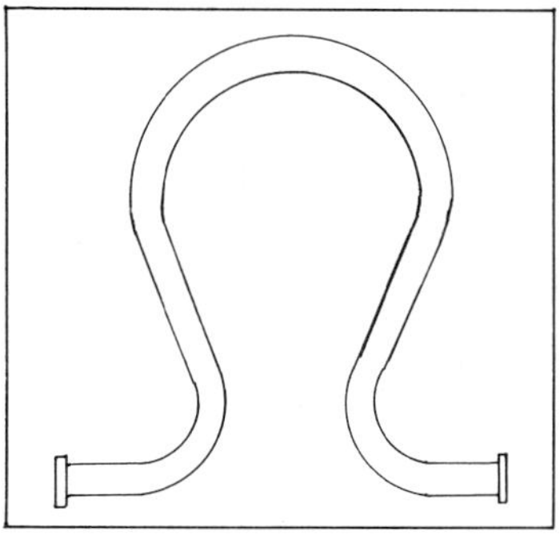

Figure 3.13 – Expansion bend

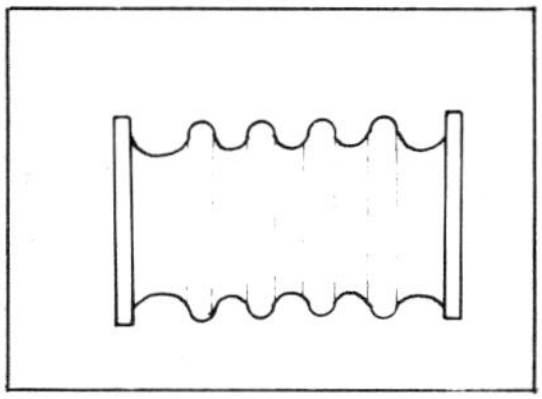

Figure 3.14 – Expansion bellows

Allowances must then be made for the pipe to expand between these points, long or slow bends will accommodate a certain amount of expansion but long straight runs must be fitted with an expansion bend or one of the other available types of expansion device (Figure 3.13, 3.14 and 3.15). The possibility of branch connections being strained by, or unduly restricting, the movement of the main must be guarded against.

Provision for draining condensate or boiler water carry-over must be made when installing steam mains, which should therefore be arranged so that there is a good fall, preferably in the direction of the steam flow, generally about ½ in in 10 ft. This fall may sometimes cause a problem if sufficient headroom is not available and it is common practice to introduce a system of relaying, the relay point providing a very suitable drain for the condensate (Figure 3.16). It is essential that steam pipes be kept free of

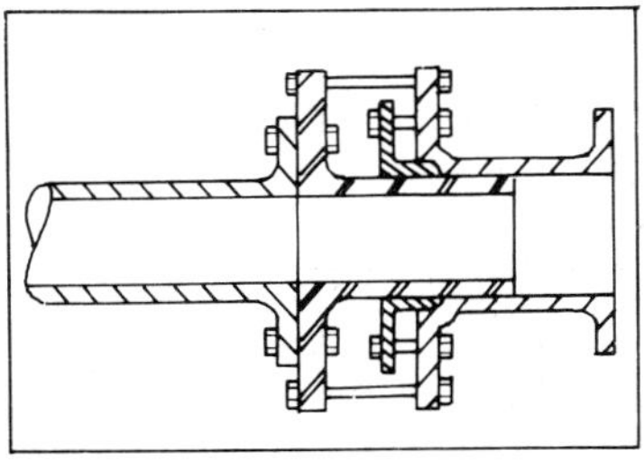

Figure 3.15 – Expansion joint

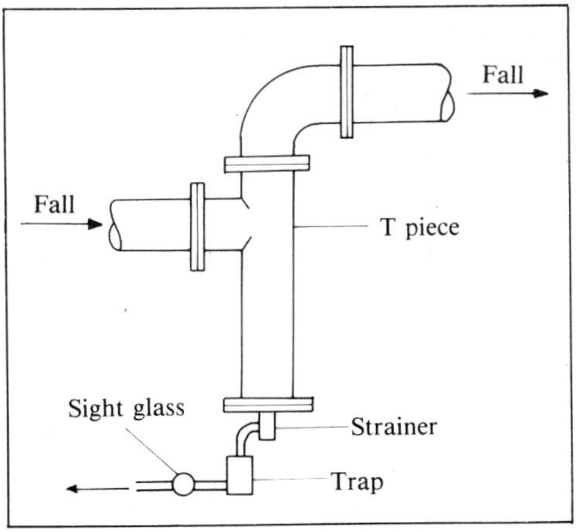

Figure 3.16 – Relay system

water, otherwise there is a danger of water hammer. This is caused by an accumulation of water being carried along the pipe with the steam and can be particularly dangerous during cold start-up, as the water can be carried along the pipe at a velocity high enough to fracture a valve or even the pipe itself.

Branch Lines
Most machinery and plant manufacturers quote steam consumption and pressures in their general specification and it is therefore not very difficult, using the method explained previously, to size accurately a branch line to feed a particular section. These connections should preferably be made to the top of the main to provide steam as dry as possible and ideally should be fitted with a valve. Expansion and draining of a branch line of, say, 2 in and above should be given careful consideration, on the general lines of the steam main.

Ring Main
When a steam line is being planned it is well worth considering a 'ring main'. In this form of main the steam from the boiler plant is usually carried in a single pipe to a point just inside the process building. Before any tappings or branches are taken off, a T piece is fitted at this point, and steam is fed in both directions into a ring of piping. There are a number of advantages in using a ring main. If the steam demand in a branch becomes heavy, not an unusual occurrence in dyeing and finishing, pressure drop is minimized as steam is being supplied to the branch from both directions. Smaller-diameter pipes can be used, which will affect the additional capital cost to some extent. If valves are fitted in the ring at strategic points, sections of the main can be shut down without interfering with the steam supply to all the machines. As the ring main is continuous there is no machine at the end of the line to be starved of steam.

Steam Traps
There are many different types of steam trap, some more suitable for a particular application than others. These include float, bucket, inverted-bucket, bimetallic, thermostatic and thermodynamic traps. It is not possible to describe here the design and applications of all the traps available, but steam-trap manufacturers are always willing to give advice on the type of trap to use for a particular purpose. I favour the inverted-bucket trap for heating batteries, jacketed vessels and steam heating coils, etc. and the thermodynamic trap for steam-pipeline draining.

There are, however, many points to note applicable to main steam traps. Strainers should be fitted on the inlet side of all steam traps to prevent the entry of dirt, rust and other foreign particles, which can block orifices and cause the trap to malfunction; some traps have strainers built into them. Sight glasses are available for fitting on the downstream side of the trap so

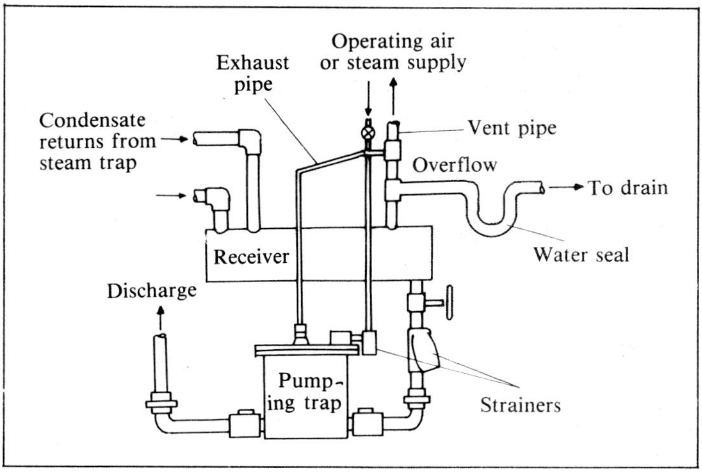

Figure 3.17 – Pumping trap

that it can be seen at any time whether or not the trap is working correctly and discharging condensate.

All drain points must be individually trapped so that each can cope with the needs of the particular point it is draining. Group trapping, i.e. several drain points laid to one trap, should be avoided.

Steam traps are constructed so that condensate can be lifted on discharge but it must be remembered that it is the steam pressure at the trap which actually carries out the lifting operation, the condensate being pushed about 2 ft up a pipe for every pound of pressure. Nevertheless, there are real and practical disadvantages in lifting condensate after the steam trap and gravity discharges are much to be preferred if the plant layout permits. When it is necessary to raise condensate to a high-level return system the best way is to use a pumping trap (Figure 3.17).

Insulation

The type and thickness of insulation must be considered at an early stage when a steam-distribution system is being designed. Steam and condensate mains should be efficiently insulated to reduce heat loss and so save fuel and running costs.

The most convenient way of insulating steam-condensate pipes is to use rigid sections of fibre-glass or mineral wool. These can be obtained in various thicknesses to suit a wide range of pipe sizes. Tables of minimum thicknesses for different temperatures are set out in British Standard Specification 5422 (1977).

The sections are closely butted together and securely fixed with metal bands. Bands, valves and fittings can be accommodated by cutting and mitring the sections as required.

Many external finishes are available to suit most situations. For interior pipework the sections are usually coated with canvas and can be painted to match the decoration scheme or the approved colour code. For exterior insulation the sections are best covered with plastic or roofing felt and reinforced with galvanized wire netting and coated with bitumen.

With energy costs rising as they are, good insulation is important; new products are continually being introduced and it would be wise to obtain advice from an insulation manufacturer or contractor.

4 MOTIVE POWER

4.1 ELECTRICITY SUPPLY

Most electricity from public generating sources is supplied at 11 000 V and an alternating frequency of 50 Hz. For industrial consumers the voltage is stepped down by transformers to 415 V at the same frequency. In the UK the supply is three-phase and neutral, giving 415 V phase to phase and 240 V phase to neutral.

Worldwide there are significant differences in the supply voltage and frequency, and machinery makers must fit motors and control equipment to suit these conditions.

4.2 ALTERNATING-CURRENT MOTORS AND STARTERS

Induction-cage Motors

The simplest, cheapest and most easily maintained type of motor to operate from this power supply is the a.c. induction-cage motor, often known as the squirrel-cage motor. The cheapest form is the ventilated type, which is usually open at both ends, cooling being effected by an air flow through the motor. It is not really suitable where there is dust, fumes or high humidity, which rules it out for dyeing and finishing processes as these conditions are present in most of these processes. The drip-proof motor is more or less protected all round, but still has ventilation openings, so it too is not to be recommended for use in steamy or dusty conditions. The best version of this type of motor for operating bleaching, dyeing and finishing machinery is the totally enclosed fan-cooled unit. This motor is completely enclosed and is dust- and moisture-proof. Cooling is by fan mounted on a rotor shaft extension at the non-driving end of the motor. This fan blows the cooling air over the outside of the motor housing and the interior is sealed from the surrounding atmospheric conditions.

The squirrel-cage motor runs at a fixed speed, determined by the number of poles in the stator and the frequency of the electrical supply. The poles, which are always in pairs, are the electromagnets formed inside the housing, usually called the stator, and the revolutions per min of the motor are given by (cycles per second $\div$ pairs of poles) $\times$ 60. Thus a motor with one pair of poles will have a theoretical speed of $(50 \div 1) \times 60 = 3\ 000$ rev./min. This is the synchronous speed for motors of this type. In practice it is found that there is a certain degree of slip and actual speeds are about 3% less.

The squirrel-cage motor derives its name from the shape of the motor, which is constructed by embedding diecast aluminium or copper bars in slots in the iron core of the rotor, the bars are short-circuited by rings of aluminium or copper at each end. There is no electrical connection to the rotor. Most squirrel-cage motors used in the textile dyeing and finishing industry are four-pole 1 500 rev./min machines. Two-pole 3 000 rev./min motors are cheaper, but this speed is on the high side for dyeing and finishing machinery. Two-speed motors are sometimes used, e.g. for driving stenter fans. These are fitted with the number of poles required for the slowest speed, e.g. six poles which will give 1 000 rev./min, but by energizing only four poles the speed will be 1 500 rev./min. The speed of squirrel-cage motors can thus be altered, but only in a very limited number of steps.

The load–speed characteristic of the squirrel-cage motor is nearly constant, with only a slight reduction in speed as the load increases, and the efficiency of motors up to about 20 h.p. (15 kW) is constant at about 90% from half load up to about 25% overload.

Switchgear and starters for three-phase induction motors are quite simple. With motors up to 20 h.p., starters are usually of the direct on-line type, when the starter contacts are closed, the full voltage is applied to the motor fields. With this method of starting the applied current is very high, up to six times the running current. Starters are usually fitted with overload devices to cope with the momentary starting current without tripping out.

On larger motors of this type, a star-delta starter is often used. With this system the current is applied to the fields in two different circuits, first in star which has the effect of reducing the applied voltage to 58% of the line voltage and increasing the current to about 1.5 times the running current. After the speed of motor has increased to about three-quarters the running speed the circuit is changed to delta, which is the normal running connection.

Squirrel-cage motors do not develop a high starting torque except with a direct on-line starter. They are not very suitable for starting under load since they only produce their power at high speeds. This is satisfactory for such items as fans, where the load in the motor rises progressively with the fan speed, but not where it is required to start up a padder or mangle nip with the pressure on, where a high starting torque is required. This drawback can be overcome to some extent by fitting one of a number of slipping coupling devices. These allow the motor to start up under no load and the drive is only taken up as the motor runs up to speed. However, special torque squirrel-cage motors are available.

Slip-ring Motors

Slip-ring or wound-rotor machines are fixed-speed motors with the same speed characteristics as the squirrel-cage motors. Their speed is governed in the same way by the number of poles in the stator.

These motors are used where a high starting torque is required without the need for excessive starting current. This is achieved by injecting current into the rotor as well as the stator windings. This current is conveyed to the rotor windings via carbon brushes running on slip rings mounted on the rotor shaft and connected to the windings. By adding an electrical resistance to the rotor for starting purposes, this type of motor can generate up to 200% of full-load torque with about twice the full-load current, as against 100% full-load torque at six times full-load current for an ordinary squirrel-cage motor.

Slip-ring motors being more complex are obviously more expensive than conventional squirrel-cage machines, but where a load has to be moved from rest with, e.g. a 50 h.p. (37.5 kW) motor, and a slipping coupling cannot be employed, with a conventional squirrel-cage motor the starting current could be up to six times the full-load current. This is not viewed very favourably by the electricity authorities, apart from the fact that it may cause a few circuit breakers to trip out.

As the foregoing motors are fixed-speed, there are limitations to their use in process machines where variation in speed is required. They can be, and often are, used in conjunction with a variable-speed gear to supply a variable-speed output, but in these instances the speed variation is achieved by hydraulic or mechanical devices, rather than by electrical means.

Variable-speed Motors

Until the development of the thyristor and other modern methods of converting a.c. to d.c., variable-speed a.c. motors were used extensively for dyeing and finishing machine drives, and indeed many of these motors are still in use.

Probably the most widely used type was the a.c. commutator type of motor, usually known as the Schrage motor. This type looks rather like a d.c. motor, in that it has a commutator and brushes for conveying current, and in fact has similar characteristics to a d.c. shunt-type motor. It also has slip rings and brushes similar to the a.c. wound-rotor machine.

In this machine the primary winding is in the rotor instead of the stator. This winding receives mains power via the slip rings. The regulating winding, which is also in the rotor, is connected to the commutator; this regulating winding acts almost like a generating winding, in that it passes current out of the rotor via the commutator and brushes to the stator windings. There are three stator windings, each of which is supplied with current from three pairs of sets of brushes radially disposed around the commutator. These brush sets can be moved around the commutator by means of a pinion and a toothed ring operated either by a handwheel, a chain pinion connected to a compensator, or a 'pony' motor and reduction gear, so that the distance between the brush sets can be varied. The function of the commutator is to deliver power from the rotor regulating windings to the stator windings at slip frequency. If the two brush sets are set so that they are short circuited, the motor will behave like an ordinary squirrel-cage (induction) motor, and the speed will be relative to the number of poles in the primary winding. When the brushes are separated, an additional voltage is injected into each stator circuit. This either increases or reduces the stator currents and slip, and the speed of the motor alters accordingly.

The speed range of the a.c. commutator motor is continuously variable at anything from 4:1 to 10:1 or even greater, and power output is relative to speed. When this type of motor is used to provide a variable-speed drive it is advisable to gear it to the machine so that the normal working speed range of the machine is as near as possible to the synchronous speed of the motor. It is not good practice to run this type of motor for long periods at the lower end of its speed range, as the high slip will generate a great deal of heat. For this reason many of these motors are fitted with their own fans for forced ventilation cooling.

The starting torque of this type of motor is high, and as speed regulation is by mechanical movement of the brushes, a simple three-phase contactor type starter is all that is required, similar to a squirrel-cage-motor starter. These motors are also quite suitable for driving the various units on a process range in series as the brush gear movements can be controlled by compensators via a chain or lever linkage, or by a pony motor and compensator-operated contactors.

As the a.c. commutator motor is rather a complicated machine to construct, it is now very expensive. In the event of a burn out, the cost of rewinding can be substantial. For these reasons and with the development of solid-state transforming and rectifying equipment, the d.c. motor has now become a more economical proposition for providing a variable-speed drive.

A number of small variable-speed drives are marketed suitable for driving inspection, rolling or other small machines. Some of the manufacturers' literature on these units can be rather misleading, they are advertised as variable-speed drives for a.c control. The motors are in fact d.c. shunt-wound machines and the control box usually incorporates a rectifier to convert the a.c. to d.c. With this system a constant voltage is fed from the rectifier to the shunt field, and a controlled variable voltage is supplied to the armature from a small transformer via a rectifier.

4.3 DIRECT-CURRENT MOTORS

Since the use of steam engines to drive textile machinery was discontinued, the direct-current motor has been the most popular method of providing a variable-speed drive for dyeing and finishing machines. In fact, in the early days of electrical power when generators were known as 'dynamos', all power supplies were d.c. Up to ten years ago some of the older dyeing and finishing plants still used a large proportion of d.c. power by rectifying the incoming public electricity supply, but eventually the cost of maintaining the older type of d.c. motors became prohibitive.

The terminology of d.c. motors is somewhat different from that of a.c. machines. The rotating part of a d.c. motor is called the 'armature' and the static part the 'field', rather than 'rotor' and 'stator' as in a.c. terminology. The basic difference between an a.c. and a d.c. machine is that the latter invariably has windings in the armature and thus a commutator and brushes are required to carry current to these windings. The d.c. machine also has windings in the static part.

Series-wound Motors

The series-wound motor has a very high starting torque, as both the armature and field current is at maximum at the moment of starting. This type is not suitable for operating under light loads as the speed is relative to the load and with no load the motor will race away. For this reason, the series-wound motor should only be used when connected to the machine by a sound mechanical coupling. As a matter of interest, being a high-torque machine, it is extensively used as a traction motor in tram cars and cranes, but rarely for driving textile machinery.

Shunt-wound Motors

The shunt-wound motor is designed to operate at a constant speed under varying loads, so long as the field is constant. The speed of this type of motor is controlled by varying the voltage applied to the armature. This is done by having a bank of resistances that can be inserted or removed by the speed controller, thus altering the voltage to the motor. From the mains a connection is taken off to the shunt field, which again is passed through a variable resistance. The speed of the motor can also be altered to a considerable extent by varying the resistance in the shunt field. These two methods of speed control can be utilized if two or more motors are required to run in tandem. If the same voltage is fed to each armature, each motor will run at more or less the same speed, and to obtain full synchronization each motor can be trimmed by independent rheostats operated by compensators.

Shunt-wound motors were used quite extensively for stenter drives when d.c. power was the normal electrical supply. A twenty-times speed variation is quite possible with this type of machine.

Compound-wound Motors

Compound-wound motors, as the term implies, have some of the characteristics of both series and shunt machines. They have a high starting torque and at the same time a limited top speed on light load.

These motors were used for driving calenders because of their ability to start up under heavy load and run at a fixed speed.

The development of new techniques in a.c. transforming and rectifying has revolutionized the use of d.c. power for variable-speed drives. Often in the past d.c. power was used in general distribution systems by employing large rectifiers supplying d.c. electricity for all power and even for lighting, which meant that d.c. motors, expensive both to buy and to maintain, had to be used in cases where the much cheaper a.c. squirrel-cage motor would have been adequate.

Nowadays, most plants will have a.c. distribution systems, and where variable speed is essential modern methods of transforming and rectifying are used to provide d.c. to drive individual machines or process ranges. Methods of rectifying a.c. to d.c. power have advanced very rapidly over the past twenty years or so; diodes and thyristors have replaced motor generators, and now even smaller semiconductors are being used and rectification is becoming more of an electronic rather than an electrical technique.

The foregoing is only a brief description of the types of d.c. motor. There were many variations, but although the physical shapes, sizes and methods of construction have changed, the operating principles remain the same. What has changed enormously and continues to change, are the methods of supplying a direct current and of controlling d.c. motors.

Direct-current Range Drives

One of the most widely used variable-speed, multi-motor driving systems is commonly known as the Ward Leonard system. This method of variable-speed control can be used for almost any number of motors running in tandem, and is used by European and North American, as well as British, machine manufacturers for driving ranges of textile processing equipment.

The Ward Leonard (WL) system in its original form was operated from a d.c. mains source, but as a.c. power replaced d.c., transforming/rectifying equipment became necessary in order to apply the system to individual machines or ranges of equipment.

To use the system, the first requirement is some means of converting a.c power to d.c. In the older WL systems this was done by using a motor-generator set. This comprised an a.c. fixed-speed motor driving a d.c. shunt-wound generator. By applying a variable exciting voltage to the shunt field, the output voltage of the generator can be controlled. This excitation voltage is supplied either from a small generator also driven by the a.c. motor, or in more modern installations by a solid-state transformer/rectifier. The variable voltage from the generator is fed to the armature of all the motors in the system so that they run at more or less the same speed. But, in a series of nips in a washing or bleaching range for instance, allowances have to be made for cloth shrinkage, for differences in the diameter of nip bowls, etc. This, of course, is the purpose of compensators between each unit of a range. In the WL driving system these compensators are mechanically linked to potentiometers, which control the supply voltage to the shunt field of each motor and the speed of each individual motor is varied accordingly. This shunt field voltage is usually supplied from an independent transforming/rectifying unit.

Compound-wound motors are used in the WL driving system, since they have the high starting torque characteristic of the series-wound machine plus the speed control associated with the shunt-wound machine.

In summary, in the WL system a common but variable voltage is applied to the armature of each motor, which has the effect of varying the speed of *all* the motors simultaneously, and a variable voltage via the compensator potentiometers is applied to the shunt field, which has the effect of varying the speed of each motor *independently* to trim each drive to the other.

While motor generators are still used in some instance as a.c.–d.c converters, these are now being replaced by solid-state converting equipment using silicon diodes or thyristors for a.c.–d.c. rectification.

Direct-current Motor Starters and Controllers

The older types of controller have practically disappeared from dyeing and finishing plant. There were two basic versions, the 'drum' type and the 'face-plate' type. Both of them are rheostatic starters and are used for inserting a variable resistance in series with the motor. In the drum type, the supply leads, motor windings and resistance banks are connected to a number of contact fingers mounted on a vertical spindle. This spindle is rotated by the handle on the top of the controller. The contact fingers are usually held on their mating static contact pieces by spring pressure so that a sequence of connections is made by rotating the handle. In a series-wound traction motor this form of controller is used for speed or torque control by inserting or removing resistance from the armature and field currents. In a fixed-speed compound-wound machine it is used simply as a graduated resistance starter.

Face-plate starters consist of a series of buttons connected to tappings on the resistance bank and a rotating arm connected to the positive lead of the supply at the fulcrum end, and fitted with copper-leaf brushes, which are moved over button contacts by an external wheel or lever, so varying the resistance on the supply to the armature. This type of starter is normally used for shunt-wound motors and has an arrangement of contacts that ensures the shunt field is fully energized for starting.

Like the older types of d.c. motor they controlled, these starter controllers are now all obsolete. Where variable resistance is required, the more modern method is to use continuously-wound resistance units with moving carbon-brush wipers to adjust the resistance in an infinitely variable or stepless manner. The wiper is usually moved by a small geared motor operated by the starting and speed-control buttons on the control panel of the machine or range of machines.

In the d.c. Ward Leonard, or 'separately excited shunt field system' as it is sometimes called, starting and speed control begins with standard three-phase a.c. supply being fed into the control box. On a modern range, starting and speed control is usually carried out from a push-button and switching station.

If the d.c. electrical supply is from a motor-generator set, start and stop buttons marked MG or Motor Generator are fitted on the panel. By pressing the start button a conventional a.c. motor starter starts up the motor of the MG set; nothing else happens at this stage. The switches for the appropriate motors are then closed. There is usually an interlocking system that ensures that these motor switches can only be closed when all the motors are at rest. Then by pressing the start button for the range, a contactor is brought in which energizes the field rectifier unit. A d.c. supply is fed simultaneously to the generator shunt field via a large variable resistance, and to all the motor shunt fields via the compensator potentiometers. At this stage the voltage to the generator field is usually minimal, so the output voltage to the motors is too low to start them turning. Then on pressing the accelerate button a small 'pony' motor operates a rheostat, which reduces the resistance or increases the voltage in the generator field and the generator starts to supply a working voltage to the motor armatures and they start up. By maintaining pressure on the accelerate button the voltage to the generator field continues to increase and therefore the voltage applied to the motor armatures and the speed of the motors continues to rise until the accelerate button is released. As the voltage to all the motors is the same, everything else being equal all the motors should now be running at the same speed, but on a linked process range the separate motors must be capable of running at slightly different speeds, for reasons already given. This is done by a secondary method of speed control, by which the resistance in the respective motor fields is varied by potentiometers operated by the cloth tension compensators.

There are a number of versions of d.c. motor-speed control for multimotor ranges, but the basic principle is the same – the overall range speed is varied by adjusting the voltage to the armatures and the individual motor speeds are trimmed by adjusting the voltage to individual motor shunt fields. Most of the versions differ only in the methods of transforming and of rectifying the mains supply brought about by continual advances in solid-state technology.

4.4 PNEUMATIC POWER AND COMPRESSED AIR

The use of compressed air is becoming more and more widespread on dyeing and finishing plant and machinery, the most typical applications being on mangle or padder nips, giant batching units and cloth-guiding equipment.

Compressed air is an excellent medium for supplying the motive power to drive machinery in difficult environments since it is impervious to steam, water or chemicals, provided the piping conveying it is made of suitable materials. It is much less of a hazard to people than steam or electricity, but unfortunately it is very expensive. Air motors are available for driving machines, but the power output of an air motor is usually less than half the power input required to drive the compressor. So the use of compressed air is usually confined to pneumatic cylinder or diaphragm operations where only a small volume of air is needed.

Compressed air is also the best, and probably the most common, source of motive power for operating pressure- and temperature-control systems on bleaching, dyeing and finishing machinery. This has now become known as the science of 'fluidics' and is becoming widely used in the control systems of all types of machinery.

One of the attractions of air-operated control systems is that, while the system may be quite complex, the basic principles of operation are relatively simple. Compressed air is supplied to the controller at a regulated constant pressure, and a pressure, temperature or electrical signal allows this air to be transmitted at a variable pressure to a servo unit, which is either piston or diaphragm operated. One of the major advantages of fluidic control over electronic control is that, by transmitting variable air pressure, a modulated rather than 'on/off' control system is possible.

The source of compressed-air power, a compressor and its associated equipment, must be selected and maintained with care. The main consideration in the selection of a compressor plant is the production of an adequate supply of compressed air at the lowest cost consistent with reliability. Important factors to be considered are:

1. Operating Pressure
   Normal industrial pressure is 100–120 p.s.i. at the compressor. Air tools and appliances generally work at 80–90 p.s.i., and for instrument control the pressure is usually around 20 p.s.i.
2. Maximum and Average Load
   Ideally, the total capacity would be based on exact knowledge of the equipment or process requirements. If this is underestimated the compressor plant will be too small and unable to maintain the required pressure in the system. Conversely if the total air consumption is greatly overestimated there may be excessive capital investment and any significant off-load running would be wasteful of energy. However, it is safer to err on the high side, as in most installations the use of compressed air will increase and soon take up any surplus capacity.

   Possible expansion should be taken into account and allowances made for leakage. Air leaks do not show up readily and it is not unusual for losses of between 25–30% of compression capacity to occur if maintenance is poor; with regular inspection and maintenance a loss of no more than 5% should be achieved. Much of the leakage occurs at hoses, couplings and valves of pneumatic equipment, and it must be remembered that, whilst the air consumption of most appliances is intermittent, leakages are continuous.

Figure 4.1 – Reciprocating compressors

Air leaks can be detected by running a compressor at the weekend or any time when the plant is stopped, and traced by ear. Another method, knowing the compressor capacity, is to charge the whole system to normal operating pressure, making sure that all compressed air is shut off. When the system is fully charged any leakage will cause the pressure to drop and the compressor to operate again. The running time should be checked with a stop-watch at least four times. If the compressor capacity is Q ft^3/min and the running time T min, the amount of air delivered is $Q \times T$ ft^3. If the total running time is t the average volume of air leakage is $(Q \times T) \div t$ ft^3/min.

Several different types of air compressor are available, the most common being the reciprocating-piston type (Figure 4.1). This type is made in a wide range of capacities and can be likened to an internal combustion engine. It is simply a piston or series of pistons reciprocated in their respective cylinders by a rotating crankshaft, usually driven by an electric motor or by an internal combustion engine if electric power is not available. The cylinder configurations of this type of compressor are many, ranging from small vertical-cylinder air-cooled machines to multiple-cylinder water-cooled units.

For high-pressure compressed-air supplies, multi-stage machines are used in which the air is compressed to the final pressure in a series of cylinders decreasing in bore as the pressure is increased. The reciprocating compressor is highly efficient both on full and on part load and can be easily controlled to suit prevailing demand. It is easily maintained and is suitable for most industrial uses.

The rotary-screw compressor is now becoming a serious competitor to the reciprocating machine in the 100–300 ft^3/min range. This type of compressor operates by the meshing of what could be termed male and female screws. These are driven by precision gearing and rotate at high speed in a very close-fitting housing. The advantage of this type of compressor is the almost complete elimination of friction, the screws and housing are so accurately made there is no metal-to-metal contact; a film of oil is always present between the moving parts. The problem with this type of machine in the early days was excessive discharge of lubricating oil. This has been overcome by the more precise construction and better oil/air techniques.

Another type of compressor quite widely used is the sliding-vane machine (Figure 4.2) in which axial vanes slide radially in a rotor revolving within an eccentric cylindrical casing. As the rotor revolves the vanes are forced outwards by centrifugal action so that they are kept in contact with the housing. Air is sucked in and entrained at the point where the rotor is farthest from the housing and discharged at the point where the rotor is nearest the housing.

Ancillary Equipment

When air is compressed, its temperature increases. This allows the now smaller volume of air to retain, mostly as vapour, the moisture originally present in it. After leaving the compressor and passing through the pipelines

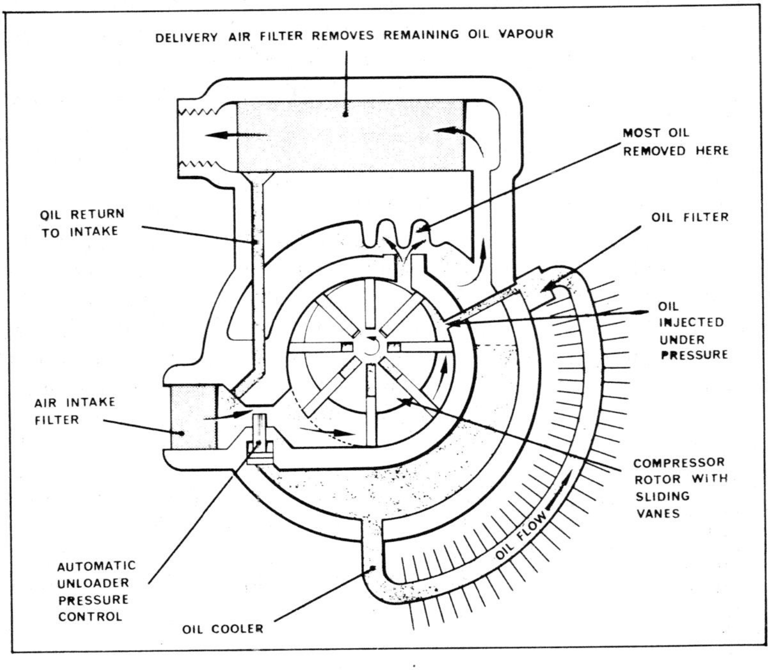

Figure 4.2 – Sliding-vane compressor

the air cools considerably and it cannot retain this moisture as vapour, which is consequently deposited as water in the pipeline. The most effective way of removing the major part of this is to cool the air immediately it leaves the compressor and to drain off the condensate.

An aftercooler is used for this purpose, fitted in the compressor discharge line before the air receiver. Two types are in common use, one using water as the cooling medium and the other using air. The former consists essentially of a chamber into which brass or copper tubes are fitted. Water flows either inside or outside these tubes and air flows in the opposite direction at the other side. The air-blast aftercooler consists of a nest of finned tubes, similar to a radiator, through which the compressed air flows and over which cold air is blown by means of a fan. A water separator should always be fitted to the air discharge side of the aftercooler.

In multi-stage compressors a proportion of the moisture is removed in the intercoolers between the compression stages.

Air Receivers

An essential part of a compressor installation is the air receiver. Air receivers absorb pulsations in the discharge air and help to precipitate oil and moisture from the air. They serve as a reservoir that compensates for temporary demands in excess of the compressor's capacity to deliver. An air receiver is a pressure vessel and must be constructed to specific standards depending on the working pressure, and is subject to inspection by the user's insurance company. It must be fitted with safety valve, pressure gauge, inspection openings and drain cock.

Drain traps are used for the draining of moisture, oil, etc. from aftercoolers and air receivers and can be effected by manual operation of a valve. However, this method can be inefficient unless a definite programme of draining is adhered to, and can be wasteful if a drain valve is left partially open. It is much better to drain automatically by fitting a suitable heavy-duty automatic drain trap (with by-pass) at all drainage points.

When planning a compressed-air system it must be decided whether to have a central compressor plant, or a number of separate compressors at the main sites of usage. There is no general rule, circumstances must be studied in each case. Points in favour of each, are as follows.

Centralized

1. Lower total compressor capacity will reduce the the cost of installation
2. Higher efficiency, larger units will give lower power costs
3. Lower supervision costs
4. Ability to control noise levels both inside and outside of the centralized compressor house.

Decentralized

1. Pipes can be shorter and smaller

2. Special foundations or housing may be unnecessary for the smaller units involved
3. Compressor and/or associated equipment failure has only localized effects
4. Auxiliary plant can be small, economical and mobile.

These comments are only general; in order to select a type of plant which will cover particular immediate and future requirements it is advisable to consult the supplier.

In some cases the compressor foundation may have to be isolated from the main building foundation so that vibration is not transmitted from the compressor to the building structure, and from heavy plant to the compressor. Where the vibrations are slight, resilient pads may be used to advantage. Heat generated by the compressor and prime mover is transmitted to the ambient air. For air-cooled units sited in enclosed rooms this heat must be removed to limit the temperature rise.

Noise from a compressor plant comes from different sources and each source has its own pattern of sound pressure. Noise can be divided into two sorts, the low-frequency sound of pulsating air intake and the higher-frequency machine noise from compressor, prime mover and fans. Local statutory regulations on noise levels should be determined, and action taken to ensure the noise levels do not exceed those stipulated. The compressor intake air must be clean and free from solid and gaseous impurities; abrasive dust and corrosive gases are particularly harmful. An air-intake filter of high capacity should be fitted to remove abrasive materials including those of small particle size. It should be able to collect large quatities of impurities without any significant decrease in filtering efficiency and air flow; normally it should be placed as close as possible to the compressor.

Where water is used for cooling compressors and ancillary equipment, it should be within the temperature and pressure levels prescribed by the compressor manufacturer and should be free from harmful impurities. The inlet temperature of the cooling water should be low in order to assist in achieving a high volumetric efficiency in the compressor and to cool the air passing through the aftercooler to a temperature adequate for effective condensation of water vapour. However, the compressor should not be overcooled, which could lead to condensation in and on the compressor. It is recommended that the cooling water be filtered. The cost of the air mains frequently represents a high proportion of the initial cost of a compressed-air installation. Using a reduced-diameter pipe may save on capital cost, but the greater pressure drop in the system results in higher operating costs, which might soon exceed the price of installing larger-diameter piping. In a new or redesigned plant early consideration should be given to the layout of the compressed-air system, otherwise the air mains may have to follow a tortuous path to avoid the existing electrical conduits, and steam, water and other pipes.

Where the points of use and of supply are relatively close, a direct single line should suffice. In this case, the diameter of the pipe must be such that the maximum flow is accommodated with no more than the maximum acceptable pressure drop. On larger installations with numerous take-off points a ring main has advantages in that the velocity at any one point of use is reduced, since the air can converge in that point from two directions; also, by correct siting of isolating valves, a section of the system can be shut down for servicing.

For ease of servicing and draining and for access when connecting to service points, the overhead main has many advantages. The primary consideration is that the main line(s) should cover the working area while being as direct as possible. Ring mains should be formed with long-sweep bends, not elbows, in order to minimize pressure losses. The main should have a fall of approximately 1 in 100 in the direction of flow, and at suitable intervals should be brought back to its original height by using long-sweep bends. By arranging drain legs, preferably fitted with automatic discharge valves, at all these points, it is possible to prevent condensate being carried to the points of use.

In some cases it is necessary to provide a large volume of air for a short period. The equipment using this air may be some distance from the compressor, and the abnormally large pipe required to carry this air with the minimum of pressure drop can be avoided by fitting an auxiliary air receiver close to the equipment in question. Branch lines are taken from the top of the ring system using long-sweep bends, and should be large enough to carry the maximum amount of air flow required with minimum pressure drop. Isolating valves should be fitted at the outlet point of the air receiver, and each branch line and drop-feed line should have an isolating valve fitted. Drop-feed lines must be taken from the top of the ring main or branch feed line using a long-sweep bend, and a drain leg with manual or automatic drain valve should be fitted at the bottom of the drop-feed lines. The feed to the appliance should be T-joined to the drop line and the required filter, regulator and lubricator fitted at this point.

Mainline pipe systems can be of steel, copper or ABS material. Generally copper or thin-wall stainless steel is not used for pipes over 1 in (2.5 cm) in diameter. Steel pipe is available in black or galvanized form; the latter, being less liable to corrode, is recommended. This type of pipe can be screwed to accept the wide range of proprietary malleable fittings. Copper tubing may be used for air mains with a nominal diameter of up to 1½ in (4 cm). The tubing and fittings chosen must be checked to ensure that they are suitable for the maximum working pressure of the system. The fittings for this type of tubing are of the compression type. ABS pipe is made from a non-corrodible and non-toxic polymeric material, which exhibits extreme toughness and ductility and is particularly suitable for conveying compressed air. However, it has temperature limitations and the manufacture should be consulted. Other thermoplastic materials such as PVC may become brittle and fracture, and should not be used in compressed-air systems. ABS pipes should not be threaded, connections to metal components should be made through adaptors or composite unions. Only jointing compounds that are compatible with ABS should be used. Since thermoplastic materials soften with increase in temperature, their pressure rating should be reduced in accordance with the manufacturer's recommendations. If flexible hose is required, good quality oil-resistant rubber or plastic should be used, suitably constructed and reinforced for the duty required of it; care must be taken as its natural flexibility causes it to sag and pipe runs often become unsightly tangles.

4.5 HYDRAULIC POWER

In the halcyon days of cotton, dyeing and finishing plants with a number of calenders often had hydraulic systems installed for pressure loading the bowls of these machines. Systems usually consisted of a motor-driven reciprocating ram-type pump, a hydraulic accumulator and a distribution network of pipes to the various machines. It is doubtful if any of these systems still exist as, where hydraulic power is required, the modern tendency is to install small hydraulic units or 'power packs' for individual machines.

In the old system, the pump would be capable of generating a pressure of several thousand pounds per square inch, the hydraulic fluid would be delivered at this pressure to the accumulator, which consisted of a long-stroking cylinder and ram of comparatively small diameter. Directly above and coupled to this vertical ram there was a large weight, usually a cylindrical tank filled with water or even scrap metal, which exerted a downward pressure on the ram determined by the weight of the loaded tank and the area of the ram. The purpose of the accumulator was to store the compressed fluid delivered by the pump and so level out the spasmodic demands on the system. It also had the effect of damping any pulsation from the pump. The fluid used in these systems was usually water and soluble oil.

What may be called the modern version of the hydraulic accumulator is the 'intensifier'. This device works on a similar principle except that the deadweight is replaced by a large compressed-air or gas-filled cylinder and piston. The intensifier also differs from the accumulator in that it does not store compressed fluid, but produces it by the application of compressed air, gas or even steam to the large piston; and the output pressure can be controlled by the regulation of the applied pressure.

The simple intensifier has to be recharged in a similar manner to an accumulator, but by making the differential pistons double acting, and using compressed air or steam to move the large piston, the intensifier acts like a pump and will generate pressure continuously in a hydraulic system.

Modern hydraulic units, or power packs as they are sometimes called, are small self-contained systems consisting of a fluid tank, usually oil, some form of high-pressure pump, and a pressure regulator. There are a number of these power packs available, most of which differ only in the type of pump used. The pumps deliver the hydraulic fluid at a high pressure and usually at constant volume. The pressure to the system is then regulated by means of a spill or leak-off valve, which returns the excess fluid to the tank or reservoir.

The type of pump used in these units will normally depend on the pressure required. The simplest type is the gear pump, which consists of a pair of meshing gears running in a close-fitting housing (Figure 4.3). As these gears rotate, the fluid is entrained and pressurized at the delivery side. This type of pump is normally used for relatively low pressures, say 500 p.s.i.

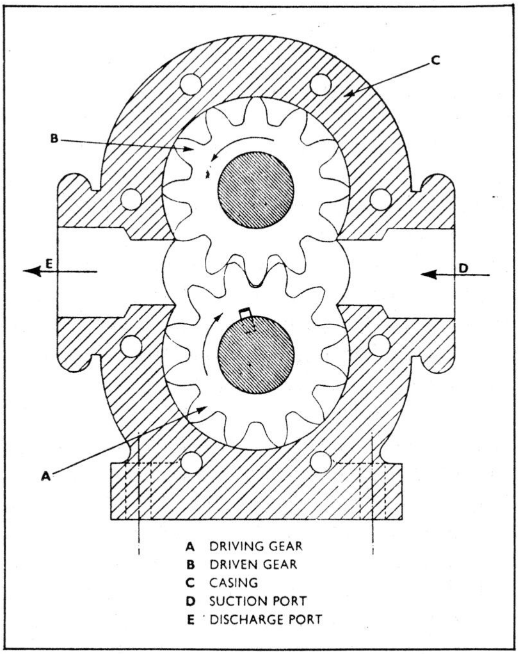

Figure 4.3 – Gear pump

The screw pump operates in a similar manner, but meshing screws or rotors are used instead of gears, and the hydraulic fluid is forced along the length of the pump (Figure 4.4). This type of pump can be used to give considerably higher pressures than the gear pump, 2 000 p.s.i. being within its capabilities. One of the main advantages of gear or screw pumps is that the fluid flow is continuous and pulse-damping equipment is not required as is the case of piston- or ram-type pumps.

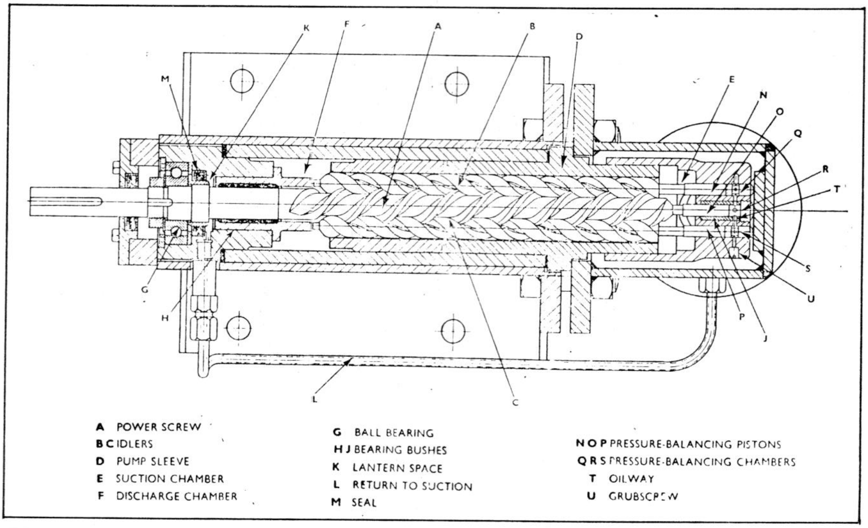

Figure 4.4

The multi-piston or ram-reciprocating type of pump is normally used for the very high pressures that are often necessary for hydraulic systems. There are numerous different designs in which the cylinders may be disposed in line, axially or radially. They all operate on the same basic principle, using reciprocating pistons or rams to deliver the fluid at high pressure, the reciprocating motion being provided by crankshafts, camshafts or swash plates (Figure 4.5). Pressures in excess of 6 000 p.s.i. are possible with this type of pump.

With the piston-type pump some form of damping is usually required, as obviously the fluid is delivered in a series of pulses. The damping device is often a small steel bottle connected to the delivery pipeline and charged with an inert gas, usually nitrogen. The compressibility of the nitrogen acts as a cushion against the pulsing of the pumps.

Apart from nip loading on calenders, the use of hydraulic power on dyeing and finishing plant is not very extensive; for mangle and padder nips, pneumatic loading is generally used.

Hydraulic Motors

Low speed combined with high torque, a common requirement in power transmission applications, is generally achieved through various forms of reduction gearing fitted to high-speed prime movers. There are, however, certain applications where this is not possible for various reasons, space limitations often being the design engineer's problem. If variable speed is required as well as speed reduction then the problem becomes very acute. It is in situations like these that the hydraulic motor often comes into its own,

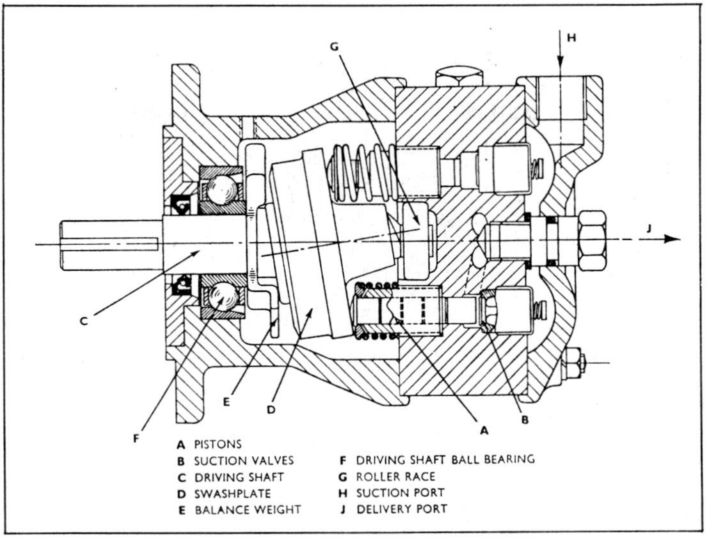

Figure 4.5 – Hydraulic pump

particularly those hydraulic motors that have good low-speed, high-torque characteristics coupled with overall high mechanical and hydraulic efficiency.

Hydraulic motors offer added advantages, such as extremely accurate control at low speeds, high power: weight ratio, universal mounting facility and an impact resistant and very robust construction.

Hydraulic motors are usually of the radial-piston type, as described in the section on variable-speed gears (Carter type) and also shown in Figure 4.6.

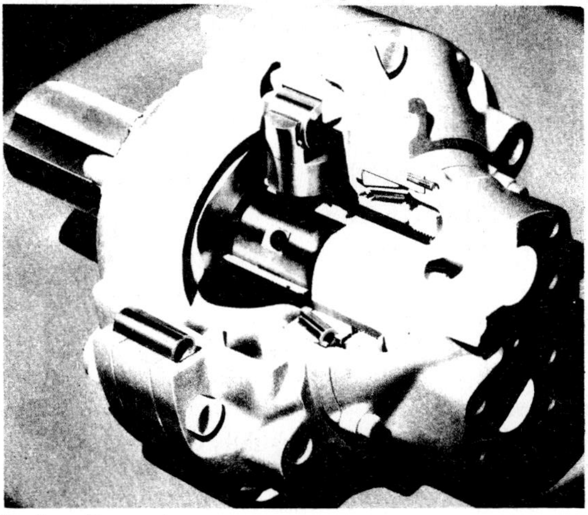

Figure 4.6 – Hydraulic motor

Figure 4.7 – Portable hydraulic power pack

A popular use is for the centre drive of a delivery batch A frame. The portable motor is clamped to the A frame shaft as required, drawing its hydraulic supply from an adjacent portable power pack (Figure 4.7).

The motor automatically reduces the revolutions of the centre-drive shaft as the batch increases in diameter; the fabric tension during winding may also be preselected and maintained.

5 POWER TRANSMISSION

5.1 VARIABLE-SPEED GEARS

The use of fixed-speed a.c. motors to drive variable-speed machines is quite common in dyeing and finishing plants. This is accomplished by interposing a variable-speed gear between the machine and the motor. The speed variation can be achieved either electrically or hydraulically.

Electrical Speed Control

Speed variation by electrical means is usually carried out by using a form of electromagnetic or eddy-current coupling. In this method the fixed-speed a.c. motor drives one half of the coupling, which might be termed the driving or input part of the coupling, and, of course, runs at a fixed speed. The

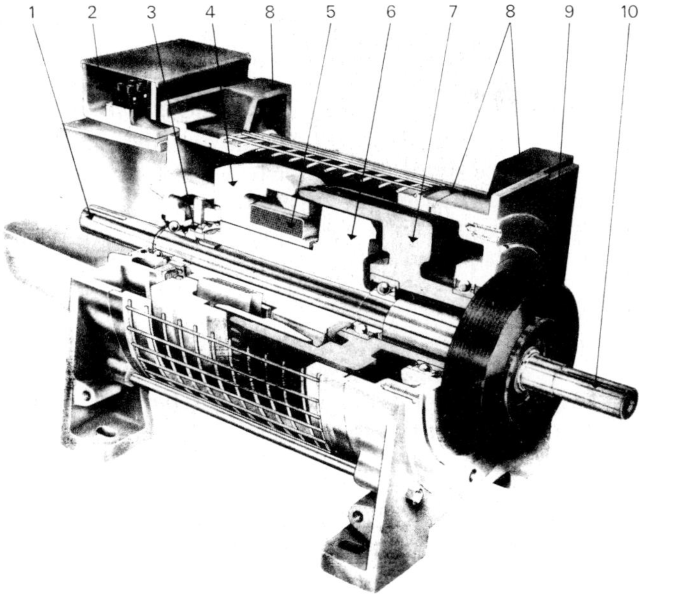

1 Shaft extension for optional brake, locking hub, pulley or half-coupling
2 Terminal box
3 A.C. governor-generator
4 Fixed coil support
5 Stationary field coil (Class E insulation)
6 Output member (rotor and through-shaft)
7 Input member (drum and quill shaft)
8 Housing assembly
9 'Poly-V' pulley
10 Output shaft

Figure 5.1 – Eddy-current speed variator

driven or output half of the coupling, although in very close proximity to the driver, has no direct mechanical connection to it. If an electrical current is now passed through electromagnetic coils in the coupling, a magnetic field is established across both halves of the coupling and the driving half will pull round the driven half at a speed and torque relative to the strength of this field. Thus by varying the input current to the coils a variable speed or torque can be obtained from the output to the machine (Figure 5.1).

This type of variable-speed drive can also be used to drive units of a processing range running in tandem. If, for example, it is required to drive a padder and a set of drying cylinders in tandem using separate driving units, the driving unit for the cylinders will be the master drive and the speed of this will be controlled by the operator from the control panel. A small tacho or signal generator is driven from the output or variable-speed shaft of the master drive. This generator produces a voltage directly proportional to the speed. The voltage is applied to the control circuitry of the padder driving unit and so synchronizes the padder drive to the cylinder drive. Trimming is done by a compensator-operated potentiometer, which also has a controlling influence on the padder drive.

For a series of drives the principle is the same, the final drive being the master and all the following drives are controlled one from the other right back to the first, or primary, drive.

This type of variable-speed unit is also very useful as an independent drive for the take-off unit of a process machine or range. For instance, on large batch drives it can be fitted as a torque or soft drive; as the drive is only transmitted through a magnetic flux, the tension between the machine and batch is completely controllable, and such driving units can actually be allowed to stall for short periods without any detrimental effects.

Mechanical Speed Control

One of the most popular, infinitely variable speed controllers that has stood the test of time is the expanding-cone pulley type. The basic construction of this consists of two pairs of cone-shaped pulleys, fixed to splined or keyed shafts, so that although the drive can be transmitted from the shafts to the pulleys or the pulleys to the shafts, the pulleys are free to move along the shafts. A pair of levers on either side of the pulleys, pivoted on fixed fulcrums located midway between the two pairs of pulleys, can be operated by one means or another so that as one pair of cones moves together the other pair moves apart.

An endless belt, which is wedge shaped to match the taper of the cones, connects and transmits the drive from one pair of cones to the other. As each pair of cones is moved together or apart relative to each other, the belt will ride up the pair that are coming together and down the pair that are moving apart, and the ratio of the speeds of the pulleys will vary accordingly. If one of the pulley shafts is driven at a constant speed by an electric motor, the other shaft will be driven via the belt and cones, at either a faster or slower speed, depending on the position of the wedge belt. When the cones in both pairs are the same distance apart, the ratio will be 1:1 and the driven shaft will rotate at the same speed as the driving shaft, but any movement of the belt up or down the cones will create a speed variation.

Versions of this type of speed variator are quite numerous and go back many years when the wedge belt was built up from leather-tipped wood lags fixed to an endless leather belt. These have now been superseded by special V belts of reinforced or synthetic-rubber material.

One well-known version of this type of speed variator uses a chain instead of the V belt. In this unit the expanding cones are made with grooves radiating from the centre to the rim and the special chain meshes into these grooves so that a positive drive is achieved, rather than the purely frictional drive of the conventional speed variator. This version of the variable speed unit is usually called a PIV (Positive Infinitely Variable) gear (Figure 5.2).

The most serious shortcoming of the expanding-cone pulley type of speed variator is the limited speed-variation range, a six to one speed variation is about the absolute limit that can be achieved and in fact a four to one varia-

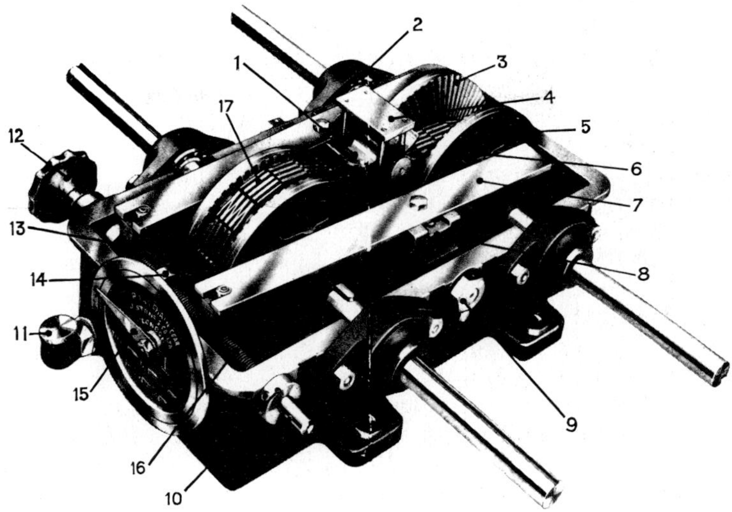

1. Shoe Assembly
2. Radial ball bearings
3. Patent P.I.V. wheel face
4. Shoe guide
5. Cast iron wheel flange
6. Ball thrust bearing
7. Operating levers
8. Chain tightening cam
9. Chain tightening screw and locking plate
10. Alignment fitting
11. Oil filler and level indicator
12. Speed control hand wheel
13. Control bracket
14. Speed limiting nut
15. Speed indicator
16. Control screw
17. Patent P.I.V. chain

Figure 5.2 – PIV speed variator

tion is more usual.

There are a number of other mechanical speed variators available, most of which are based on metal-to-metal friction devices, a form of driving that is not generally regarded as good engineering practice. With drives based on metal-to-metal friction it is difficult to avoid rapid wear.

Infinitely variable speed transmission by mechanical means is a very difficult proposition. If it were not so, gear levers would have disappeared from the motor car long ago!

Hydraulic Speed Control

Probably the best known of hydraulic variable-speed gears in the textile dyeing and finishing industry is the Carter gear (Figure 5.[illegible]).

The main components of this gear are a radial, variable stroke, piston-type hydraulic pump and a hydraulic motor of similar construction. The pump and motor are mounted on separate coaxial shafts and both these units are enclosed in a cylindrical housing, rather like an electric motor with shafts at either end. The input or pump shaft is driven by an ordinary fixed-speed induction motor either by direct coupling or by V pulleys and belts.

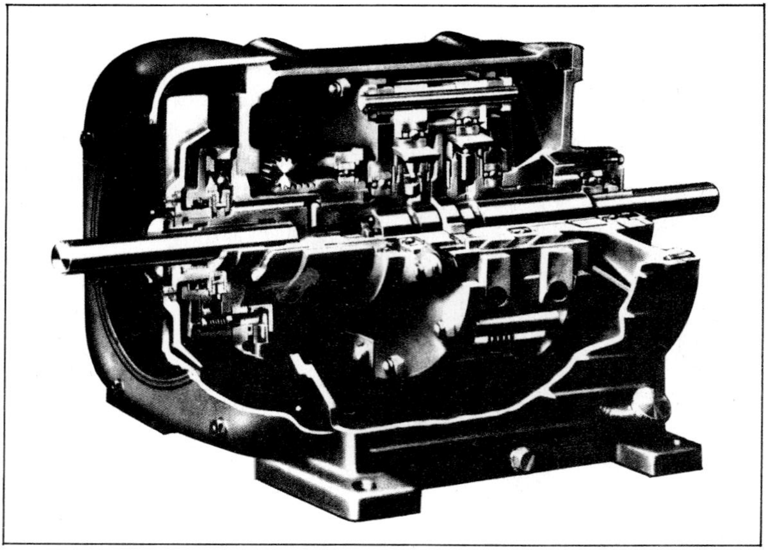

Figure 5.3 – Hydraulic speed variator

Both the pump and motor run in an oil bath, which serves both as a lubricant and a transmission fluid. A control shaft, which can be operated either manually or electrically, projects from the gearbox. The inner end of this shaft operates eccentric elements to vary the stroke of the pump and motor units. The variable stroke of the pump determines the volume of oil delivered to the motor unit, the speed of which is relative to that volume, so that while the speed of rotation of the pump or input unit is constant, the speed of the motor or output unit will vary.

The Carter gear can be used to drive machines or ranges in tandem by using mechanical or electrical linkages from compensators to the control shaft of the gear.

The Carter gear is now probably the only hydraulic variable-speed gear that is still in use for dyeing and finishing machinery drives. The Vickers hydraulic gear which was used as a variable-speed drive, mainly on roller-printing machines, operated in a similar principle except that the pump and motor units were separate and coupled by pipework, and while there are a number of these still in use, it appears they are no longer manufactured.

The Carter gear will transmit the full-rate torque from zero up to the maximum speed of the input shaft, but the makers usually recommend that the gear should not be run continuously at less than, say, 4% of maximum speed. This gives a variable speed range of 25 to 1, which is normally adequate for any drive.

Possibly because of the high cost, hydraulic variable-speed gears now appear to be losing ground to modern d.c. electrical methods of speed variation (see the section on electric motors).

5.2 REDUCTION GEARS

Most dyeing and finishing machine drives incorporate some form of reduction gear. Motor revolutions have to be reduced to equivalent cloth speeds and some machines require high torque at low speeds, which again is achieved by the use of a reduction gear.

Many of these reduction gears are incorporated in the motor so that motor and gearbox are constructed as a single unit. This type of gear is normally made coaxial, so that the low-speed output shaft is on the same axis as the motor shaft and the speed reduction is obtained by the use of a train of toothed pinions and spur wheels. As all the gears are running in an enclosed box and immersed in lubricating oil, this type of reduction gear requires very little maintenance, apart from checking the oil level from time to time (Figure 5.4).

Another form of spur and pinion reduction gear is the torque-arm reducer (Figure 5.5). This type of reduction gear is designed to be mounted on the slow-speed driven shaft so that it does not require fixing to the floor or to a baseplate. Some machine makers favoured this type of reduction gear for padder and mangle drives, the gear being mounted on the bottom-bowl shaft extension and the drive to the input shaft by V belts from the motor. A torque arm links the unit to a fixed anchorage (otherwise the unit would attempt to revolve round the shaft). The torque arm incorporates a turnbuckle for adjusting belt tension. The only problem with this type of drive to a nip is that if the bottom bowl has to be taken out for regrinding or

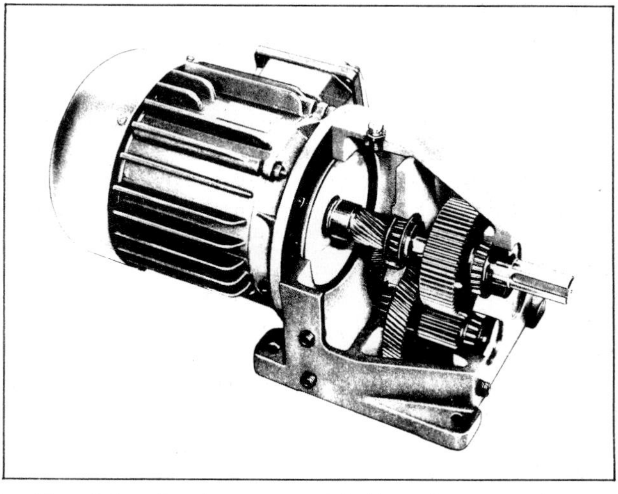

Figure 5.4 – Coaxial reduction gear with flange-mounted motor

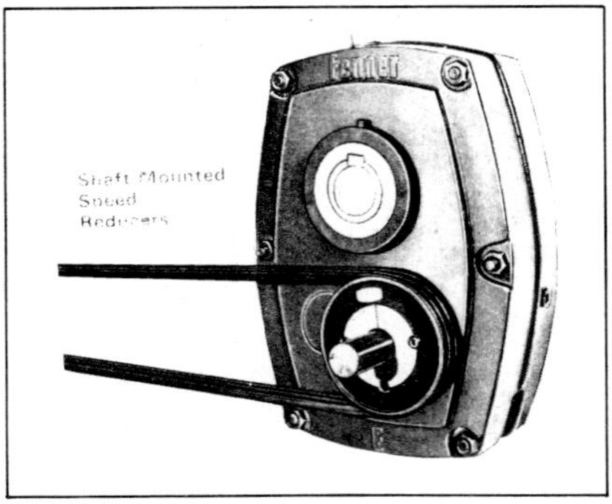

Figure 5.5 – Torque-arm reduction gear

replacement of a bearing, the torque-arm reduction gear has also to be removed. Probably the main advantage of the torque-arm reduction gear is that no couplings are required between the motor and the machine. There is very little to go wrong with the gear if the oil level is regularly checked, an annual oil change is more than adequate.

Probably the most common type of reduction gear used on machine drives is the worm and worm-wheel gear (Figure 5.6). On this gear the input and output shafts are at right angles to each other, the input or worm shaft being driven by the motor and the output or worm-wheel shaft drives the machine either by a coupling or by a roller-chain drive. Again this type of reduction gear requires very little maintenance other than oil checks and changes. Reduction ratios of up to 70:1 are available.

The purpose of all these reduction gears is to reduce the motor speed down to a reasonable cloth speed of the process machinery. The type of gear

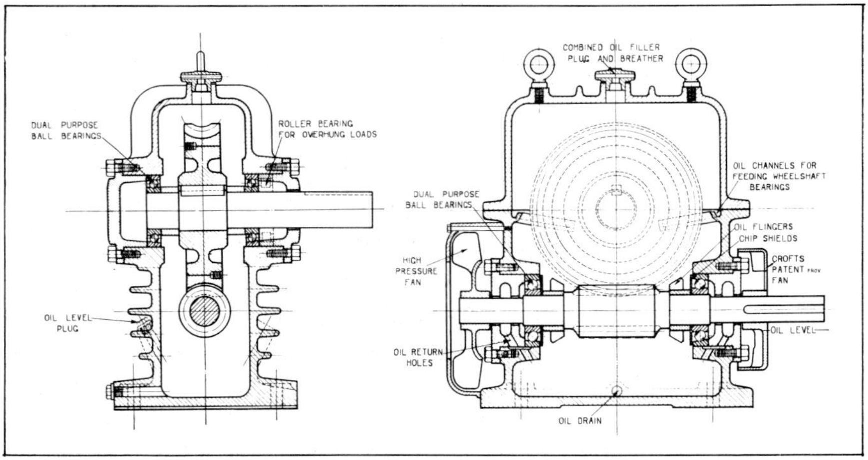

Figure 5.6 – Worm reduction gear

used on any particular application is usually decided by the machine manufacturer, cost being a major consideration. For instance, it is obvious that for a mangle or a padder drive any of these three types of reduction gear could be used. In my opinion the worm reduction box has many advantages.

Reduction Gears and Mechanical Variable-speed Gears in Combination
It has been noted that many textile drives require variable speed with high torque (twisting power) at relatively low speeds.

The reduction from motor speed to machine speed is relatively large and often greater than the maximum reduction ratio possible in the variator itself. Thus, when the two units are necessary to make up the drive, the mechanical variator is always installed between the motor and the reduction gearbox. The reason for this is that for a given amount of power a variator in this position does not have to transmit the full torque required at the machine shaft, and may therefore be relatively small and cheap. The final reduction gearbox is mechanically less complex and cheaper to produce in the necessary rugged form.

5.3 DRIVING MECHANISMS

Belt Drives
Where relatively fast motors are used for machine driving and only nominal speed reductions are required, for instance if an a.c. motor running at 1 440 rev./min is required to drive a shaft at say 360 rev./min, a 4:1 reduction, a belt drive would normally be used for such a drive.

V belts are the most common form of belt drive now in use. Such drives are simple to install, economical and clean, and if the alignment of the pulleys and the belt tension are correct, there is very little to go wrong. V belts have been around a long time, and an almost unlimited range of pulleys and belts are available for different loadings up to 200 h.p. (150 kW) and speed ratios up to 8:1.

Power transmitted depends upon the friction between the belt and pulley. The belt does not rest in the bottom of the groove but is wedged between the sides. The groove angle of the pulley is usually between 32 and 40°.

Early V belts were made of natural rubber, reinforced with cotton or rayon cord. In those days the design of a drive was relatively simple, all that was required to be known was the power to be transmitted and the speed of the driving and of the driven shaft, but by the use of synthetic rubber with nylon, polyester and even metal reinforcing there is now a multitude of different V belts available and drives have become much more compact.

The availability of a wide variety of V belts, while being quite a boon to the machine designer and manufacturer, can be a problem for the plant engineer. In a dyeing and finishing plant the age range of the machinery can be twenty years or more, so it may be that the older machines have what can be called the standard V belt drives, and more modern machines with any of a number of compact V belt drives, none of which are interchangeable. This fact is of particular importance if a new plant is being assembled, and standardization of spares is desirable.

V belt drives cannot be considered to be completely positive; they slip under excessive overload. This is a factor in their favour, as if a machine stalls due to a wrap-round or for any other reason less damage will be done by belt slipping than would be the case with a more positive drive, which might result in a burnt-out motor or a few bent rollers.

For more positive drives, or where synchronization is required, a 'timing' belt drive may be used. This is a flat belt with inverted teeth formed on the inner or pulley face of the belt. These teeth mesh with grooves across the face of the pulleys so that a positive drive is obtained. This type of drive has taken over from chain drives to a large extent, a prime example being camshaft drives on car engines. Speed ratios of 8:1 and above, and power ratings of up to 100 h.p. (75 kW) are possible with 'timing' belt drives. They are quiet in use and of course need no lubrication.

Flat-belt drives using leather, Balata, cotton-duck or even camel-hair belts are now largely obsolete. They were used mainly for line-shaft drives. No doubt there are quite a number of orthodox belt drives still in use but it is doubtful if many machines are now being fitted with this form of drive.

Although the old type of flat-belt drive has gone out of fashion, there is a new form of flat belt that is proving a worthy successor to it. This is the reinforced or 'sandwich' belt. The most usual form of this type of belt is an inner and outer layer of chrome-tanned leather, between which a series of extruded nylon strips are bonded. The leather faces provide the gripping surfaces and the nylon strips the reinforcement (Figure 5.7). This type of belt is particularly useful where both sides are required to drive, e.g. in the sinuous drive to a stack of drying cylinders or for quarter-twist drives, where V belts cannot be used.

In my plant all the drying cylinder stacks are driven with sandwich belts; these have proved very successful. They are much cleaner than chain or spur-wheel drives. Several have been in use for eight years or more and are still in excellent condition. There is little or no stretching of this type of belt, so that after the initial tensioning no further adjustments are required.

Allowances for cloth shrinkage can also be made with this form of flat-

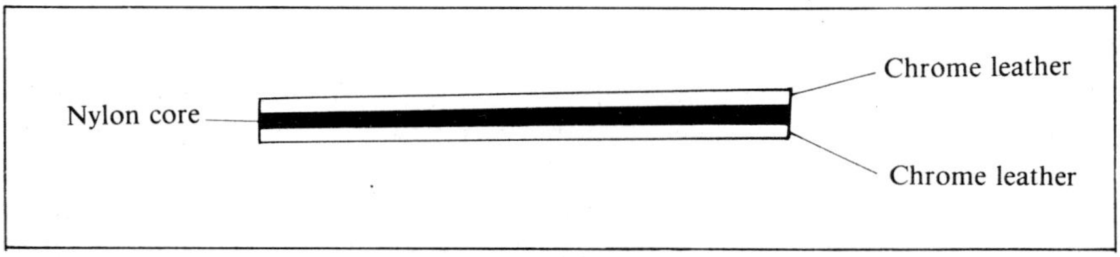

Figure 5.7 – Modern sandwich belt

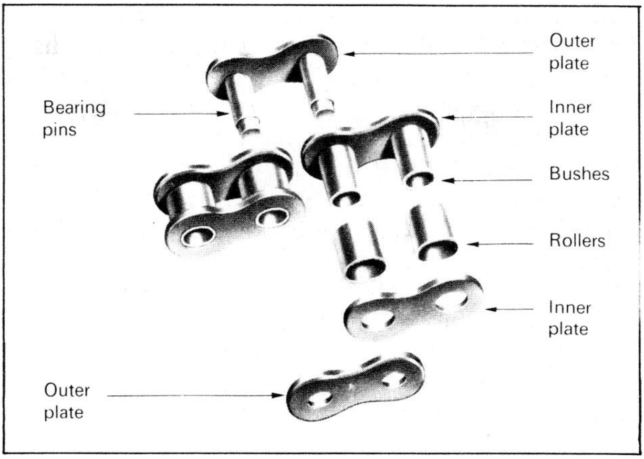

Figure 5.8 – Construction of simple roller cahin

belt drive by varying the diameter of the driving pulley. This is not possible with chain or spur-wheel drives, as these can only be altered by a minimum of one tooth, which may be too much.

Chain Drives

For a positive or synchronous drive, timing belts are excellent within their limitations, but for a heavy, relatively slow-running drive the roller chain is far superior. With roller chains reduction is usually restricted to a maximum of 6:1 in a single stage and up to 600 h.p. can be transmitted. When fitted, the roller-chain drive needs some centre distance adjustment, or more usually an adjustable jockey pinion.

As with other types of drive, some kind of guard is needed for safety reasons, indeed this is now mandatory and, as at all but the lowest power or speed chains require some form of lubrication, the guard or chain case must be oil tight.

Roller chains are available from 5/32 in (4 mm) pitch, for low power and instrument drive, to 3 in (7.5 cm) pitch, for really heavy drives. They are manufactured in most pitches in the simple, duplex or triplex form complete with a wide range of wheels and pinions, and a wide variety of connecting links are available for joining the ends of the chain (Figures 5.8 and 5.9).

Choosing the Right Drive

There are no hard and fast rules when choosing a drive for any particular purpose and the decision is usually made by the machine designer. In the dyeing and finishing industry the choice might be as follows.

For straightforward motor-to-shaft drives where the motor speed is 720 rev./min or higher, the slight slippage is not detrimental, a V belt drive is the most economical. For a similar drive, but where speed synchroniza-

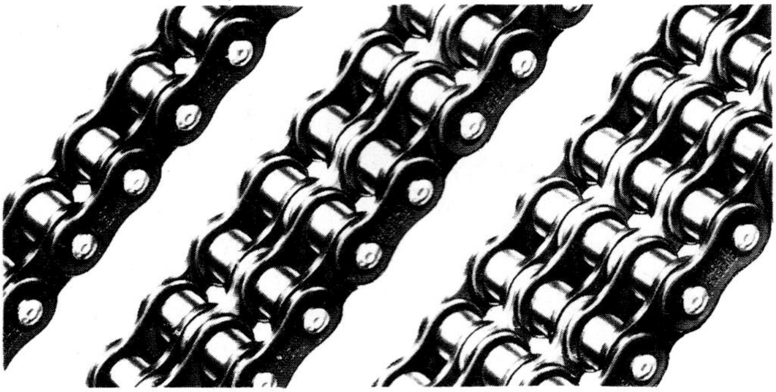

Figure 5.9 – Roller chains

tion is essential, a 'timing' belt drive might be considered.

For a positive drive at any speed, and where the transmission can be adequately lubricated and preferably totally enclosed, and particularly for low-speed drives, a roller-chain drive will be the most appropriate.

For high-speed twisted drives or where a drive is required from both sides of the belt, the reinforced sandwich belt is the answer.

There is a variety of other belt and chain drives available, the choice of which is largely a matter for the machine designer, but to date these have not yet replaced to any great extent the traditional forms of transmission.

5.4 BEARINGS AND BEARING MATERIALS

In textile dyeing and finishing machinery, probably the most common piece of mechanical equipment is the roll or roller. These include any cylindrical piece of equipment used for conveying, squeezing or drying cloth, e.g. padder and mangle rolls, calendar bowls and drying cylinders, as well as the multitude of trolley or guide rollers found in various process machines and ranges. The feature common in all these rolls is that whether driven or free running, they all have to be supported by a bearing at each end.

The design and construction of these bearings are most important. In the textile bleaching, dyeing and finishing industry, bearings are required to operate under a wide variety of conditions, including exposure to temperatures in excess of 200°C, boiling water, highly corrosive chemicals and steam, and under very high stresses and loadings.

They must be reliable, as the failure of one small bearing can mean a shut-down of a large machine or process range.

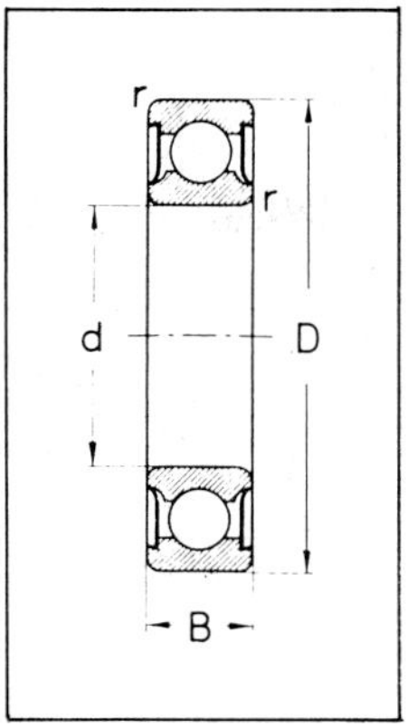

Figure 5.10 – Standard single-row bearing with two seals

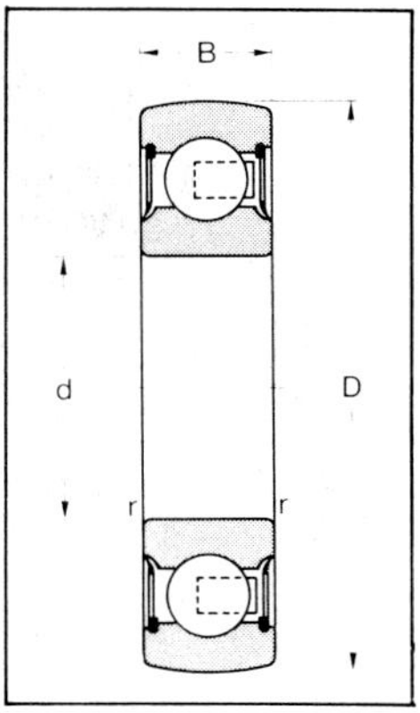

Figure 5.11 – Single-row bearing with spherical outside diameter and two seals

Ball and Roller Bearings

Probably the most common type of bearing in use is the ball bearing, which of course can only be used in locations where it can be lubricated with grease or oil. Many of the smaller and light-duty ball bearings can be supplied prepacked with lubricant and sealed. These are known as 'lubricated for life' bearings, and provided the working conditions are not too arduous, they perform very well, but they are certainly not suitable in every instance on dyeing and finishing plant and machinery, particularly when conditions are hot, wet or steamy. In fact the life of the grease can determine the life of the bearing.

The simplest and cheapest ball bearing is the single-row bearing, i.e. one which has one row of balls and is usually supplied in its basic form for fitting in suitable recesses in the machine or mounted in a variety of housings bolted to the machine. This type of bearing is designed to accommodate both radial and thrust loads and is often used in the prelubricated and sealed form (Figure 5.10).

Single-row bearings are not designed as self-aligning bearings, and where any reasonable degree of misalignment has to be catered for, they are usually mounted in a spherical housing or cartridge so that the complete bearing can swivel to accommodate any misalignment of the shaft or spindle (Figure 5.11).

Where loadings are higher, double-row ball bearings are generally used. Typical uses in the dyeing and finishing industry are with light to medium mangle and padder rolls, drying cylinders, dye jigs, and winches, etc. Special lubricants are available to suit the conditions under which the bearing has to function.

Several different types of double-row bearings are available, but by far the most popular is the self-aligning bearing. The outer race of this type is

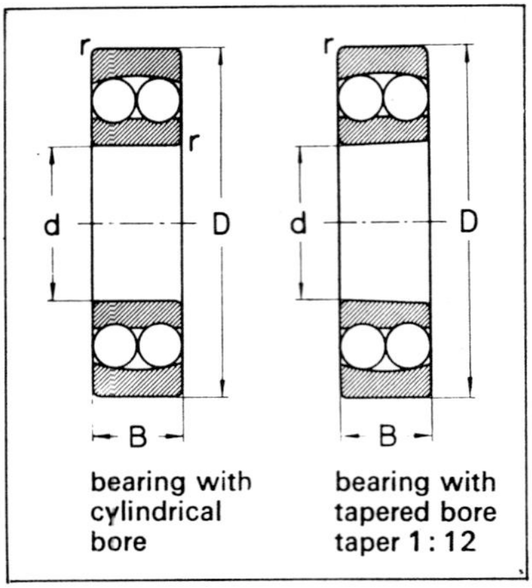

Figure 5.12 – Double-row self-aligning ball bearings

made spherical on the inside diameter so that the inner race and balls will run quite freely, even with some misalignment of the shaft or spindle (Figure 5.12).

Double-row bearings are not normally available in the prelubricated and sealed form, as the fitting of seals would restrict self alignment, also in many instances this type of bearing is used where special lubricants must be used. This type of bearing is also normally fitted in the standard regreaseable plummer block, which is the bearing and housing unit used extensively in cylinder drying ranges, in external bearings in bakers and steamers, and in most locations where working conditions are difficult.

For heavier loads and more precise applications, the roller bearing is generally preferred, because the load capacity is greater for a given diameter. Typical uses include medium to heavy mangle and padder nip rolls and calender bowls. This type of bearing is available in a rigid form with parallel cylindrical rollers or as a self-aligning bearing with barrel-shaped rollers in a spherical outer race.

Ball and roller bearings are supplied either to the machine manufacturer for fitting in prepared housings, or as complete pillow or plummer blocks for bolting to the machine. The pillow block unit is usually fitted with the prelubricated and sealed bearing, and the plummer block with a regreaseable bearing.

Generally speaking, the pillow block is quite suitable for conditions that can be described as normal. It has its limitations, a fact that is not always appreciated by machine manufacturers. The pillow block is much cheaper than the plummer block and, apart from the standard pedestal type of housing, the same type of ball bearing can be supplied in a variety of housings easy to mount in the machines.

One of the major problems with the prelubricated type of bearing unit

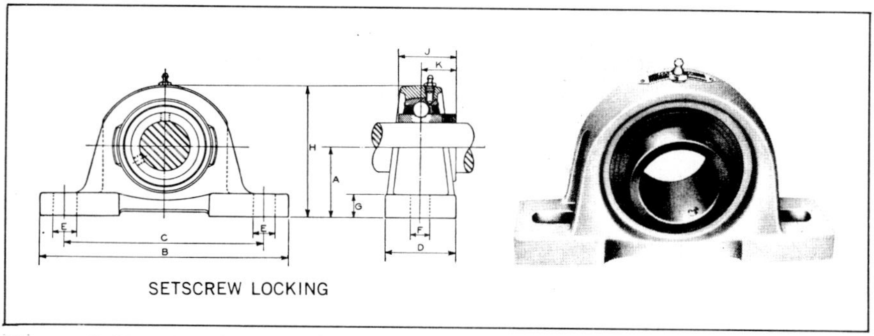

Figure 5.13 – Ball bearing pillow block

arises when it is fitted in locations that are wet, steamy or hot, not uncommon in dyeing and finishing plant. Under these conditions it is often necessary to use special lubricants such as water-repellent or high-temperature greases, so the bearing has to be fitted at the manufacturing stage, which makes them special, and if the lubricant does not have a long life regreasing can be a problem.

Pillow blocks (Figure 5.13) are designed to be regreaseable, but the lubricant has to be pumped through a tiny hole in the outer race, which under

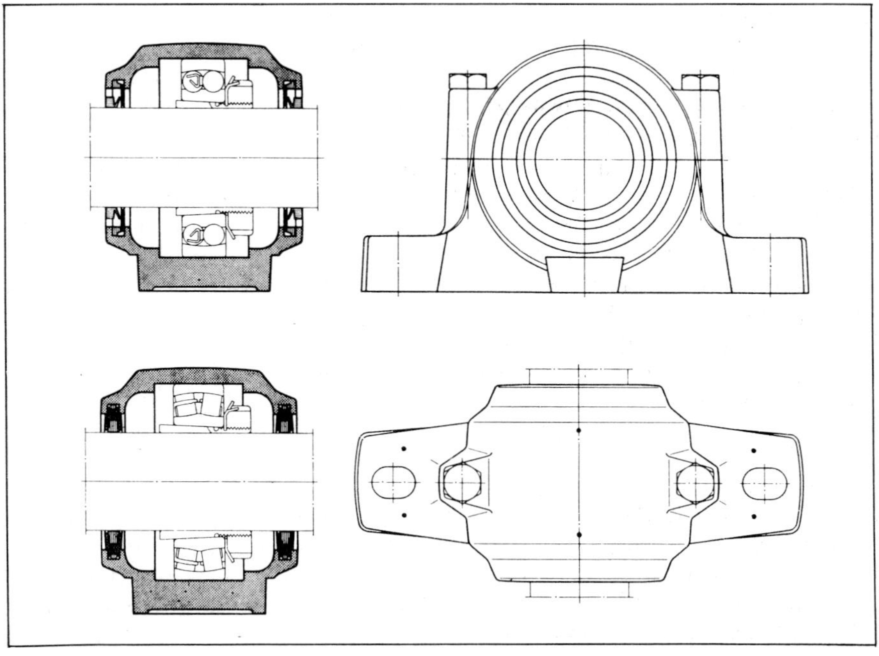

Figure 5.14 – Ball and roller bearing plummer blocks

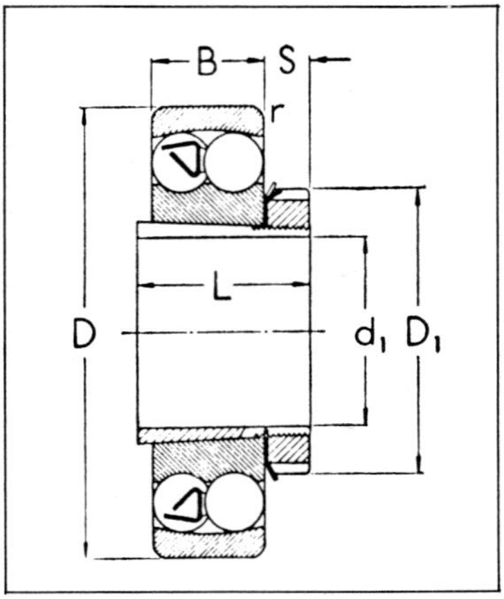

Figure 5.15 – Bearing with taper bore and adapter sleeve

high-temperature conditions soon becomes blocked with carbonized grease.

For difficult applications where steam, water or high temperatures require the use of a special lubricant it is much better to use a plummer block (Figure 5.14). In the plummer block the ball or roller bearing is not sealed, but is completely surrounded by the lubricant. The quantity of grease needed is much greater than in the pillow block, so there is less danger of the bearing running dry. The lubricant is applied by the user of the bearing rather than by the manufacturer. If at any time the lubricant requires changing, or a different lubricant is to be tried, the plummer block can quite easily be flushed out and repacked.

In short, if a bearing unit is required to operate under conditions that make it necessary to use a special lubricant, the plummer block is preferable to the pillow block. Although more expensive, it is soon found that the extra cost is offset by the reduction in down time.

There are a number of methods of fitting ball or roller bearings. Where the shaft spindle can be turned to a fine tolerance, the bearing will usually be an interference fit and is either pressed onto the shaft or spindle, or warmed up in hot oil, causing it to expand enough to allow a push fit on the shaft. Bearings for electric motors and gear boxes, for instance, are usually fitted in this way, as the rotating parts can be manufactured to fine limits.

In textile processing machinery, bowl and roller ends are not usually made to such fine limits to accept an interference fit. On smaller bearings, the commonest method is to have the inner race of the bearing extended, drilled and tapped for grub screws, so that the bearing is just a push fit on the shaft and clamped by means of the grub screws (see Figure 5.13).

The third method, generally used on larger machines and where more precision is required, is to use a tapered sleeve between the inner ring of the bearing and the shaft (Figure 5.15). The inner ring has a tapered bore and

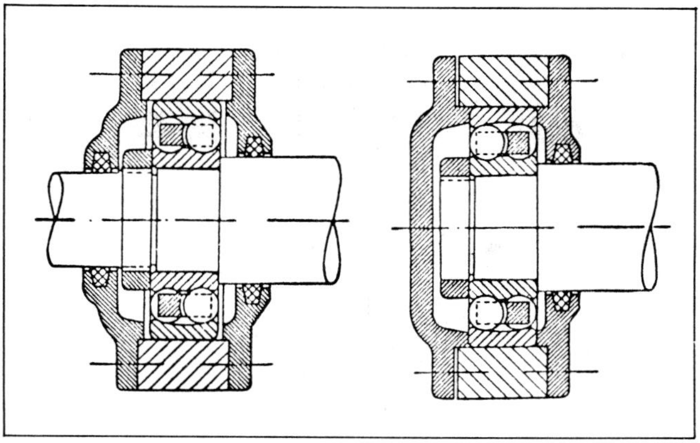

Figure 5.16 – Self-aligning ball bearings in a common shaft, showing endwise location of the outer rings, that on the right being free to move within its housing

the sleeve has a matching taper on the outside and a parallel bore on the inside. The sleeve, which is split to allow for compression, is then either pushed or pulled into the bearing by means of a threaded ring and clamped to the shaft by compression.

The use of ball and roller bearings at elevated temperatures, e.g. in hot flues, heat-setting stenters and bakers, where working temperatures can exceed 200°C, is a factor machine makers sometimes overlook. Bearings expand when heated, and as a standard bearing is made to very fine limits, at high temperatures this expansion will often result in a premature bearing failure. Ideally, bearings should be sited outside the maximum temperature zones.

To overcome this problem, bearings are manufactured with different degrees of clearance between the balls or rollers and the inner and outer races. The clearance required depends on the operating temperature, and the former is usually indicated on the bearing, either by a number from one to five followed by the letter C, or by one to five dots, following the actual bearing number in both cases. Bearing manufacturers' literature usually indicates the clearance required and the best bearing to use for a specific working temperature. Lengthwise expansion and contraction of a shaft or roller due to temperature must be provided for. This is done by firmly clamping the outer ring of one bearing within its housing, the outer ring of the opposing bearing being free to move endwise in its housing (Figure 5.16) in which the left-hand bearing is clamped and the right-hand bearing is free.

Other problems arise when ball or roller bearings cannot be used, e.g. for rollers in wash and chemical becks and troughs, inside steamers, in dye jigs and winches, in padder and filling mangle troughs. Stainless steel ball bearings have been tried for these applications, but have never been very successful and the manufacturing costs are prohibitive.

In such circumstances, bearings are required that will operate satisfactorily with 'lubricants' such as strong and dilute acids and alkalis, hydrogen peroxide, various dye solutions, a large variety of fillings and resins, steam and hot and cold water. In many instances, the same troughs or becks may be used for completely different chemicals, depending on the process, so bearings are required that are impervious to all these liquors and suitable for use at temperatures in excess of 100°C.

Plastic Bearings

The use of plastics has gone a long way to solving the problems arising when bearings have to function in water, steam and chemical solutions.

Phenolic, epoxide and polyamide resins are all used as the basis of bearings for use under these conditions or even for dry running. Nylon makes quite a satisfactory wet or dry bearing within the limitations of its resistance to chemicals and heat. It will not stand up to some chemicals used in scouring and bleaching, nor is it suitable for use at temperatures above 80°C.

Reinforced phenolic resin is used extensively for liquor- or steam-lubricated bearings. This material is available with a variety of laminates, paper, asbestos, glass fibre and several different fabrics. The reinforcing medium determines the chemical- or heat-resisting properties of the material. Resin-bonded asbestos laminates are about the best all-round materials for resisting the effects of heat and chemicals. Manufacturers' literature usually tabulates the properties of the different grades.

Plastic bearing materials do not perform very well under dry-running conditions, they are much better with some form of lubricant, even if it is only steam. The addition of graphite or PTFE has improved the dry-running qualities of these materials considerably, but these bearing materials can cause considerable damage to a stainless steel journal or spindle end. Stainless steel is not the best material for use as a bearing journal, but because of its resistance to corrosion and chemicals it is used almost universally in the construction of textile and processing machinery.

Reinforced plastic materials are not very suitable for precise applications, but precision bearings are not necessary for submerged rollers in wash becks or dye and filling troughs, or for internal steamer rollers. Some of these bearing materials, although highly resistant to chemical attack, tend to swell in steam or even in boiling water, particularly if certain chemicals are also present. Scouring and bleaching steamers where caustic soda or hydrogen peroxide are used are typical examples of difficult conditions.

Where these conditions prevail it is advisable to make the bearings a very loose fit on the journal or spindle. On a 1½ in (35 mm) diameter roller spindle, a clearance of 20 thou (0.5 mm) would not be excessive. If these sort of clearances are not allowed it will often be found that plastic bearings will quickly bind on the spindle and create excessive drag on the roller.

In less severe and colder conditions, the bearing clearance can be reduced, but never to the same limits as with ball, roller, or plain metal bearings. These are the reasons why the use of plastic bearings should be limited to the less precise applications.

Carbon Bearings

One of the best materials for hot, steamy and even dry-running conditions is solid graphite or carbon. This material will stand up to most chemicals and conditions encountered in scouring, bleaching and dyeing and retain good dimensional stability. It does not damage shafts or spindles to the same extent as plastic bearing material, and its temperature stability is far greater than is required in textile processing. If used in completely dry conditions they may squeak badly.

Graphite-impregnated PTFE is also available as a bearing material. This is a very good combination as both materials are virtually chemically inert, but like all PTFE products it is quite expensive.

One shortcoming of solid graphite and carbon is that it is quite a brittle material and is readily damaged by shock loading, which can occur with a wrap-round or a break-out, particularly if the bearing block is made completely of carbon.

This can be overcome by making the bearing block from one of the reinforced plastic materials and fitting it with a carbon bush or liner. The bush need only be about 1/5 in (5 mm) thick and the bearing-block material need only be selected for its chemical- or heat-resisting properties, not for its effectiveness as a bearing material. Making the bearing unit in this manner allows the plastic block or housing to act as a shock absorber protecting the carbon bush.

The same remedy can be adopted when using the expensive graphite-impregnated PTFE, but for a different reason. A bearing unit in which only the bush or liner is made of costly material is much cheaper. Also when wear does take place, only the bush requires replacing, the housing block can be reused.

No particular suppliers of these specialized bearing materials have been mentioned. Several manufacturers produce similar products, most of whom advertise in one or other of the technical journals.

Plain Bearings

Long before ball and roller bearings were invented, the surfaces of metals and alloys were used as bearings, and indeed they still are used where ball, roller or plastic bearings cannot be employed. In many instances where a split bearing or a half bearing is required, e.g. for calendar bowls, plain bearings are used.

The most common metals employed are the alloys of copper, zinc and tin. Gunmetal is probably the oldest alloy still in use as a bearing material, and a similar alloy with the addition of phosphorus, known as phosphor bronze, is an excellent bearing material.

The use of plain bearings is essentially confined to applications where oil or grease lubrication can be applied, but special oil-retaining or dry-lubricant-impregnated bronze alloys are available that will run with the minimum of lubrication. The bearings are made by pressing metal powders into the required shape and sintering at a temperature below the alloy's melting point to form the finished component. The pores in the metal then either act as tiny oil-retention reservoirs, or are filled under pressure with a dry lubricant such as molybdenum disulphide, graphite or PTFE.

Such materials are used for light-duty bearings and in locations where there is not enough space for a ball or roller bearing.

Chapter 8

Economics – Investment in Plant and Machinery

H CATLING

INTRODUCTION

It is a human weakness to devote attention to the various aspects of a topic in proportion to the intrinsic interest of each, rather than to their importance in relation to the topic as a whole. Where the topic is the economics of investment in industrial plant, this weakness has the effect of concentrating attention on those aspects that most readily lend themselves to precise quantification. It would, perhaps, be better to speak of pseudo-precise quantification, as the economic equations involved in plant selection are so complex that it is generally necessary to make a number of largely arbitrary assumptions about some extremely critical factors.

Once this has been done and the assumptions have been accepted, it is only too easy to develop an apparently precise economic model, unaware of the fact that the reliability of one or more of the assumptions is insufficient to justify the use of the model in certain critical areas of decision making.

In general, investment in plant and machinery is made in the expectation that the result will be a financial return to the company sufficient to warrant the expenditure involved. Such investment is more amenable to quantitative evaluation than is, say, investment in process research or in machinery development. In particular, the capital requirements for an investment in machinery can be predicted with relatively high accuracy. The greatest problem lies in estimating the value of the return that may be expected from the investment. The nature of the benefits the new plant will confer can be broadly identified and qualitatively appraised without undue difficulty, but it is a very different matter to express this in significant terms with the necessary degree of accuracy. Many pitfalls may be met. Unforseen technological developments, changes in consumer requirements and emergence of competition from an unexpected quarter are some of the factors that can materially affect the return which an investment will yield. Although the impact of many of them may be determined only on the basis of probability, they may, nevertheless, be of critical importance in an appraisal.

When even one factor is known only in terms of its probability the process of decision making inevitably becomes more complex. With only one such factor present, and that of an essentially 'go/no go' nature, it may be possible to use a simple numerical criterion to avoid the need to exercise a personal, subjective judgement. For example, a company may decide that for given relationships between the reward for success and the penalty of failure a particular level of risk is acceptable. Using such a criterion an absolute limit to the maximum level of risk could be set. This approach will generally suffice where only one major risk factor is involved, but we must not lose sight of the fact that we have not in reality eliminated subjective judgement. Such judgement was necessarily employed in formulating the risk criterion and also in determining the maximum allowable failure penalty.

In making investment decisions concerning major projects it is often found that there are appreciable uncertainties in a number of important factors. The interaction of these uncertainties greatly complicates matters and although theoretically it may be possible to devise a framework of formal logic to provide unequivocal answers, it is generally impracticable and may well be grossly misleading to do so. The unpalatable fact is that there is no way in which fundamental uncertainties may be eliminated and the treatment of uncertainty is a crucial activity in investment appraisal. Because of its importance we will return to this topic later and discuss some of the more important areas of uncertainty and how they may best be dealt with.

This chapter gives an outline of economic-appraisal techniques for use in the preparation of investment plans for textile finishing, along with the broad strategies of the deployment of fixed-capital resources.

1 COST STRUCTURE OF THE INDUSTRY

Proposals for investment in plant cannot be appraised by technologists or economists alone. Guided by a neutral, patient and painstaking chairman, the conception, development and final appraisal of a major plant-investment proposal might conceivably be undertaken by a team composed of pure technologists and pure economists. A far more effective team would be one in which the technologists were familiar with the detailed cost structure of processing operations and the economists appreciated the technological considerations.

The first stage of any plant-investment proposal is to identify and to quantify the opportunity to profit provided by it. This stage is essentially commercial, it is concerned with market situations, appraisal of competition and a full appreciation of company resources, both financial and physical. To play an effective part economists and technologists alike should be completely familiar with the cost structure of processes in their own organization and should have, as background, a broad understanding of the general cost structure of each major sector of the industry of which the company is a part.

Tables 1 and 2, which are based on an AITIT (Association Internationale de la Teinture et de l'Impression Textiles) survey carried out in 1976, give

TABLE 1

Cost Components of Textile Finishing in the UK as a Percentage of Total Costs in Each Industry Sector

Costs	Yarn dyeing	Bleaching	Piece dyeing	Printing	All sectors
Salaries and wages*	40.3	40.1	46.4	37.9	40.9
Dyes and chemicals	28.8	19.8	18.7	25.8	22.7
Energy	13.1	14.9	12.2	12.4	13.0
Maintenance	5.1	6.5	5.7	5.0	5.5
Depreciation	3.0	2.7	3.5	4.9	3.8
Finance**	4.6	7.1	5.4	5.1	5.3
Other costs	5.1	8.9	8.1	8.9	8.8
Total	100	100	100	100	100

*Including social charges

**The total cost of all non-equity funds, i.e. all funds employed in the business apart from shareholders' funds

Based on an AITIT survey of 1976

TABLE 2

Cost Components of Textile Finishing in the EEC as a Percentage of Total Costs in Each Industry Sector

Costs	Yarn dyeing	Bleaching	Piece dyeing	Printing	All sectors
Salaries and wages*	41.2	39.3	42.6	37.9	40.4
Dyes and chemicals	27.1	21.8	20.3	27.8	23.4
Energy	13.7	14.1	12.5	11.6	12.8
Maintenance	4.7	6.8	5.3	4.9	5.5
Depreciation	3.3	2.8	3.3	5.4	3.7
Finance**	4.3	6.7	4.8	5.1	5.3
Other costs	5.7	8.5	11.2	7.3	8.9
Total	100	100	100	100	100

*Including social charges

**The total cost of all non-equity funds, i.e. all funds employed in the business apart from shareholders' funds

Based on an AITIT survey of 1976

basic information on costs in the UK and the EEC respectively, by means of which a proposal in the conception stage may viewed broadly.

In each table the costs in four major sectors of textile finishing are broken down into seven components, each expressed as a percentage of the total cost. The first thing to note is that the 'all sectors' figures show no significant differences between the UK and the EEC, and there are no striking differences between the four sectors surveyed. In all sectors salaries and wages was the greatest single category, typically accounting for about 40% of all costs. Dyes and chemicals, at around 23%, was always the second-ranking category, with energy in third place at around 13%. With one exception, all of the 'also rans' were appreciably lower than 10%.

Of the 'also rans' the most important in the present context is depreciation. It averaged approximately 3.75%, variation between sectors being remarkably small, ranging only from 2.7% for bleaching in the UK to 5.4% for printing in the EEC. However, if we regard depreciation as a measure of the level of investment in plant and machinery, these figures appear, at first sight, to be rather low in relation to other costs. The reason for this is that, for the most part, they refer to established works, which have been continuously updated and selectively re-equipped over a long period and, for accountancy purposes, the value of plant and machinery purchases has been progressively depreciated (in accountancy terms) over an equally long period. At any time, but particularly when currency inflation is high, the accountant's figure for depreciation is by no means a completely satisfactory measure of the technological situation.

An alternative measure applicable to relatively stable industries such as textiles is the annual rate of investment in fixed capital, preferably expressed in relation to strictly contemporary operating costs. Although considerable variations can occur from one year to another, such a figure averaged over about ten years is a good indicator of a company's attitude and practice in relation to investment in plant and machinery.

Using two such measures, Table 3 shows the position of finishing in relation to other major sectors of the textile industries of the UK. It gives investment in plant and machinery over the ten-year period 1966 to 1975 expressed as (a) percentage of net output and (b) percentage of labour costs, over the same years, for four sectors of the cotton and allied textile industries. These measures are generally more significant in comparing the importance of machinery costs with other costs. Spinning, weaving and finishing are not very different as far as the relationship between machinery costs and net output is concerned – all are very much more capital intensive than is garment making in this respect. In relation to labour costs, however, spinning and weaving are roughly twice as capital intensive as is finishing, and from four to five times as capital intensive as garment making.

Another measure of considerable interest to the technologist is the relationship between annual expenditure on dyes and chemicals and the average annual expenditure on plant and machinery. For the textile finishing in-

TABLE 3

Fixed-capital Investment in the Cotton and Allied Textile Industries of the UK during the Period 1966–75

	As % of net output	As % of wages cost
Spinning	14.2	43.3
Weaving (including preparation)	15.8	37.5
Finishing	13.8*	19.4
Garment making (shirts and overalls)	5.5	8.9

*Net output excludes the value of dyes and chemicals (see Tables 1 and 2)

Based on Business Monitor census of production returns

dustry as a whole the value of gross output* during the five years to 1976 averaged 78.5% above net output*, with most of the difference attributable to purchases of dyes and chemicals. In round figures the average expenditure on dyes and chemicals exceeds net output by 75% and expenditure on plant and machinery averages about 6% of dye and chemical costs.

To summarize this quick glance at the statistics of the operating economics of the industry as a whole, we can say that the figures suggest that there is generally considerable scope for higher capital expenditure on equipment for saving labour, dyes and chemicals. The case for capital expenditure with the object of saving energy was not then so clearly manifest in that energy costs were not overwhelmingly important. The rapidly escalating cost of fuels is now changing the situation. It has aroused a great deal of technical interest in low-energy processes and also in the design and construction of cheaper and more efficient heat exchangers. As a result, much new equipment and a growing fund of energy-saving expertise is beginning to offer many attractive opportunities for capital investment in this area of cost saving.

2 INVESTMENT-APPRAISAL CRITERIA

In making a financial appraisal of a proposal to purchase plant and machinery we are forced to do arithmetic with many disparate elements. Reduced to its essentials, we are attempting to relate a financial liability incurred today to the present value of the net future earnings of a fixed-capital asset. The magnitude of the financial liability may be fairly reliably assessed but the value of future earnings is a vastly different matter.

It is first necessary to make a number of very critical assumptions concer-

*Gross output is the value of the total sales. Net output is gross output less the cost of material and service purchases.

ning the initial earning capacity of the plant and the way in which its earning capacity will decline with time, until the day comes when it will work no more. Its working life may cease for a number of reasons. It may be worn out, it may be technically obsolete and hence unprofitable to operate, or it may be that the product with which it is concerned is no longer required. This aspect of investment appraisal calls for a great deal of skill and judgement plus the use of a crystal ball – or at least a little future-scenario writing – but however thoroughly it is done it can only be regarded as being in the nature of a forecast based on a number of factors of varying likelihoods. Nevertheless it is the best that can be done and, so long as the critical assumptions are clearly stated and their relative importance assessed, it can be a valuable aid to corporate decision making.

At this point the question arises as to how the outcome of the appraisal may be expressed. There are many devices of varying complexity by means of which the results of an appraisal may be expressed in a simple form. The more popular of these are discussed below.

Crude Pay-back Period

Probably the first approach to the problem was the natural and instinctive concept of a pay-back period. Primitive man, before making himself a simple tool, would intuitively, perhaps even subconsciously, ask and answer the question, 'How long will it be before the time saved by the use of this tool is equal to the time expended in making it?'. This approach was valid then and is equally valid today. The knowledge that in *x* months or years the saving in production cost resulting from the purchase of a new machine will have equalled the cost of the machine gives one an instinctive feeling on whether the projected purchase is worthwhile. In its simplest form this device is known as the crude pay-back period approach. The cost of the machine is divided by the number of months or years that will have elapsed before the new machine has 'payed for itself'. Where this period is short, e.g. no more than two years, this very simple device is good enough. More usually investment in textile finishing machinery is made on narrower margins, and more complex methods are needed.

Discounted Pay-back Period

For terms longer than two years the weakness of the crude pay-back period approach is the fact that the pound of tomorrow will not equal in value the the pound of today, and the promise of a pound in five years time is worth a great deal less than a pound today. It is necessary, therefore, to discount the expected future earnings from a proposed machinery purchase to their value today. The principle on which this may be done is very simple.

The concept used is that the present value of a sum of money receivable at some future date equals the sum of money which if invested today would be exactly the same as the future sum at the date when that sum will be receivable. Although the concept is simple the arithmetic involved in using it can be very tedious. This arises because compound interest calculations necessarily involve geometric progressions, developed from the basic expression for the final value of a sum invested for a number of years at a given rate of return. This is:

$$s_n = s_o(1+r)^n$$

where s_n = final sum
s_o = original sum
r = interest rate, expressed as a fraction
n = period of investment, in years.

As our present concern is with the determination of the duration of a period, it is readily apparent that this is not the sort of calculation that can be done in a moment on the back of an envelope. When the basic expression given above is used to calculate a pay-back period directly it results in the rather more complex formula:

$$n = \frac{\log s - \log(s - rc)}{\log(1+r)}$$

where n = pay-back period, in years
s = annual return on use of plant
r = interest or discount rate, expressed as a fraction
c = cost of plant.

Evaluation of n is not unduly time consuming, either using this formula or by interpolation within present-value tables based on the formula. It is, however, limited to the situation in which the return from the plant becomes available immediately and remains substantially unchanged throughout the pay-back period.

This is seldom the real-life situation. More usually there is a lengthy period of installation and commissioning, followed by a settling-down period during which full potential utilization is reached. From this point onwards the net real saving from the new plant declines in value, slowly at first, then with increasing speed as normal wear and tear, technical obsolescence and changes in market requirements inevitably erode the margin of advantage relative to one's competitors. The situation can be catered for only by means of a series of step-by-step calculations, taking account of each complete year (or appropriate shorter period) from the date of purchase onwards. The following example indicates the method of working.

Consider an investment of £10,000 to be made in plant offering savings in manufacturing cost in relation to plant at present in use. It is estimated that six months will elapse after purchase before the plant becomes fully operational. During the final six months of the first year there will be a saving of

TABLE 4

Discounted Recovery of a Fixed-capital Investment of £10,000

		Present values* (at date of purchase)		
Year	Net earnings for the year (£)	£1	Net earnings for the year (£)	Cumulative total (£)
1	800**	0.902	722	722
2	4,000	0.837	3,347	4,069
3	3,600	0.743	2,675	6,744
4	3,200	0.659	2,111	8,855
5	2,800	0.586	1,640	10,495
6	2,400	0.520	1,248	11,743
7	2,000	0.462	924	12,667
8	1,600	0.410	656	13,323
9	1,200	0.364	437	13,760
10	800	0.325	260	14,020
11	400	0.286	114	14,134
12	0			

*Discounted quarterly in arrears at 12% per annum

**Accruing during the final 6 months

4 years is too short leaving a deficit of £1,145

5 years is too long giving a surplus of £495

$$\text{Actual recovery period} = 4 + \left(\frac{1.145}{1.145 + 495}\right) \simeq 4.7 \text{ years}$$

£1,800 in production costs against which must be charged non-recurring commissioning expenses of £1,000; consequently during the first year the net return will be £800. As this will accrue during the last six months of the year it must be discounted by, in rough terms, three-quarters of a year's interest in order to arrive at its value at the time of purchase of the plant.

Throughout the second year the plant will be fully operational and will show a net saving of £4,000, spread uniformly through the year. Thereafter, erosion of net earning capacity will proceed at the rate of 10% per annum until in the twelfth year after purchase the plant will have completely lost its premium earning capacity in relation to competitors. Table 4 illustrates this hypothetical case worked out on the assumption that savings will become available quarterly in arrears and will be discounted at the rate of 12% per annum.

The computation shows that on these assumptions the pay-back (or capital-recovery) period will be a little less than five years. The use of the simpler crude pay-back period method, assuming an immediate and continuing saving of £4,000 per annum, would have given the misleading answer that the new plant would have paid for itself in two and a half years. Discounted pay-back applied without taking account of commissioning delay and costs, and assuming no erosion of savings with time, would have indicated a period of almost three years – a little nearer to the truth but still seriously misleading.

Total Discounted Return

No one buys a plant merely to allow it to pay for itself. The ultimate justification for all investment is the making of a profit, and from this point of view the pay-back period method of investment appraisal is seen as being a negative approach to decision making. Total discounted return is a device by which the discounting technique outlined above is employed to show the net profitability of a proposed investment. The principle can readily be seen by going back to the example considered above and outlined in Table 4.

The assumptions made in the appraisal of this project include termination of profitable working life at the end of the eleventh year. In the event, it is likely that some measure of updating will be an economic way of extending the profitable working life of the plant. However, we must complete our appraisal within the constraints provided by the original assumptions, plus a further assumption on the scrap value of the plant. Let us assume that £500 will be realized on the sale of the plant in the middle of the twelfth year. Discounted to the time the investment was made, this is worth about £135 and may properly be added to the total in the final column of Table 4.

Thus the total discounted return is forecast to be £14,269 in money available at the time of purchase of the plant. This means that the proposed investment in plant, in return for acceptance of the attendant commercial risks, will give, over the lifetime of the plant, exactly the same financial return as would be obtained from a high-security investment of £14,269 in, for example, Government stock returning 12% per annum.

DCF Rate of Return

This is a method by which the performance of a proposed investment may be expressed by a single figure. Like the two previous methods, it uses the concept of discounting future earnings to give present values, but does this from a different viewpoint. In the total discounted return method the present value of the net return that may be expected is computed using a postulated discount rate based on the rates of return available from gilt-edged investments. The DCF rate of return method uses the same basic assumptions with regard to the initial cost and forecast earnings of the proposed investment, but does not assume a value for the discounting rate. Instead it computes the discounting rate, which when applied in the manner already described will show the present value of future earnings to be exactly equal to the cost of the investment.

The use of this figure, often loosely referred to simply as 'the DCF return', has now been fashionable for more than a decade. It has the considerable merit that it is based on sound economic principles and is perhaps the most convenient single-element measure of financial performance. It is worth looking more closely at the exact meaning of the figure. The DCF return, as defined above, is in fact the rate of return on outstanding capital (i.e. capital not yet recovered) during each year of the forecast life of an investment. Thus it is, throughout the life of the plant, the true annual rate of return on invested funds over and above full recovery of the capital. It can therefore be used directly to relate the forecast financial return of the proposed use to any alternative use of company funds.

Arithmetically it is not a figure that can be calculated directly, but quite small desk-top computers can be programmed to derive the figure by an iterative process. To illustrate the essentials of the method of calculation a worked example is given in Table 5. The forecast conditions used are broadly those used earlier to illustrate discounted pay-back period and total discounted return in Table 4. In Table 5 the discounted returns for each year of the life of the project have been calculated and summed for two discount rates, 20% and 24%, exactly as for a determination of total discounted return. In this example a 20% discount rate gives a TDR surplus of £399, and a 24% discount rate gives a deficit of £829. Since at the correct discount rate the TDR is zero, the correct figures lie somewhere between 20% and 24%. From this point it is sufficiently accurate for most purposes to proceed by simple linear interpolation. Thus in this case

$$\text{Correct DCF rate} \simeq 20+4 \,.\, \frac{399}{399+829}$$

$$= 21.3\%$$

With so specific a measure of performance available, it is tempting to use it as the sole criterion in decision making, but it must be appreciated that it is no more reliable than the essentially subjective assumptions of forecast plant performance used in a computation. Where doubt exists in relation to a particular assumption it can be very rewarding to explore the sensitivity of DCF rate to changes in the value of that assumption. In a qualitative way one may discern the order of sensitivity to different assumptions by inspection of cumulative discounted return tables. From Tables 4 and 5 one can see immediately that the earnings during the last few years of the life of the plant have little effect on total real earnings and, conversely, it is clear that untoward delay or additional expense in commissioning can be disastrous. As a broad generalization for most investments in textile finishing plant, it may be said that the returns during the first four or five years are of paramount importance and every effort should be made to reduce uncertainty over this period, particularly over the first two years.

TABLE 5

Calculation, by Interpolation, of the DCF Rate of Return on a Fixed-capital Investment of £10,000

		Present values* (at date of purchase)			
		Discounted at 20% per annum		Discounted at 24% per annum	
Year	Net earnings for the year (£)	£1	Net earnings for for the year (£)	£1	Net earnings for the year (£)
1	800	0.833	666	0.806	645
2	4,000	0.694	2,776	0.650	2,600
3	3,600	0.579	2,084	0.524	1,886
4	3,200	0.482	1,542	0.423	1,354
5	2,800	0.402	1,126	0.341	955
6	2,400	0.335	804	0.275	660
7	2,000	0.279	558	0.222	444
8	1,600	0.232	371	0.179	286
9	1,200	0.194	233	0.144	173
10	800	0.161	129	0.116	93
11	400	0.135	54	0.094	37
12	500**	0.112	56	0.076	38
		Total	10,399		9,171

*Discounted annually in arrears

**Plant sold as scrap

20% is too low giving a net return of £399

24% is too high giving a net return of −£829

$$\text{Correct rate} = 20 + \left(\frac{399}{399 + 829} \,.\, (24 - 20)\right) \simeq 21.3\%$$

3 SENSITIVITY ANALYSIS

A qualitative appreciation of the extent to which the outcome of an investment appraisal is sensitive to the values of the assumptions made is generally sufficient to enable a manager with flair and experience to make a sound judgement. It is, however, unsuitable as an input to a formal appraisal methodology. When preparing proposal appraisals for consideration by a committee or board, it is necessary to quantify sensitivity and present it in a easily understood form.

There are no quick and easy ways for doing this satisfactorily. Basically this is because of non-linearity of the relationships with which we are here concerned, and of the mutual interactions between them. Conceptually, in most situations, the simplest approach is the most reliable and the most informative.

Firstly, each assumption should be examined and the extreme values determined, the most favourable and the least favourable, which are fully credible in the light of all available information and informed opinion. It is important to ensure at the outset that the concept of fully credible (as distinct from remotely possible) extremes is fully appreciated by all the people consulted and from whom expert opinions are solicited. Having assembled this array of extreme values, the original computation of the chosen criterion of profitability is repeated using the alternative extreme values of each assumption in turn, at the same time retaining the original most-likely values for all other assumptions. These computations reveal the order of sensitivity of each assumption and enable the assumptions of high criticality to be earmarked for further study.

The next step is to investigate the nature and extent of the mutual interaction of the assumptions. This is done by repeating the appraisal procedure, using all possible combinations of the alternative values of those assumptions to which the results of the appraisal have been found to be sensitive. For all those assumptions that have been shown to be of low criticality, only the original most-likely values need to be taken. Even so, the amount of work involved in computation can become formidable, and the problems of presentation become increasingly complicated as the number of alternative values to be taken into account increases. For example, where the result is materially influenced by the value chosen for each of four assumptions over the whole range of full credibility, it is necessary to carry out the appraisal computation 2^4 times, in addition to the original computation based on the most-likely values, a total number of 17 computations. The labour involved in the purely routine computations is trivial in relation to what is at stake in any sizeable investment proposal, so all assumptions of a significant degree of criticality should be included in the sensitivity analysis.

When the necessary computations, each set out to show year-by-year net earnings and discounted cumulative totals in the manner of Table 4, have been completed they should be studied in order to give a broad understanding of the interactions and the non-linearities involved. An analysis of this nature gives a far deeper awareness of the opportunities and risks in fixed-capital investment than is possible using only blind computer print-outs of appraisal criteria. However, for presentation to a committee or working party, a much more concise, although necessarily less informative, statement is needed.

A convenient and substantially complete digest of the outcome of a sensitivity analysis can be put on a single sheet, in the form illustrated by Table 6. This summarizes an analysis of a proposal broadly similar to that already used in describing the basic appraisal criteria. It refers to a common situation, in which the proposal is investment in new plant offering the prospect

TABLE 6
Sensitivity Analysis of Project XYZ

No.	Assumption	Extreme values* M**	Extreme values* L**	Most-likely values
1	Cost of installing and commissioning	£1,500	£5,000	£2,000
2	Time to being fully on stream (months)	3	12	6
3	Initial rate of earning (per annum)	£5,500	£2,500	£4,000
4	Rate of decline of earning rate (per annum)	6%	16%	10%

Assumption 1	Assumption 2	Assumption 3	Assumption 4	DCF rate of return (%)
M	M	M	M	44.3
M	M	M	L	39.8
M	M	L	M	38.0
M	M	L	L	32.1
M	L	M	M	34.7
M	L	M	L	29.1
M	L	L	M	30.2
M	L	L	L	26.4
L	M	M	M	22.7
L	M	M	L	19.3
L	M	L	M	15.9
L	M	L	L	11.1
L	L	M	M	12.3
L	L	M	L	6.6
L	L	L	M	– 2.0
L	L	L	L	– 7.4
Most-likely values				21.3

*Extremes of complete credibilty
**M = most-favourable value, L = least-favourable value

of savings in manufacturing costs. The direct cost of the plant is known precisely and this assumption need not be subject to sensitivity analysis. Four other assumptions:

1. Cost of installation and commissioning
2. Time before coming fully operational
3. Initial annual rate of earning when fully operational
4. Rate of decline of annual rate of earning

can be seen to be critical to the outcome. For each of these assumptions most-favourable and least-favourable extreme values have to be determined and are stated at the head of Table 6, together with corresponding most-likely values used in the original appraisal. In this presentation, DCF rate of return is the preferred criterion. The values of this are for each of the sixteen possible combinations of most-favourable and least-favourable assumptions. They are then tabulated in descending order of importance of single assumptions. The single-value criterion obtained using most-likely values for all assumptions is also given.

The value of this table to a committee or a working party is somewhat debatable. It depends a great deal on the composition of the committee and on its terms of reference. Nevertheless, however closely the figures are studied there comes a point at which subjective judgement must be exercised. In some circumstances, having the whole gamut of fully credible outcomes conveniently shown may be useful, but often it will be sufficient to consider only the two extreme criteria given by the conjunction of all most-favourable and all least-favourable assumptions respectively, together with the single criterion obtained when all the most-likely assumptions have been made.

More elaborate schemes involving the allocation of probabilities of occurrence to each alternative assumption have been devised and can be used. The value of all such schemes is greatly limited. Each alternative assumption must be regarded as a unique event, and, without a background of experience of the frequency of occurrence of similar events in similar circumstances on many occasions, subjective judgement must be used to determine the probability of occurrence of each alternative assumption made.

Once these probabilities have been determined, probability theory can be used to compute the overall probabilities of a range of possible conjunctions of different alternative assumptions. This can be done with great accuracy, but it must not be forgotten that the value of the result is wholly dependent on the soundness of the judgement applied to the initial determination of each of the individual-assumption probabilities used.

4 BASES OF ASSUMPTIONS

We now have adequate means for determining the true return on an investment, subject to the acceptance of a number of essential assumptions on the expected performance of the projected new plant. We also have means for treating the extent to which an appraisal is sensitive to particular assumptions. It remains to establish a methodology for determining the assumptions.

This is very dangerous ground. The fact must be faced that the most important assumptions are essentially forecasts, dependent to a greater or lesser extent on subjective judgement, and for this reason complete reliability is not attainable. Nevertheless, for major investment proposals, the discipline of a formal forecasting procedure can greatly reduce the risks involved.

In recent years the methodology of technological forecasting has been given a great deal of attention and much of what has been written on the topic is applicable to our present concern. The techniques developed range from the coldly mathematical to the highly intuitive, but it is impossible to recommend a universal approach suitable for all investment situations.

For our present purpose, appraisal of whether investment in plant and machinery for textile finishing is worthwhile, we need only a limited time horizon. We do not need to see with any real precision the situation 25, or even 15, years ahead. The examples of Tables 4 and 5 are typical of finishing equipment investment, in that earnings of the plant during the first five years of operation are of the greatest importance, whereas direct earnings to be received after more than ten years can do little to justify present capital expenditure. Thus, if the proposed investment is concerned with well-established products and processes, simple extrapolation of present trends, both commercial and technical, will suffice. When a radically new product or a major process innovation is under consideration, it is necessary to be much more imaginative and the appropriate tool may well be scenario writing – a technique by which a range of possible future situations may be visualized and imaginatively developed.

Whatever the tools the forecaster uses, at the end of the day he must provide the firm figures for each assumption that the appraisal technique requires. Often he will be able to do this with an entirely adequate degree of confidence. For example, although the working life of a machine may not be predictable with any confidence beyond ten years, this is generally unimportant because the present value of such distant earnings is low. Conversely, although the viability of a project may be critically dependent on the time spent and the costs incurred in installation and commissioning, when purchasing new machinery of established design from a reputable maker the accuracy with which these factors can be forecast will be sufficiently precise.

This is not so when considering a major process or product innovation involving machinery of novel design. Both installation and commissioning will be fraught with difficulties, and it will be necessary to identify and consider their physical nature, which initially is only vaguely apprehended. The likely interaction of the difficulties must then be studied to produce alternative scenarios for the possible course of the installation and commissioning phase of the venture. Only then will it be possible to know and understand the whole gamut of possible outcomes of this critical step in the implementation of the investment proposal.

When the study has been completed we are still not out of the wood. We are faced with the problem of expressing the results of the study in terms that will allow the application of our investment-appraisal criteria. Clearly, a new single-figure presentation cannot deal adequately with the results of a study that has shown, for example, that if all goes well the cost of commissioning will be £5,000 over a period of three months, but a concatenation of possible troubles, of varying probability, may cost as much as £100,000 over a period of two years before the plant will be fully operational. Using the two extreme values to show the best and worst possible outcomes is the simplest solution, but it is more helpful to the decision maker to supplement this with figures for the most-likely outcome and to give some estimate of the probability of occurrence of a number of less-likely outcomes.

Estimation of degrees of probability in such a situation generally involves an amount of work disproportionately high in relation to the reliability of the basic premises and to the practical value of the probability figures. A list of possible outcomes and their associated probability figures can be more confusing than helpful to a board of directors and a low-key graphical presentation is often more appropriate. A rough plot of probability against time and cost gives the information clearly without implying an unwarranted degree of accuracy.

Installation and commissioning costs and delays are examples of an area in which the determination of a single most-likely value for a possible outcome can be difficult so that its use in investment appraisal may be dangerously misleading. The earning potential of a new process can never be accurately predicted. The introduction of a novel process, with substantially lower operating costs than the conventional process, can be expected to increase profit per unit of output and increase the company's market share only if the company's major competitors make no changes. At one end of the range of possibilities, they may be so disheartened by the advantage the innovation has given that they withdraw, leaving the company with a clear field and hence a better than expected return. Alternatively, at the other extreme, they might fight back by the purchase of second-generation equipment for carrying out the same new process, and thus effectively leapfrog the original advance. They are then likely to have a plant more reliable and cheaper to operate, and the company will then find itself with lower margins and a smaller market share than before.

This raises a very important factor in relation to the whole rationale of fixed-capital investment appraisal. It is rare to find that the equipment bought today is identical with that which would have been bought only ten years ago. There is continuous development of products, processes and machinery for carrying out processes while, at the same time, market preferences and the relative costs of all inputs to processing and manufacture are also changing. As a consequence, all machinery eventually becomes obsolete. In general, those responsible for the management of fixed assets should always be asking not 'For how long can I continue to operate my plant profitably?' but, more pertinently, 'At what point in time will it be most profitable to purchase new plant?'. This means that the profitability of one's plant should be continuously under review. This can be conveniently done using the defender-and-challenger concept.

Present equipment is regarded as the defender and all available alternatives are appraised as challengers. After directly comparable measures of performance have been evaluated, the cost, during the next accounting year, of not replacing the defender by the leading challenger is assessed and related to the whole cost of such replacement. The first figure, which may be referred to as the next-year inferiority of the defender, can be computed with fair reliability, as the present operating costs and returns of the equipment in use are known and the expected costs and returns using the new equipment are not subject to any long-term uncertainties. The whole cost of replacement is more difficult to determine. In addition to the need to discount future earnings to present values, it is necessary to make assumptions about the rate at which the new equipment – which after installation will have become the new defender – will deteriorate with respect to future challengers. This can be done by extrapolation from historical data relating to the rates of performance improvement for broadly similar equipment during the recent past. Although this cannot be done with confidence for more than, say, five years into the future, this is generally not a serious limitation, since the first three or four years of operation are overwhelmingly important in terms of capital recovery.

Inherent in the defender-and-challenger concept is the progressive erosion of the earning capacity of all plant, as a result of machinery developments and other changes occurring continuously. Acceptance of this concept and its implications does nothing to simplify fixed-asset management. On the contrary it raises some formidable difficulties. It does, however, reduce the risks of complacency in regard to plant-acquisition policy, which can so easily undermine the financial soundness of a company.

Although obsolescence is a continuing process, the effect of complacency on the competitive position of a company can be most clearly appreciated by examining the effect of one significant change in technology. Consider a single innovation that gives a material reduction in process or product cost or an increase in product value. The first company to adopt the innovation may do so hesitantly, but once it can be seen to be successful others will follow, and the pace of adoption will quicken as confidence in the new process grows. For a time the innovators will enjoy an advantage over their competitors, which may be exploited by taking a greater profit or by gaining a greater share of the market. Over a period of years the change will proceed, until virtually the whole industry is using the new process, and as this takes place the advantage to both early and late adopters will be progressively eroded. The last company to take up the new process will gain no positive advantage, it will merely shed an encumbrance and incur a financial liability.

Three options are open to the last company. It may:

(a) Withdraw from that sector of the industry
(b) Continue activity but in a poor and declining competitive position
(c) Adopt the process.

Option (c) is the only one consistent with company survival in the medium to long term, and even this is conditional upon the financial resources being sufficient to meet the cost of the equipment and of the upheaval of installation and commissioning. It is true that these costs will be no higher than they were for those who adopted the process earlier, but, whereas they were able to meet them out of higher profits arising from the competitive advantage gained over companies still operating conventional plant, the last company will reap no such benefit. Installation of the new plant will do no more than give operating-cost parity with the rest of industry.

In practice the situation is usually less clear cut. The experience of the first users smooths the path of those who follow, thereby ameliorating the situation of the stragglers, but even so being the last can seldom be justified.

5 METHODS OF FINANCING

Having decided that new investment in plant and machinery is in the best interests of a company, the decision must then be made as to the most appropriate way to finance the investment. Possible ways range widely from short-term borrowing, medium- and long-term borrowing and the issue of preference shares, to the use of equity funds. These possibilities are listed in ascending order of risk to the provider of the funds and of cost to the (equity) shareholders. The apparent anomaly of it being cheaper to borrow money than to use one's own money requires a word of explanation. The basic factor that produces this situation is logically fundamental to the whole concept of industrial enterprise, if we regard such enterprise as an activity involving risk and demanding courage on the part of those who engage in it, i.e. the equity shareholders. This means that for the system to be viable it is a mandatory condition that investment in industrial equity must, in the long term at least, yield a higher return than is available from low-risk investment.

In practice there are a number of factors combining to complicate the issue. Any one of the available options may become the most appropriate at a particular time or to a particular company. Such factors as the current rate and expected future rates of currency inflation, bank interest rates and the general state of trade can have an overwhelming effect on the cost and availability of funds from the various sources, both generally and in relation to specific uses of funds. The situation is further complicated by a country's tax structure, in that it is a common arrangement for the servicing of loans to be an allowable pre-tax expense, whereas it is usual for returns to shareholders to be made after tax, and interaction between differential tax rates can materially alter the balance.

Briefly, the situation may be summarized by saying that no one method of financing new investment can be recommended as being the best in all situations. For each situation it is necessary to consider the position in some detail, and as an aid to such consideration, relevant features of the major alternative ways of financing are discussed below.

Short-term Borrowing

A short-term bank loan or a bank overdraft is a very flexible way of financing additional investment. An overdraft is particularly flexible, because of the way charges are computed over very short periods (often daily) on the outstanding balance, and the ease with which the outstanding balance may be reduced or increased from time to time. This makes it possible for a company to reduce financing cost whenever funds become available and, in effect, provides a very convenient way of getting a relatively high return on any temporary surplus of funds that may accumulate for any reason. Short-term loans of one to three months duration, renewable by mutual agreement, are rather less flexible but offer broadly the same advantages to the borrower.

Short-term lending, either by term agreement or in the form of an overdraft within an agreed maximum credit line, is also flexible, and of relatively low risk from the viewpoint of the lending bank. If difficulties arise the bank can withhold renewal of a short-term loan or withdraw an overdraft facility. This largely offsets the fact that in the event of liquidation the bank has no prior claim on specific assets and ranks after secured creditors such as debenture holders. For this reason short-term loans and overdrafts are generally regarded as giving good security to a vigilant lender, and as such are available at relatively low cost. However, availability is very seriously limited by the prospective lender's view of the financial soundness of the prospective borrower. Although many other factors are involved, in particular the 'track record' of the prospective borrower, the ratio of current assets to current liabilities is often used as a single-figure measure of the credit-worthiness of a company. As a broad generalization this ratio should not be appreciably lower than 2:1.

Medium- and Long-term Borrowing

The general basis of most medium- and long-term loans is repayment on a specified date and at a fixed rate of interest, payable periodically throughout the term of the loan. To reduce the risk inherent in the provision of funds for industrial enterprises, it is not uncommon for these loans to be secured on specific assets and with a clear agreement as to the conditions under which foreclosure of the mortgage would take place. Other ways in which the risk to the lender may be reduced include the limitation of the borrower's right to incur further debt of equal or earlier priority, agreement that payment of due interest shall constitute a first charge on the income of the enterprise, or a limitation of dividend payments by the company in the event of earnings failing to meet an agreed criterion.

There are many variants on this theme, but all are essentially mortgage agreements and the most common mechanism of implementation is by means of mortgage debentures. The borrowing company will negotiate an appropriate debenture trust deed, with a bank (or other organization) acting as trustee for the individuals or companies who will become the debenture holders. Debentures will then be offered and issued, not necessarily at par, but with a yield interest rate judged to be such as will attract funds to the extent required, having regard to all relevant considerations at the time.

Preference Shares

The distinguishing feature of preference shares is that their holders have claim to a share of profits that has priority over the corresponding claim of ordinary shares. This claim is, however, limited to an agreed fraction of the face value of the shares, whereas the participation of ordinary shares in profit is unlimited. Furthermore, the claim may only be met from profits (or surplus in the event of liquidation) when all creditors, including loan holders, have been paid in full.

From the borrower's point of view, funds raised by the issue of preference shares have an advantage over true borrowings in that the dividend (which is equivalent to the agreed interest rate payable on a loan) is payable only subject to the enterprise having made profits sufficient to permit payment. In the case of cumulative preference shares, unpaid dividends are not forfeit, but are deferred until profitability is sufficient to permit payment being made, and no dividends may be paid to ordinary shareholders until all arrears of accumulated preference-share dividends have been met.

Preference shareholders are, in a number of ways, treated as part owners of a company. In particular, the fact that they share in the profits of a company, rather than receive due interest as do loan holders, means that the payment of preference-share dividends is not truly a business expense. Logically therefore, preference-share dividends, unlike loan interest, are not allowable pre-tax expenses.

For both parties preference shares suffer from the weakness of low flexibility. Where the shares are irredeemable there is, of course, no flexibility at all, but it is possible to have shares redeemable at a fixed date or at the company's option. In the former case flexibility is neither better nor worse than for fixed-term loans, and in the second case there is a small degree of flexibility available to the company only. The combination of low-flexibility and low-priority status relative to loans makes preference shares a higher-risk form of investment. As a consequence, the cost of funds raised in this way tends to be higher than the cost of borrowing.

Equity Funds

Investment in equity entails high risk. The earnings of an enterprise become available to the equity holders only when all business expenses, including interest payable to other suppliers of funds, have been met. Similarly, when the enterprise is ultimately wound-up only the residue of funds, realized after the due claims of all creditors, loan holders and employees have been met, is available to the equity holders. On the credit side, however, there is no upper limit to the dividend that may be paid to equity holders, and when an enterprise prospers greatly the return will equally greatly exceed that available to low-risk investors. Hence, in general, it is more expensive to finance fixed-capital acquisitions by equity funding than by the other methods discussed above. This broad generalization is certainly applicable when considering the creation of new equity. In so far as existing equity has accepted the risks involved in the enterprise, in the expectation of receiving a return higher than can be obtained with greater security, the sharing of this expectation with others must effectively reduce the value of it to existing equity holders.

In practice the creation of new equity on the open market is only resorted to where the sums involved are large, otherwise the promotional costs are likely to be prohibitively high. The creation of new equity by means of a rights issue is much more common and is usually less costly. In this operation new ordinary shares are created and offered to existing holders of ordinary shares at a discount compared to the price of ordinary shares at the time of issue. The number of these privilege shares that any one shareholder is allowed to buy is strictly related to the amount of his existing holding. A shareholder not wishing to buy is entitled to sell his rights for whatever price they may command. These provisions ensure equality of treatment. No hardship is suffered by any industrial shareholder, so long as he either buys his quota of privilege shares or sells his entitlement to buy. He is automatically compensated for the reduction in market value of his original holding, which must be expected to occur as a result of the operation. Thus, at low risk and without injury to existing shareholders, the primary object of providing new equity funds has been achieved. Although not an inexhaustible source of funds, this is a ready means of increasing equity capital by a substantial fraction. The cost of the operation depends on many factors, not least the judgement exercised in determining the extent and price of the issue in relation to the price and market strength of the company's shares at the time the issue is made.

An entirely different approach to the financing of fixed-capital projects out of equity funds is by the use of retained funds. Broadly these will principally comprise provision for depreciation and retained profits. As depreciation is usually an allowable pre-tax business expense but profit is not, the conceptual difference between the two categories must not be lost sight of in computing the true cost to shareholders of using these funds for fixed-capital acquisitions. The intricacies of the situation depend on the detailed structure of company and personal taxation in a country, so this must be taken fully into account in assessing the true cost of this method of funding to both the company and to individual shareholders. A considerable complication in such assessment is that although retained funds are truly shareholders' funds, they are undistributed shareholders' funds, and the two broad categories, depreciation provision and retained profits, usually have a distinctly different legal and taxation status.

The extended use of revenue funds for the financing of fixed-capital developments may, over a long period, lead to a situation in which the nominal value of issued ordinary shares is small in relation to the value of the fixed assets of the company and also in relation to market valuation. No legal difficulty arises on this account, but there are a number of practical considerations that may make this inconvenient at times. The situation may be prevented or rectified quite simply by means of a bonus-share issue. This means creating a number of new ordinary shares and distributing them as a free issue to existing shareholders in strict proportion to their existing shareholding. The net result is a formal conversion of revenue reserves into capital funds. Again there are tax considerations, as a result of which the assessment of real cost is complicated, but once these are appreciated accurate assessment is only a matter of arithmetic.

6 EFFECTS OF INFLATION

Mild inflation of a roughly predictable character, such as has occurred from time to time over many centuries in many countries, presents no serious difficulties. The same is true of mild deflation, a less-common phenomenon, but, nevertheless, one that has occasionally persisted for several successive decades. Severe inflation, occurring spasmodically at high rates, such as has been experienced worldwide during the 1970s is a very different matter. It is seriously disruptive.

Although the major disruptive effects could be greatly reduced – if not entirely eliminated – by fiscal means, few governments have been moved, or have been in a strong enough position, to deal radically with the problem. An important consideration here is that the principal effect of inflation is to benefit the borrower at the expense of the lender. As in most countries the state is a very considerable borrower it would be unrealistic to expect radical government action.

Inflation and the Return on Investments

So far as the return on fixed-capital investments is concerned, inflation has only second-order effects. Constant, uniform inflation has no effect whatever on the true pay-back period or on the true DCF return on an investment made in plant or machinery. The basic principle involved here is that an investment being made today can only be properly appraised in terms of the present value of its future earnings. Consider the situation in which a particular investment is solely dependent for its justification on a single earning at a specified future date which, discounted without allowance for possible inflation, has a present value of £1,000. If, by the time the earning is made, general inflation of 50% has taken place, has the investment thereby become justified more fully? Of course not; although the actual earning will have been increased to £1,500, in terms of currency at the time of investment it is worth only £1,000.

The effect of inflation on working-capital needs is also a second-order factor in investment appraisal and may affect the choice of plant. The working capital needed to fund work in progress is directly and continuously subject to inflation as inflation proceeds. It is necessary, therefore, for a company to increase current funds continuously to meet the rising 'value' of work in progress. As, in most countries, this is not an allowable pre-tax expense it is a factor that may greatly reduce profitability in times of rapid inflation. Under these circumstances, new plant that offers a reduction in operating costs but involves an increase in the amount of work in progress may lose its attraction when the cost of financing a continuously growing work-in-progress fund is taken into account.

Differential inflation poses real problems and offers attractive opportunities for exploitation to those able to forecast the pattern of variability with confidence. For example, where the price of a particular class of machinery or a particular category of raw materials has lagged behind other prices, it could be opportune to invest or increase stocks before prices catch up with (or possibly go ahead of) general inflation. This is, however, speculative trading rather than manufacture.

Inflation and the Financing of Investment

There are no sovereign solutions to the considerable problems that arise in the financing of investment during times of severe inflation. Accounting techniques are adequate for appraising the pros and cons of the various means by which investment may be financed for any known pattern of future inflation throughout the course of a transaction. The great difficulty lies in forecasting future rates of inflation. Long-term borrowing (and lending) is clearly fraught with risk, but short-term transactions are also subject to very considerable risks as a result of the severe variations in inflation rate and base lending rates occurring from time to time as a direct consequence of government actions taken as control measures.

Given a constant inflation rate, interest rates for loans and the returns to be expected on equity funds will settle down to reflect fairly accurately the changing value of a currency. In a rapidly changing situation, they more truly reflect the prevailing view of the likely future pattern of inflation. Experience during the recent past in the UK has shown that this view has often been seriously wrong. For the most part, behaviour has been consistent with a belief that 'this cannot go on for much longer in our country'. In the event, inflation has far outstripped the predictions upon which interest rates and expectation of dividends have been based.

In the present situation (early 1982) money is a commodity of very variable price, which is essential to the undertaking of an industrial enterprise. During times of severe, spasmodic inflation the market price of this commodity is subject to violent fluctuations against which there is no secure hedge. Fortunes can be made by playing the market, but equally disastrous losses may also be incurred and it is hardly the business of a textile finisher to participate in such hazardous games.

As the true cost of any method of funding capital investment is dependent on the pattern of inflation rates during the term of the funding transaction, it is necessary to make some assumptions of future rates of inflation if a rational choice is to be made. Clearly, this cannot be done with anything approaching absolute confidence, particularly when considering longer-term transactions. A rational technique would have much in common with that described for sensitivity analysis. It would not eliminate risk, but would at least indicate the degree of risk.

The first step is to obtain full-credibility estimates of what rates of inflation will be throughout the term of each form of funding transaction under consideration. Even for short terms the range of estimates may be very wide indeed, but so long as care has been taken to ensure that they are true full-credibility estimates by persons knowledgeable in these matters, this is the best that can be done. In addition estimates of most-likely values must be obtained from each respondent.

The next step is to calculate the cost of financing the project by each method under consideration, using first the most-likely values of future inflation. Then the costs are calculated, assuming the extreme values of the full-credibility estimates. This will show the extent of the risk involved in each method of funding and the relative sensitivity of each variation in inflation rates. From this point onwards it will be necessary to resort again to subjective judgement between, for example, a method of funding giving lower cost under the most-likely forecast conditions but at high risk, and a higher-cost method of lower risk.

So far we have spoken of cost of financing without referring to fundamental difficulties inherent in the concept of a 'true cost', by which transactions of very different duration may be directly compared. Perhaps the simplest practical way in which this may be done is by calculating the present value of all payments to be made during the term of the transaction, discounting only for currency inflation. To do this, the chronological order of all payments throughout the term is tabulated. Next the forecast inflation indices, relative to the time of commencement of the transaction, are calculated; these are expected to be applicable at the time of each payment. Each future payment may then be discounted to a present-value-of-cost figure and the total of these discounted payments may be regarded as the present value of the whole cost of the transaction.

Table 7 has been prepared to show this method of working applied to determination of the present value of the cost of a loan of £10,000, repayable in five instalments payable annually in arrears, with interest at 20% per annum charged on the outstanding balance, assuming inflation to proceed at a constant rate of 12.5% per annum for the whole term. It will be seen that the present value of the total cost is £11,727. Assuming zero inflation, the total present value of the cost of an otherwise identical transaction at an interest rate of 5.75% per annum would be £11,725. It may be said that for both lender and borrower an interest rate of 20% per annum with inflation at 12.5% per annum is, in gross terms, substantially equivalent to 5.75% per annum interest rate with zero inflation for a transaction of this nature over a five-year term.

TABLE 7

Effect of Inflation on True Interest Rate

End of year	Repayment (£)	Interest (£)	Total (£)	Inflation index	Present* value (£)
1	2,000	2,000	4,000	112.5	3,556
2	2,000	1,600	3,600	126.6	2,844
3	2,000	1,200	3,200	142.4	2,247
4	2,000	800	2,800	160.2	1,748
5	2,000	400	2,400	180.2	1,332
			Present value* of the total cost		£11,727

*Discounted for inflation only

Assumptions: Loan of £10,000 repayable in five instalments payable annually in arrears with interest at 20% per annum on the outstanding sum. Inflation at a constant rate of 12.5% per annum.

With zero inflation the total cost of an otherwise identical transaction at an interest rate of 5.75% per annum would be £11,725. Thus for both lender and borrower 20% per annum with inflation at 12.5% per annum is substantially equivalent to 5.75% with zero inflation for a transaction of this nature over a five-year term.

Applying the same approach to appraisal of a mortgage debenture of £10,000, redeemable after ten years with interest charged at 13% per annum, we find that with inflation at 12.5% throughout the ten years the present value of the total cost is £10,279. This is rather more attractive than the shorter-term loan at 20% per annum repayable in the five annual instalments just evaluated. But if we consider the effect a reduction in the rate of inflation during the later years of the transaction will have on the present value of the total cost we get a different answer. On the assumption that inflation will be at 12.5% per annum during the first five years and at 2.5% per annum during the second five years of the term, the present value of the total cost works out at £12,923. It is clear that, in the light of the great variability in both inflation rates and interest rates in many countries during recent years, completely objective appraisal of the relative merits of different methods of funding fixed-capital investment is not really possible.

7 INSTALLATION AND COMMISSIONING

The cost of installing and commissioning plant and machinery must be regarded as fixed-capital investment. From the accountant's point of view, since these costs add to the value of the equipment purchased, they rank as fixed assets and must be depreciated over the expected life of the plant in exactly the same way as must the equipment itself. It follows, therefore, that account must be taken of this aspect of installation and commissioning costs when comparing the expected return on alternative investment proposals. With plant of established design this presents no difficulty, but where there is an element of novelty it may be difficult to predict the cost and duration of installation and, more particularly, of commissioning, where substantial costs may be incurred as a result of initial troubles leading to spoilt goods or the production of seconds. In this situation we are again reduced to the use of forecasting techniques. Those discussed under 'Sensitivity Analysis' are applicable.

For all but the crudest methods of investment appraisal it is necessary to forecast the direct cost and the duration of installation and commissioning. Direct cost is very simply dealt with. Forecast values of direct costs are merely a straightforward addition to the cost of purchasing the new asset. They have no effect on the expected earnings of the asset during its operating lifetime. In contrast, the time taken to install and commission can significantly affect the total earning capacity by two clearly differentiated mechanisms.

Firstly, the time interval between nominal purchase of a machine and when it begins to earn money is a factor that directly reduces the present value of the expected future earnings of the machine.

Secondly, if purchase of a new machine is justified on the grounds that, because it is a technical leader, it will reduce operating costs and thereby increase net earning capacity in relation to machinery being operated by competitors, then delay in commissioning will erode the expected margin of advantage in two ways. Firstly, delay will reduce the differential between the new machine and machinery used by the average competitor, who must be assumed to be progressively improving his technology, both initially and throughout the lifetime of the machine. Secondly, delay will reduce the technically viable life of the machine as development by others inexorably advances the relevant technology.

The choice of method of financing expenditure on installation and commissioning may depend on the taxation position, but as it is essentially fixed-capital expenditure the considerations discussed under 'Methods of Financing' are broadly applicable. Some scope for judgement exists in determining the point at which commissioning should properly be regarded as complete. Up to that point expenditure is clearly of a capital nature, possibly attracting grants or special capital allowances, but essentially needing to be met out of equity funds or by borrowing. After commissioning is normally complete, costs attributable to shortcomings in the new equipment, or lack of familiarity with it, may often arise and be regarded as teething troubles. For the most part they may be borne as operating costs and as such would rank as pre-tax-allowable expenses.

8 ECONOMIC OPERATION OF PLANT

In conducting our personal affairs most of us instinctively draw a very clear distinction between capital expenditure and current expenditure. Except for the most trivial items, the acquisition of fixed-capital assets is carried out only after a more or less searching, albeit informal, value analysis has been made. On the other hand, a great part of our current expenditure is undertaken largely as a matter of routine, much of our day-to-day spending being sanctioned only by custom.

An important factor leading to a similar situation in industry is that current expenditure is like water under the bridge, whereas fixed-capital expenditure is of the nature of the bridge itself. In a works, electricity consumed and process labour employed are both intangible, but each piece of installed machinery is a lasting memorial to a management decision. Machinery seen to be working intensively is high justification of the decision, but machinery standing idle damns the decision makers. This leads to a compulsion on the part of management to operate machinery as intensively as possible, and the more expensive and the more recent the machinery the stronger the compulsion. In this climate of judgement it is only too easy to equate intensive operation with most-profitable operation. It is by no means always so. It can be a rewarding discipline to keep the cost structure of each major process continuously under review, to ensure that processing is being carried out as economically as possible with the available plant and machinery.

Considering the relative magnitude of the major components of finishing costs puts the situation into perspective. Costs are set out in Tables 1 and 2, expressed as a percentage of total costs, for each of the four principal finishing processes. Table 1 refers to the UK industry and Table 2 to the industry in the EEC. The differences between the two are very small indeed.

Over all processes, both in the UK and the EEC generally, depreciation is only about 3.75% of total cost. For individual processes it ranges from 2.7% for bleaching in the UK to 5.4% for printing in the EEC. With salaries and wages making up a little over 40% of the total, dyes and chemicals about 23% and energy about 13%, it would appear unlikely that intensive operation of plant could offer useful economies. This is true with respect to direct economies from reduction of depreciation charges per unit of throughput, but some associated economies accruing when plant is operated more intensively generally make multi-shift working attractive in economic terms.

With the exception of salaries and wages, all costs are reduced, per unit of throughput, by more-intensive working. There are marginal savings in the cost of dyes and chemicals from reduced stockholding and from reduced wastage of some liquors. Very substantial savings in energy costs arise as a result of (a) less wastage of thermal energy on starting-up and shutting-down of machines, and (b) lower unit electricity costs consequent on improvement in the ratio of average usage to maximum demand. The latter factor applies whether electricity is bought from the regional supply authority or generated on site. Although the cost of maintenance may be reduced slightly, this will generally be too small to be significant.

The most striking reduction is in depreciation costs. Whether the plant is running for 24 hours each day or is completely idle, depreciation continues at the same rate, as technical obsolescence inexorably takes its course and ultimately the plant ceases to be technically viable. Thus the depreciation charge per unit of throughput is inversely proportional to the number of hours per year during which the plant is working. Although finance and other costs increase in absolute amount with more-intensive working, the increase is generally less than directly proportional to the increase in throughput and there is a net reduction in both of these costs per unit of throughput.

The interaction of the various factors discussed above is too complex for it to be possible to make completely general recommendations about whether more-intensive working is worthwhile. It will always be necessary to consider each of the component costs individually and totally. As an example, in Table 8 the effects of changing from two-shift working (80 hours per week) to three-shift working (120 hours per week) for a bleachworks operating under average UK conditions are set out, and the method of working explained. Although the modifying factors tabulated in column 2 are estimates in relation to typical UK conditions, very different values may be appropriate in specific circumstances. Determination of these factors is, of course, a critical operation in carrying out this sort of appraisal.

The net result in this example is a 6.9% reduction in cost per unit of production. The principal contributors to this result are an increase of 1.60% attributable to salaries and wages and a reduction of 4.77% attributable to energy. Finance and other costs together contribute a further reduction of 3.68%. The contribution from reduced depreciation is a reduction of only 0.89%.

Although the total reduction in cost per unit of production may be important in relation to the competitive position of the company, the potential increase in total earnings that the increased productive capacity can bring is of greater importance. For example, consider the total effect on our hypothetical bleachworks. Suppose that the turnover with two-shift working is £100,000 per annum and the total operating cost is £90,000 per an-

TABLE 8

Effect of Night-shift Working on Costs

The situation

Present working is 2 × 8 hours/day, 5 days/week.

Proposed working is 3 × 8 hours/day, 5 days/week.

Method of computation

1. Tabulate present components of cost as a percentage of total cost (Column 1).
2. Consider each cost component separately and determine the average cost per hour worked under the proposed conditions as a fraction of the present cost per hour. Tabulate these fractions (Column 2).
3. For each component multiply the percentage figure of Column 1 by the fraction of Column 2, tabulate (Column 3).
4. Total the percentage figures of Column 3 to give the cost per unit of throughput for three-shift working as a percentage of the corresponding cost for two-shift working.

	Column 1	Column 2	Column 3
Salaries and wages	40.1	1.04	41.70
Dyes and chemicals	19.8	0.98	19.40
Energy	14.9	0.68	10.13
Maintenance	6.5	1.00	6.50
Depreciation	2.7	0.67	1.81
Finance	7.1	0.85	6.04
Other costs	8.9	0.85	7.57
Total	100		93.15

Thus under these circumstances three-shift working reduces the costs per unit of throughput by 6.9%.

num, giving gross earnings of £10,000 per annum. With three-shift working the turnover may be increased to £150,000 per annum with total operating costs increased to $3/2 \times (93.15/100) \times$ £90,000 = £125,750 per annum, giving gross earnings of £24,250 per annum.

Thus, in this example, the gross earnings of a company have been increased by more than 140% without any increase in fixed-capital requirement. The result may not always be so favourable, but it can be informative, and possibly rewarding, to carry out proper studies of the way in which the level of the plant utilization can affect earnings and profitability in specific circumstances.

9 MAINTENANCE, SPARES AND RELIABILTY

To a greater or lesser extent all machines require maintenance and spares during their working life; the penalty of inadequate maintenance and non-availability of spares is unreliability. As this can be very expensive in terms of reduced throughput and increased production of second-quality or even damaged goods, responsibility for maintenace should be regarded as an important management function. Because machine maintenance has little in common with actual plant operation, it is generally preferable to establish maintenance as an independent specialist function with clearly defined terms of reference. These must recognize that machinery is not an end in itself but a means of carrying out a range of processes, and, as such, maintenance practices and schedules must interfere as little as possible with machine operation, without unduly jeopardizing machine life and performance.

This is not a new concept. Since man first began to use cutting tools it has always been appreciated that a balance must be struck between the effort devoted to sharpening a tool and the consequent saving of effort in use. The increasing complexity of our tools tends to obscure the concept, and, although it has never been wholly lost sight of, often only lip service is paid to it. The stumbling block to full recognition is the consequent need for the application of quite refined cost-benefit-analysis procedures to maintenance. Many of the problems are peculiar to maintenance work. In particular, identifying and quantifying the benefits resulting from the application of specific maintenance procedures (or the penalties of not applying them) can be quite complex. Acceptance of this fact, together with an appreciation of the increasing dependence of industry on machinery, has led to a great surge of interest in the economic management of machine maintenance under the title of 'terotechnology'. The conferring of a new title on an old but neglected topic has not of itself solved any problems but, coinciding with the coming of low-cost computers, it has stimulated interest in the development of formal methodologies for optimizing maintenance practices.

In a large plant it is worthwhile to have a small team concerned only with management of maintenance. It may be led by an engineer fully conversant with cost-accounting procedures or by a cost accountant with a strong aptitude for engineering. In smaller establishments it is sufficient for one man to accept continuing responsibility for the implementation of schedules prepared by specialist consultants, subject to occasional review of the situation by such consultants. A similar arrangement is recommended for individual premises of multi-plant companies, except that here the ideal arrangement is to have a specialist operational-research team at headquarters rather than to employ outside consultants.

In a newly commissioned plant, guidance on what maintenance procedures and schedules are necessary, together with a recommended inventory of spares, will be available from the machinery makers. These procedures and schedules should not be regarded as sacrosanct and used blindly but should be refined in the light of operating experience. Operating conditions vary greatly from plant to plant and, in addition, all machinery tends to become less reliable and increasingly prone to produce sub-standard goods as it becomes older. To counter this the performance of all machinery in the plant should be kept continuously under review. Account should be taken of the cost of down-time and loss of revenue arising from the production of second-quality goods, throughout the lifetime of all important machines. These costs and losses of revenue should then be related to maintenance and spares-inventory costs with the object of determining optimum procedures, schedules and inventory levels. Although this cannot be done with great precision, it will generally be found possible, by quantitative analysis of operating experience and extrapolation of trends, to effect useful economies in this area of plant management.

Major reconditioning may restore the original performance of the machine in all important respects or, in the light of experience and with the availability of better engineering materials and components, it may be possible to update a machine to give a better-than-new performance. Even so, the alternative of replacement should not be overlooked. In cost-benefit terms there are a number of problems in deciding when maintenance costs have become excessive and replacement of a machine, or even a complete process line, is the more economic alternative.

The basic criterion for the repair-or-replace decision is very simple indeed. The moment when the increasing cost of maintaining and operating the existing machine exceeds the cost of buying, maintaining and operating a new machine is the moment when replacement should be made. In practical terms annual review of the situation is usually sufficient. Quantification of this criterion is bedevilled with difficulties. Common complications are that the new machine might produce a slightly different product, offer a higher level of operator amenity or be less flexible in operation than the existing machine. Such factors are easily appreciated in qualititative terms, but it is desirable that their net value or cost should be quantified to a reasonable degree. It must be remembered that maintenance and spares are pre-tax operating costs but that replacement is a capital acquisition, which requires different funding. Finally there is the problem, already discussed under 'Investment Appraisal', of equating criteria of different sorts, in this case relating expected operating costs during the coming year to that portion of the net cost of acquiring a new fixed asset properly chargeable to that year.

It should be noted that in the common situation, where maintenance and operating costs of an old machine are increasing year-by-year, it is misleading to compare forecast average costs of the old and the new over a future period of years. This will lead to premature scrapping of the old. Ideally one would replace the old by the new as soon as the review shows that the increasing cost of the old equals the cost of the new, but not before.

Index

D

E

F

G

H

N

O

P

R

S